THE OXFORD HANDBOOK OF

PHILOSOPHY OF COGNITIVE SCIENCE

THE OXFORD HANDBOOK OF

PHILOSOPHY OF COGNITIVE SCIENCE

Edited by
ERIC MARGOLIS,
RICHARD SAMUELS,
AND STEPHEN P. STICH

OXFORD
UNIVERSITY PRESS

OXFORD
UNIVERSITY PRESS

Oxford University Press, Inc., publishes works that further
Oxford University's objective of excellence
in research, scholarship, and education.

Oxford   New York
Auckland   Cape Town   Dar es Salaam   Hong Kong   Karachi
Kuala Lumpur   Madrid   Melbourne   Mexico City   Nairobi
New Delhi   Shanghai   Taipei   Toronto

With offices in
Argentina   Austria   Brazil   Chile   Czech Republic   France   Greece
Guatemala   Hungary   Italy   Japan   Poland   Portugal   Singapore
South Korea   Switzerland   Thailand   Turkey   Ukraine   Vietnam

Copyright © 2012 Oxford University Press

First issued as an Oxford University Press paperback, 2017

Published by Oxford University Press, Inc.
198 Madison Avenue, New York, New York 10016
www.oup.com

Oxford is a registered trademark of Oxford University Press

Library of Congress Cataloging-in-Publication Data
The Oxford handbook of philosophy of cognitive science / edited by
Eric Margolis, Richard Samuels, and Stephen P. Stich.
p. cm.
ISBN 978-0-19-530979-9 (hardcover : alk. paper); 978-0-19-084621-3 (paperback : alk. paper)
1. Philosophy and cognitive science.   I. Margolis, Eric, 1968–
II. Samuels, Richard (Richard Ian)   III. Stich, Stephen P.
IV. Title: Handbook of philosophy of cognitive science.
BD418.3.O93 2012
128—dc22   2011010135

Contents

Contributors

Colin Allen, Department of History and Philosophy of Science and Cognitive Science Program, Indiana University

Peter Carruthers, Department of Philosophy, University of Maryland

B. Jack Copeland, Department of Philosophy and Turing Archive for the History of Computing, University of Canterbury

Stephen Crain, Macquarie Centre for Cognitive Science, Macquarie University

Lisa Damm, Department of Philosophy, University of California, San Diego (PhD)

John M. Doris, Philosophy-Neuroscience-Psychology Program and Philosophy Department, Washington University in St. Louis

Frances Egan, Department of Philosophy and Center for Cognitive Science, Rutgers University

Daniel M. T. Fessler, Department of Anthropology and Center for Behavior, Evolution, & Culture, University of California, Los Angeles

Susan A. Gelman, Department of Psychology, University of Michigan

Alvin I. Goldman, Department of Philosophy and Center for Cognitive Science, Rutgers University

Steven Gross, Department of Philosophy, Johns Hopkins University

Rick Grush, Department of Philosophy, University of California, San Diego

Ben Jeffares, postdoctoral fellow, Department of Philosophy, Victoria University of Wellington

Joshua Knobe, Program in Cognitive Science and Department of Philosophy, Yale University

Stephen Laurence, Department of Philosophy and Hang Seng Centre for Cognitive Studies, University of Sheffield

Edouard Machery, Department of History and Philosophy of Science, University of Pittsburgh

Eric Margolis, Department of Philosophy, University of British Columbia

Christopher Mole, Department of Philosophy, University of British Columbia

Shaun Nichols, Department of Philosophy and Cognitive Science Program, University of Arizona

Casey O'Callaghan, Department of Philosophy, Rice University

Gualtiero Piccinini, Department of Philosophy, Department of Psychology, and Center for Neurodynamics, University of Missouri-St. Louis

Paul Pietroski, Department of Linguistics and Department of Philosophy, University of Maryland

Jesse J. Prinz, Philosophy Program, The Graduate Center, The City University of New York

Diane Proudfoot, Department of Philosophy and Turing Archive for the History of Computing, University of Canterbury

Georges Rey, Department of Philosophy, University of Maryland

Richard Samuels, Department of Philosophy, Ohio State University

Lawrence A. Shapiro, Department of Philosophy, University of Wisconsin

Kim Sterelny, Philosophy Program, Research School of Social Sciences, The Australian National University; and Philosophy Department, Victoria University

Stephen P. Stich, Department of Philosophy and Center for Cognitive Science, Rutgers University

Peter M. Todd, Cognitive Science Program, Department of Psychological and Brain Sciences, and School of Informatics and Computing, Indiana University

Robert Van Gulick, Department of Philosophy, Syracuse University

Elizabeth A. Ware, Psychology Department, Viterbo University

Jonathan M. Weinberg, Department of Philosophy and Cognitive Science Program, University of Arizona

THE OXFORD HANDBOOK OF

PHILOSOPHY
OF COGNITIVE
SCIENCE

....

CHAPTER 1

INTRODUCTION: PHILOSOPHY AND COGNITIVE SCIENCE

....

RICHARD SAMUELS, ERIC MARGOLIS, AND STEPHEN P. STICH

THE past few decades have witnessed dramatic growth in the cognitive sciences. New experimental techniques have given us unprecedented access to the inner workings of the mind and brain, and this has led to a massive expansion in theorizing, covering everything from the mechanisms of memory and attention to the cultural processes that shape moral reasoning. This same period has also witnessed a significant shift in philosophy in that many philosophers have become substantially more attuned to developments in the cognitive sciences. This change is reflected in both the issues that philosophers study and the intellectual tools upon which they draw. Moreover, philosophers influenced by the development of cognitive science very typically view their work as having an inherently interdisciplinary dimension, and consider research emanating from such fields as linguistics, neuropsychology, and cognitive anthropology as having genuine philosophical significance. What is more, these changes in cognitive science and philosophy are not unrelated. The influence has been mutual, and the intersection between the two—what philosophers sometimes call *the philosophy of cognitive science*—has become an extraordinarily active area of enquiry.

This handbook focuses on topics in the philosophy of cognitive science and brings together a range of chapters that represent the state of the art in this active field. But our aim is not to be comprehensive. There are simply too many topics in the philosophy of cognitive science—too much fruitful ongoing research—for any single volume to encompass it in satisfactory fashion. Instead, we have organized the volume around a representative sample of important topics. Some of these have been central to cognitive science since its inception; others, though vital to our understanding of the mind, have only recently gained the attention they deserve. Each chapter focuses on one such topic and provides an introduction to—and an illustration of—research on that topic.

We are inclined to think that the best way to learn about the philosophy of cognitive science is to work through a number of case studies. It is here that we encounter the issues that have fascinated philosophers and scientists alike—in some cases, for centuries or even millennia. Still, a preliminary overview of the field may be helpful. Though there are no doubt many ways to carve up the terrain, we think that the following five broad categories of issues cover much of what is usually taken to fall within the domain of the philosophy of cognitive science:

- *Traditional philosophical issues about the mind*: Long-standing philosophical problems that concern the nature of the mind.
- *Meta-theoretic issues*: Issue concerning the practice of cognitive science and its foundational assumptions.
- *Conceptual issues:* Issues concerning the explication and clarification of the core concepts of cognitive science.
- *First-order empirical issues:* Issues about mental phenomena and behavior of the sort that comprise the primary focus of psychology, linguistics, and allied disciplines.
- *Traditional philosophical issues that are not ostensibly about the mind*: Long-standing issues that are most naturally construed as falling within areas of philosophy other than the philosophy of mind (e.g., ethics, aesthetics, epistemology, philosophy of religion, and metaphilosophy).

We do not suppose that these categories are mutually exclusive or that, in practice, they are easy to delineate. Rather they are often intimately intertwined in the research of philosophers of cognitive science. Nevertheless, it is useful in the first instance to separate them out and to look at each in turn. And this is exactly what we propose to do in the rest of this chapter, in each case, illustrating the broad category with examples from this volume.

1. TRADITIONAL PHILOSOPHICAL ISSUES ABOUT THE MIND

There are many issues of long-standing interest to philosophers that also fall within the purview of cognitive science. The most obvious of these concern fundamental questions about the mind. Indeed, the hope that cognitive science might facilitate progress on these perennial questions is perhaps the central historical reason for philosophical interest in the cognitive sciences. In our view, this hope has not been a forlorn one. On the contrary, the bridge between the philosophy of mind and the cognitive sciences has invigorated traditional philosophical debates about the mind and has resulted in considerable progress and insight. In many ways, this is one of the great success stories of contemporary philosophy. Some examples are in order.

1.1. Cognition in a Physical World

One primary objective of cognitive science has been to articulate detailed, empirically plausible answers to a range of questions that fall within the orbit of perhaps the most venerable of all philosophical issues about the mind: the mind-body problem. In most general terms, the mind-body problem is that of characterizing the relationship between mental phenomena on the one hand, and physical phenomena on the other. A critical aspect of this general problem is how minds such as ours—with their rich suite of cognitive capacities—fit into a world whose fundamental laws and entities are entirely physical (Fodor 1975; Newell 1980; Pylyshyn 1984; Anderson 2007). Human beings perceive, reason, and make decisions, and, as a consequence of such mental activities, produce behaviors that exert an influence on the world. But how this could be—how mere physical beings could exhibit such cognitive capacities—has seemed to many theorists to be a source of profound puzzlement.

   One response to the mind-body problem is to maintain, as Descartes did, that a human being's mental life depends on the existence of an immaterial entity that is not subject to physical law. But such a position is implausible in the light of scientific developments over the past three hundred years or so. Whatever the explanation for our cognitive and behavioral capacities, many now suppose that it cannot be one that requires a special immaterial component—something akin to the traditional notion of a soul. Here is where cognitive science enters the picture. Because cognitive scientists seek to specify *mechanistic* explanations of psychological phenomena, cognitive science offers the prospect of an empirically informed approach to addressing issues about how such mental capacities as perception, reasoning, and memory could result from the activity of mere physical systems. The successes and failures of research programs within cognitive science can thus be seen as informing our assessment of different answers to the question of how minds like ours manage to exist in a world whose ultimate constituents are physical.

   Over the past few decades, a number of different research programs have emerged from the cognitive sciences that explore divergent proposals about the fundamental nature of the mind. Many of these are discussed in the following chapters. For example, Gualtiero Piccinini focuses on perhaps the most enduring proposal of this sort—*computationalism*. As Piccinini makes clear, all computationalists endorse the general claim that certain mental capacities can be explained computationally. What they disagree about is how to construe the nature of computation and how to characterize the type of computation that is most relevant to the explanation of our cognitive capacities. Some, for example, maintain that cognition depends on classical symbol manipulation, while others advocate connectionist computation. In addition to providing a survey of these options, Piccinini looks at the difficult questions of what it *means* to say that a physical system is a computational system, and how the distinction between digital and analog computation should be drawn.

Although computationalism of some sort is the most prominent view of cognition among cognitive scientists, it is not the only one. Another family of views, often subsumed under the general heading of *embodied or embedded* accounts of the mind, is the focus of Larry Shapiro's contribution to the present volume. Shapiro clarifies the historical roots of the embodied cognition approach and notes the influence of Gibsonian theories of perception on its development. In addition, he explicates the various respects in which such views diverge from more traditional computational accounts of cognition and provides an overview of some of the evidence that has been marshaled in support of embodiment.

One point of disagreement between traditional computationalism and alternative views, such as those often advocated by embodiment theorists, concerns the nature and role of mental representation. According to most computationalists, representations have an essential role to play in cognition. In contrast, many embodiment theorists urge that much of cognition can be explained in nonrepresentational terms. This point of divergence forms a focal point for Shapiro's chapter. But issues about representation also figure prominently elsewhere in this volume. For example, Frances Egan's chapter on representationalism reviews the various senses in which the human mind may be understood to be an information-processing system, and examines whether the best way to reconstruct this idea requires the postulation of mental representations. Though mental representations play an important role in much theorizing about the physical basis of cognition, as Egan and Shapiro both point out, there are lively debates in the philosophy of cognitive science concerning whether mental representations are truly needed to serve this role.

1.2. The Problem(s) of Consciousness

The primary focus of cognitive science has been the explanation of cognitive capacities. As a consequence, the aspects of the mind-body problem about which cognitive science has had most to say concern the physical realization of cognitive capacities. Yet as Thomas Nagel and others have noted, the biggest mystery associated with the mind-body problem—what seems to make it especially bewildering—is *consciousness*. More specifically, the problem of consciousness is to explain how a physical system could exhibit what philosophers have variously called phenomenal consciousness, the "what it is like" feature of experience or qualia (Nagel 1974; Chalmers 1996). Here it is rather less clear what, if anything, cognitive science has to offer.

Until quite recently consciousness had been, at most, a topic of only fringe interest for scientists generally and cognitive scientists in particular. But in the past decade or so, there has been a resurgence of interest in consciousness, and a number of new philosophical theories have emerged that explicitly draw upon developments in the cognitive sciences. In his chapter, Robert Van Gulick provides a field guide to this exciting area of enquiry, distinguishing among the heterogeneous phenomena subsumed by the word *consciousness*, setting out the main empirically informed theories of consciousness, and describing some of the more interesting scientific results that may also be of interest to philosophers.

1.3. The Nature of Thought

Another class of philosophical issues about the mind that has been revitalized by research in cognitive science concerns the nature and preconditions for thought. A number of the chapters in this volume are concerned with aspects of this issue.

Language and Thought. Peter Carruthers's chapter addressees the relationship between language and cognition. This relationship has been for many centuries a focal point of philosophical debate. Philosophers have been interested in the extent to which thought *depends* on the possession of a natural language (or vice versa). According to one prominent family of views, widespread in twentieth century philosophy, the possession of a natural language is a necessary condition for thought of any sort (Davidson 1975; Dummett 1993). In contrast, other philosophers maintain either that the dependency goes the other way, from thought to natural language (Fodor 1975), or that the two are on par with each other (Brandom 1994).

Of course, philosophers are not the only ones with interests in the relationship between thought and language. At least since the time of Whorf and Vygotsky, psychologists and anthropologists have also had a deep interest in this relationship, and in recent decades there has been a concerted effort to assess the extent to which our cognitive capacities depend on the prior possession of natural language. Carruthers explores the implications of some of this research and argues that it supports the conclusion that natural language is implicated in some of the most distinctively human of our cognitive capacities, including the capacity to perform exact arithmetic calculations and to engage in sophisticated executive tasks that involve the manipulation of goals and attention.

Thought and Concepts. Another philosophical issue concerning the preconditions for thought arises in the study of concepts. If we assume that thoughts are composed of concepts, concept possession is clearly essential to having thoughts. But what concepts themselves are and what is required for concept possession are enormously controversial matters. For example, different proposals vary in the claims they make about the ontological status of concepts (whether concepts are abstract entities or mental representation), the structural properties of concepts (including whether word-size concepts have any structure at all), and the acquisition of concepts (whether there are any innate concepts and how concepts can be learned).

In their chapter, Laurence and Margolis focus on an aspect of the controversy about concept possession that they dub the problem of the *scope of the conceptual*. Many philosophers routinely distinguish between two types of states or content—conceptual and nonconceptual—and also maintain a sharp divide between creatures that are supposed to have concepts and those that do not. Laurence and Margolis survey the main arguments that have been marshaled by philosophers to support these distinctions and argue that they turn on a surprisingly large number of different ways of delimiting the scope of the conceptual, and that few of the arguments that are meant to support one way or another of drawing this distinction stand up to scrutiny.

Evidence that bears on the nature of concepts has come from a variety of different regions of cognitive science—linguistics, cognitive psychology, and neuroscience, among others. But one area of enquiry that has proven especially valuable to the study of concepts is developmental psychology. In their chapter, Gelman and Ware focus on the view known as *psychological essentialism*, which owes a great deal to work in developmental psychology. This is the view that much of cognition depends on concepts that function like natural kind terms in that they partition the world into categories whose members are represented as sharing an underlying essence that both determines category membership and explains the surface commonalities between members. Gelman and Ware argue that, in some cognitive domains—for example, folk biology—a satisfactory account of development requires that we adopt such a view of concepts.

1.4. More Specific Mental Phenomena

So far we have focused on traditional philosophical questions about the mind that have quite a broad scope—questions that concern thought, cognition, or consciousness per se. But in addition to these very general issues, philosophers have long been interested in a range of relatively specific mental phenomena such as attention, memory, and perception. Clearly, such phenomena are also of interest to cognitive scientists, and many of the chapters in the present volume seek to integrate—and draw out the philosophical implications of—recent empirical research on these topics.

An illustration of this approach to the philosophy of cognitive science can be found in Casey O'Callaghan's chapter on perception and multimodality. For a host of different reasons, including its centrality to our knowledge of the world, philosophers have always been interested in the nature of perception. But as O'Callaghan observes, philosophical discussions of perception have tended to be highly vision-centric, focusing on vision to the near complete exclusion of other perceptual modalities. Moreover, even when other perceptual modalities do receive attention, there is a widespread tendency to treat each modality (olfaction, audition, etc.) as if it were largely isolated from the others. O'Callaghan maintains that this *unimodalism* is a mistake. Contrary to widespread opinion, the connections between distinct perceptual modalities are rich and complex, and an account of perception that ignores this fact will, at best, be incomplete and, at worst, seriously misrepresent the nature of perception.

Another chapter that considers a specific mental phenomenon of traditional interest to philosophers is Jesse Prinz's chapter on the emotions. Though Prinz focuses primarily on issues about the individuation (or counting) of emotions, he does so with an eye to assessing competing theories of what emotions are. More specifically, he argues that many accounts of the emotions yield exceedingly implausible answers to the question of how many emotions we have, either implying that we have far fewer than is in fact plausible or far too many. Prinz then goes on to argue that his preferred account of the emotions—the somatic appraisal theory—yields a vague though plausible answer to the "how many" question.

A final example of a chapter that focuses on a specific mental phenomenon is Christopher Mole's on attention. Philosophers have seldom studied attention for its own sake. Rather, what interest it has garnered has, in large measure, been motivated by a concern to understand other phenomena (e.g., free will or consciousness). As Mole notes, however, attention has had a rather more central place within cognitive science than it has had in the history of philosophy, and largely as result of Donald Broadbent's influence has been the subject of intensive empirical enquiry. Mole argues that the best accounts of attention to have emerged from this research tradition—the so-called biased competition models—promise to have important implications for both the methodology of cognitive psychology and the metaphysics of attention.

2. META-THEORETIC ISSUES

The philosophy of cognitive science is not just empirically informed philosophy of mind; it is also part of the philosophy of science. Like the philosophy of biology or the philosophy of physics, it takes a region of science as an object of enquiry and addresses issues about its nature and practices. In the present case, the object of enquiry is the confederation of disciplines that make up cognitive science (cognitive psychology, linguistics, cognitive neuroscience, artificial intelligence, etc.) considered separately and in toto. Sometimes the goal of enquiry is largely descriptive—to articulate aspects of the scientific practices that are typical of cognitive science. But more commonly the goals are at least partially normative. Philosophers of cognitive science seek an understanding of how the science *ought* to be done. Again, some illustrations are in order.

2.1. Foundational Assumptions

One class of meta-theoretic questions that has figured prominently in the philosophy of cognitive science concern how best to characterize the foundational assumptions of the cognitive sciences. Such issues arise throughout the present volume, and some of them have already been mentioned. One very widely held view, for example, is that the information processing that goes on in the brain depends on the presence of mental representations, and that cognitive processes are, in some sense, computational processes. As noted earlier, both of these ideas receive extended treatment in the chapters by Egan and Piccinini.

Another example of a foundational assumption that has received considerable attention in the philosophy of cognitive science is what we might call the *mechanistic assumption*. According to this very widely held view, the mind is indeed a mechanism of some sort—roughly speaking, a physical device decomposable into functionally specifiable subparts. Moreover, given this assumption, a central goal for cognitive science is to characterize the nature of this mechanism, or to provide an account of our *cognitive architecture*. A huge amount of cognitive science is concerned in one way or another with the attainment of this goal. Issues that flow from the mechanistic assumption recur throughout this volume. For example, much debate about the nature of cognitive mechanisms has occurred in the context of debates over modularity. In his chapter, Samuels clarifies the notion of a module as it figures in recent disputes about whether the mind is massively modular. In the course of this discussion, he considers a variety of issues about the individuation of cognitive mechanisms as well as the relative plausibility of different accounts of the sorts of mechanisms that comprise the human mind.

2.2. Relationships between Disciplines

A second kind of meta-theoretic issue of widespread interest to philosophers of cognitive science concerns the relationships between, and potential contributions of, the various disciplines that comprise cognitive science. Cognitive science is, of course, a multidisciplinary field of research, but there remains enormous disagreement regarding the relevance and precise role of these various disciplines within the overarching program of cognitive science. Again, such issues are well represented in the present volume.

One issue that has been the focus of much heated debate concerns the role of evolutionary biology in the development of a science of cognition. At one extreme, some highly influential theorists maintain that evolutionary biology is of little or no value to the project of cognitive science (Fodor and Piattelli-Palmarini 2010). At the other extreme, there are researchers, especially evolutionary psychologists, who maintain that cognitive science is a branch of biology, and that evolutionary considerations furnish us with both a unified methodology for the study of cognition and the means to generate plausible hypotheses about the structure of the mind (Tooby and Cosmides 1992; Pinker 1997; Carruthers 2006). These are issues that Jeffares and Sterelny take up in their chapter on evolutionary psychology. In doing so, they reject both of these two extremes and set out a pluralistic conception of how evolutionary considerations can inform research in cognitive science that draws on a variety of evolutionary sciences, including behavioral ecology, physical anthropology, and population genetics.

Evolutionary biology is not the only discipline whose role within cognitive science has been a source of dispute. Another such discipline, one that Copeland and Proudfoot discuss in their chapter, is the area of computer science known as artificial intelligence (AI). Historically, AI has been hugely influential in the development of cognitive science. Indeed, on the conception of cognition that emerges from the seminal work of Alan Turing, Allen Newell, Herbert Simon, and David Marr, the goals of cognitive science and of AI are virtually identical: to provide computational accounts of cognitive phenomena, such as reasoning, problem solving, and vision (Marr 1982; Newell and Simon 1976; Turing 1951). Yet the significance of AI to the study of mind has also been the subject of concerted criticism, both on the basis of empirical considerations and on broadly a priori grounds. In their chapter on AI, Copeland and Proudfoot clarify the foundational assumptions of AI and consider some of the more prominent objections that have been leveled against it. In addition, they take a look at recent arguments that purport to show the technological inevitability of AI.

A final (and surprising) example of a discipline whose significance to the study of cognition has been much disputed is neuroscience. On its face, this seems very surprising indeed. After all, few would deny that brains have a central role to play in cognition! Nonetheless, as Grush and Damm make clear in their chapter on cognition and the brain, there has been a widespread tendency within cognitive science to argue that brain science has little to tell us about cognition per se. Moreover, even those cognitive scientists who do not accept such arguments have tended until relatively recently to make little use of neuroscientific results in the construction of their theories. Grush and Damm explore some of reasons for this tendency, and argue that cognitive science has been dominated by overly simplistic models of cognition, neural function, and their interrelation. For all that, they conclude by sounding a note of cautious optimism: we may be, as they say, finally on the verge of rendering tractable the mystery of how brains give rise to cognition.

3. CONCEPTUAL ISSUES

A third broad class of issues in the philosophy of cognitive science concerns the clarification and explication of core theoretical concepts. Cognitive scientists certainly make heavy use of a number of concepts that generate knotty theoretical and philosophical problems. Some of these have already been touched on in Section 2. For example, the notions of cognition, computation, representation, and consciousness have all been the subject of efforts at conceptual clarification, and the chapters in this volume that make central use of these concepts all engage in efforts to clarify them. In matters philosophical, issues about the meanings of our concepts are seldom far from the surface.

It is perhaps worth giving a more extended example of a concept that has received its share of attention from philosophers of cognitive science: innateness. Of
course, innateness has been a source of bafflement since well before the inception of
cognitive science. As far back as the eighteenth century, David Hume complained
that the concept permitted those incautious enough to use it to "draw out their
disputes to a tedious length, without ever touching the point in question" (Hume
1975, 18). Yet despite the frequent pronouncements of its demise, innateness continues to play a central role in much of cognitive science, especially when it comes to
developmental issues. Largely for this reason, philosophers of cognitive science have
expended considerable effort trying to clarify the concept and its role in scientific
theorizing. In their chapter, Gross and Rey provide a survey of this research. They
consider the main attempts to explicate *innateness* and argue that all are subject to
serious problems. Even so, Gross and Rey maintain that the lack of a good analysis
of this concept does not pose an insurmountable problem for its continued use in
cognitive science, and with this idea in hand they discuss ongoing disputes concerning the extent to which aspects of the human conceptual system are innate.

We have briefly mentioned some issues in the philosophy of cognitive science
that are readily subsumed under the general rubric of conceptual issues. But in case
it is not already obvious, we should stress that most philosophers of cognitive science think that there is *much more* for philosophers to do than *just* clarify concepts.
They thus assign a markedly different significance to conceptual issues than that
proscribed by some very influential views of philosophical methodology, such as
those associated with the so-called linguistic turn (see, e.g., Dummett 1996; Grice
1987; Jackson 1998; Moore 1966). According to these views, *the* central goals of philosophy in general—and the philosophy of special sciences in particular—are conceptual in character. In one familiar view of this kind, philosophy is in the business
of providing classical definitions, that is, sets of individually necessary and jointly
sufficient conditions, which capture the meanings of our words or the concepts
they express. In a related view widely associated with the work of Wittgenstein, the
analysis of our concepts is viewed as a means by which to expose the chimerical
character of philosophical problems and show that they are not genuine problems
about the nature of the world, but just pseudo-problems that result from linguistic
or conceptual confusion.

Though there remains considerable disagreement about the nature and goals of
philosophical enquiry—issues that we cannot begin to address here—the above
sorts of views are not widespread among philosophers of cognitive science. For one
thing, it has proven extraordinarily hard to provide sets of necessary and sufficient
conditions that adequately track pretheoretic intuitions about the application of
words, and indeed much research in cognitive science appears to cast doubt on such
projects (Fodor et al. 1980). Moreover, efforts to show that problems about the mind
are the product of linguistic confusion have not met with much success. On the
contrary, what the development of cognitive science appears to suggest is that many
of the most enduring philosophical issues about the mind—the kinds of issues

discussed earlier in this Introduction—are genuine ones that can be the subject of fruitful, empirically informed research.

Of course, this does not mean that careful attention to psychological concepts has *no* role to play in the philosophy of cognitive science. How we understand the ways of framing our intellectual projects has genuine implications for how research proceeds. (What could be more obvious?) And clarifying such notions as consciousness, representation, and innateness can be helpful, if for no other reason than because their uses are multifarious, and failures to draw the right distinctions can lead to confusion. As such, many of the activities that one might label as "conceptual analysis" may still have a role to play in the philosophy of cognitive science. Nonetheless, all this would typically be viewed as a preliminary step to addressing substantive issues about mental phenomena themselves.

4. First-Order Empirical Issues

One consequence of rejecting the view of philosophy as an exercise in conceptual clarification is to also reject a sharp distinction between philosophy and workaday scientific research. If philosophical questions are intimately concerned with substantive empirical issues, then the boundary between philosophy and science is unlikely to be a clear one. This attitude is very much reflected in the recent work of philosophers of cognitive science, where contributions have been made—sometimes in collaboration with scientists—to a range of ongoing empirical debates about the mind that are largely alien to traditional philosophy of mind. Such research is well represented in the present volume.

4.1. Theory of Mind

Some of the empirical issues on which philosophers of cognitive science have focused concern quite specific mental capacities. For instance, philosophers have been very active in research regarding what is often called *theory of mind*—our ordinary yet remarkable ability to interpret one another as psychological agents whose behavior can be predicted in terms of our beliefs, desires, and other mental states. Initial philosophical interest in this topic was motivated by its prima facie connections to debates in mainstream philosophy of mind (e.g., those concerning eliminativism and commonsense functionalism) (e.g., Lewis 1972; Churchland 1981). But recent attention has focused more on empirical issues about the cognitive mechanisms that are responsible for our theory of mind capacities. Alvin Goldman's contribution to the present volume provides an overview of research in this vein and defends a strong version of one particular approach to theory of mind abilities—the so-called *simulation theory*.

4.2. Language

Another empirical issue that has been of great interest to philosophers concerns the structures and mechanisms underlying our linguistic capacities. Of course, issues about language have occupied philosophers since antiquity, and the philosophy of language remains an active area of research. But the study of language has also played a pivotal role in both cognitive science and the philosophy of cognitive science. This is, in part, due to the profound influence of Noam Chomsky's early work—for example, his critique of behaviorism and the development of the generative linguistics—on the development of cognitive science (Chomsky 1957, 1959). But in addition, linguistics has proven to be an enormously fertile region of enquiry, and research within this field has often provided models for the pursuit of research on other, nonlinguistic domains of cognition.

Though debates over the nature and role of language recur throughout this volume, they receive their most concerted treatment in Pietroski and Crain's chapter on the language faculty. Pietroski and Crain focus on facts about how children acquire languages as a way of gaining insight in the domain of language. In doing so, they argue for a broadly Chomskyan conception of language according to which humans possess a language-specific system—a language faculty—that is governed by innate, though logically contingent principles that are unique to natural language.

4.3. Culture and Cognition

A third example of an issue that has become a focus for research among philosophers of cognitive science concerns the relationship between cognition and culture. Though philosophers have long been interested in culture, early philosophical interest in cognitive science tended not to assign a central role to cultural transmission in mental life. As a consequence, issues about the psychological mechanisms responsible for culture—and the concomitant influence of cultural factors on cognition—were not ones that received significant attention. In recent years, however, the relationship between culture and cognition has increasingly become a focus for research in cognitive science. No doubt, there are many reasons for this development, but one is simply the recognition that social learning and cultural transmission are pivotal to many of the most distinctive cognitive feats of our species, such as the development of science and technology. There is also increasing awareness of the significance of cumulative cultural evolution (Richerson and Boyd 2005). It is thus the burden of any satisfactory account of the human mind to characterize the processes and mechanisms that figure in the cultural transmission of knowledge. This burden is equally shared by theorists whether they endorse more nativist or more empiricist conceptions of cognitive development.

In their chapter on culture and cognition, Daniel Fessler and Edouard Machery provide an overview of recent work on the relationship between culture and cognition. They begin with a body of research that has a long tradition: the investigation of features of the mind that are universal, and of features of mind that vary across cultures. Underlying these features, they argue, are a mix of mechanisms of cultural transmission, including quite a few that may be domain-specific. Finally, Fessler and Machery note the effects of features of cognition on culture, that is, the influence that the structure of our minds has on how cultural information is retained and transmitted. Though the investigation of such issues is still fairly new, there is every indication that issues about the cognitive basis of culture will have increasing influence on theorizing in the philosophy of cognitive science—and rightly so.

5. Traditional Philosophical Issues That Do Not Ostensibly Concern Mental Phenomena

The kinds of issues we have considered so far are all ones that are either ostensibly about mental phenomena or about the nature and practices of cognitive science. This is how one would expect things to be. But there is a further kind of topic of interest to philosophers of cognitive science that does not readily fall within either of these categories. This last general category in our taxonomy concerns traditional philosophical issues that are most naturally construed as belonging to areas of philosophy that are not, in the first instance, about the mind at all, but where research from cognitive science may nonetheless be brought to bear.

5.1. Rationality

One family of such issues concerns the nature of rationality. On the face of it, such issues are *normative* ones. Some of these issues—roughly, the ones that concern how we ought to revise our beliefs—are traditionally viewed as the proper subject matter of epistemology. In contrast, those issues about rationality that concern how we ought to *act* are traditionally viewed as the proper subject matter of ethics. Yet in recent years there has been an increasing appreciation of the fact that empirical research in cognitive science may have implications for traditional normative philosophical issues.

Two contributions to the present volume follow this trajectory. The first—Allen, Todd, and Weinberg's chapter on the topic of reasoning and rationality—argues that a range of deeply entrenched assumptions about the mind underwrite a pervasive "Residual Cartesianism" about rationality that treats internal, individualistically characterizable mental states as the primary targets of normative assessment. Though they do not purport to show that this view is wrong, the authors of this chapter do maintain that research from cognitive science—on judgment and decision making in particular—suggest an alternative and more plausible *ecological* approach to rationality in which environmental and social factors are essential to an adequate characterization of rationality.

A second contribution to the present volume that focuses on issues of rationality is Doris and Nichols's chapter on moral psychology. The conclusions they reach are largely consonant with those of Allen, Todd, and Weinberg. In particular, they maintain on broadly empirical grounds that an individualistic "Cartesian" conception of practical rationality—as something that cognizers do best in isolation from peers and other social influences—is misguided. In its stead, they propose that an empirically defensible philosophical account of practical rationality will be more collaborativist than individualist.

5.2. Metaphilosophical Issues

Philosophers are inveterate navel-gazers, and the nature of philosophy itself quite often becomes an object of enquiry for them. Thus another kind of topic not ostensibly about the mind, but very much of interest to philosophers of cognitive science, is metaphilosophical issues about the nature of philosophy itself. In particular, philosophers have been greatly concerned with characterizing its proper domain of enquiry and the methods that ought to be used in theorizing about this domain.

There is a sense in which much philosophical attention to cognitive science has, since its inception, been built on contentious metaphilosophical commitments. Most obviously, the very idea that empirical research is important to the resolution of long-standing philosophical matters involves a departure from one traditional view of philosophy as a purely a priori discipline. Rather, the philosophy of cognitive science is invested with a commitment to a version of *methodological naturalism,* or the view that (many) philosophical issues ought to be addressed by methods that render philosophy continuous with the scientific enterprise. But in recent years there has emerged from the philosophy of cognitive science a more radical critique of certain metaphilosophical traditions in philosophy. In particular, there is a family of views, often associated with *experimental philosophy,* which recruits the empirical methods of cognitive psychology in order to show that commonplace philosophical reliance on intuition is fundamentally misguided. In the present volume, Joshua Knobe provides a survey of some of the empirical studies to have emerged from experimental philosophy, as well as a discussion of the metaphilosophical implications of this work.

6. Conclusion

In the forgoing sections we have sought to provide an overview of the kinds of topics that comprise the philosophy of cognitive science. The conception of philosophy more or less implicit in this field of enquiry is, we think, one that exhibits a number of important commitments. First, it is naturalistic in its methods, at least in the sense that empirical research of the kind produced within the sciences is taken to be of direct relevance to the resolution of issues within philosophy. Second, the kind of research that is pursued is pluralistic in a variety of respects. The kinds of *issues* that comprise its subject matter are, as we have seen, multifarious. But also—perhaps in part because of this—there is pluralism with respect to *methods*. The traditional philosophical project of clarifying concepts is very much in evidence. But so too is the invocation of empirical results, the widespread use of inferences to the best explanation, and an openness to empirical speculation. The philosophy of cognitive science even has its share of philosophers who directly engage in designing and implementing their own experiments when the need arises. Of course, the philosophy of cognitive science as represented in this volume is built on the methodological assumption that the deliverances of cognitive science can shed considerable light on issues about the nature of the mind, and perhaps even on a broader range of philosophical questions, including how philosophy itself should be practiced. As with most methodological commitments, there is no guarantee that this one will prove fruitful. But we think that the sorts of research presented in this volume provide ample grounds for optimism. We hope you agree.

REFERENCES

Anderson, J. (2007). *How Can the Human Mind Occur in the Physical Universe?* New York: Oxford University Press.

Brandom, R. (1994). *Making It Explicit: Reasoning, Representing, and Discursive Commitment.* Cambridge, MA: Harvard University Press.

Carruthers, P. (2006). *The Architecture of the Mind: Massive Modularity and the Flexibility of Thought.* New York: Oxford University Press.

Chalmers, D. (1996). *The Conscious Mind: In Search of a Fundamental Theory.* New York: Oxford University Press.

Chomsky, N. (1957). *Syntactic Structures.* The Hague: Mouton.

———. (1959). A review of B. F. Skinner's *Verbal Behavior, Language* 35(1): 26–58.

Copeland, B. J. (ed.) (2004). *The Essential Turing.* Oxford: Oxford University Press.

Churchland, P. (1981). Eliminative materialism and the propositional attitudes. *Journal of Philosophy* 78(2): 67–90.

Davidson, D. (1975). Thought and Talk. In D. Davidson (1984), *Inquiries into Truth and Interpretation.* Oxford: Oxford University Press.

Dummett, M. (1993). *The Seas of Language.* Oxford: Oxford University Press.

———. (1996). *Origins of Analytical Philosophy.* Cambridge, MA: Harvard University Press.

Fodor, J. (1975). *The Language of Thought.* Cambridge, MA: Harvard University Press.

Fodor, J., Garrett, M., Walker, E., and Parkes, C. (1980). Against definitions. *Cognition* 8(3): 263–67.

Fodor, J., and Piattelli-Palmarini, M. (2010). *What Darwin Got Wrong.* London: Profile Books.

Grice, P. (1987). Conceptual Analysis and the Province of Philosophy. In P. Grice (1989), *Studies in the Way of Words*. Cambridge, MA: Harvard University Press.

Hume, D. (1777/1975). *Enquiries Concerning Human Understanding and Concerning the Principles of Morals*, edited by L. A. Selby-Bigge (3rd edition revised by P. H. Nidditch). Oxford: Oxford University Press.

Jackson, F. (1998). *From Metaphysics to Ethics: A Defence of Conceptual Analysis*. Oxford: Oxford University Press.

Lewis, D. (1972). Psychophysical and theoretical identifications. *Australasian Journal of Philosophy* 50: 249–58.

Marr, D. (1982). *Vision. A Computational Investigation into the Human Representation and Processing of Visual Information*. San Francisco: W. H. Freeman.

Moore, G. E. (1966). *Lectures on Philosophy*, edited by Casimir Lewy. London: George Allen and Unwin.

Nagel, T. (1974). What is it like to be a bat? *Philosophical Review* 83: 435–50.

Newell, A. (1980). Physical symbol systems. *Cognitive Science*, 4(2): 135–83.

Newell, A., and Simon, H. (1976). Computer science as empirical inquiry: Symbols and search. *Communications of the Association for Computing Machinery* 19(3): 113–26.

Pinker, S. (1997). *How the Mind Works*. New York: W. W. Norton.

Pylyshyn, Z. (1984). *Computation and Cognition. Toward a Foundation for Cognitive Science*. Cambridge, MA: MIT Press.

Richerson, P., and Boyd, R. (2005). *Not by Genes Alone: How Culture Transformed Human Evolution*. Chicago: University of Chicago Press.

Tooby, J., and Cosmides, L. (1992). The Psychological Foundations of Culture. In J. Barkow, L. Cosmides, and J. Tooby (eds.), *The Adapted Mind: Evolutionary Psychology and the Generation of Culture*. New York: Oxford University Press.

Turing, A. (1951). Can Digital Computers Think? In B. J. Copeland (ed.), *The Essential Turing*. Oxford: Oxford University Press.

Whorf, B. (1956). *Language, Thought, and Reality: Selected Writings of Benjamin Lee Whorf*, edited by J. Carroll. Cambridge, MA: MIT Press.

Vygotsky, L. (1986). *Thought and Language*, edited by A. Kozulin. Cambridge, MA: MIT Press.

CHAPTER 2

....

CONSCIOUSNESS AND COGNITION

....

ROBERT VAN GULICK

CONSCIOUSNESS has only recently again become a major topic of serious inquiry in the scientific study of mind and cognition. During most of the twentieth century, consciousness was regarded as a marginal or even disreputable notion, but that was not so at the very outset of modern empirical psychology. The methods deployed during the first few decades of laboratory research from the end of the nineteenth into the early twentieth century relied heavily on introspection, and the nature and properties of conscious experiential states were a primary object of investigation. However, the rise of behaviorism brought a general anti-mentalism, with consciousness being regarded as especially suspect. The move away from behaviorism and toward a more cognitive approach in the 1960s lead to a renewed acceptance of mentalistic concepts and models, but consciousness still remained a largely neglected fringe topic until the later 1980s. Since that time, there has been an explosive growth of research on consciousness spurred in part by the advent of powerful new neuro-imaging technology. This renewed interest in consciousness has brought a flood of relevant data and a proliferation of theories. The scientific study of consciousness is still very much in its infancy, and there is as yet no consensus about its nature, substrate, function or role in our cognitive and mental life. All these issues are under active inquiry, and new theories, models and methods of investigation are being regularly introduced.

The following eight sections provide an overview of some of the major issues and directions at the current time in what is a rapidly evolving field of research. Section 1 gives a survey of ten distinct concepts of consciousness, each of which has some relevant application. The general question of the relation between consciousness and cognition is discussed in Section 2, which provides a general framework for the survey of theories of consciousness in the next three sections: philosophical theories in Section 3, cognitive theories in Section 4, and neurobiological theories in Section 5. Section 6 addresses some general issues about the aims and methods of consciousness research, and the last two sections survey some specific research programs and results in two major topic areas: consciousness and perception in Section 7, and consciousness and memory in Section 8.

1. DIVERSE CONCEPTS OF CONSCIOUSNESS

The words *conscious* and *consciousness* are used in a variety of ways both in research and everyday speech. None of the many senses of those terms is standard or privileged. Thus a diversity of usage is entirely appropriate. However, to avoid ambiguity and confusion, it is necessary to distinguish among those different meanings and be clear about one's intended use.

The adjective "conscious" is heterogeneous not only in its meaning but in its range of application. Sometimes it is applied to whole organisms or persons and at other times to particular mental states or processes. One can ask not only whether an iguana or a person under light sedation is conscious, but also whether a given perception or memory is conscious. Thus a standard division is drawn between the various senses of so-called "creature consciousness" which apply to *whole organisms* and the diverse ways in a specific *mental state or process* may be conscious, so-called "state consciousness." Within each of those two main categories, a variety of senses can be distinguished. Given their special concern with conceptual matters, philosophers have done a lot to clarify and categorize the different senses of consciousness in each of these two main families. At least five types of creature consciousness are typically distinguished.

1.1. Sentience

An organism may be said to be conscious simply in so far it can sense and perceive its environment and has the capacity to respond appropriately. However, what counts as sensing, perceiving, or responding in the relevant regard is not always clear. Some organisms surely pass the threshold; mammals, birds, and reptiles all seem obviously conscious in the sentient sense, and the same applies to cephalopods such as squid and octopuses. On the other hand, plants and oysters seem not to qualify even though they do in some limited respects receive and act on information from their environments. But what of caterpillars and jellyfish? On which side of the divide do they fall? In the absence of any theoretically supported rationale, there seems to be no principled way to draw the boundary.

1.2. Wakefulness and Arousal

A second sense of creature consciousness requires not merely the capacity to sense or perceive, but the current active use of those capacities. In this sense, a cat in deep sleep or a person under general anesthesia would not count as conscious. The sleeping cat is sentient—it has the capacity to sense—but it is not using that capacity, and thus not conscious in this second sense of being awake and alert. Again there are clear cases and unclear border cases. I am surely conscious in the awake sense as I write this sentence, as are you as you read it. But what about times when one is dreaming, drifting into sleep? What about a person under hypnosis or light sedation, or in a fugue state? Again there are many possible ways to draw the border, and opting for one rather than another seems arbitrary in the absence of supporting theoretical reasons of the sort that we yet lack.

1.3. Self-Awareness

A more demanding notion of creature consciousness requires that organisms be not only aware but self-aware. As well as being actively and sentiently aware of its world, such a creature must also be reflectively aware of its own awareness and of some significant subset of its own mental states and processes. Self-awareness comes in degrees and varies along multiple dimensions. One person or organism might be more self-aware that another by being aware of a *greater portion* of its mental states and processes or by being *more fully and richly informed* about them.

Like externally directed awareness, self-awareness can vary in its sophistication. It can be explicit and propositional, implicit and nonconceptual, or something in between. Which creatures count as conscious in the self-aware sense will again depend on where and how the threshold is drawn. Requiring explicit propositional self-awareness might well exclude human infants and even toddlers as well as most nonhuman animals, though perhaps chimps and other apes might qualify. However, if implicit nonconceptual self-awareness sufficed, then most mammals might count as conscious creatures in the self-aware sense, and perhaps many nonmammals as well. Squid and octopi and many birds might meet the standard.

1.4. Phenomenal and Qualitative Experience

Rather than focusing on cognitive abilities, conscious creatures might be defined as those that have an experiential life. Following a famous suggestion of the philosopher Thomas Nagel (1974), an organism is conscious if there is something that is it like to be that organism (i.e., if it has a subjective experiential inner point of view). To use Nagel's prime example, we may not know what is like for a bat to experience its world through echolocation, but we are confident that it is like something from the bat's point of view. Bat experience has a qualitative or phenomenal character, even if it involves qualia that we are unable to comprehend just as a person born blind might be unable to comprehend what is it like for us to see or experience the red of a ripe tomato.

1.5. Transitive Consciousness

Organisms are sometimes said to be *conscious of* various items or objects, for example, "The robin is *conscious of* the cat watching it from the window." Consciousness in this sense is understood as an intentional relation between the organism and some object or item of which it is aware. Thus it often referred to as *transitive consciousness* to indicate that it involves a relation to some object at which the consciousness is directed.

State consciousness also comes in many forms, and at least five types are commonly distinguished, reflecting to some degree the various notions of creature consciousness.

1.6. States of Which One Is Aware

Conscious mental states might be regarded as simply those of which one is aware of being in. On this reading, what makes a desire or memory conscious is simply the fact that one is aware of having the relevant desire or memory. By contrast, unconscious desires or memories are those of which one is not aware. David Rosenthal's (1997) higher-order notion of state consciousness essentially involves self-awareness in this sense. However, he identifies the conscious state not with the higher order awareness but with the first order mental object of awareness. An unconscious desire is transformed into a conscious one not by any intrinsic change in the desire itself, but by one's becoming aware that one has it.

1.7. Qualitative States

Conscious mental states are sometimes identified as those that involve so-called "qualia" or "raw feels," the qualitative experiential aspects such as the taste of a pineapple, the experienced blue of the sky, or the throbbing pain of a toothache.

1.8. Phenomenal States

In a similar but not necessarily identical sense, conscious states might be regarded as those that have phenomenal properties or phenomenal character.

The term *phenomenal* refers to how things appear in experience, and the terms *qualia* and *phenomenal property* are sometimes used interchangeably. However, others use the term *phenomenal* to refer not merely to raw feels but to the structure of experience (e.g., to the fact experience involves the awareness by a subject of a world of objects located in space and time). Taken in this richer sense, the phenomenal is not exhausted by the qualitative.

1.9. Access Consciousness

Following the philosopher Ned Block (1995), states might be regarded as conscious in a quite different sense in so far as their contents are accessible for report and application in a variety of ways. In this sense, which has relevance to much empirical work on consciousness, a state is *access conscious* if the information or content it carries is generally accessible by other modules or processes, such that the person can report the content of that state and use it in a variety of ways to guide internal or external behavior. For example, the difference between conscious and unconscious perception is a matter of the sort of access that the perceiver has to the content of what is perceived. What is unconsciously perceived may have some effects on behavior, but it is far less accessible for use and report than what is consciously perceived. The notion of access consciousness thus may be a graduated one rather than an all-or-nothing matter.

1.10. Narrative Consciousness

In humans at least, our mental life is accompanied an ongoing interpretative account of our feelings, thoughts, and actions—the so-called "stream of consciousness" or "self-narrative" (Dennett 1991). Conscious mental states might be identified with those that get included in the narrative or stream. How accurately that narrative reflects the reality of our mental states and processes is open to debate (Gazzaniga 1988) but insofar as it delimits how we experience ourselves, it provides one plausible means of distinguishing conscious from nonconscious states; those in the narrative stream count as conscious and others do not.

These ten senses or concepts of consciousness do not exhaust the possible types of creature consciousness or state consciousness. Nor should they be regarded as mutually exclusive; indeed there may be important theoretical links among them, and some combinations may be interdependent. Nonetheless, they do provide a basic starting point for sorting out the variety of senses in which organisms or states and processes are described as conscious in everyday life and scientific research.

2. THE RELATION BETWEEN CONSCIOUSNESS AND COGNITION

From the cognitive science perspective, the relation between consciousness and cognition is a central issue. How are the two related? Is one primary and the basis for the other? If so, which is basic, and which is derived? Are they two distinct and independent mental phenomena that perhaps interact, or is one a form or subtype of the other? (Van Gulick 1995).

Recent theorizing, as surveyed below, offers a diversity of answers to these questions. Some opt for a one-way dependence. However, among those who do, there is disagreement about the direction of dependence. Some regard consciousness as more basic, and others conversely analyze consciousness in terms of cognition or representation. According to some, (Searle 1990; Horgan and Tienson 2002) there can be no genuine mentality or cognition without subjective consciousness, which is thus the more basic feature of mind. On the other hand, representationalists such as the philosopher Michael Tye (1995, 2000) or higher-order theorists such as David Rosenthal (1997) analyze and explain consciousness in various ways in terms of representation and cognition, which are thus more basic from their perspectives. Subtype views also have proponents especially among those empirical researchers who take many types of cognitive activity—perception, memory, or action control—to occur in both conscious and nonconscious forms. Such theorists accept the reality of unconscious cognition, but they aim to discover how it differs from conscious forms of cognition (Merikle, Smilek, and Eastwood 2001). Even if they do not aim to fully explain consciousness in terms of cognition, they hope to discover how conscious cognition differs from unconscious cognition.

To sort through these options, one needs to consider some of the specific theories of consciousness that have been proposed. Although there is some overlap, they can be grouped in three main categories: philosophical theories, cognitive theories, and neurobiological theories.

3. Philosophical Theories of Consciousness

3.1. Representational Theories

While it is generally agreed that conscious states, or at least most conscious states, have representational or intentional properties, representationalist theories go farther in claiming that conscious states have no mental properties other than their representational properties. Thus two conscious states that are the same in all representational respects would be of exactly the same mental type (Harman 1990; Tye 1995, 2000). Representationalist theories aim to fully explicate the nature of consciousness in terms of representational features. Representation per se is obviously not sufficient for consciousness insofar as there can be unconscious or nonconscious mental states with representational content. Thus representationalist theories aim to explicate the nature of conscious mental states in terms of their involving a distinctive range of contents, their using distinct modes or formats to represent their contents, or their having distinctive functional roles associated with those contents.

The representationalist approach is motivated in part by a desire to avoid any commitment to qualia as mental but nonrepresentational features of conscious states. In so far as such qualia are thought to pose an intractable obstacle to any physicalist account of consciousness, their elimination would be a big plus for physicalism. Representationalists often appeal to the supposed "transparency of perceptual experience" in support of their view (Harman 1990; Tye 2000). According to the transparency thesis we are not aware of—and indeed cannot become aware of—any features of our perceptual experience other than how the experience represents the world as being. Experience is "transparent" in that "we look right through" it to the world. All we are supposedly aware of in perception is the represented world; we are not aware of any properties of the process or vehicles of representing, but only of what is represented.

Much of the dialectic between representationalists and their critics centers on the disputed interpretation of cases in which the critics allege that the representational and mental natures of conscious states diverge. Some involve pairs of states that are alleged to differ mentally despite being the same in all representational respects, that is, despite sharing their contents and all their other representational features. Supposed cases of this first sort include imaginary inverted qualia examples (Block 1996) and cases of objects (such as trees) visually represented as being the same size but viewed at different distances and thus differing in visual appearance (Peacocke 1983). Other cases allegedly involve the converse, conscious states that are mentally the same in some key respects, such as their associated qualia, despite their representational differences. The philosopher Ned Block has offered the "Inverted Earth" thought experiment (Block 1996) as a case of this second sort, and ambiguous reversing percepts such as those of the famous duck-rabbit might also qualify. Representationalists have responded, and the interpretation of such cases remains an open controversy. Thus so too does the status of representationalist theories in general.

3.2. Higher-Order Theories

Higher-order theories analyze consciousness as a form of self-awareness. Conscious mental states and processes are simply those we are immediately aware of being in. According to higher-order theories, the process of transforming a unconscious mental state M (e.g., an unconscious memory or desire) into a conscious one is not a matter of changing the state M itself, but of a generating a higher-order state H whose content is that one is in M. By becoming aware of M in that way, we make M a conscious state (e.g., a conscious memory or desire) (Rosenthal 1986, 1997). To use the searchlight metaphor, H is the spotlight and M is conscious because it is in the beam.

Higher-order theories come in several forms. Some treat the requisite higher-order states as perception-like, and thus the process of generating such states is a kind of inner perception or perhaps introspection (Lycan 1996). Others explicate the process in terms of higher-order thoughts and eschew any model of inner perception (Rosenthal 1986, 1997 Gennaro 1995). While most higher-order theories require the presence of an actual occurrent higher-state to confer conscious status on the mental state at which it is directed, others require only the disposition to give rise to such a higher-order awareness (Carruthers 2000). On these latter theories, a state M's having the mere disposition to produce a higher-order thought or perception of itself suffices to make M conscious.

Higher-order theories deal well with the notion of consciousness as self-awareness, but less obviously suffice in accounting for other forms of consciousness such as those associated with phenomenal or qualitative notions. Indeed one standard criticism of higher-order theories is that it would be easy to build systems that had the relevant meta-states but that nonetheless lacked any what-it-is-ness or any phenomenal or qualitative inner life (Chalmers 1996).

Simply giving an organism or system the power to monitor and represent its own representational states seems insufficient to produce any consciousness in those latter subjective senses. Other critiques focus on the possibility of higher-order illusion or hallucination. If the conscious-making process is a matter of generating the higher-order state, what happens in cases in which no actual mental state corresponds to the higher-order thought or perception? If one is not really having a perception of blue, but has a higher-order thought that one is, can there be any blue what-it-is-likeness or blue phenomenality given the absence of any perception with the relevant property? Higher-order theorists have responded in detail to these criticisms (Lycan 2006; Rosenthal 1997; Carruthers 2000), and lively debate continues on the viability of the higher-order approach.

The space of reflexive theories has expanded in recent years to include what are sometimes called "same-order" theories of consciousness. Like standard higher-order theories, these theorists regard consciousness as having an essential reflexive or self-representational aspect. However, rather than locating that reflexive or metamental content in a distinct higher-order thought or perception, they analyze it as a component of the intentional content of the conscious state itself (Kriegel 2006). Some such models explicate the relevant reflexive content as the consequence of the states being recruited into a globally integrated complex of state that forms the continuously shifting substrate of conscious experience. These latter models thus in some ways aim to combine elements of higher-order theory with features of global integration theories.(Van Gulick 2004, 2006).

3.3. Multiple Drafts Model

The philosopher Daniel Dennett (1991) has offered the so-called "multiple drafts model" of consciousness. The model rejects any notion that conscious mental states are those that occur in some privileged location or module of the brain or in some special format. There is no inner observer, no self who watches the flow of consciousness on an inner screen in what Dennett calls the "Cartesian Theater" of the mind, a metaphor which he claims is implicit in much of our confused thinking about consciousness.

According to Dennett, contentful representations are continuously being generated and revised at multiple locations through the brain. The difference between those that are conscious and those that are not is a graduated matter of how much influence, persistence, and effect on other contents they have throughout the system. Using another metaphor, he describes consciousness as "fame in the brain." There is no sharp line between those states or contents that are consciousness and those that are not; it is simply a matter of relative influence and impact, including on what we are likely to say or do. Dennett denies the existence of qualia and any suggestion that conscious states differ from nonconscious ones by possessing distinctive qualitative or phenomenal properties. In the human case, influential contents often get incorporated into the more or less serial narrative that emerges from the interpretative activity of the brain. Thus the multiple drafts model also has relevance to the notion of narrative consciousness.

3.4. Qualia Realist Theories

A diverse array of other philosophical theories agree on regarding qualia and phenomenal properties as both real and mentally basic, where "basic" implies not being fully explicable in terms of representational properties or higher-order awareness. Sometimes the point is asserted as a belief in qualia as "mental paint," a metaphor meant to capture the idea that our mental states involve introspectible media of representation as well as contents (Block 1996). Others simply assert that consciousness is an intrinsic property of our mental states, indeed one that is required for any genuine mental representation and thus incapable of being defined in terms of it (Searle 1990, 1992).

Most supporters of such theories are nonetheless physicalists and regard such properties—qualia or intrinsic consciousness—as fully caused and constituted by their underlying neural substrates in ways as yet to be scientifically discovered (Searle 1992). However, some qualia realists doubt that these properties can be understood as physical, and they thus embrace some form of mind-body dualism, or at least a dualism that admits the reality of both physical and nonphysical properties (Chalmers 1996).

4. COGNITIVE THEORIES OF CONSCIOUSNESS

Though cognition plays a role is some philosophical theories of consciousness, such as the multiple drafts and representationalist models, it figures even more prominently in a number of psychological theories of consciousness. Psychological theories in many cases also address the question of how consciousness relates to other aspects of standard cognitive models such as attention, executive control, working memory, and the distinction between automatic and controlled processing.

4.1. Global Workplace Theories

Global workplace models, as initially developed by the psychologist Bernard Baars (1988), employ a theater metaphor of the mind in which multiple specialized processors or modules compete for the attention of the "audience." Because the audience, like focal attention in most cognitive models, is capacity-limited only in that the outputs of a subset of processors will be conscious at any given moment. The success of a processor in gaining the audience's attention will depend both on its own level of activity and its ability to form mutually supporting coalitions with other processors based in part on the correspondence or coherence among the contents of their respective outputs. At any given moment, a single coalition typically dominates in a more or less "winner take all" fashion, and the content of consciousness shifts continuously as one coalition replaces another.

Employing another metaphor, the contents associated with the momentary dominant group are said to be "broadcast" throughout the system and thus widely available for use by other modules, on a par with the role sometimes attributed to working memory. By being broadcast, those contents also have the potential to recruit other processors into the global mutually reinforcing coalition. Global workspace models also provide a role for context and systemic priorities such as those that influence attention. The tendency of the system as a whole to settle into one or another dominant state is modulated in part by general background factors, including expectations, current goals, and past experience. Dennett's (1991) multiple drafts model is inspired in part by the global workspace theory with which it shares many features. Some neurobiological theories of consciousness have also drawn on global workspace theory, which continues to have wide influence in the field.

4.2. Intermediate Level Representation Theories

The intermediate level representation model has been offered as a partial account of consciousness (Jackendoff 1987; Prinz 2000, 2005). Its primary concern is the contents of conscious experience. It draws upon the common view that perceptual processing proceeds through a succession of content levels that deal with progressively more sophisticated and abstract contents. For example, early or low level vision is often taken to deal with features of the retinal image, middle or intermediate vision with the geometrical properties of the external scene, and late or high level vision to be concerned with computing the categorical and specific identities of objects and persons. Similarly the auditory processing of speech deals first with acoustic properties, then phonological features, and finally with meaning.

The central claim of the intermediate level theory is that the contents of conscious experience are limited to those at the intermediate level. What is represented in visual experience are the features computed at the middle level of processing and content: the colors, shapes, sizes, positions, distances, textures, and motions of the perceiver's visual surroundings. Under this theory, consciousness represents neither low level features associated with the visual image itself nor high level features associated with the identity or category of objects, people, and actions. Such information can often be inferred from what is consciously represented, and often is inferred quite automatically, but it is not actually represented in consciousness.

If one is looking at a mobile phone, all that is consciously represented according to the intermediate level theory are the shape, size, color, and location of the phone and its parts. One quickly recognizes it as a phone, but that categorical nature is not itself consciously represented. The claim is controversial and rejected by those who hold a "thick" notion of conscious content according to which the awareness of a object's kind and identity is reflected in conscious experience itself rather than being restricted to cognitive states that merely accompany conscious experience without themselves being conscious (Siewert 1998; Strawson 1994, 2006).

However, even if conscious content is restricted to the intermediate level, representation of such content is obviously not sufficient for consciousness. Such contents can be unconsciously processed. According to the theory, the extra requirement is the addition of attention. Such contents become conscious only when they are the objects of attention, which increases their activation and makes them available to working memory. According to Prinz (2005), they need not actually be stored in working memory. The requirement is only that they be modulated by attention in ways that make them available to working memory in order to be conscious. The viability of the intermediate level theory thus depends in part on our evolving understanding of attention and working memory.

4.3. Information Integration Theories

Cognitive theories of a very different sort focusing on the integration of information have been proposed by Gerald Edelman and Giulio Tononi (2000), with further development by Tononi (2004). The key idea behind this approach is that conscious mental states are distinguished by the very high degree to which they integrate diverse items of information. Integration thus involves both diversity or differentiation and unity. Many distinct and diverse contents are nonetheless bound together in a coherent way. Such integration results in a very high information value for conscious contents. For example, if one considers all the information carried by a momentary visual experience, this constitutes an enormous reduction in uncertainty relative to the vast class of all possibilities that could be represented by other visual experiences.

Tononi (2004) has provided a specific mathematical means of measuring the degree of integration in such systems, which he labels *phi*. Without giving the details here of how phi is computed, the outcome is that complex systems with highly differentiated but interconnected contents will show a high phi value, while simpler systems with less differentiated contents or less interconnection will have lower phi values. In a complex dynamic system such as the brain there will be multiple shifting subsystems or coalitions, each with its respective transient phi values. At any given moment, the coalition or subsystem with the highest phi value will be conscious, and the content of consciousness will correspond to the informational content of that subsystem.

Since phi can take a continuous range of values, the model regards consciousness as also admitting of degrees. Insofar as the phi values associated with the dynamic coalitions of a frog's brain show a lower phi value than those associated with a human brain, the frog would have a lower degree of consciousness. Nor is consciousness in this view limited to brains or comparable biological systems. Any information system—whether biological, electronic, or otherwise—that exhibits differentiation and integration sufficient to receive a phi value will exhibit a comparable degree of consciousness. In terms of an underlying neural mechanism, the integration theory appeals to the reentrant circuits of the thalamo-cortical system originally identified as a possible substrate of consciousness by Edelman (1990). At this point it is not possible in practice to measure or compute the phi values of actual brain systems. The theory nonetheless presents an intriguing abstract framework for distinguishing conscious from nonconscious mental processes.

5. Neurobiological Theories of Consciousness

Most philosophical and cognitive theories of consciousness such as those surveyed above assume that it has a neural substrate. However their explanatory focus is on higher level features realized by those neural mechanisms rather than on the detailed nature of the substrate itself. By contrast, neurobiological theories put more emphasis on the identity and nature of the neural structures and processes. Nonetheless, they are typically associated or motivated at least in part by their affinity with one or another higher level functional model of consciousness. There are too many such theories to cover all of them here, but a selection of three should provide a representative sample.

5.1. Neural Synchrony Theories

The neurophysiologist Wolf Singer has proposed a model of perceptual consciousness that relies heavily on neural synchrony, that is, on the synchronous phase-locked firing among assemblies of neurons (1999). Singer's model incorporates at least two types of neural representation within perceptual systems. Some neurons on the ascending pathway code information in a local fashion by serving as highly selective detectors that fire only in response to the input patterns produced by very specific stimuli. A second coding system, which Singer sometimes refers to as involving meta-representations, consists of complex assemblies linked to those local detectors or even built out of them. Information in this second system is encoded more globally by unified overall patterns of activity. It is here that neural synchrony plays its role. Arrays of local neurons are bound together into a coherent representational assembly by their synchronous and mutually reinforcing firing patterns (Crick and Koch, 1990). The relevant oscillations are typically taken to be in the 40 Hz range, which thus provides the interval during which the contents of conscious perception can shift.

5.2. Neuronal Global Workspace

Neural models proposed by Dehaene and Naccache (2001) are closely related to cognitive theories of the global workspace. Like the cognitive theories, these neuronal models distinguish between specialized processors or modules and more global processing systems, identifying the latter with assemblies of long axon neurons in prefrontal, cingulate, and parietal cortex. Through their descending pathways these long-range neurons amplify or suppress the activity of other neurons located within more specific processors, and thereby coordinate highly flexible globally coherent processing to carry out complex and effortful cognitive tasks. It is the outcomes of these processes that provide the contents of consciousness, though only some and not all such contents become conscious, with attentional mechanisms playing a selective role.

5.3. Local Recurrence Models

A more localist theory of perceptual consciousness has recently been proposed by Victor Lamme (2006). He notes that though activation spreads forward through the entire visual processing system in less than 100 ms, perceptual experience does not occur unless there is a subsequent recurrent spread of activation back from areas associated with higher processing into earlier areas associated with more low level processing. If the backward spread is prevented by one or another method, perceptual awareness does not occur. Lamme distinguishes this backward spread within the visual processing system from more global spreading that may make contact with the sort of prefrontal areas associated with neuronal global workspace models. He argues that interaction with those frontal areas may be necessary for visual attention, but not for mere visual awareness or visual experience (2006).

Lamme thus draws on Block's (1995) distinction between phenomenal consciousness and access consciousness to suggest that the local recurrence within the visual systems may be sufficient for phenomenal visual consciousness, but that more global involvement with frontal systems may be needed for access consciousness and for the ability to make verbal reports on the contents of one's visual experience.

These three are just a few of the many neurobiological theories of consciousness that have been offered, but they provide a good illustration of how work on neural models interconnects with more functional or high level models of the sort that would be of most immediate relevance to the study of cognition.

6. Aims and Methods of Consciousness Research

There have been long-running debates about how consciousness should be studied, including those concerning the value of first person versus third person approaches. Because much of our conscious mental life is directly accessible through introspection or other forms of self-awareness, it offers a means of observation not generally available in the sciences. We cannot directly observe our genes or our immune systems, so there is no alternative in such cases to gathering data through external third person means. Consciousness by contrast offers at least the possibility of collecting data directly from the first person perspective that each of us bears to our own conscious mental processes and states.

As noted above, introspective methods played a major role in early psychological research, but then fell out of favor. In part this was due to difficulties about how to interpret introspective reports. The data generated by introspective means are not the raw facts of consciousness but the reports that subjects give about their conscious experience. Even when researchers serve as their own subjects, they must still convert their first-person observations into reports in order for them to be used as scientific data.

The unavoidable reliance on such reports introduces a potential problem about the meanings that different first-person observers give to the words they use to report on their experiences. Given the impossibility of checking those reports against the private facts of consciousness that they purport to describe, there is the danger that different reporters mean different things by their descriptions. Nor is this an idle worry since real controversies arose about just such concerns, as in the early debates concerning the existence of imageless thoughts, which were regularly reported by subjects in some laboratories but uniformly denied by those in others.

Moreover, introspection has long been regarded as fallible and open both to biases and the possibility of confabulation. Subjects may sometimes sincerely but inaccurately report the mental states they believe they should be in rather than those they actually are in (Gazzaniga 1988). Given all these worries, there has been a general reluctance to rely on introspection, and a corresponding preference for third-person methods.

Nonetheless first-person awareness presents a rich source of information about our conscious mental lives, and it would be unwise to simply disregard it. Nor need first-person methods be limited to those of causal self-observation by untrained observers. The philosophical tradition of phenomenology has developed rigorous and systematic methods for first-person observation, and based on that data has produced a rich body of theory about the structure and organization of consciousness (Husserl 1913, 1929 Merleau-Ponty 1945). Indeed empirical researchers have recently gained important insights about the nature of cognition and consciousness from phenomenological theories concerning diverse topics such as the role of implicit background knowledge, embodied cognition, and our experience of time (Varela, Rosch, and Thompson 1991; Thompson 2007).

The conflict between first-person and third-person methods has echoed in contemporary debates concerning the respective merits of subjective and objective criteria for determining when a mental process or percept is conscious. Such criteria are especially important for contrast studies that aim to compare the functionality or neural substrates of conscious processes with those of unconscious processes (Merikle, Smilek, and Eastwood 2001). Contrast studies thus require clear criteria for deciding when a process or percept is conscious.

Subjective criteria rely primarily on subject reports. A percept or memory counts as conscious if and only if the subject reports being aware of the relevant percept or memory. If he denies awareness, then it counts as unconscious. Objective criteria by contrast rely on behavioral performance factors other than reports. For example, having been briefly shown a target stimulus under near threshold conditions, the subject may then be asked to pick or guess the identity of the target from a display that includes both the target and a number of previously unviewed distractors. If the subject can correctly guess or pick out the target from among the distractors, then he is judged to have consciously perceived it on the initial presentation.

Subjective and objective criteria often coincide (Merikle, Smilek, and Eastwood 2001). When a subject is unable to perform the identification or other behavioral task, he normally will also have had no experience of the target that he would subjectively report. And if the subject were to report a subjective experience of the target despite failure on the behavioral task, the relevant behavioral criterion would not normally be regarded as a valid measure of lack of awareness. The difficulty lies with cases in which subjects perform above chance on the behavioral task but fail to report any subjective experience of the target. If the positive behavioral performance is taken to indicate conscious perception in such cases, as it sometimes is by those who favor objective measures, it may lead to any overly expansive view of conscious perception and a correspondingly too-limited view of the scope of unconscious perceptual processing (Merikle, Smilek, and Eastwood). Thus what such research reveals about the relative scope and nature of unconscious and conscious processing may well depend on whether one uses an objective behavioral measure or subjective report criterion for determining when processing is conscious.

Most research studies on consciousness can be grouped into a few main families based on the type of questions they aim to answer and the variables they rely on to investigate those issues.

As to their goals, some studies aim to discover the neural correlates of consciousness and conscious processing, with the ultimate goal being to explain the neural basis or substrate of consciousness. Other studies are more concerned with uncovering the functional differences between conscious and unconscious processing, and with the consequences such differences have at the level of cognitive or behavioral capacities. The two sorts of questions are distinct but often linked. As was seen above in the survey of theories, models of the functional organization of consciousness can both guide and be constrained by models of its neurobiological basis. There is a mutual interplay: discoveries about the kind of processing that consciousness involves can provide insight into its neural basis as well as the converse.

As to the means of investigating these goals, the most common methods involve difference or contrast studies that attempt to discover the functional or neural differences between cases of conscious and unconscious processing.

7. Conscious and Unconscious Perception

The contrast method is most easily understood with respect to perceptual cases. A typical study would aim to produce paired cases where a given stimulus is consciously perceived in one case and perceptually processed without conscious awareness in the other. The two cases might then be compared using various neural imaging techniques—fMRI, PET, EEG, MEG—to determine differences in the neural structures and patterns of activity in the two cases. Alternatively one might investigate differences in the impact or effects of the perceptual processing in the two cases. For example, one might attempt to discover to which features of the stimulus the subject's behavior is sensitive in the nonconscious case, and how those features compare with the ones to which she is sensitive in the conscious case. How do the results of the two sorts of processing affect subsequent memory? Or what sort of factors (including other contents) can affect the outcome of processing in the unconscious cases versus the conscious case?

The contrastive cases are sometimes produced by varying the stimulus conditions, for example, by shortening the duration of the exposure or decreasing the luminance contrast to the point where conscious awareness of the stimulus is lost. The use of backward masking, pioneered by Anthony Marcel (1983), is especially effective in blocking conscious awareness. If the stimulus is presented for a short duration such as 40 ms and then followed immediately by another stimulus, subjects report no awareness of the initial stimulus. However, behavioral tests show that the initial masked stimulus has indeed been processed, though just which features of it get detected remains in dispute.

If for example the masked stimuli are words, subjects will show a bias in favor of the target on a subsequent word stem completion task despite reporting no awareness of having seen any initial stimulus at all. Given word "motor" as the masked target, subjects are more likely to complete the stem "mot-" as "motor" rather than as "moth," "mother," or "motel" in comparison with matched controls who had not been exposed to the target. Some have claimed that the masked targets are processed all the way up to the semantic level and that they have found behavioral effects to support such high level unconscious processing, but others who have not gotten such results claim that unconscious processing is restricted to more structural features of the stimulus.

Other techniques to produce contrast cases manipulate attention. For example, subjects who are instructed to carry out a given attention-demanding task will often report not seeing a stimulus presented to focal vision well above threshold in duration and contrast if it not of the sort for which they are looking (Rock and Mack 1998). One explanation of such cases is that conscious perceptual awareness requires attention, and given that attention has been directed to the selective detection of other stimulus features, the target stimulus is not consciously processed.

Attention deficits can also prevent conscious processing in neuropathological cases of neglect or extinction. In neglect cases, imaging studies show that the stimulus appears to be processed up to the high level features, yet the patients report no awareness of the stimuli and typically show no behavioral evidence of having perceived it (Driver and Mattingley 1998). However, such effects do sometimes implicitly manifest themselves (e.g., a patient may show avoidance behavior in response to the presentation of an aversive visual stimulus of which he denies any awareness). In extinction cases, the patient reports seeing the stimulus if it is shown to his neglected visual field without any competing stimulus in his non-neglected field, but does not see it if a competing stimulus is simultaneously presented to his non-neglected field. The latter stimulus extinguishes all perception of the former, most likely by capturing all of the patient's attention.

So-called "blindsight" (Weiskrantz 1986) also provides cases of visual processing in the absence of conscious awareness. These patients have damage to cortical areas such as V1 that are involved with early visual processing. They report no visual awareness of stimuli presented to the relevant portions of their visual field. Yet if asked to guess or randomly point to a direction where a stimulus might have been presented, they do so accurately. They sometime show other residual visual capacities such as the ability to correctly guess the orientation of a grating, to guess whether a moving stimulus was presented as well as its direction, and in some cases to produce appropriate hand motions to grasp presented objects.

These patients do not make any of these responses unless cued to do so since they deny having any visual awareness of the stimuli, and they typically perform most accurately if they are instructed to just guess or act in whatever way they choose rather than being asked to try to respond the stimulus. Though there is some debate about the basis of the preserved unconscious visual processing in blindsight cases, the most accepted hypothesis is that they derive from other secondary visual pathways, such as that from the thalamus to the superior colliculus that are involved in orientation responses and that do not support conscious visual perception (Weiskrantz 1986).

Direct brain manipulations are also sometimes used to produce contrast cases. For example, transcranial magnetic stimulation (TMS) can be used to disrupt the sequence of perceptual processing. Evidence about which disruptions prevent conscious perception and which do not can help indicate which neural structures and processing stages are involved in conscious perception (Logothetis and Schall 1988).

The distinction between conscious and unconscious perception has also figured prominently in research on what has come to be called the "two visual streams," which refers to two anatomically distinct processing pathways that receive inputs from early visual areas (Goodale and Milner 1992, 2004). One pathway, the ventral stream, radiates to largely temporal areas, and appears to be concerned with object recognition, categorization, and conceptualization. The other pathway, the dorsal stream, links to largely parietal regions that are concerned with the control of action. Patients with selective damage to one cortical area and not the other show dissociated patterns of visual deficits. For example, those with temporal damage but intact parietal areas cannot correctly describe the orientations of objects or slots, but they are able to correctly insert the objects in the relevant slots and to shape their hands correctly for grasping. Those with the converse pattern of damage can give correct descriptions but are unable to use that information to guide motor responses, a condition referred to as optic ataxia. Goodale and Milner (2004) argue that only the ventral stream processing supports visual consciousness and that the dorsal stream does not do so. The matter remains under debate and investigation.

8. Conscious and Unconscious Memory

Although perception has been the primary focus of research investigating conscious and unconscious cognition, major work has also been done with regard to memory. As with perceptual studies, an important goal has been to discover the differences, both functional and neural, between conscious and unconscious forms of memory.

Given everyday common sense notions of memory, the concept of unconscious memory might seem contradictory. If memory of an event means recalling and being able to report on it or describe it, then how could such an activity be unconscious? Of course, memory researchers use the term *memory* to refer to many other ways in which information about the past may be stored and have a relevant impact on subsequent behavior, and unconscious memory may be possible in these other senses. Somewhat surprisingly, cognitive psychologists were slow to acknowledge the existence of unconscious memory, perhaps because they wished to avoid association with popular Freudian notions of repressed memories of the sort supposedly uncovered by psychoanalysis. However, in recent decades that has changed, and there has been active research and theorizing about the nature of unconscious memory processes.

The conscious versus unconscious distinction in contemporary memory research largely tracks that between what is called "explicit memory" and "implicit memory" (Schacter 1987). Explicit memory is typically manifest by the ability to recall or recognize previously observed items. Implicit memory by contrast normally can exist in the absence of any such recall or recognition abilities. It most typically manifests itself in priming effects, that is by enhancing or speeding the performance of tasks with regard to previously viewed items, but which do not involve actually reporting or recognizing those items as having been previously presented.

Amnesic patients have played an important role in establishing the existence of implicit memory in the absence of explicit memory. Amnesic patients have suffered some form of brain damage, such as bilateral lesion of the hippocampus, that has left them unable to form and recall new explicit memories of events for items they have clearly and consciously perceived. As shortly as a few minutes later they will deny knowledge of the experience and fail to show any memory of the relevant items on recall or recognition tasks. However, if they are tested indirectly, as on a word stem completion task, they will show priming effects such as an enhanced likelihood of completing a stem with a word that they recently viewed, despite having no explicit memory of having seen it or even of having viewed a list of words.

Implicit memory is of course not restricted to amnesic patients, and under suitable conditions normal subjects will also show implicit memory effects in the absence of explicit memory. The distinguishing mark of implicit memory is the effect of information about past experience on current behavior in the absence of any awareness of the past experience, and it is that absence of awareness that justifies labeling it as unconscious memory.

There are functional differences between the two types of memory. Factors that affect one do not necessarily affect the other. For example, the depth of processing that the subject applies to the stimulus on initial exposure is known to affect explicit memory. The deeper the initial processing, the greater is the likelihood of subsequent recall. For example, words are more likely to be remembered if at initial exposure the subject had to answer a semantic question about the word—such as "Does it refer to an animate thing?"—than if the task involved answering a merely phonological question, such as "Does it rhyme with 'cake'?" Depth of processing however has no effect on implicit memory performance.

There are many competing theories to explain implicit or unconscious memory and its persistence in the absence of explicit memory. Some regard implicit and explicit memory as two separate systems with distinct neural bases (Schacter and Tulving 1994), though there do not seem to be any cases in which implicit memory is lost while explicit memory is preserved, which would seem possible if they really were distinct and independent neural systems. Others view the two as based in a single memory system but explain the dissociation in performance as the result either of the different sorts of information required for implicit versus explicit tasks or differences in the ways in which the two sorts of tasks access that single memory system. For example, Roediger and McDermott (1993) argue that implicit tasks are bottom-up perception driven and require information only about the surface features of the remembered items such as the structures of the words, while explicit memory tasks are conceptual driven and require access to information about meaning, semantics, or gist. The loss of explicit memory could thus result from the loss of the relevant sorts of information. Alternatively the dissociation might turn on how the information is accessed. Jacoby (1991) has argued that explicit memory tasks typically involve controlled attentional processing while implicit memory tasks are largely automatic and unconscious. The latter form of access might persist even if the former is lost.

With memory as with perception, the distinction between conscious and unconscious processing is now regarded as a serious research topic: one that may shed important light on the nature and neural bases of cognition and consciousness and their relation to other important mental phenomena such as attention and self-awareness.

9. CONCLUSION

Consciousness research is now well established as an important if less than mature field of empirical investigation and theorizing. Some current theories and models of consciousness will no doubt be discarded as the field develops, but some may provide a route to a genuine understanding about the nature and function of consciousness. New models and theories will certainly also be needed, and even basic and fundamental questions remain unclear or deeply puzzling. Consciousness is a diverse and complicated aspect of our mental life, and understanding it will likely require a plurality of methods, concepts, theories, and models. Real progress has been made in the last two decades, but as yet our understanding of consciousness is limited. There is much work to be done, and the field is at an exciting stage. The next few decades will likely provide surprising data and novel models, both of which will transform our understanding of consciousness and its role in perception, memory, and the whole of our cognitive life.

REFERENCES

Baars, B. (1988). *A Cognitive Theory of Consciousness*. Cambridge: Cambridge University Press.

Block, N. (1995). On a confusion about the function of consciousness. *Behavioral and Brain Sciences* 18: 227–47.

————. (1996). Mental paint and mental latex. *Philosophical Issues* 7: 19–49.

————. (2007). Consciousness, accessibility, and the mesh between psychology and neuroscience. *Behavioral and Brain Sciences* 30: 481–99.

Carruthers, P. (2000). *Phenomenal Consciousness*. Cambridge: Cambridge University Press.

Chalmers, D. (1996). *The Conscious Mind*. Oxford: Oxford University Press.

Craik, F., and Tulving, E. (1975). Depth of processing and the retention of words in episodic memory. *Journal of Experimental Psychology* 104(3): 268–94.

Crick, F., and Koch, C. (1990). Toward a neurobiological theory of consciousness. *Seminars in Neuroscience* 2: 263–75.

Dehaene, S., and Naccache, L. (2001). Towards a cognitive neuroscience of consciousness: Basic evidence and a workspace framework. *Cognition* 79: 1–37.

Dennett, D. (1991). *Consciousness Explained*. Boston: Little, Brown.

Driver, J., and Mattingley, J. (1998). Parietal neglect and visual awareness. *Nature Neuroscience* 1: 17–22.

Edelman, G. (1990). *The Remembered Present: A Biological Theory of Consciousness*. New York: Basic Books.

Edelman, G., and Tononi, G. (2000). *A Universe of Consciousness: How Matter Becomes Imagination*. New York: Basic Books.

Gazzaniga, M. (1988). *Mind Matters*. New York: Houghton Mifflin.

Gennaro, R. (1995). *Consciousness and Self-Consciousness: A Defense of the Higher-Order Thought Theory of Consciousness*. Amsterdam: John Benjamins.

Goodale, M., and Milner A. (1992). Separate visual pathways for perception and action. *Trends in Neuroscience* 15: 20–25.

————. (2004). *Sight Unseen: An Exploration of Conscious and Unconscious Vision*. Oxford: Oxford University Press.

Harman, G. (1990). The intrinsic quality of experience. *Philosophical Perspectives* 4: 31–52.

Horgan, T., and Tienson, J. (2002). The Intentionality of Phenomenology and the Phenomenology of Intentionality. In D. Chalmers (ed.), *Philosophy of Mind: Classical and Contemporary Readings*. Oxford: Oxford University Press.

Husserl, E. (1913/1931). *Ideas: General Introduction to Pure Phenomenology* (*Ideen au einer reinen Phänomenologie und phänomenologischen Philosophie*), translated by W. Boyce Gibson. London: George Allen and Unwin.

————. (1929/1960). *Cartesian Meditations: An Introduction to Phenomenology*, translated by Dorian Cairns. The Hague: Martinus Nijhoff.

Jackendoff, R. (1987). *Consciousness and the Computational Mind*. Cambridge, MA: MIT Press.

Kriegel, U. (2006). The Same-Order Monitoring Theory of Consciousness. In U. Kriegel and K. Williford (eds.), *Self-Representational Approaches to Consciousness*. Cambridge, MA: MIT Press.

Jacoby, L. (1991). A process dissociation framework: Separating automatic from intentional uses of memory. *Journal of Memory and Language* 30: 513–41.

Lamme, V. (2006). Towards a true neural stance on consciousness. *Trends in Cognitive Sciences* 10: 494–501.

Logothetis, N. K., and Schall, J. D. (1989). Neural correlates of subjective visual perception. *Science* 245(4919): 761–763.

Lycan, W. (1996). *Consciousness and Experience*. Cambridge, MA: MIT Press.

Marcel, A. (1983) Conscious and unconscious perception: experiments in visual masking and word recognition. *Cognitive Psychology* 15: 197–237.

Merikle, P., Smilek, D, and Eastwood, J. (2001) Perception without awareness: perspectives from cognitive psychology. *Cognition* 79: 115–34.

Merleau-Ponty, M. (1945/1962). *Phenomenology of Perception (Phénoménologie de lye Perception)*, translated by Colin Smith. London: Routledge and Kegan Paul.

Nagel, T. (1974). What is it like to be a bat? *Philosophical Review* 83: 435–50.

Peacocke, C. (1983). *Sense and Content*. Oxford: Oxford University Press.

Prinz, J. (2000). A neurofunctional theory of visual consciousness. *Consciousness and Cognition* 9: 243–59.

———. (2005). A neurofunctional theory of consciousness. In A. Brook and K. Akins (eds.), *Cognition and the Brain: Philosophy and Neuroscience Movement*. Cambridge: Cambridge University Press.

Rock, I., and Mark, A. (1998). *Inattentional Blindness*. Cambridge, MA: MIT Press.

Roediger, H., and McDermott, K. (1993). Implicit Memory in Normal Human Subjects. In F. Boller, F. Grafman, and J. Grafman (eds.), *Handbook of Neuropsychology*. Amsterdam: Elsevier.

Rosenthal, D. (1986). Two concepts of consciousness. *Philosophical Studies* 49: 329–59.

———. (1997). A Theory of Consciousness. In N. Block, O. Flanagan, and G. Guzeldere (eds.), *The Nature of Consciousness*. Cambridge, MA: MIT Press.

Schacter, D. (1987). Implicit memory: History and current status. *Journal of Experimental Psychology, Learning, Memory and Cognition* 13: 501–18.

Schacter, D., and Tulving, E. (1994). Current Thinking on Memory Systems of 1994. In D. Schacter and E. Tulving (eds.) *Memory Systems*. Cambridge, MA: MIT Press.

Searle, J. (1990). Consciousness, explanatory inversion, and cognitive sciences. *Behavioral and Brain Sciences* 13: 585–642.

Searle, J. (1992). *The Rediscovery of the Mind*. Cambridge, MA: MIT Press.

Siewert, C. (1998). *The Significance of Consciousness*. Princeton, NJ: Princeton University Press.

Singer, W. (1999). Neuronal synchrony: A versatile code for the definition of relations? *Neuron* 24: 49–65.

Strawson, G. (1994). *Mental Reality*. Cambridge, MA: MIT Press.

———. (2006). Intentionality and Experience: Terminological Preliminaries. In D. Smith and A. Thomasson (eds.), *Phenomenology and Philosophy of Mind*. Oxford: Oxford University Press.

Thompson, E. (2007). *Life in Mind*. Cambridge, MA: Harvard University Press.

Tononi, G. (2004). An information integration theory of consciousness. *BMC Neuroscience* 5:42, http://www.biomedcentral.com/1471-2202/5/42.

Tye, M. (1995). *Ten Problems of Consciousness*. Cambridge, MA: MIT Press.

———. (2000). *Color, Content and Consciousness*. Cambridge, MA: MIT Press.

———. (2002). Blurry images, double vision and other oddities: New problems for representationalism?. In Q. Smith and A. Jokic (eds.), *Consciousness: New Philosophical Essays*. Oxford: Oxford University Press.

Van Gulick, R. (1995). How should we understand the relation between intentionality and consciousness?. *Philosophical Perspectives* 9: 271–289.

———. (2004). HOGS (Higher-Order Global States)—An Alternative Higher-Order Model of Consciousness. In R. Gennaro (ed.) *Higher-Order Theories of Consciousness*. Amsterdam and Philadelphia: John Benjamins.

———. (2006) Mirror-Mirror, Is That All?. In U. Kriegl and K. Williford (eds.), *Self-Representational Approaches to Consciousness*. Cambridge, MA: MIT Press.

Varela, F., Thompson, E, and Rosch, E. (1991). *The Embodied Mind: Cognitive Science and Human Experience*. Cambridge, MA: MIT Press.

Weiskrantz, L. (1986). *Blindsight: A Case Study and Its Implications*. Oxford: Oxford University Press.

CHAPTER 3

....

REASONING AND RATIONALITY

....

COLIN ALLEN, PETER M. TODD,

AND JONATHAN M. WEINBERG

REASONING and rationality have long been thought of as uniquely human charac-teristics, made possible by the special power of the human mind. The "modern" philosophy of Descartes took those special powers to lie beyond what can be ac-complished by mechanisms composed of mere matter; according to him, reason-ing, understanding, and thought were the operations of a mind wholly distinct from material body. Cognitive scientists and like-minded philosophers have thor-oughly rejected Descartes's substance dualism, but other concomitants of that dualism are still active. Guilt-by-association is just as a bad an argument in sci-ence as it is in court, so it is possible that these other views deserve to survive the death of their metaphysical progenitor. Nonetheless, our purpose in this chapter is to review the evidence against them while promoting a more "ecological" ap-proach to rationality.

Let us consider a suite of five views on the topic of reasoning and rationality that might be called "Residual Cartesianism" (cf. Clark 1987).[1] First, this perspective would say that reasoning should be explained and evaluated at the level of the *indi-vidual* cognitive agent (individualism). Second, and more specifically, it should be explained in terms of processes *internal* to that agent (internalism). Third, the appropriate norms for such evaluation are given in mathematical and symbolic terms, such as the predicate calculus or Bayes' theorem, and normatively apply to agents entirely independently of those agents' particular biological makeup or

[1] Note that our use of this phrase is a convenient label for some received views that, though Cartesian in flavor in various ways, may not be views that were actually held by Descartes.

environmental circumstances (rationalism). Fourth, we should expect all non-pathological humans to share the *same* basic competence in reasoning, and as such, we can evaluate humanity itself as basically rational or irrational, depending on whether the processes that implement that competence do or do not comply with the appropriate norms (universalism; although the descriptive and normative aspects of universalism could come apart, traditionally they have not been separated). Fifth, because reasoning is at the core of rationality and because the processes of reasoning are understood as essentially enmeshed with language, only humans reason; only humans are rational (human exceptionalism). On this final point, Descartes was the inheritor of a tradition going back at least to Aristotle.

It is easy to see how these pieces of Residual Cartesianism fit together. Given rationalism, it can seem that individualism and internalism must be true as well—the truths of mathematics and logic are unaffected by whatever variation there is in the local, environmental circumstances of different agents, and thus the normative quality of an agent cannot depend on its environmental particulars, including its relationships to other agents. It is, moreover, hard to deny the value of the mathematical frameworks that rationalists draw upon. Those who would resist Residual Cartesianism do not want to find themselves needing to deny, say, the validity of *modus tollens*. A similar argument would lead from rationalism to universalism, unless one wished to court a form of relativism about logic. And those who would deny human exceptionalism cannot ignore the fact that only humans have made their logical and mathematical principles explicit and no (other) animal seems capable of doing so.

There are also powerful philosophical intuitions that pull in the direction of Residual Cartesianism. Indeed, one recent prominent argument in epistemology, owed to Stewart Cohen (1984), asks us to again consider the universe that Descartes describes in his *Meditations*, in which an evil, but powerful entity is hell-bent on deceiving a group of beings. They are in an environment for which their modes of reasoning radically fail to be truth-conducive, yet they are using the same modes of reasoning that *we* do, and "from the inside" their world seems the same to them as ours does to us. Whereas Descartes's aim is to make plausible the skeptical view that we know nothing, Cohen's aim is to challenge the "reliabilist" view that rationality requires reliable access to the environment; since such matters of reliability are outside of the agent, this picture of rationality is inconsistent with the internalism of Residual Cartesianism. This "new evil demon" problem is meant to put a pointed question to the externalist reliabilist: who are we to say that we are, in any sense, more rational or justified in our reasoning than our poor deceived doppelgangers? Cohen's underlying idea here is that the norms of epistemic evaluation should not depend on what possible world we happen to find ourselves in. Maybe we were lucky enough to be born into a world in which our reasoning capacities are reliable, and our deceived brethren were spectacularly unlucky: the intuition that Cohen is exploring is that whether one is or is not rational should not be a matter of luck. By endorsing internalism, Cohen exemplifies that strand of Residual Cartesianism.

On internalist views, rationality is largely a matter of consistency or coherence among a set of beliefs. Philosophers and logicians raised in the tradition of Cartesian epistemology typically see themselves as engaged in the *normative* task of articulating the *right* procedures or processes to guarantee internal consistency. Psychologists, many of whom are well trained in formal logic and philosophy, see themselves as engaged in the *descriptive* task of discovering the *actual* processes used by human reasoners. These projects converge on the *evaluative* task of assessing whether the actual processes used by humans are the right ones (Samuels, Stich, and Faucher 2004) and on the *prescriptive* task of saying how to improve human reasoning (Baron 2008).

This focus on *process* is, according to Kacelnik (2006), the defining characteristic of a conception of rationality that is widely shared by philosophers and psychologists—what he calls "PP-rationality." Kacelnik identifies two other concepts of rationality, the E-rationality of economists and the B-rationality of biologists. These are less concerned with process and more with outcomes: maximization of expected utility in the case of E-rationality, and biological fitness in the case of B-rationality. The addition of game theory to the analytical tools of economists and biologists adds a social dimension to rationality. In this way, the internalist strand of Residual Cartesianism is challenged, for what tends to maximize utility or fitness can no longer be considered outside the context of what is going on outside the agent.

Even where philosophers and psychologists have in some ways diverged from this picture, Residual Cartesianism has continued to haunt their views, an unexorcised ghost in the cognitive scientific machine. For example, one might think that the proliferation, nearly to the point of dogma, of *dual-process* accounts of reasoning might represent a chance to break with the monolithic Cartesian approach.[2] Such accounts distinguish between two systems of reasoning. System 1 is evolutionarily old, based on associative and/or affective mechanisms; it is shared with other animals, operates unconsciously and mandatorily, and consists of processes that are generally conducive to attaining goals within specific domains. System 2 is evolutionarily recent, based on sequential processing, distinctively human and domain general; it operates consciously and with deliberate effort, and is specifically engaged in abstract reasoning guided by normative rules, such as *modus ponens*. People produce different patterns of performance on the same task under different circumstances, often with one set of responses very uniformly observed under conditions where they are making decisions quickly, unreflectively, or under high cognitive load, with the other set being those responses that people more commonly produce under conditions that provoke greater deliberation, or allow them to bring explicit training to bear.

Some dual-process theorists have taken this descriptive distinction between two types of cognition and postulated a normative distinction as well. Evans and

---

[2] In addition to Evans and Over (1996), discussed below, see Reber (1993), Sloman (1996) (and Gigerenzer and Regier's (1996) comment), the papers in Chaiken and Trope (1999), and Evans (2003). Early versions of this thesis are also present in James (1890) and Neisser (1963).

Over (1996) distinguish two flavors of rationality to go with the two types of system:

> Rationality$_1$: cognition or behavior that is "generally reliable and efficient for achieving one's goals."
>
> Rationality$_2$: cognition or behavior where "one has a reason for what one does sanctioned by a normative theory." (Evans and Over 1996, 8).

The idea, then, is that System 1 is a fairly good source of much of our success in terms of rationality$_1$, at least on the cheap, but only System 2 will support the full-blown epistemic normativity of rationality$_2$. Thus Residual Cartesianism still finds a home even in these more contemporary theories in two ways: first by positing a special subset of our cognition for which its claims hold, and second by privileging the rationality of that subset as *the* normative form of rationality.

Reflecting a broad consensus among psychologists, Evans and Over regard the set of processes responsible for rationality$_2$ as constrained by working memory and other resource limits. Of course, even Descartes recognized that human rationality is bounded—he supposed that angels and the Deity have superior intellects, after all. But the standard of unbounded reasoning in accordance with the principles of logic and mathematics provided the ideal measure against which human shortcomings were to be evaluated. Neither psychologists nor economists appeal any more to supernatural beings for normative standards. However, another sign of Residual Cartesianism is that philosophers, psychologists, and economists continue to adopt models of rational decision making which assume standards of reasoning that people fail to meet even though they may in fact be acting rationally when considered from the ecological perspective (to be developed below). The rash of books on how "irrational" people are show that Residual Cartesian ideas are still current.[3]

1. FROM BOUNDED RATIONALITY TO ECOLOGICAL RATIONALITY

That people are not perfect reasoners is, of course, not news. The concept of bounded rationality (Simon 1982, 1990) has been widely discussed. Simon introduced two important bounds that human cognition must operate within: purely external constraints, such as the costs of searching for information in the world, and purely internal ones, such as limits on the amount of information humans can store and process, and the speed with which we can do that processing. These constraints led

---

[3] E.g., *Nudge: Improving Decisions about Health, Wealth, and Happiness* (Thaler and Sunstein 2008); *Kluge: The Haphazard Construction of the Human Mind* (Marcus 2008); *Predictably Irrational: The Hidden Forces that Shape Our Decisions* (Ariely 2008); *How We Know What Isn't So: The Fallibility of Human Reason in Everyday Life* (Gilovich 1993); *Inevitable Illusions: How Mistakes of Reason Rule Our Minds* (Piattelli-Palmarini 1996); and *Stumbling Upon Happiness* (Gilbert 2006).

Simon to propose that humans must employ *bounded* rationality, in contrast to the unbounded forms that many psychologists and economists assume. Many researchers following Simon focused on one or the other of these two sources of constraints on human reasoning and decision making, external and internal. This led some on the one hand to see bounded rationality as the attempt to do as well as possible given the demands of the external world—the notion of optimization under constraints, often pursued by economists (e.g., Stigler 1961) and biologists (e.g., Stephens and Krebs 1986). On the other hand, many psychologists and behavioral economists interpreted bounded rationality as the suboptimal outcome of the limited cognitive system, hobbled with internal constraints when it faces a complex environment, leading to cognitive biases and illusions (Kahneman, Slovic, and Tversky 1982; Conlisk 1996).

We think that this literature has failed to appreciate Simon's point sufficiently. He saw bounded rationality as emerging through the close interaction of *both* of these sets of constraints: "Human rational behavior...is shaped by a scissors whose two blades are the structure of the task environments and the computational capabilities of the actor" (Simon 1990, 7; see Todd and Gigerenzer 2003 for a discussion). While this perspective may be somewhat agnostic as to just how well these two blades fit together, and hence the nature of the rational behavior that they shape, Gigerenzer and colleagues (Gigerenzer, Todd, and the ABC Research Group 1999; Todd and Gigerenzer 2007) have put forth the idea that the two sources of constraints on decision making can come to be finely honed and closely matched to each other, yielding *ecological rationality*: the ability to make good decisions with often simple mental mechanisms whose constrained internal structure can exploit the external information structures available in the environment. This approach to rationality directly confronts aspects of Residual Cartesianism (dis)embodied in the principles of internalism and rationalism.

1.1. Ecological Rationality

Ecological rationality comes about through the coadaptation of minds and their environments. The internal bounds comprising the capacities of the cognitive system can be shaped by evolution, learning, or development to take advantage of the structure of the external environment (Todd 2001). Likewise, the external bounds, comprising the structure of information available in the environment, can be shaped by the effects of minds making decisions in the world, including most notably in humans the process of cultural evolution (Boyd and Richerson 1985; Richerson and Boyd 2005). This coadaptation results in decision mechanisms that are fit to particular environments, and to the particular adaptive goals that humans (or other species) have in those environments. Sterelny points out (2006, 305) that "what counts as efficient information use is specific to organism and environment." According to proponents of ecological rationality, efficient information use is to be understood causally and constitutively as an adaptive interaction between organism and environment.

Ecological rationality is thus a binary relationship—a particular decision mechanism cannot be said to be ecologically rational (or not) in itself, but only when assessed in a particular environment. This means that external correspondence criteria are what matter for judging ecological rationality: the extent to which the mechanism leads to adaptive behavior in a specific environment. Internal coherence criteria used in other definitions of rationality, such as making logically consistent choices, do not take the environment into consideration and so are not involved in this conception of rationality. (This does not mean that outcomes of ecologically rational mechanism are never appropriately described in terms of the rules of logic—just that minds did not evolve specifically to implement those rules.)

Both the internal constraints on decision making (including limited computational power and limited memory in the organism)—and the external ones (including limited time, and costs of gathering information in the environment)—push toward simple cognitive mechanisms for making decisions quickly and without much information. Simon (e.g., 1990) spoke of the need for approximate decision methods, such as *satisficing*, using aspiration levels to tell when a good-enough solution has been found. Gigerenzer et al. (1999) propose that the mind draws on an adaptive toolbox of simple, fast, and frugal heuristics or rules of thumb. This simplicity is actually enabled by the fit of the heuristics to particular environments: by being able to rely on the presence of useful structure in the information in the world, the heuristics themselves can be simpler, essentially letting the world do some of the work.

Simplicity through environmental fit, though, makes it unlikely that the heuristics in the adaptive toolbox will work well in other environments. Ecological rationality does not lead one to expect decision processes that are appropriate in all possible new environments, whereas traditional conceptions of rationality (e.g., first order logic, or Bayes' rule) are specified abstractly so as to apply regardless of content. Ecological rationality is domain-specific or structure-specific rather than domain-general. However, simple heuristics are often robust to a range of changes in particular environments, more so than complex mechanisms, because they avoid overfitting (Gigerenzer and Brighton 2009).

As an example of a robust simple heuristic that is ecologically rational in particular environments, consider the well-studied *take-the-best* heuristic (Gigerenzer and Goldstein 1996; Gigerenzer et al. 1999). Take-the-best makes decisions between pairs of objects on the basis of features or cues of each object. It searches for cues in order of their validity—that is, their rate of making correct decisions in similar contexts—and stops searching as soon as it finds a cue that distinguishes between the objects (which may be the first cue, or the second, or one further down the list); this is the single "best" available cue that the heuristic takes to make a decision.

Take-the-best and other so-called *one-reason decision heuristics* are frugal in that they do not look for any more information than is needed to make an inference, and they are fast because they do not involve any complex computation—not even the multiplication and addition required by weighted additive mechanisms that underlie the traditional standard rational approach to decision making. Take-the-best will not do well compared to weighted additive or tallying mechanisms in environments where the distribution of cue importance is uniform (i.e., the available pieces of information are roughly equal in their usefulness). But many environments, such as those of consumer choice or mate choice, are characterized instead by a distribution of cue importance that falls off rapidly (a "J-shaped" distribution), so that the most influential cue is considerably more important than the second, which is considerably more important than the third, and so on. In such environments, take-the-best is ecologically rational, and is able to outperform the weighted additive model, particularly when generalizing to new decisions (Gigerenzer et al. 1999; Hogarth and Karelaia 2006).

People have been found to use simple heuristics in ecologically rational ways in numerous studies, for instance, employing one-reason decision mechanisms when information is costly (Newell, Weston, and Shanks 2003) or must be sought in memory (Bröder and Schiffer 2003), or when the decision must be made under time pressure—all these are situations in which it is advantageous to limit information search. Furthermore, people are sensitive to the structure of information in the environment, for instance learning appropriately to employ either take-the-best or a weighted additive mechanism depending on which will be more accurate (Rieskamp and Otto 2006). The processes of figuring out which heuristic to apply and when, along with the application of the heuristic itself, are not necessarily conscious processes; Gigerenzer (2007, 19) captures this possibility by saying "The intelligence of the unconscious is in knowing, without thinking, which rule is likely to work in which situation." Exactly how people determine which type of environment they are in, and then which heuristics will be appropriate to apply, remain open questions in the ecological rationality framework. This strategy selection process may be a point at which Residual Cartesianism has some appeal. As Hurley and Nudds (2006, 13) put it, "We may use reasoning to select which heuristics to use, even if using heuristics is not itself a kind of reasoning; such metacognition may be enough to make the use of heuristics a rational process, or part of one." Another example, that does not appeal to meta-cognition, is provided by Lee and Cummins (2004) whose Bayesian model of evidence accumulation is intended to explain the use of different decision strategies. But proponents of ecological rationality expect that strategy selection itself can be done in a simple, fast, and frugal manner (or in different ways in different situations), such as through reinforcement learning (Rieskamp and Otto 2006) or innate association.

The study of ecologically rational mechanisms (Todd and Gigerenzer 2007) begins with identifying important decision tasks (e.g., on psychological or evolutionary grounds) and specifying the structure of information in the environment that can be exploited in making decisions. Next, computational models of candidate heuristics can be proposed that are psychologically plausible, based on human competences. These can be tested via simulation in various environments to see when they work, and then via experimentation or observation to see when people (or other animals) actually use the proposed heuristics. This research program differs from that of (1) the heuristics-and-biases program, in emphasizing explicit computational models of heuristics; (2) evolutionary psychology, in considering individually and culturally learned heuristics as well as evolved ones; and (3) behavioral ecology, in starting from the bottom up with psychologically plausible building blocks, rather than top-down beginning with optimal models.

1.2. Social Rationality

Of the most important environments that organisms must navigate, many are social, where part of the environment is made up of other members of their species. Some researchers have argued that rational agents in social environments gain an advantage from representing the intentional states of others and modeling their reasoning about goals and desires so as to incorporate this into their own reasoning about attaining personal goals and desires. The social intelligence hypothesis (Byrne and Whiten 1988; Whiten and Byrne 1998) goes even further in stating that the need to navigate social environments has been the major selective force in apes and humans, driving the evolution of social intelligence and explicit reasoning about the minds of others. But reasoning about the contents of other minds must also be done within the kinds of constraints identified by Simon. Consequently, researchers in the ecological rationality framework have proposed that social rationality, like other forms of ecological rationality, can be achieved through the appropriate use of simple heuristics.

Sterelny (2003, 2006) opposes this view. He argues that social environments are often hostile, wherein "competitive interactions...degrade agents' information environments" (2006, 297), rendering fast and frugal heuristics insufficient. Whether this is true in general remains to be shown; there are certainly specific counterexamples. One set of counterexamples stems from the fact that even in some competitive interactions, both parties may have aligned incentives—vying deer stags, for instance, may want to assess each other without fighting, and can do so by using a one-reason rule of thumb that evaluates a sequence of cues (parallel walking for visual size assessment, strength of roar, etc.) until finding the first cue that discriminates between the two rivals (Hutchinson and Gigerenzer 2005, 117). Furthermore, there are heuristic approaches to dealing with "unfriendly" asocial environments (e.g., Fasolo, McClelland, and Todd 2007), and these may apply in hostile social situations as well.

It is also important to note that not all social interaction is hostile, competitive, or unfriendly: agents' goals may be aligned in whole (e.g., in cooperative settings) or in part (e.g., in mate choice), or agents may be making decisions that are about the social world but are not strategic (e.g., estimating the prevalence of some feature among the population). In each of these cases, researchers have found simple heuristics at work, for instance for combining information to make a group decision (Reimer and Katsikopoulos 2004), for finding a suitable mate who is also agreeable (Todd and Miller 1999), or for assessing how common a disease is in one's social circle (Hertwig, Pachur, and Kurzenhäuser 2005). These heuristics and others provide a positive answer to Hurley and Nudds' question (2006, 9–10) of whether social rationality can be achieved by simple mechanisms "that reliably generate behaviour fit to achieve one's ends in social environments and that are unlikely to be undermined by deception and the manipulation of information."

1.3. Cognitive Niche Construction

Human exceptionalism is the strand of Residual Cartesianism that puts the greatest focus on language and symbolic reasoning as the basis for human rationality. It is undeniable that abstract reasoning is greatly enhanced by language and a general facility with symbolic forms of representation. But how independent is symbolic reasoning from the heuristics of ecological rationality? One possibility is Evans and Over's view that symbolic reasoning is handled independently by System 2, given input from System 1. Another possibility is that symbolic reasoning consists in the application of ecologically rational heuristics to the symbolic and linguistic structures that we create. For instance, properties of symbol structures might themselves be targets for heuristic decision making. In a recent study, Landy and Goldstone (2007) reported that experimental subjects competent in algebra could nevertheless be misled by formulas in which spacing cues were discordant with the rules that govern the precedence of arithmetic operators. Landy and Goldstone also demonstrated that similarly competent subjects introduce spacing into equations that they themselves write out in a way that typically supports their understanding of operator precedence. When parentheses are included, these are the best cue for guiding the perceptual groups to which the operators apply, but when they are absent, subjects fall back on the next best cue, which happens in their normal environments to be the spacing between symbols, even though the use of spacing is only implicit in the practice and teaching of arithmetic and algebra. This can be likened to an application of the take-the-best heuristic.

The invention of symbolic systems shows how humans deliberately and creatively alter their environments to enhance learning and memory and to support reasoning. Nonhuman animals also alter their environments in ways that support adaptive behavior. Loose parallel, or deep similarity? Analogies between human and animal behavior are easy to find. But it is much harder to track the shifting border in the dispute over whether animals have "genuine" rationality or merely behave "as if" rational (Hurley and Nudds 2006). Even if an ecological conception of rationality obviates the problem of determining whether animals have the mental states required by internalists, careful study of the actual interactions between animals and their environments is needed to show whether the putative similarities between humans and animals justifies speaking of animal rationality.

The past few decades have seen a remarkable expansion in the number of studies of animal cognition. Memory, self-recognition, tool use, culture, communication, and metacognition have been studied in a widening range of species and habitats and in increasingly sophisticated laboratory experiments. This is not the place to survey all of these developments (see Andrews 2008 for a philosophically motivated review). Rather, we wish to highlight a couple of examples in support of a more general point about animal rationality. Of course it is possible to defend human exceptionalism by emphasizing the very real differences in information and reasoning afforded by human language. But human accomplishments are built on top of thousands of years of cultural development and the accumulation of material culture. Human-enculturated chimpanzees and bonobos, specifically those trained in symbolic communication, seem capable of cognitive feats that those living in more natural conditions cannot (Premack 1986; Boysen and Capaldi 1993; Savage-Rumbaugh and Lewin 1996). Nevertheless, the more ephemeral and less noticeable interactions between animals and their environments may provide the ground-level scaffolding for human cultural achievements and the apes that have been enculturated by humans. Vigo and Allen (2009) argue that once the structural relations among stimuli are properly measured, it is possible to show that human reasoning is grounded in processes of categorization that may be shared with language-less animals.

Stigmergy, an important mechanism for swarm intelligence, is the product of interactions among multiple agents and their environments. It is enhanced through cumulative modification of the environment by individuals. For instance, a foraging ant returning from a food source leaves a pheromone trail. A single ant's trail has little power to impact the colony, but when reinforced by other ants on the same trajectory, the behavior of the entire colony can be affected. In the right environment, such mass action can result in efficient exploitation of resources, with nearer food sources more likely to be harvested first. The environmental modifications supporting stigmergy often have a short half-life, as with fast-evaporating volatile pheromones, but can also involve longer-term modifications (e.g., deer trails) that may be used by multiple generations. The creation of enduring symbols, not just in the abstract but in their actual concrete realizations and applications, is a major cultural accomplishment that makes unique forms of stigmergy possible. In this way, humans create "cognitive trails" in the information environment (see Cussins 1992 who introduces the term *cognitive trails*, giving it a related but more elaborate meaning than our use here). The collective intelligence of humans has been enhanced by the development of symbolic markings, beginning perhaps with a few lines etched on stone axes and continuing through cave wall paintings to the twenty-first century's vast repositories of digitized data and the technology that makes it possible to access them. Whether it involves following pheromone trails, road signs, or the algebraic rules learned by attending to the symbol-laden presentations of teachers, the ecological rationality of doing so is a matter of using fast and frugal heuristics to fit behavior to an informational environment that is simultaneously being transformed. The correspondence between human and animal rationality is not a loose analogy but a deep biological similarity, albeit the degree of stigmergy reaches its most hypertrophic form in the human species.

2. Variation of Reasoning and Rationality across Individuals

2.1. Cultural Differences

A further tenet of Residual Cartesianism is universalism—all human minds reason in fundamentally similar ways (*modulo* instances of disease, injury, etc.), and are properly evaluable in the same normative terms. To reject the latter part of that claim would lead to a kind of epistemic relativism, and though that is a view that has been defended on occasion (Barnes and Bloor 1982; Rorty 1991 (though Rorty contests the term *relativism*); for more recent defenses, see Stich 1990; Hales 2006; Weinberg 2007; Goldman (forthcoming)), it remains controversial at best. But the former part can be seen as a special instance of what Paul Griffiths (1997) has labeled "the doctrine of the monomorphic mind" (122), and a number of recent studies challenge it.[4]

One of the more famous results that has been taken in past decades to show that humans are, on the whole, irrational is the "fundamental attribution error" (Ross 1977), also known as the "correspondence bias": a tendency to explain others' behavior too much in internal terms, and insufficiently in contextual terms. Although the findings have been robust (see, e.g., Gilbert and Malone 1995 for a review)—and perhaps they even partially explain the attraction of the internalist strand of Residual Cartesianism!—they have only very recently been examined for cross-cultural validity. Choi and Nisbett (1998) have shown that in task conditions in which situational factors are made salient, East Asian subjects are not prone to making this error, even though Western subjects are not any more likely to take these factors into account than they are when the situational factors are less salient. Based on these findings and others like them, Nisbett et al. (2001) have hypothesized that there are robust East/West cultural differences in reasoning, such that one can describe Easterners as operating within a "holistic system of thought" and Westerners within an "analytic system of thought." The former, contrasted with the latter, displays a greater sensitivity to field or context (hence the reduced risk of committing the fundamental attribution error), is more tolerant of contradictions, and endorses reasoning based more on similarity- or exemplar-based categories than rule-based categories. There is no clear sense in which either system is normatively superior to the other. For example, both systems deviate from traditional rationality$_2$ accounts of the treatment of evidence: when subjects are presented with a strong argument for one proposition and a weaker argument for a proposition inconsistent with it, the presence of the weak argument leads Western subjects, surprisingly, to increase their confidence in the first

---

[4] See in particular Henrich et al. (2010) for a powerful survey of recent work indicating that subjects from "WEIRD" societies—Western, Educated, Industrialized, Rich, and Democratic—may be demographic outliers across a large range of psychological dimensions.

proposition, whereas the presence of the strong argument for the first proposition leads Eastern subjects, equally surprisingly, to increase their confidence in the second proposition: "[Easterners] actually found the less plausible proposition to have more merit when it had been contradicted than when it had not...a normatively dubious inference but utterly different in kind from that of [Westerners]." (302).

There is also some evidence of cross-cultural variation in explicitly epistemo-logical judgments. Weinberg, Nichols, and Stich (2001), following the lead of the work just discussed, surveyed subjects on a number of traditional philosophical thought-experiments concerning whether hypothetical cases do or do not count as instances of knowledge. Among the differences they found, some concerned a ver-sion of the "Gettier case" in which an agent has good reasons for a belief, and the belief is true, but nonetheless the belief intuitively fails to count as an instance of knowledge. (For example, someone may believe correctly that a colleague currently owns an American car, because the believer is aware of the Buick that colleague has driven for years—but the believer is *not* aware that the Buick is no more, and that the only reason the belief is correct is that the colleague has very recently, and secretly, bought a Pontiac.) Or, at least, that is the "intuitive" status standardly reported in the (primarily Western) analytic epistemology literature. And Weinberg et al.'s Western subjects overwhelmingly shared that judgment, with a very large majority refraining from attributing knowledge. Their East and South Asian subjects were radically more split on the matter, with slightly over half attributing knowledge. It may be that cul-turally variable preferences for similarity-based over rule-based categorical reason-ing can explain this difference. The moral, though, is that there is extant evidence that styles of both cognition and metacognition may vary along cultural lines, un-dermining the tenet of universalism. The descriptive element of universalism is un-dermined directly. The normative element is threatened on a more methodological basis by challenging one of the chief sources of evidence philosophers appeal to for determining the relevant norms: intuitions about what is or is not a case of knowl-edge, or epistemically justified belief, and so on. If different groups have divergent intuitions about various cases, this putative source of evidence for our norms ap-pears to be prone to errors in hitherto unexpected ways.

2.2. Rationality in Groups

We have seen one kind of challenge to the individualism of the traditional view: that an organism's reasoning can only be evaluated in terms of its relations to its envi-ronment. One can also challenge the presupposition that it is only the individual organism, as an epistemic agent, that can be evaluated in terms of rationality. Groups of agents can be considered in rational terms as well, where that rationality cannot be simply reduced to the question of whether the group's members are rational when considered individually (Hutchins 1995; Wilson et al. 2004). The simplest way this may be done is if a group comprised of agents is evaluated according to the same norms of rationality that apply to its individual agents. Classic approaches in social choice theory already attempt to analyze the rationality of group decisions in terms of joint satisfaction of a collection of individually held rational preferences, but by grounding success of the group decision procedure in questions about the maximization of individual preferences, the apparent rejection of the individualism of residual Cartesianism only goes part way. One can go a bit further and consider groups themselves as having beliefs, drawing inferences, making decisions, and so on (Gilbert 1989), then one can ask whether the rules, heuristics, and so on that they use in doing so are normatively appropriate.

But groups need not work just as if they were individuals, and they also need not be evaluated as such; their plural composition makes possible an epistemic pluralism as well. Groups can sustain a "cognitive division of labor" (Kitcher 1993), and the environment of a group may render some forms of cognition rational that would not be rational outside of that environment. It may be epistemically valuable for the group on the whole if individual members are permitted, and even encouraged, to cognize in different ways. Sometimes this is the only way for science to move forward in its own piecemeal way. As Philip Kitcher has observed, "for example, in the resolution of the Darwinian debate, and even more, in the triumph of Lavoisier's new chemistry, it was important that opposing points of view were kept alive and that the objections they generated were used to refine the ultimately successful positions" (Kitcher 1993, 344; see also Solomon 2001; Longino 2002; Sunstein 2003).

Lu Hong and Scott Page (2004) have recently provided a formal treatment of the epistemic value of diversity. Interestingly, groups comprised of uniform "expert" artificial agents—that is, agents that performed maximally well as individuals on the problem-solving task—were regularly outperformed by groups comprised of a diverse collection of artificial agents that nonetheless were individually less successful than the individuals in the expert group. Hong and Page write, "[A] random collection of agents drawn from a large set of limited-ability agents typically outperforms a collection of the very best agents from that same set. This result is because, with a large population of agents, the first group, although its members have more ability, is less diverse. To put it succinctly, diversity trumps ability" (16386). Different agents bring different "perspectives" and "heuristics" (16385): ways of representing the problem space, and rules for navigating that space. Even a very good agent will sometimes get trapped in a local optimum, lacking either a way of viewing that position that allows it to "see" that a better outcome is possible, or a tool for getting to that better outcome. A team of diverse agents, under the right circumstances, can find optima that an agent that is individually superior to every member of the team could not find. Diversity of reasoning might, therefore, be not just descriptively correct, but prescriptively correct as well.

2.3. Individual Differences

In addition to differences across groups, there are important differences in reasoning across individuals within groups. Stanovich and West (2000) document a wide range of tasks on which, while the modal performance deviates significantly from what is standardly taken to be the normative response, nonetheless some subjects do give the endorsed answers. On such tasks such as syllogistic reasoning and nondeontic Wason card selection, a distinct minority of subjects give the answers that the original experimenters specified as the correct answers, whereas the majority of subjects gave answers that are labeled as instances of irrational "belief bias" or "matching bias." Stanovich and West argue that it is not just that for any such task a minority of subjects give the answers that are standardly labeled as normatively correct—their findings suggest that it is the *same* minority that will do so on each. These subjects scored highly on several measures of cognitive ability, such as their SAT scores. Such individual differences thus provide evidence inconsistent with the universalism of Residual Cartesianism: some perfectly healthy people really just do think in ways differently from how others do. That these differences are systematic, and not random, precludes explaining these differences away as merely differing rates of performance errors (Stein 1996; Stich 1990). This is further reinforced by that fact that these high cognitive ability subjects did not show significantly different performance on such tasks as the false consensus effect (Ross, Greene, and House 1977) and other "my side" biases, nor on noncausal base rate problems (Stanovich and West 1998).

Stanovich and West try to extract normative consequences from these individual differences by means of Slovic and Tversky's (1974) "understanding-acceptance principle": when there is a covariation between subjects' having a deeper understanding of a norm or axiom and their acceptance of it, then we thereby have evidence of its correctness as well. This principle is itself intuitively plausible, and could be seen as a special version of the more general principle of trusting the judgments of those who show more of the hallmarks of expertise in any given area—perhaps one of the best ways we have of getting "oughts" from "is's." The principle becomes even more compelling when we have not just individual judgments considered *seriatim*, but a manifold of covariant judgments of the sort that Stanovich and West have uncovered. Furthermore, within the framework of Residual Cartesianism the inference can look almost mandatory, for if there is only one maximally correct form of human rationality, then it would be bizarre *not* to take the more enlightened people to be closer to the One True Norm.

Our suspicions about Residual Cartesianism, however, also lead us to worry that normative matters are less tidy than Stanovich and West picture them. One can see a little bit of pressure on this picture in Stanovich and West's paper, when they consider how System 1 cognition might be more attuned to "evolutionary rationality" while System 2 cognition's job is to subserve "normative rationality" (2000, 662). As with Evans and Over, their choice of vocabulary clearly indicates that they take the normative question to be settled. But it is not clear that it is, in fact, normatively rational for a person with a well-tuned set of System 1 resources, but a comparatively impoverished set of System 2 resources, to rely as primarily on System 2 as Stanovich and West seem to suggest that person ought. The argument here is similar to that offered by Hilary Kornblith (1999) to the effect that a person who is susceptible to reasoning that is mere rationalization may be cognitively better off ignoring reasons proffered by his interlocutors than he would be if he tried hard to take those reasons seriously and respond to them. A form of relativism presents itself—the norms that are well-followed by those with ample System 2 resources may simply not be appropriate for those without such resources, especially when their intuitive cognition is more or less adequate to their cognitive lives and the environments in which they find themselves.

This last issue—of the ecological rationality of intuitive cognition—plays a central role in Stanovich and West's account. They claim that people's System 1 cognition is apt for richly embedded social and personal contexts, ripe with the cues that this system can make excellent use of, but that system-2 comes into play precisely for decontextualized tasks in which more impersonal rules may yield better results (2000, 659). This suggests a different possible relativism, not in terms of the psychological resources of the agents, but of the environments in which they find themselves: agents who operate in ecologically appropriate environments may do better to simply trust to system 1, and the "override" function of System 2 (662) may be more likely than not to lead the agents astray, but agents operating in less cognitively friendly terrain may do well to have a higher distrust of their intuitions.

3. CONCLUDING REMARKS

We have identified five strands of Cartesian thought (individualism, internalism, rationalism, universalism, and human exceptionalism) that still have a grip on philosophical and psychological theorizing about rationality. Each of these five strands of Residual Cartesianism faces strong challenges. In this chapter we have presented the ecological approach to rationality as a viable alternative to the traditional approach, and outlined the challenge it provides to the five elements of Residual Cartesianism. Some philosophers may wish to dismiss this approach on the grounds that the kinds of intuitive processes it proposes belong only to System 1 processes, reserving the honorific term "rational" for System 2 processes. In the cognitively unfriendly domain of a philosophy seminar it may seem that intuitive forms of decision making are not rational, but this we suggest is an artifact of the situation that is not devoid of irony given the centrality of intuitions to philosophical argumentation. Going forward, the task for defenders of ecological rationality is to describe in detail the features of information environments that allow an organism to select the right heuristic for the situation.

REFERENCES

Andrews, K. (2008). Animal Cognition. In Edward N. Zalta (ed.), *The Stanford Encyclopedia of Philosophy (Winter 2008 Edition)*, http://plato.stanford.edu/archives/win2008/entries/cognition-animal.

Ariely, D. (2008). *Predictably Irrational: The Hidden Forces That Shape Our Decisions.* New York: Harper.

Baron, J. (2008). *Thinking and Deciding,* 4th ed. New York: Cambridge University Press.

Barnes, B., and Bloor, D. (1982). Relativism, Rationalism and the Sociology of Knowledge. In M. Hollis and S. Lukes (eds.), *Rationality and Relativism.* Oxford:.Blackwell.

Boyd, R., and Richerson, P. (1985). *Culture and Evolutionary Processes.* Chicago: University of Chicago Press.

Boysen, S. T., and Capaldi, E. J. (eds.) (1993). *The Development of Numerical Competence, Animal and Human Models.* Hillsdale, NJ: Erlbaum.

Bröder, A., and Schiffer, S. (2003). "Take the best" versus simultaneous feature matching: Probabilistic inferences from memory and effects of representation format. *Journal of Experimental Psychology: General,* 132: 277–93.

Byrne, R.W., and Whiten, A. (eds.) (1988). *Machiavellian Intelligence: Social Expertise and the Evolution of Intellect in Monkeys, Apes and Humans.* Oxford: Oxford University Press.

Chaiken, S., and Trope, Y. (1999). *Dual-Process Theories in Social Psychology.* New York: Guilford Press.

Choi, I., and Nisbett, R. E. (1998). Situational salience and cultural differences in the correspondence bias and in the actor-observer bias. *Personality & Social Psychology Bulletin* 24: 949–60.

Clark, A. (1987) Being there: Why implementation matters to cognitive science. *Artificial Intelligence Review* 1(4): 231–44.

Cohen, S. (1984). Justification and truth. *Philosophical Studies,* 46(3): 279–95.

Conlisk, J. (1996). Why bounded rationality? *Journal of Economic Literature* 34: 669–700.

Cussins, A. (1992). Content, embodiment, and objectivity. *Mind* 101: 651–88.

Evans, J. S. (2003). In two minds: Dual-process account of reasoning. *Trends in Cognitive Sciences* 7(10): 454–59.

Evans, J. S., and Over, D. E. (1996). *Rationality and Reasoning.* Hove, UK: Psychology Press.

Fasolo, B., McClelland, G. H., and Todd, P. M. (2007). Escaping the tyranny of choice: When fewer attributes make choice easier. *Marketing Theory* 7: 13–26.

Gigerenzer, G. (2007). *Gut Feelings: The Intelligence of the Unconscious.* New York: Viking Press.

Gigerenzer, G., and Brighton, H. (2009). Homo Heuristicus: Why biased minds make better inferences. *Topics in Cognitive Science* 1: 107–43.

Gigerenzer, G., and Goldstein, D. G. (1996). Reasoning the fast and frugal way: Models of bounded rationality. *Psychological Review* 103: 650–69.

Gigerenzer, G., and Regier, T. P. (1996). How do we tell an association from a rule? *Psychological Bulletin* 119(1): 23–26.

Gigerenzer, G., Todd, P. M., and the ABC Research Group. (1999). *Simple Heuristics That Make Us Smart.* New York: Oxford University Press.

Gilbert, D. (2006). *Stumbling on Happiness.* New York: Knopf.

Gilbert, D., and Malone, P. (1995) The correspondence bias. *Psychological Bulletin* 117(1): 21–38.

Gilbert, M. (1989). *On Social Facts.* London: Routledge.

Gilovich, T. (1993). *How We Know What Isn't So: The Fallibility of Human Reason in Everyday Life.* New York: Free Press.

Goldman, A. (2009). Epistemic Relativism and Reasonable Disagreement. In R. Feldman and T. Warfield (eds.), *Disagreement.* Oxford: Oxford University Press.

Griffiths, P. (1997). *What Emotions Really Are.* Chicago: University of Chicago Press.

Hales, S. (2006). *Relativism and the Foundations of Philosophy*. Cambridge, MA: MIT Press.

Henrich, J., Heine, S., and Norenzayan, A. (2010). The weirdest people in the world? *Behavioral and Brain Sciences*, 33: 61–83.

Hertwig, R., Pachur, T., and Kurzenhäuser, S. (2005). Judgments of risk frequencies: Tests of possible cognitive mechanisms. *Journal of Experimental Psychology: Learning, Memory, and Cognition* 31: 621–42.

Hogarth, R. M., and Karelaia, N. (2006). Take-the best and other simple strategies: Why and when they work "well" in binary choice. *Theory and Decision* 61: 205–49.

Hong, L. and Page, S. (2004). Groups of diverse problem solvers can outperform groups of high-ability problem solvers. *PNAS* 101(46): 16385–16389.

Hurley, S., and Nudds, M. (eds.) (2006). *Rational Animals?* Oxford: Oxford University Press.

Hutchins, E. (1995). *Cognition in the Wild*. Cambridge, MA: MIT Press, Bradford Books.

Hutchinson, J. M. C., and Gigerenzer, G. (2005). Simple heuristics and rules of thumb: Where psychologists and behavioural biologists might meet. *Behavioural Processes*, 69: 97–124.

James, W. (1890). *The Principles of Psychology*, vol. 1. New York: Cosimo.

Kacelnik, A. (2006). Meanings of Rationality. In S. Hurley and M. Nudds (eds.), *Rational Animals?* Oxford: Oxford University Press.

Kahneman, D., Slovic, P., and Tversky, A. (eds.). (1982). *Judgment under Uncertainty: Heuristics and Biases*. Cambridge: Cambridge University Press.

Kitcher, P. (1993). *The Advancement of Science*. New York: Oxford University Press.

Kornblith, H. (1999). Distrusting reason. *Midwest Studies in Philosophy* 23(1): 181–96.

Landy, D., and Goldstone, R. L. (2007). How abstract is symbolic thought? *Journal of Experimental Psychology: Learning, Memory, and Cognition* 33(4): 720–33.

Lee, M. D., and Cummins, T. D. R. (2004). Evidence accumulation in decision making: Unifying the "take the best" and the "rational" models. *Psychonomic Bulletin and Review* 11: 343–52.

Longino, H. (2002). *The Fate of Knowledge*. Princeton, NJ: Princeton University Press.

Marcus, G. (2008). *Kluge: The Haphazard Construction of the Human Mind*. New York: Houghton-Mifflin.

Neisser, U. (1963). The multiplicity of thought. *British Journal of Psychology* 54: 1–14.

Newell, B. R., Weston, N. J., and Shanks, D. R. (2003). Empirical tests of a fast-and-frugal heuristic: Not everyone "takes-the-best." *Organizational Behavior and Human Decision Processes*, 91: 82–96.

Nisbett, R., Peng, K., Choi, I., and Norenzayan, A. (2001). Culture and systems of thought: Holistic versus analytic cognition. *Psychological Review* 108(2): 291–310.

Piattelli-Palmarini, M. (1996). *Inevitable Illusions: How Mistakes of Reason Rule Our Minds*. Hoboken, NJ: Wiley.

Premack, D. (1986). *Gavagai! or the Future History of the Animal Language Controversy*. Cambridge, MA: MIT Press.

Reber, A. S. (1993). *Implicit Learning and Tacit Knowledge*. Oxford: Oxford University Press.

Reimer, T., and Katsikopoulos, K. (2004). The use of recognition in group decision-making. *Cognitive Science* 28: 1009–29.

Richerson, P., and Boyd, R. (2005). *Not by Genes Alone: How Culture Transformed Human Evolution*. Chicago: University of Chicago Press.

Rieskamp, J., and Otto, P. E. (2006). SSL: A theory of how people learn to select strategies. *Journal of Experimental Psychology: General* 135: 207–36.

Rorty, R. (1991). *Objectivity, Relativism, and Truth: Philosophical Papers*, vol. 1. Cambridge: Cambridge University Press.

Ross, L. (1977). The intuitive psychologist and his shortcomings: Distortions in the attribution process. In L. Berkowitz (ed.), *Advances in Experimental Social Psychology* vol. 10. New York: Academic Press.

Ross, L., Greene, D. and House, P. (1977). The false consensus effect: An egocentric bias in social perception and attribution processes. *Journal of Experimental Social Psychology*, 13(3): 279–301.

Samuels, R., Stich, S., and Faucher, L. (2004). Reason and Rationality. In I. Niiniluoto, M. Sintonen, and J. Wolenski (eds.), *Handbook of Epistemology*. Dordrecht, Netherlands: Kluwer.

Savage-Rumbaugh, S., and Lewin, R. (1996). *Kanzi: The Ape at the Brink of the Human Mind*. New York: Wiley.

Simon, H. A. (1982). *Models of Bounded Rationality*. Cambridge, MA: MIT Press.

Simon, H. A. (1990). Invariants of human behavior. *Annual Review of Psychology* 41: 1–19.

Sloman, S. A. (1996). The empirical case for two systems of reasoning. *Psychological Bulletin* 119(1): 3–22.

Slovic, P., and Tversky, A. (1974). Who accepts Savage's axiom? *Behavioral Science* 19: 368–73.

Solomon, M. (2001). *Social Empiricism*. Cambridge, MA: MIT Press.

Stanovich, K. (1999). *Who is Rational? Studies of Individual Differences in Reasoning*. Mahwah, NJ: Erlbaum.

Stanovich, K., and West, R. (1998). Individual differences in rational thought. *Journal of Experimental Psychology: General* 127(2): 161–88.

——. (2000). Individual differences in reasoning: Implications for the rationality debate? *Behavioral and Brain Sciences* 23(5): 645–65.

Stein, E. (1996). *Without Good Reason*. Oxford: Oxford University Press.

Stephens, D. W., and Krebs, J. R. (1986). *Foraging Teory*. Princeton, NJ: Princeton University Press.

Sterelny, K. (2003). *Thought in a Hostile World*. New York: Blackwell.

——. (2006). Folk Logic and Animal Rationality. In S. Hurley and M. Nudds (eds), *Rational Animals?* Oxford: Oxford University Press.

Stich, S. (1990). *The Fragmentation of Reason*. Cambridge, MA: MIT Press.

Stigler, G. J. (1961). The economics of information. *Journal of Political Economy* 69: 213–25.

Sunstein, C. (2003). *Why Societies Need Dissent*. Cambridge, MA: Harvard University Press.

Thaler, R., and Sunstein, C. (2008). *Nudge: Improving Decisions about Health, Wealth, and Happiness*. New York: Penguin.

Todd, P. M. (2001). Fast and frugal heuristics for environmentally bounded minds. In G. Gigerenzer and R. Selten (eds.), *Bounded Rationality: The Adaptive Toolbox*. Cambridge, MA: MIT Press.

Todd, P. M., and Gigerenzer, G. (2007). Environments that make us smart: Ecological rationality. *Current Directions in Psychological Science* 16(3): 167–71.

Todd, P. M., and Gigerenzer, G. (2003). Bounding rationality to the world. *Journal of Economic Psychology* 24: 143–65.

Todd, P. M., and Miller, G. F. (1999). From pride and prejudice to persuasion: Satisficing in mate search. In G. Gigerenzer, P. M. Todd, and the ABC Research Group (eds.), *Simple Heuristics That Make Us Smart*. New York: Oxford University Press.

Vigo, R., and Allen, C. (2009). How to reason without words: Inference as categorization. *Cognitive Processing* (in press), http://dx.doi.org/10.1007/s10339-008-0220-4.

Weinberg, J. (2007). Moderate epistemic relativism and our epistemic goals. *Episteme* 4(1): 66–92.

Weinberg, J., Nichols, S., and Stich, S. (2001). Normativity and epistemic intuitions. *Philosophical Topics* 29: 429–59.

Whiten, A., and Byrne, R.W. (eds.) (1997). *Machiavellian Intelligence II: Extensions and Evaluations*. Cambridge: Cambridge University Press.

Wilson, D., Timmel, J., and Miller R. (2004). Cognitive cooperation: When the going gets tough, think as a group. *Human Nature* 15: 225–50.

CHAPTER 4

....

MASSIVE MODULARITY

....

RICHARD SAMUELS

Cognitive scientists disagree on many issues, but one very widespread commitment is that the mind is a *mechanism* of some sort: roughly speaking, a physical device decomposable into functionally specifiable subparts. On this assumption, a central project for cognitive science is to characterize the nature of this mechanism—to provide an account of our *cognitive architecture*—which specifies the basic operations, component parts, and organization of the mind. As such, this project is (albeit in modern, mechanistic guise) an attempt to answer issues that have been central to philosophy at least since Plato. The recognition of this fact—as well as the foundational character of the issues and arguments involved—has meant that philosophers have been actively involved in contemporary discussions of cognitive architecture.

Though the overarching project of specifying a cognitive architecture spans many different topics and regions of enquiry, one central cluster of issues focuses on the extent to which our minds are *modular* in organization. It is this cluster of issues that I focus on here. Specifically, I discuss the question of whether the human mind is *massively modular*: roughly, whether our minds—including those "central" regions responsible for reasoning and decision-making—are largely or entirely composed of a great many specialized cognitive mechanisms or modules. This question represents the confluence of many issues of central theoretical import to philosophy and cognitive science, including issues about the scope and limits of computational explanation, the role of evolutionary theorizing in understanding the mind, and the extent to which our psychological capacities are innately specified. In large measure because of this, the issue of massive modularity has come to mark a major fault line dividing different approaches to the study of human cognition, and has attracted both prominent advocates—Leda Cosmides, John Tooby, Steven Pinker, Peter Carruthers, and Dan Sperber, to name a few—and its share of influential detractors (e.g., Jerry Fodor and Stephen J. Gould).

The present chapter is not the place to provide a comprehensive survey of the debate surrounding massive modularity (MM). Its goals are more limited. First, in Section 1, it explains what is at issue between advocates and opponents of MM, and spells out the hypothesis itself in more detail. Second, it sketches some of the more prominent arguments for MM. In particular, in Section 2, it considers some well-known arguments from evolution, and in Section 3, arguments from computational tractability. Finally, in Section 4, it considers what many regard as the most serious theoretical challenge for MM—the *problem of flexibility*—to explain our cognitive-behavioral flexibility within the restrictions imposed by a modularist conception of cognitive architecture (Carruthers 2006).

1. What Is at Issue?

To a first approximation, massive modularity is the hypothesis that the human mind is largely or entirely composed from a great many modules. More precisely, MM can be formulated as the conjunction of three claims:

- *Composition*: The human mind is largely or entirely composed from modules.
- *Plurality*: The human mind contains a great many modules.
- *Central Modularity*: Modularity is found not merely at the periphery of the mind but also in those *central* regions responsible for such "higher" cognitive capacities as reasoning and decision making.

In what follows I assume advocates of MM are committed to the conjunction of these claims. Even so, each is amenable to a variety of different interpretations. More needs to be said if we are to get clearer on what is at issue.

1.1. Composition Thesis

MM is in large measure a claim about the kinds of mechanisms from which our minds are composed—viz. it is largely or even entirely composed from *modules*.[1]

---

[1] There is a familiar notion of modularity, sometimes called Chomskian modularity, on which modules are not mechanisms but systems of mental representations—bodies of mentally represented knowledge or information—such as a grammar or a theory (Segal 1996; Samuels 2000; Fodor 2000). Paradigmatically, such structures are truth-evaluable in that it makes sense to ask of the representations if they are true or false. Moreover, they are often assumed to be innate and/ or subject to informational constraints (e.g., inaccessible to consciousness). Although Chomskian modules are an important sort of cognitive structure, they are not the ones most relevant to the sort of position advocated by massive modularists. This is because advocates of MM appear to assume that modules are a species of cognitive *mechanism* (Sperber 2002; Sperber and Hirschfeld 2007; Cosmides and Tooby 1992; Carruthers 2006).

But this is vague in at least two respects. First, it leaves unspecified the precise extent to which minds are composed from modules. In particular, this way of formulating the proposal accommodates two different positions, which I call *strong* and *weak* massive modularity. According to strong MM *all* cognitive mechanisms are modules. Such a view would be undermined if we were to discover any non-modular cognitive mechanisms. By contrast, weak MM maintains only that the human mind is *largely* modular in structure. In contrast to strong MM, such a view is clearly compatible with the claim that there are some non-modular mechanisms. So, for example, a proponent of weak MM can readily posit non-modular devices for reasoning and learning.

A second crucial respect in which the Composition Thesis is vague is that it leaves unspecified what modules *are*. For present purposes, this is an important matter since the interest and plausibility of the thesis turns crucially on what one takes modules to be.

1.2.1. *Robust Notions of Module*

Though there are many notions of modularity in play within cognitive science,[2] perhaps the most well-known and most demanding is due to Fodor (1983). On this view, modules are functionally characterizable cognitive mechanisms that tend to possess the following features to some interesting degree:

- Domain-specificity: They operate on a limited range of inputs, defined by some task domain such as vision or language processing;
- Informationally encapsulation: They have limited access to information in other systems;
- Innateness: They are unlearned components of the mind;
- Inaccessibility: Other mental systems have only limited access to a module's computations;
- Shallow outputs: Their outputs are not conceptually elaborated;
- Mandatory operation: They respond automatically to inputs;
- Speed: Their operation is relatively fast;
- Neural localization: They are associated with distinct neural regions;
- They are subject to characteristic and specific breakdowns; and
- Their developmental trajectories exhibit a characteristic pace and sequence.

This full-fledged Fodorian notion has been highly influential in many areas of cognitive science (Garfield 1987); but it has not played much role in debate over MM,[3] and for good reason. The thesis that minds are largely or entirely composed of Fodorian modules is obviously implausible. Indeed, some of the entries on Fodor's list—relative speed and shallowness, for example—make little sense when applied to central systems (Carruthers 2006; Sperber and Hirschfeld 2007). And even where

---

[2] The following discussion is by no means exhaustive. For more detailed discussions of different notions of modularity see Segal (1996); Samuels (2000); and Carruthers (2006).

[3] Incidentally, not even Fodor adopts it in his recent discussions of MM (Fodor 2000, 2008).

Fodor's properties can be sensibly ascribed—as in the case of innateness—they carry a heavy justificationary burden that few seem inclined to shoulder (Baron-Cohen 1995; Sperber 1994).

In any case, there is a broad consensus that not all characteristics on Fodor's original list are of equal theoretical import. Rather, domain-specificity and informational encapsulation are widely regarded as most central. Both concern the architecturally imposed[4] informational restrictions to which cognitive mechanisms are subject—the range of representations they can access—though the kinds of restriction involved are different.

Domain-specificity is a restriction on the representations a cognitive mechanism can take as *input*—that "trigger" it or "turn it on." A mechanism is domain-specific (as opposed to domain-general) to the extent that it can only take as input a highly restricted range of representations.[5] Standard candidates include mechanisms for low-level visual perception, face recognition, and arithmetic.

Informational encapsulation is a restriction on the kinds of information a mechanism can use as a *resource* once so activated—paradigmatically, though not essentially, information stored in memory. Specifically, a cognitive mechanism is encapsulated to the extent that it can access, in the course of its computations, less than all of the information available to the organism as a whole (Fodor 1983). Standard candidates include mechanisms such as those for low-level visual perception and phonology that do not draw on the full range of an organism's beliefs and goals.

Though there are many characteristics other than domain-specificity and encapsulation that have been ascribed to modules, when discussing more robust conceptions of modularity I will focus on these properties. This is both because they are widely regarded as the most theoretically important features of Fodorian modules, and because—as we will see—they are central to the topics to be considered here.

---

[4] To claim that a property of a cognitive mechanism is *architecturally* imposed minimally implies the following. First, they are relatively enduring characteristics of the device. Second, they are not mere products of *performance* factors, such as fatigue or lapses in attention. Finally, they are supposed to be *cognitively impenetrable* (Pylyshyn 1984). To a first approximation: they are not properties of the mechanism that can be changed as a result of alterations in the beliefs, goals, and other representational states of the organism.

[5] Two comments. First, it should go without saying—though it will be said anyway—that the notion of domain-specificity *admits of degree* and that researchers who use the notion are interested in whether we possess mechanisms that are domain-specific to some *interesting* extent. The same points also apply to the notion of informational encapsulation. Second, there is a range of different ways in which theorists have proposed to characterize types or domains of representations. For example, in one common view, domains of representations are *content* domains: sets of representations that are characterized in terms of what they are about, or what they mean (Fodor 1983). On another view, domains of representations are individuated by formal properties of representations (Jackendoff 1992; Barrett and Kurzban 2006). In this view, the representations that comprise a domain share various formal, non-semantic properties. For further discussion of issues about the nature and individuation of domains see Sperber 1996; Fodor, 2000; Samuels, 2000; and Barrett and Kurzban 2006.

That said, it is important to stress that not all those interested in modularity assume the centrality of these notions.

1.1.2. A Minimal Functional Notion of Module

According to another, *minimal* conception of modules that has become increasingly commonplace in cognitive science—especially among advocates of MM—modules are just distinct, functionally characterized cognitive mechanisms of the sort that correspond to boxes in a cognitive psychologist's flow diagram (Fodor 2005). In a recent paper, Barrett and Kurzban (2006) summarize and endorse this growing consensus:

> We similarly endorse the view espoused by many evolutionary psychologists that the concept of modularity should be grounded in the notion of *functional special-ization* (Barrett 2005; Pinker 1997, 2005; Sperber 1994, 2005; Tooby and Cosmides 1992) rather than any specific Fodorian criterion. Biologists have long held that structure reflects function, but that function comes first. That is, determining what structure one expects to see without first considering its function is an ap-proach inconsistent with modern biological theory. The same holds true, we argue, for modularity. (Barrett and Kurzban 2006)

Of course, there is nothing wrong per se with adopting such a conception of modu-larity. Indeed, one obvious virtue is that it renders MM more plausible. But it does so at the risk of leaching the hypothesis of its content, thereby rendering it rather less interesting than it may initially appear to be. For in the context of cognitive sci-ence, the idea that minds are composed of functionally specifiable mechanisms has near universal acceptance.[6] So, if being a module *just is* being a functionally specifi-able mechanism, then the thesis that minds are composed of modules is just the consensus view.

1.2. Plurality Thesis

Still, it does not follow, as many have claimed, that no distinctive version of MM can be formulated with the minimal notion of modularity (Fodor 2005; Prinz 2006). In particular, some proponents of MM maintain that their thesis is interesting not because it implies that minds are composed of minimal modules, but because it implies what I earlier called the Plurality Thesis: the view that minds contain a great many cognitive mechanisms or modules (Carruthers 2006).[7]

Is this an interesting thesis? Clearly, if formulated in terms of a robust notion of modularity, the Plurality Thesis is quite radical since many deny that

---

[6] This is so even for fans of empiricist and domain-general accounts of cognitive processes. After all, a domain-general learning mechanism is still a functionally specifiable device.

[7] It may also be that functional specialization admits of degree, and that an interesting version of MM could maintain that minds are largely or entirely composed of *highly* specialized mechanisms.

domain-specific and/or encapsulated devices have a substantial role to play in our cognitive economy. But things are less clear if one adopts the minimal notion. According to some advocates of MM, such a thesis would still be interesting since many deny that there are *lots* of such minimal modules. Carruthers, for example, maintains that such a claim is rejected by "those who...picture the mind as a big general-purpose computer with a limited number of distinct input and output links to the world" (Carruthers 2006). But on reflection this cannot be quite right. Big general-purpose computers are not simple entities. On the contrary, they are almost invariably decomposable into a large number of functionally characterizable sub-mechanisms.[8] So, for example, a standard von Neumann-type architecture decomposes into a calculating unit, a control unit, a fast-to-access memory, a slow-to-access memory, and so on; and each of these decomposes further into smaller functional units that are themselves decomposable into sub-mechanisms, and so on. As a consequence, a standard von Neumann machine will typically have hundreds or even thousands of distinct functionally characterizable sub-components.[9] Thus is would seem that even radical opponents of MM can endorse the sorts of Plurality Thesis advocated by Carruthers and others. Indeed, some have argued that this is little more than a consequence of the consensus view in cognitive science—viz. that cognitive mechanisms are hierarchically decomposable into smaller systems (Fodor 2005).

Still, there is an important distinction between this anodyne version of plurality and the sort of view that is characteristic of MM, even on the minimal conception of modules. To a first approximation, non-modularists, such as those who construe the mind as a big general-purpose computer with a limited number of distinct input and output links, are committed to a plurality of functional modules because, qua mechanists, they are committed to the idea that complex mechanisms are decomposable into simpler parts. On this view, there will be large numbers of parts at lower levels in the decomposition. But there will also be some relatively abstract level of description at which there is only a small number of devices. Roughly, on such views the highest level of analysis will be one in which all the parts are organized into a relatively small number of cognitive mechanisms. In contrast, advocates of MM deny that there is *any such level* of composition. Rather, they maintain that even at the highest levels of description, the human mind will resemble a confederation of hundreds or even thousands of functionally dedicated devices—a cheater detection mechanism, a theory of mind

---

[8] Indeed this is more or less guaranteed by the widespread assumption that the functional decomposition of a "large" system will typically have many levels of aggregation (Simon 1962). I return to this point below.

[9] A similar point applies to the sort of radical connectionism on which the mind is characterized as one huge undifferentiated neural network. This is often—and rightly—seen as the antithesis of MM (Pinker 1997), and yet it is committed to a vast plurality of mechanisms. After all, each node in a neural network is a mechanism, and in any version of the connectionist story, there will a great many such nodes.

device, a frequentist module, and so on—that in no interesting sense compose to form some larger single unitary mechanism. In short, all mechanists about cognition are committed to a plurality of cognitive mechanisms because they are committed to functional decomposition. Call this *decompositional* plurality. But only advocates of MM are committed to what we might call a *compositional* plurality: the existence of a large number of mechanisms that cannot be *composed* further. It would thus seem that, contrary to what many have claimed, an interesting version of MM could be formulated in terms of the minimal, functional notion of a module.[10]

1.3. Central Modularity

Let us turn to the final thesis that comprises MM:

> *Central Modularity*: Modules are found not merely at the periphery of the mind but also in those *central* regions responsible for such "higher" cognitive capacities as reasoning and decision making.

This does not strictly follow from the claims discussed so far since one might deny that there are any central systems for reasoning and decision making. But this is *not* the view that advocates of MM seek to defend. Indeed, a large part of what distinguishes MM from the earlier, well-known modularity hypothesis defended by Fodor (1983) and others is that the modular structure of the mind is not restricted to input systems (those responsible for perception, including language perception) and output systems (those responsible for producing behavior) (Jackendoff 1992). So, for example, it has been suggested that there are modules for such central processes as social reasoning (Cosmides and Tooby, 2000), biological categorization (Pinker 1994), and probabilistic inference (Gigerenzer et al. 1999).

How interesting is the Central Modularity thesis? This depends on the notion of modularity involved, but also the kind of plurality that is at stake. Start with versions of the thesis formulated with the minimal notion of a module. If the claim is merely that there are central, functional modules, then Central Modularity is merely the consensus view in cognitive science. Similarly, if the claim is merely that there are *lots* of central, functional modules, then once more it is hard to discern any interesting and distinctive position. But if the kind of plurality involved is not merely decompositional, but *compositional* in character, then we appear to have a position that is rather more worthy of attention:

---

[10] It should be noted that the present discussion presupposes answers to some genuine but largely unaddressed questions about the individuation of cognitive mechanisms. In particular, it is far from clear when two or more mechanisms are themselves parts of some larger mechanism. For some discussion of such issues, see Lyons 2001.

Central Compositional Modularity: Central cognition depends on a great many functional modules that are not themselves composable into "larger" more inclusive systems.

This *would* be a distinctive version of Central Modularity. Not because it maintains that central cognition depends on functional modules, or because it assumes the existence of many such mechanisms, but because it implies a kind of decentralized or *confederate* view of central cognition: one on which our capacity for thought, reasoning, judgment, and the like depends on the interaction of a multitude of distinct mechanisms. This is one way to articulate an interesting version of Central Modularity without recourse to a Fodorian conception of modules. Moreover, it is a suggestion that comports well with views articulated by some prominent advocates of MM. So, for example, it appears to capture what Tooby and Comsides have in mind when they liken our cognitive architecture to "a confederation of hundreds or thousands of functionally dedicated computers (often called modules)" (Tooby and Cosmides, 1995, xiv) What makes their position interesting is not merely that there are lots of such devices, but that they comprise a loose confederacy of subsystems as opposed to, say, an all-encompassing unitary central executive.

Let us now consider versions of Central Modularity formulated in terms of a more robust conception of modularity. Here, the degree to which one's version of Central Modularity is interesting will depend on both (1) the extent to which central cognition is subserved by domain-specific and/or encapsulated mechanisms, and (2) how many such modules there are. Both these questions could be answered in a variety of different ways. At one extreme, for example, one might adopt the following relatively weak claim:

Weak Central Modularity: There are a number of domain-specific and/or encapsulated central systems, but there are also non-modular—domain-general and unencapsulated—central systems as well.

Such a proposal is not without interest. But it is not especially radical in that it does not stray far from the old-fashioned peripheral modularity advocated by Fodor. Moreover, as we will see in Section 4, it does not raise the sorts of deep theoretical problems that plague other, stronger versions of MM. At the other extreme one might maintain:

Strong Central Modularity: All central systems are domain-specific and/or encapsulated, and there are a great many of them.

This is a genuinely radical position since it implies that there are no domain-general, informationally unencapsulated central systems. But this Strong Central Modularity is also implausible, for as we will see in later sections, there are no good reasons to accept it, and some reason to think it is false.

2. MASSIVE MODULARITY AND EVOLUTION

Discussions of MM are closely tied to claims about the evolutionary plausibility of different architectural arrangements. Specifically, many have argued that MM is plausible in the light of quite general considerations about the nature of evolution. Though this is not the place to discuss such arguments in detail, what follows aims to provide a flavor of the evolutionary motivations for MM. In doing so, I discuss briefly two prominent arguments for MM.[11] (For more detailed discussion of such arguments see Tooby and Cosmides 1992; Sperber 1994; Samuels 1998; Fodor 2000; Buller 2005; and Barrett and Kurzban 2006.)

2.1. Evolvability

One common argument for MM derives from Simon (1962)'s seminal discussion of evolutionary stability (Carston 1996; Carruthers 2006). According to Simon, for an evolutionary process to reliably assemble complex functional systems—biological systems in particular—the overall system needs to be *semi-decomposable*: hierarchically organized from components with relatively limited connections to each other. Simon illustrates the point with a parable of two watchmakers, Hora and Tempus, both highly regarded for their fine watches. But while Hora prospered, Tempus became poorer and poorer and finally lost his shop. The reason:

> The watches the men made consisted of about 1000 parts each. Tempus had so constructed his that if he had one partially assembled and had to put it down—to answer the phone, say—it immediately fell to pieces and had to be reassembled from the elements.... The watches Hora handled were no less complex... but he had designed them so that he could put together sub-assemblies of about ten elements each. Ten of these subassemblies, again, could be put together into a larger subassembly and a system of ten of the latter constituted the whole watch. Hence, when Hora had to put down a partly assembled watch in order to answer the phone, he lost only a small part of his work, and he assembled his watches in only a fraction of the man-hours it took Tempus. (Simon 1962)

The obvious moral—and the one Simon invites us to accept—is that evolutionary stability requires that complex systems be hierarchically organized from dissociable subsystems, and according to many, this militates in favor of MM (Carston 1996, 75).

Though evolutionary stability may initially appear to favor MM, one concern is that the argument only supports the familiar mechanistic thesis that complex machines are hierarchically assembled from (and decomposable into) many

---

[11] There are other less plausible arguments for MM, which due to space limitations, will not be considered here. For further discussion of other arguments for MM see Tooby and Cosmides (1992); Sperber (1994), and Samuels (2000).

subcomponents. But this clearly falls short of the claim that all (or even any) are domain-specific or encapsulated. Rather it supports at most the sort of banal Plurality Thesis which I earlier referred to as decompositional plurality: one that is wholly compatible with even a Big Computer view of central processes. All it implies is that if there are such complex central systems, they will need to be hierarchically organized into dissociable subsystems—which incidentally was the view Simon and his main collaborators endorsed all along (Simon 1962; Newell 1990).

2.2. Task Specificity

Another well-known kind of evolutionary argument, widely associated with the work of the evolutionary psychologists Leda Cosmides and John Tooby, purports to show that once we appreciate the way in which natural selection operates and the character of the cognitive problems that human beings confront, we will see that there are good reasons for thinking that our minds contain a large number of distinct, modular mechanisms.

In brief, the argument is this: Human beings confront a great many evolutionarily important cognitive tasks whose solutions impose quite different demands. For example, the demands on vision are distinct from those of speech recognition, of mindreading, cheater detection, probabilistic judgment, grammar induction, and so on. Further, it is unlikely that there is be a single general inference mechanism that could perform all these cognitive tasks, and even if there could be such a mechanism, it would be systematically outperformed by a system comprised of an array of distinct mechanisms, each of whose internal processes were specialized for processing the different sorts of information in the way required to solve the task (Carruthers 2006; Cosmides and Tooby 1994; Tooby and Cosmides 1992). But if this is so, then we should expect the human mind to contain a great many functionally specialized cognitive mechanisms since natural selection can be expected to favor superior solutions over inferior ones. In short: we should expect minds to be massively modular in their organization.

Though there is a lot to say about this argument, I will restrict myself to two brief comments. (See Samuels 1998; Buller 2005; Fodor 2000; and Carruthers 2006 for further discussion.) First, if the alternatives were MM or a view of minds as comprised of just a single general-purpose cognitive device, then MM would be the more plausible. But these are clearly not the only options; on the contrary, there are lots of different options. For example, opponents of MM might deny that central systems are modular while still insisting there are plenty of modules for perception, motor control, selective attention, and so on. In other words, the issue is, not merely whether some cognitive tasks require specialized modules, but whether the sorts of tasks associated with central cognition—paradigmatically, reasoning and decision making—require a proliferation of such mechanisms.

Second, it is important to see that the addition of functionally dedicated mechanisms is not the only way of enabling a complex system to address multiple tasks. An alternative is to provide some (small set of) relatively functionally nonspecific mechanism with the requisite bodies of information for solving the tasks it confronts. This is a familiar proposal among those who advocate non-modular accounts of central processes. Indeed, advocates of non-modular reasoning architectures routinely assume that reasoning devices have access to a *huge* amount of specialized information on a great many topics, much of which will be learned but some of which may be innately specified (Newell 1990; Anderson and Lebiere 2003). Moreover, it is one that plausibly explains much of the proliferation of cognitive competences that humans exhibit throughout their lives—for example, the ability to play chess, or reason about historical issues as opposed to politics or gene splicing or restaurants. To be sure, it *might* be that each such task requires a distinct mechanism, but such a conclusion does not flow from general argument alone. For all we know, the same is true of the sorts of tasks advocates of MM discuss. It may be that the capacity to perform certain tasks is explained by the existence of specialized mechanisms. But how often this is the case for central cognition is a largely open question that is not adjudicated by the argument from task specificity.

3. COMPUTATIONAL TRACTABILITY AND RELEVANCE

A second family of arguments for MM focuses on a range of problems that are familiar from the history of cognitive science: problems that concern the computational tractability of cognitive processes. Though such *intractability arguments* vary considerably in detail, they share a common format. First, they proceed from the assumption that cognitive processes are classical computational ones—roughly, algorithmically specifiable processes defined over mental representations. This assumption has been criticized in many quarters, but it has widespread acceptance in the context of the present debate, and for this reason I assume it here. Second, given the assumption that cognitive processes are computational ones, intractability arguments seek to undermine non-modular accounts of cognition by establishing the following *Intractability Thesis*:

> IT: Non-modular cognitive mechanisms—in particular mechanisms for reasoning and other central processes—are *computationally intractable* in roughly the sense that they require more time or cognitive resources—for example, memory and processing power—than humans can reasonably be expected to possess.

But if this is so, and if the human mind is, as many cognitive scientists suppose, a computational system of some kind, then it follows that the mind is composed of modular cognitive mechanisms. After all, a model of cognition that requires resources that we do not possess is simply not one that can accurately characterize the architecture of our minds.

3.1. Informational Impoverishment

Why accept the Intractability Thesis? One well-known argument for IT, often associated with the work of Cosmides and Tooby, proceeds from the assumption that a non-modular mechanism—one that is task nonspecific or domain-general "lacks any content, either in the form of domain-specific knowledge or domain-specific procedures that can guide it towards the solution of problems" (Cosmides and Tooby 1994, 94). As a consequence, it "must evaluate all the alternatives it can define" (94). But as Cosmides and Tooby observe, such a strategy is subject to serious intractability problems, since even routine cognitive tasks are such that the space of alternative options tends to increase exponentially. Non-modular mechanisms would thus seem to be computationally intractable: at best intolerably slow, and at worst incapable of solving the vast majority of problems they confront.

Though frequently presented as an objection to non-MM accounts of cognitive architecture, this argument is really only a criticism of theories that characterize cognitive mechanisms as suffering from a particularly extreme form of informational impoverishment. Any appearance to the contrary derives from the stipulation that domain-general mechanisms possess no specialized knowledge. But this conflates claims about the need for informationally rich cognitive mechanisms—a claim that is not denied—with claims about the need for modularity, and though modularity is one way to build specialized knowledge into a system, it is not the only way. As noted earlier, another is for non-modular devices to have access to bodies of specialized knowledge. Indeed, it is commonly assumed by non-modular accounts of central processing that such devices have access to huge amounts of information. This is obvious from even the most cursory survey of the relevant literatures. Fodor (1983), for example, maintains explicitly that non-modular central systems have access to huge amounts of information; Gopnik, Newell, and many others who adopt a non-modular conception of central systems maintain this as well (Gopnik and Meltzoff 1997; Newell 1990). The argument currently under discussion thus succeeds only in refuting a straw man.

3.2. Relevance Problems

Non-modularists can avoid the conclusion of Cosmides and Tooby's argument by positing relatively task nonspecific mechanisms that have access to lots of information. Yet it is precisely the assumption of informational richness that generates the most well-known tractability problems for non-modular accounts of cognition: what have historically been construed as versions of the frame problem, though are perhaps more accurately characterized as *relevance problems* (Pylyshyn 1987; Ford and Pylyshyn 1996; Samuels 2010). Roughly put, such problems conform to the following general schema:

> *Relevance Problems*: Given a task, T, and computational system S, how does S determine (with reasonable levels of success) from all the available information which is *relevant* to the specific task at hand? (Glymour 1987).

Such problems can arise in the performance of many different tasks, including planning, decision making, pragmatics, perception, and so on. But perhaps the most well-known—and notoriously difficult to address—is a kind of relevance problem that arises in the context of belief revision, what might be called *problem of relevance in update*:

> *Relevance in Update*: Given some new information, how do we determine (with reasonable levels of success) which of the representational states we possess are *relevant* to determining how to update our beliefs?

Does this problem undermine non-modular, computational accounts of cognition? Presumably it is a very hard research topic for cognitive science. Among other things, it requires the specification of tractable, psychologically plausible computational processes that manage to successfully recruit those representations relevant to the task at hand. But the fact that the problem constitutes a hard research topic is not, by itself, reason to reject non-modular views. Rather, it is merely the specification of one central part of the problem of explaining belief revision, and moreover, a part of the problem that presumably modular and non-modular views alike need to address. What is required to turn this into an *objection* to non-modular, computational accounts is an argument for the claim that that non-modular accounts cannot plausibly accommodate the sort of relevance-sensitivity characteristic of human cognition. In what follows, I consider two arguments of this sort.

3.2.1. *Exhaustive Search*

One might think that in order to identify those items of information relevant to the task at hand, a non-modular central system would need to perform exhaustive searches over our beliefs. But given even a conservative estimate of the size of any individual's belief system, such a search would be unfeasible in practice. In this case, it would seem that non-modular reasoning mechanisms are computationally intractable.

Though it is unclear that anyone really endorses this argument, some have found it hard not to view advocates of non-modular central systems as somehow committed to exhaustive search (Carruthers 2004; Glymour 1985). Yet this view is incorrect. What the non-modularist does accept is that unencapsulated reasoning mechanisms have access to huge amounts of information—paradigmatically, all the agent's background beliefs. But the relevant notion of access is a modal one. It concerns what information—given architectural constraints—a mechanism *can* mobilize in solving a problem. In particular, it implies that any background belief can be used, not that the mechanism in fact mobilizes the entire set of background beliefs—that is, that it engage in exhaustive search.

3.2.2. *Inferential Holism and the Intractability of Unencapsulated Processes*

A second closely related intractability argument focuses on the apparent implications of the assumption that modular mechanisms are paradigmatically encapsulated. Though the argument has been formulated many times over (see Carruthers 2006; Samuels 2005; Barrett and Kurzban 2006), one relatively plausible rendering of the argument proceeds from the observation that much human reasoning is holistic in character. In contrast to the argument from exhaustive search, the sort of holism at issue is *not* that all—or even most—of our beliefs *actually* figure in any specific instance of reasoning. Instead, the sort of holism at stake here is modal in character. What it amounts to is that under the appropriate conditions—especially those involving different background beliefs—the relevance of a belief to a reasoning task *can* vary dramatically. Slightly more precisely:

> *Inferential Holism:* Given appropriate background beliefs, (almost) any belief can be rendered relevant to the assessment of (almost)[12] any other belief.[13]

To take a fairly simple example:[14] On the face of it, the current cost of tea in China has little to do with whether my brother's baby in England will cry on Saturday morning. But suppose that I believe my brother has stocks invested in Chinese tea, that he reads the business section of the newspaper every Saturday morning, and that on reading bad financial news he tends to fly into a rage. Given these background beliefs, it seems that beliefs about the current cost of tea in China may well be relevant to beliefs about whether my brother's baby will cry on Saturday morning. *Mutatis mutandis* for other beliefs. Or so it would seem. In which case, it would seem that under the appropriate conditions, a given belief can be relevant to the assessment of (almost) any other.

How is the apparent holism of human inference related to issues of modularity? One connection is this: If our capacity for belief revision depends on some kind of domain-general inference system, then such a system will need to be highly unencapsulated. Otherwise the mechanism in question could not explain the holistic character of much human inference. But, the argument continues, such unencapsu-

---

[12] Clearly, this could do with refinement. So, for example, few beliefs will presumably be relevant to the assessment of logical beliefs—e.g., that if P, then P.

[13] Or to use Fodor (1983)'s terminology: belief revision processes are *isotropic*.

[14] The example is based on a case used in Copeland (1993), which, in turn, was based on an example from Guha and Levy (1990).

lated processes would be computationally intractable. They would require more time and resources than we, in fact, possess. In which case, it cannot be that we possess unencapsulated reasoning mechanisms.

What are we to make of this argument? The first premise is plausible. If belief revision is holistic and depends on a single mechanism, then the mechanism would need to be unencapsualted. The problem is with the second premise. There is a long story to tell here. (See Samuels 2005.) But the short version is that tractability does not require encapsulation. As with most real-world computational applications—Web search engines, for example—there may be heuristic and approximation techniques that permit feasible computation: techniques that often, though not invariably, identify a substantial subset of those representations that are relevant to the task at hand. Of course, this would not be an option if we maintained that, when reasoning, we are *guaranteed* to identify relevant beliefs. But there is no reason whatsoever to suppose that this claim is true. Indeed, one very clear moral from the last four decades of research on human judgment and reasoning is that such standards of accuracy are misplaced.[15] I conclude, then, that the present argument fails.

3.3. The Locality Argument

A final kind of intractability argument that I consider here—one that has been hugely influential in recent debate—is due to Jerry Fodor (2000, 2008). Fodor's argument is a complex one, but the core idea can be framed in terms of a tension between two claims.[16] The first is that classical computational processes are local in roughly the following sense: what computations apply to a particular representation is determined solely by its constituent structure—that is, by how the representation is constructed from its parts (2000, 30). To take a very simple example, whether the addition function can be applied to a given representation is solely determined by whether it has the appropriate syntactic structure—for example, whether it contains a permissible set of symbols related by "+."

The second claim is that much of our reasoning is global in that it is sensitive to context-dependent properties of the entire belief system. In arguing for this, Fodor focuses primarily on abductive reasoning (or inference to the best explanation). Such inferences routinely occur in science and, roughly speaking, consist of coming to endorse a particular belief or hypothesis on the grounds that it constitutes the best available explanation of the data. One familiar feature of such inferences is that the relative quality of hypotheses are not assessed merely in terms of their ability to fit the data, but also in terms of their simplicity and conservativism. According to Fodor, however, these properties are not intrinsic to a belief or hypothesis but are

---

[15] See, for example, Pohl (2005) for discussion of the myriad errors that we make in reasoning.

[16] For more detailed discussion of the argument see Ludwig and Schneider (2007) and Samuels (2010).

global characteristics that a belief or hypothesis possesses by virtue of its relation-ship to a constantly changing system of background beliefs.

The problem, then, is this: If classical computational operations are local, how could global reasoning processes, such as abduction, be computationally tractable? Notice that if the above is correct, a classical abductive process could not operate merely by looking at the hypotheses to be evaluated. This is because, by assumption, what classical computations apply to a representation is determined solely by its constituent structure, whereas the simplicity and conservativism of a hypothesis, H, depends not only on its constituent structure but its relations to our system of back-ground beliefs, K. In which case, a classical implementation of abduction would need to look at both H and whatever parts of K determine the simplicity and con-servativism of H. The question is: How much of K needs to be consulted in order for a classical system to perform reliable abduction? According to Fodor, the answer is that lots—indeed, very often, the totality—of the background will need to be ac-cessed, since this is the "only guaranteed way" of classically computing a global property. But this threatens to render reliable abduction computationally intracta-ble. As Fodor puts it:

> Reliable abduction may require, in the limit, that the whole background of
> epistemic commitments be somehow brought to bear on planning and belief fixa-
> tion. But feasible abduction requires in practice that not more than a small subset
> of even the relevant background beliefs are actually consulted. (2000, 37)

In short: if classicism is true, abduction cannot be reliable. But since abduction presumably is reliable, classicism is false.

If sound, the above argument would appear to show that classicism itself is untenable. So, why would anyone think it supports MM? The suggestion appears to be that MM provides the advocate of CTM with a way out: a way of avoiding the tractability problems associated with the globality of abduction without jettisoning CTM (Sperber 2005; Carruthers 2006). Fodor himself put the point as well as anyone:

> Modules are informationally encapsulated by definition. And, likewise by
> definition, the more encapsulated the informational resources to which a compu-
> tational mechanism has access, the less the character of its operations is sensitive
> to global properties of belief systems. Thus to the extent that the information ac-
> cessible to a device is architecturally constrained to a proprietary database, it
> won't have a frame problem and it won't have a relevance problem (assuming that
> these are different); not, at least, if the database is small enough to permit approx-
> imations to exhaustive searches. (2000, 64)

The modularity of central systems is thus supposed to render reasoning processes sufficiently local to permit tractable computation.

There are a number of serious problems with the above line of argument. One that will not be addressed here concerns the extent to which MM provides a satis-factory way of shielding computationalism from the tractability worries associated with globality. What will be argued, however, is that although simplicity and conservativism are plausibly context dependent, Fodor provides us with no reason whatsoever to think that they are global in any sense that threatens non-modular versions of computationalism.

First, when assessing the claim that abduction is global, it is important to keep firmly in mind the general distinction between normative and descriptive-psychological claims about reasoning: claims about how we ought to reason, and claims about how we actually reason. This distinction applies to the specific case of assessing the simplicity and conservativism of hypotheses. On the normative reading, assessments of simplicity and conservativism ought to be global: that is, normatively correct assessments ought to take into consideration one's total background epistemic commitments. But of course it is not enough for Fodor's purposes that such assessments ought to be global. Rather, it needs to be the case that the assessments humans make are, in fact, global—and there is no reason whatsoever to suppose that this is true.

A comparison with the notion of consistency may help to make the point clearer. Consistency is frequently construed as a normative standard against which to assess one's beliefs (Dennett 1987). Roughly, all else being equal, one's beliefs ought to be consistent with each other. When construed in this manner, however, it is natural to think that consistency should be a global property in the sense that any belief ought to be consistent with the entirety of one's background beliefs. But there is absolutely no reason to suppose that human beings conform to this norm, and some reason to deny that we do. So, for instance, there is good reason to suppose that reliable methods of consistency checking are computationally too expensive for creatures like us to engage in, if consistency is construed as a global property of belief systems (Cherniak 1986). Moreover, this is so in spite of the fact that consistency really does play a role in our inferential practices. What I am suggesting is that much the same may be true of simplicity and conservativism. When they are construed in a normative manner, it is natural to think of them as global properties, but when construed as properties of the beliefs that figure in actual human inference, there is no reason to suppose that they accord with this normative characterization.

Second, even if we suppose that simplicity and conservativism are global properties of actual beliefs, the locality argument still does not go through, since it turns on the implausible assumption that we are guaranteed to make successful assessments of simplicity and conservativism. Specifically, in arguing for the conclusion that abduction is computationally unfeasible, Fodor relies on the claim that "the only guaranteed way of Classically computing a syntactic-but-global property" is to take "whole theories as computational domains" (2000, 36). But guarantees are beside the point. Why suppose that we always successfully compute the global properties on which abduction depends? Presumably we do not. And one very plausible suggestion is that we fail to do so when the cognitive demands required are just too great. In particular, for all that is known, we may well fail under precisely those circumstances the classical view would predict—namely, when too much of a belief system needs to be consulted in order to compute the simplicity or conservativism of a given belief.

3.4. Modularity and Tractability

Even if intractability arguments for MM are not decisive, it is important to stress that modularity does provide a number of resources for addressing tractability problems. First, where a mechanism is functionally specialized or domain-specific, it becomes possible to utilize a potent design strategy for reducing computational load: namely, to build into the mechanism substantial amounts of information about the problems that is it supposed to address. This might be done in a variety of ways. It might be only implicit in the organization of the mechanism, or it might be explicitly represented; it might take the form of rules or procedures or bodies of propositional knowledge and so on. But however this information gets encoded, the key point is that a domain-specific mechanism can be informationally rich and, as a result, capable of rapidly and efficiently deploying those strategies and options most relevant to the domain in which it operates. Such mechanisms thereby avoid the need for computationally expensive search-and-assessment procedures that might plague a more general-purpose device. For this reason, domain specificity has seemed to many a plausible candidate for reducing the threat of combinatorial explosion without compromising the reliability of cognitive mechanisms (Sperber 1994; Tooby and Cosmides 1992).

Second, encapsulation can help reduce computational load in two ways. First, because the device only has access to a highly restricted database or memory, the costs incurred by memory search are considerably reduced since there just is not that much stuff over which the search can be performed. Second, by reducing the range of accessible items of information, there is a concomitant reduction in the number of relations between items—paradigmatically, relations of confirmation and relevance—that can be computed.

Yet one might reasonably wonder what all the fuss is about. After all, computer scientists have generated a huge array of methods—literally hundreds of different search and approximation techniques—for reducing computational overheads (Russell and Norvig 2003). What makes encapsulation of particular interest? Here is where the deeper explanation comes into play. Most of the methods that have been developed for reducing computational load require that the implementing mechanisms treat the assessment of relevance as a computational problem. Roughly, they need to implement computational procedures that select from the available information some subset that is estimated to be relevant. In contrast, encapsulation is supposed to obviate the need for such computational solutions. According to this view, an encapsulated device (at least paradigmatically) only has access to a very small amount of information. As a consequence, it can perform a (near) exhaustive search on whatever information it can access, thereby avoiding the need to assess relevance. There is a sense, then, in which highly encapsulated devices avoid the relevance problem altogether (Fodor 2000).

4. PROBLEMS OF COGNITIVE-BEHAVIORAL FLEXIBILITY

So far I have considered some prominent arguments for MM and found them wanting. I now consider a family of challenges for massive modularity that concern the apparent flexibility of human behavior and cognition. Section 4.1 spells out three sorts of representational flexibility that are alleged to pose a problem for MM, at least in its more radical forms. Next, Section 4.2 highlights some closely related problems that behavioral flexibility pose for MM. Finally, Section 4.3 reviews briefly some possible responses to these problems.

4.1. Representational Flexibility

Perhaps the most commonly posed flexibility worries for MM concern various kinds of *representational* plasticity that human thought appears to exhibit, but that are not readily accommodated within a MM framework.

Representational Integration. A first kind of flexibility concerns our capacity to freely combine conceptual representations across different subject matters or content domains. That is, we exhibit what Carruthers (2006) calls content flexibility. So, for example, it is not merely that we can think about colors, about numbers, about shapes, about food, and so on. Rather we can have thoughts that concern all these things at once—for example, that we had two roughly round red steaks for lunch. But if this is so, then the natural explanation of this capacity is that there are cognitive mechanisms that are able to combine representations from different cognitive domains (Fodor 1983, 102). In this case, it would seem that there must be at least some domain-general cognitive mechanisms.

Content General Consumption. Not only can we freely combine concepts, we can also *use* the resulting representations in theoretical and practical inference to assess their truth or plausibility, but also to assess their impact on our plans and projects (Fodor 1983; Carruthers 2006). But if this so, then there must be mechanisms that can utilize such complex, novel representations. And the obvious explanation for this capacity is that we possess domain-general cognitive mechanisms—for example, for planning and belief revision—that can take representations as input more or less irrespective of their content.

Inferential Holism. A third kind of representational flexibility concerns the range of information that we can bring to bear on solving a given problem. As noted in Section 3, human reasoning appears to exhibit a kind of holism or isotropy (Fodor 1983). Given *surrounding conditions*—especially background beliefs—the *relevance* of a belief to the theoretical or practical tasks in which one engages can change dramatically. Indeed, it would seem that given appropriate background assumptions, almost any belief can be rendered relevant to the task in which one engages (Copeland 1993). But if this is so, then the obvious explanation is, as Fodor noted long ago, that we possess central systems that are unencapsulated to an interesting degree.

What do these considerations show? First, they clearly do *not* show that there are no modular central systems. This is because even if the explanation of representational flexibility requires the existence of some non-modular central systems, this would be wholly consistent with the existence of other central systems that are modular in character. In other words, the above considerations are wholly compatible with what I earlier called weak MM: the thesis that central cognition depends on both modular mechanisms and domain-general, unencapsulated ones. Second, the above considerations are also compatible with the sort of *compositional MM* formulated using the minimal notion of a module. This is because such a thesis does not require what the above considerations render implausible—that all modules are domain-specific and/or encapsulated—and this is simply because, in the minimal sense of modularity, domain-general, unencapsulated mechanisms *are* modules.

So, we should be cautious not to interpret the present considerations as undermining all versions of MM. Nevertheless, taken together the above kinds of representational plasticity do provide prima facie reason to suppose that there are cognitive mechanisms that are domain general and unencapsulated. This is because the assumption that there are such mechanisms yields the simplest and most natural explanation of the kinds of flexibility outlined above. To that extent, then, the existence of representational flexibility renders Strong MM implausible.

Advocates of Strong MM have sought to provide accounts of the above kinds of flexibility—accounts that eschew any commitment to the sorts of non-modular mechanisms posited by Fodor and others. If such proposals could be made to work, then the argument from representational flexibility would be significantly weakened. In Section 4.3 I briefly review some of these modularist proposals. But first we need to consider a closely related kind of flexibility problem that an adequate version of MM must address.

4.2. Behavioral Flexibility and Flow of Control

The worries considered so far concern the apparent flexibility of our representational capacities. But there is another very closely related kind of worry that concerns a striking fact about the character and range of our cognitive-behavioral repertoire. To a first approximation:

> *Flexibility Thesis*: We are capable of performing an exceedingly wide—perhaps unbounded—range of tasks in a context-appropriate fashion.

According to some critics, the worry about MM, at least in radical form, is that it lacks the resources to account for this kind of flexibility.

Some comments are in order. First, though there are many issues of detail regarding precisely how best to formulate the Flexibility Thesis, the general idea has very widespread acceptance. Indeed, it has a heritage that goes back at least as far as Descartes; it is widely endorsed by cognitive scientists (Newell 1990; Anderson and Lebiere 2003); and it has seemed irresistible to those who study either the anthropological record (Richerson and Boyd 2006) or the contrasts between human behavior and that of other primates (Whiten et al. 2003).

Second, though the Flexibility Thesis is logically distinct from the sorts of representational flexibility mentioned in Section 4.1, it is important to stress that on many extant accounts of cognition, the two are very intimately related. Specifically, one very common reason for invoking flexible, representation-rich processes is to explain the highly variable yet context-appropriate character of human behavior. Crudely put: on one very common view of cognition—one that many modularists endorse—human behavior is flexible in large measure because it causally depends on flexible representational processes (Newell 1990; Pylyshyn 1984).

Third, the fact of behavioral flexibility is, of course, not merely an explanatory challenge for modular theories of cognitive architecture, but a serious explanatory challenge for any account of cognition (Newell 1990). Indeed, it is arguably just the problem of explaining intelligent behavior. Nevertheless, some critics maintain that behavioral flexibility poses quite specific and serious challenges for advocates of MM because their position appears to preclude the sorts of explanations that most plausibly explain the character and range of human behavior: that is, those that posit domain-general, functionally nonspecific mechanisms.

One central virtue of domain-general, functionally nonspecific mechanisms is that they can underwrite the performance of a great many tasks. They are, in Descartes's memorable phrase, "universal instruments." Advocates of a thoroughgoing MM cannot, of course, avail themselves of such mechanisms. But neither can they plausibly suppose that we possess a specific module for each task we can perform. As Descartes observed, the range of tasks that we can perform is simply *too great* for such a proposal to be at all plausible.[17] How, then, can advocates of MM explain the range of tasks that we are capable of performing?

It would seem that there is only one available option. Advocates of MM are committed to providing what might be called a *confederate* account of cognitive flexibility: one on which flexible behavior is, as Pinker puts it, the product of "a network of subsystems that feed each other in criss-crossing but intelligible ways" (Pinker 2005. See also Pinker 1994 and James 1890). But merely pointing this out is not, of course, an explanation of our cognitive-behavioral flexibility so much as a statement of the problem given a commitment to MM. The challenge for advocates of MM is to sketch the *right sort* of plurality and "criss-crossing" between mechanisms, and this would require an account that addresses at least the following problems.

---

[17] As Fodor once pointed out, sometimes we manage to balance our checkbooks, but it is not at all likely that there is a modular device for doing that!

First, on the assumption that behavioral flexibility causally depends on flexible representation-rich processes, such an account would need to handle the sorts of flexibility mentioned in Section 4.1., that is:

Integration Problem: Advocates of MM need to explain how novel, cross-domain representations can be produced.
Consumption Problem: Advocates of MM need to explain how novel, cross-domain representations could be utilized in reasoning, decision making, and other cognitive processes.
Holism Problem: Advocates of MM need to explain how some inferential processes could exhibit their characteristic holism.

To avoid positing non-modular mechanisms, a thoroughgoing MM would need to explain such phenomena as a product of the collaborative activity of multiple modules.

Second, because MM is committed to a confederate account of behavioral flexibility, advocates of MM also need to address a problem about the *flow of control* that is often ignored in discussions of MM. If solutions to the problems we confront frequently depend on the collaborative interaction of a host of modules, there needs to be some account of how the right module "gains control" of the process at the right time. This is because on such a model, a correct or appropriate outcome will occur only if an appropriate module is activated at the right time in the process. So, advocates of MM need to address what might be called the *allocation problem*:

Allocation Problem: Advocates of MM need to characterize the *control structures* that ensure that representations are allocated to the relevant modules at the right time.

Issues about flow of control are commonplace in computer science, and there are many ways to organize a computational system in order to address such issues. In the case of thoroughgoing versions of MM, however, the allocation problem has seemed especially pressing because it has proven hard to think of plausible control structures that could enable cognitive-behavioral flexibility without compromising the assumption that our minds contain only modular systems.

Though there are many variants of the allocation problem, it would be useful to start with one especially well-known version, discussed by Fodor (2000). What Fodor purports to show is that the allocation problem poses a kind of *logical* problem for strong versions of MM. Specifically, he argues that, on pain of regress, solving the allocation problem requires that there exist at least some domain-general mechanisms. In this case there is, according to Fodor, a sense in which the hypothesis of a completely modular architecture—one that eschews domain-general mechanisms entirely—is "self-defeating."

To appreciate Fodor's version of the allocation problem, we need to focus on the question of whether the mechanisms responsible for allocating representations to modules are themselves domain-specific. Fodor maintains that there are really

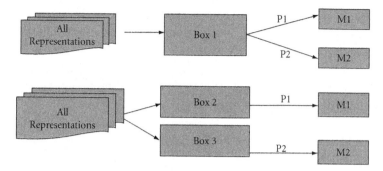

Figure 1. Fodor's input problem.

only two options, which are represented schematically in Figure 1. According to the first option, represented by Box 1, the allocation mechanism is relatively domain-general, in which case it is able to perform its allocating function because it can access both those representations that should be allocated to M1 and those that should be allocated to M2. (Think of someone passing apples to one friend and oranges to another. They need to have access to both apples and oranges to perform that task.) But the problem with this option is that the allocating mechanism (Box 1) is not itself domain-specific, and so (strong) MM is false. The second option, represented by Box 2 and Box 3 in the diagram, is that allocation mechanisms are no more domain-general than the modules to which they allocate representations. In this arrangement, the existence of allocation mechanisms does not violate MM by assuming the existence of non-modular devices. But according to Fodor, it is now unclear how the allocation mechanisms could, themselves, have been allocated the relevant representations. If we suppose that it was another domain-specific alloca-tor, then regress ensues, and if we suppose that the allocator is domain-general, then we once more violate the assumptions of strong MM. On the face of it, then, alloca-tion poses a serious challenge for strong versions of MM.

4.3. Massively Modular Architectures for the Explanation of Cognitive-Behavioral Flexibility

What sort of massively modular architecture could address the various problems of flexibility and allocation outlined above? At this time, the issues remain largely open, and extant proposals are pitched at a very abstract—sometimes metaphori-cal—level. Nonetheless, I now propose to consider some of the suggestions that have been floated in recent years.

4.3.1. Weak Massive Modularity

One response, mentioned earlier, would be to acknowledge the need for at least some domain-general and/or unencapsulated mechanisms. Such positions are commonplace in recent cognitive science among theorists who are otherwise quite sympathetic to modular accounts of cognition. Thus, for example, Susan Carey and

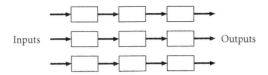

Figure 2. Pipeline architectures.

John Anderson both endorse versions of this weak MM position, and moreover they do so in large measure because it helps handle the sorts of problems mentioned earlier (Anderson 2007; Carey 2009).

Weak MM is a plausible position for it has the resources to accommodate the empirical evidence for modularity while also allowing for aspects of cognition—various kinds of learning, analogical inference, planning, and so on—that do not seem modular in character. Nonetheless, for some advocates of MM, such a position may seem unattractive on broad theoretical grounds. As we saw in Sections 2 and 3, it is quite common to maintain that, for evolutionary and computational reasons, non-modular mechanisms are implausible, and consequently that some more thoroughgoing version of MM is required. For such theorists it would be implausible to suppose that non-modular devices have a major role to play in human cognition.

4.3.2. *Pipeline Architectures*

Suppose that one seeks a thoroughgoing—or strong—MM. How might one address the problems of representational flexibility and allocation? One possibility would be to advocate what are sometimes called *pipeline architectures.* The general idea is that the modules within such a system are organized in a lattice-like fashion so that their interconnections satisfy two conditions:

a) Information flow is unidirectional: Once information enters a device in a given layer, n, it cannot subsequently enter another device in n, or a device in any layer prior to n. Rather, information is automatically routed to a device to some subsequent layer of the system.
b) Uniqueness: Information processed by one module can be routed to (at most) one other module.

On these assumptions, then, the overall system can be schematically represented as a set of parallel pipelines, each composed of a number of interconnected modular processing units. (See Figure 2.)

It is important to stress that no one has ever seriously defended pipeline architectures in the simple form presented here.[18] Nonetheless, it will be instructive to

---

[18] It is worth noting, however, that there are various influential proposals that come very close. In particular, the Subsumption Architecture advocated by Rodney Brooks and his collaborators bears striking similarities, and raises very similar problems. For further discussion see Barrett (2006), Brooks (1991), Kirsch (1991), and Hurley (2001).

consider them since simple pipeline architectures possess a number of properties relevant to our present discussion, which can help clarify the problems that flexibility poses for MM.

First, such architectures enforce a strong kind of MM because they ensure that different mechanisms have access to different, *non-overlapping pools* of information. Indeed they ensure that modules satisfy exceedingly strong conditions on both domain–specificity and informational encapsulation.[19] Modules within such a system will be domain-specific because they receive inputs from at most one other system. And since the information that any module receives is simply its input, each module will also be encapsulated. Thus pipeline architectures are both strongly modular and satisfy Fodorian conditions on modularity.

Second, there is a sense in which pipeline architectures evade the sorts of allocation worries discussed earlier. Since modules within such an architecture are triggered by their inputs and pass information uniquely and unidirectionally, the flow of control within a pipeline architecture is rigid and inflexible. For example, if the first module in a pipeline, P, is activated by a sensory input, then every subsequent module in P will also be activated, and modules that are not in P will not be activated by the sensory input. One way to put the point is that in such a view there is no *computational* problem of allocation; rather, allocation is brute-causal and hardwired.

Third, the previous observation is important for understanding Fodor's allocation problem. This is because it highlights that strong MM per se is not self-defeating—or at least not for the reasons that Fodor provides. Fodor's problem turns on the putative fact that regress ensues unless one posits domain-general allocation mechanisms. But within a pipeline architecture the regress of allocation is halted by the first module in the pipeline—we might suppose a sensory mechanism of some sort. Thus, the dilemma that Fodor seeks to generate for MM—either regress of allocation or domain-general allocators—never gets off the ground.

Fourth, it is important to see that pipeline architectures only succeed in resolving Fodor's puzzle at a serious cost. Specifically, the proposed solution implies that modules in different pipelines cannot interact—that many configurations of intermodular interaction are impossible—and this, in turn, imposes serious limitations on the sorts of flexibility that can be accommodated by such a confederate system. First, representations in different pipes cannot be freely combined, in which case a pipeline architecture will not solve the integration problem. Second, since pipeline architectures cannot combine representations from distinct domains, they cannot explain our apparent capacity to *use* cross-domain representations in our practical and theoretical inferences. In other words, they cannot offer a solution to the consumption problem for MM.[20] Third, since pipeline architectures enforce a rigid

---

[19] Of course, this assumes the (obvious) fact that no sensory mechanisms are domain-general.

[20] Indeed there is a sense in which such systems do not confront a consumption problem since there are no cross-domain representations to be consumed.

distinction between informational pools, such a system cannot exhibit inferential holism. Finally, since behavioral flexibility is supposed to depend on the above sorts of representational flexibility, pipeline architectures preclude the kinds of inter-modular interactions that seem required to produce novel, flexible behavior. In short: though pipeline architectures evade the allocation problem that Fodor poses, they do so at the cost of completely failing to accommodate the sorts of flexibility that advocates of MM need to explain.

Finally, the above discussion suggests an interesting connection between the problems posed by allocation and the problems that cognitive flexibility poses. Recall: it is precisely because pipeline architectures enforce a rigid division between pipelines of modules that they both evade Fodor's allocation problem and fail to exhibit representational flexibility. But what this suggests is that, within MM archi-tectures, problems of allocation or control are closely related to the system's capac-ity to exhibit various kinds of representational flexibility. Roughly put, the more flexibility the system exhibits, the more serious we should expect allocation prob-lems to be. More specifically, the above discussion suggests that Fodor's version of the allocation problem—that of requiring domain-general control structures—is one that only arises for modular systems that exhibit the appropriate kinds of rep-resentational flexibility, and the more flexibility exhibited, the more need there will be for such control structures. I return to this issue below. But for now let us con-sider another possible approach to the problems of flexibility and allocation.

4.3.3. *Enzymatic Computation*

Recently, Clark Barrett has presented a proposal that is intended to address worries about allocation at the same time as it explains aspects of cognitive flexibility (Bar-rett 2005). Barrett's point of departure is Fodor's version of the allocation problem. As such, his proposal might be viewed merely as an attempt to resolve the kind of logical problem that, according to Fodor, allocation poses for massively modular architectures. Alternatively, it might be construed as trying to satisfy the stronger demand of providing an *empirically plausible* model for how a massively modular system might exhibit flexibility. In what follows I outline Barrett's proposal and then considers each of these options.

In developing his view, Barrett takes enzymatic systems in biochemistry as a model for how a modular mind might be organized. Broadly speaking, enzymatic systems possess two kinds of properties that make them appropriate as a model of cognitive modularity. The first class of properties is those that allow enzymes to function as specialized computational devices. Specifically:

a) Enzymes accept information of a particular kind, generally in the form of chemical substrates with particular properties that meet the binding specificity criteria of the enzyme in a "lock and key" fashion.
b) They perform specific operations on the information they admit, catalyzing reactions that produce reaction products with different properties than the input substrates.

c) The reaction products produced by enzymes are in a format useable by other systems, thereby allowing for complex cascades of activity.

A second class of properties possessed by enzymatic systems that Barrett thinks make them appropriate as a model of cognitive modularity concerns the environment in which interactions between enzymes and substrate occur. Specifically, such interactions occur in "open" systems (solutions) in which all substrates are accessible, in principle, to all enzymes. Thus in such enzymatic systems one has *access generality*—where all information (substrates) are available to all processing mechanisms (enzymes)—with *processing specificity*: where each kind of enzyme only performs highly specific operations on a very specific range of substrates—viz. those that satisfy the binding criteria of the enzyme. Importantly for Barrett's purposes, enzymatic systems achieve this combination of access generality and processing specificity without the need for a mechanism that delivers substrates to enzymes. Thus within such systems there is no rigid routing (à la the pipeline model) or domain-general "meta" device for allocation (à la Fodor). Thus, according to Barrett, enzymatic systems provide both (1) an existence proof of naturally occurring modular systems that avoid the sorts of allocation problems Fodor raises, and (2) a model of how the flow of control within cognitive systems might occur.

What should we make of Barrett's proposal? First, let us consider it as a response to the putatively logical problem of allocation that Fodor poses for MM. Barrett is correct that flow of control within a MM cognitive system could, in principle, operate in the same way as enzymatic systems do. Specifically, it is possible to envisage a system comprised of process-specific mechanisms operating in an access general environment in such a way that specialized mechanisms gain access to relevant inputs without the need for non-modular allocation devices. As such, Barrett's proposal provides a way to resolve the logical problem Fodor seeks to generate for MM, and to that extent the proposal is successful.[21]

How does Barrett's enzymatic proposal fare as an *empirically plausible* model of the mind's organization? Here I am rather less sanguine. Construed literally, the enzymatic model is deeply implausible, and construed as mere metaphor it is utterly unclear that it can be cashed out without reintroducing precisely the sorts of domain-general control structures that it seeks to avoid.

In order to see why, on a literal construal, the enzymatic model is implausible, we need to get clearer about why enzymatic systems—real enzymatic systems—do not require routing mechanisms or meta-control devices. The central problem of control within a modular cognitive architecture is the problem of enabling the right mechanism to access the right representations at the relevant time. There is an analogous problem for enzymes. In order for enzymes to produce their products, they must bind the relevant substrate. How does that occur? Of course, there is no "routing system" that brings the relevant substrate to the right enzyme. Instead, enzymes

---

[21] Though since the simpler pipeline model achieves the same result, it is unclear that this success is of any great significance.

interact with substrates via a process that crucially depends on chance collision. To a first approximation, dumb luck (random probability) is a central component of the story of how enzymes come to have their characteristic effects within an open system. Where solution conditions are fixed, the rate at which enzymatic processes occur is a function of enzyme and substrate concentrations. Where either concentration is high, catalytic reactions occur rapidly because there is a higher probability that individual enzymes will encounter substrates that satisfy their binding conditions. By contrast, where concentrations are low, rates of reaction are low because compatible proteins and substrates rarely encounter each other. And, of course, where there is just one instance of an enzyme and one instance of an appropriate substrate in a sea of other substrates, the probability of a relevant collision at any particular point in time is exceedingly low.

The moral should be clear for discussions of cognitive architecture. Enzymatic systems *do* possess properties that are closely analogous to those possessed by putative cognitive modules. Moreover, enzymes are able to do their job—convert substrate into products—in the absence of any routing or meta-device for allocation. Thus the possibility of enzymatic computation shows, contrary to Fodor, that there is no *logical* problem of allocation within a massively modular system. But much more would need to be said in order to render the enzyme metaphor plausible as an empirical model of control flow. This is because, when applied in a literal fashion to cognitive systems, the proposal yields a conception of control flow on which appropriate cognitive processing—as opposed to lots of fruitless failed interactions between modules and representations—is simply the product of random interactions in an environment that contains high concentrations of modules of the same type and high concentrations of representations of the same type. And this is not even remotely plausible as a story for how cognition works.[22]

Of course, Barrett is well aware that, construed literally, enzymatic systems are not a plausible model for how modular cognitive systems interact. Indeed, it is a point he stresses repeatedly. As a consequence, he instead treats talk of enzymatic systems as a metaphor in need of further development. But now the problem is that it is far from clear how this can done while still preserving the idea that the human mind has a MM architecture that manifests flexibility and yet avoids the need for domain-general allocation devices.

In developing his view, Barrett tends to draw on examples of computational models that bear an abstract resemblance to enzymatic systems. In particular, he is fond of developing the enzymatic metaphor with reference to the sorts of blackboard architectures that exhibit a kind of open access combined with processing specificity (Hayes-Roth 1985). But the problem with this is that computationally well-specified backboard architectures—as opposed to breezy descriptions of the

---

[22] Though there is not enough space to consider the issue in detail here, one possible way to respond to the present worry would be to hypothesize that (1) the informational repository is open but highly structured, and (2) modules tend to be located in close proximity to those areas of the repository that contain information that satisfies their input conditions.

rough idea—incorporate precisely the sort of domain-general allocation mechanism that Barrett seems so keen to avoid. Specifically, though such architectures are comprised of an open source repository of information (the blackboard), and multiple (software) modules—often called "knowledge sources"—they also have a *control shell* as a core component. What the control shell does is use generic control knowledge in order to make runtime decisions about the course of problem solving. In other words, it is an allocation device that determines which of the available modules gets to perform its computations at a given time on information in the blackboard (Corkill 2003). But this is precisely the kind of domain-general control device that is so conducive to Fodor's presentation of the allocation problem, and so at odds with the spirit of the enzymatic model. In short, if the blackboard architecture is what one gets when the enzymatic metaphor is spelled out, then strong MM must be rejected.

5. CONCLUSION

This chapter sought first to clarify MM and distinguish a range of importantly different versions of the thesis. Second, it critically assessed some of the more prominent arguments for MM—arguments that purport to show that it is plausible either on evolutionary grounds or on grounds of computational tractability. Finally, it introduced some of the problems that cognitive-behavioral flexibility appears to pose for MM, at least in its strongest forms. The foregoing discussion of flexibility clearly does not preclude the possibility of an empirically plausible strong MM. Among other things, there are other important modularist proposals—such as those proposed by Sperber (2005) and Carruthers (2006)—that have not been discussed here, and a comprehensive assessment would require due consideration of these proposals. But neither was the discussion intended to ground such a strong conclusion. Rather, the goals were threefold. The first was to flag some of the kinds of flexibility that appear to pose problems for MM. The second was to highlight that, in the absence of any well-specified and plausible modularist account of flexibility, positing non-modular devices appears to yield the most natural and plausible account of flexibility. The final goal was to suggest that in going beyond mere metaphor and vague suggestion, extant modularist proposals that seek to accommodate cognitive flexibility appear to risk reintroducing precisely the sorts of domain-general, non-modular mechanisms that they seek to banish. For all that has been said, this might reflect mere contingent features of extant proposals. But another and rather more intriguing possibility is that there are deep and systematic connections between manifesting human levels of cognitive-behavioral flexibility and the need for domain-general mechanisms. Though this is not the place to spell out the argument, I suspect that this is what is really going on.

REFERENCES

Anderson, J. R. (2007). *How Can the Human Mind Occur in the Physical Universe?*
New York: Oxford University Press.

Anderson, J. R., and Lebiere, C. L. (2003). The Newell test for a theory of cognition.
Behavioral & Brain Science 26: 587–637.

Baron-Cohen, S. (1995) . *Mindblindness: An Essay on Autism and Theory of Mind.*
Cambridge, MA: MIT Press.

Barrett, H. C. (2005). Enzymatic computation and cognitive modularity. *Mind & Language*
20: 259–87.

Barrett, H. C., and Kurzban, R. (2006). Modularity in cognition: framing the debate.
Psychological Review 113: 628–47.

Brooks, R. A. (1991). Intelligence without reason. *Proceedings of the 12th International
Conference on Artificial Intelligence* 569–95.

Buller, D. (2005). *Adapting Minds.* Cambridge, MA: MIT Press.

Carey, S. (2009). *The Origin of Concepts.* New York: Oxford University Press.

Carruthers, P. (2003). On Fodor's problem, *Mind and Language,* 18(5): 502–23.

———. (2004). Practical reasoning in a modular mind. *Mind and Language* 19: 259–78.

———. (2006). *The Architecture of the Mind: Massive Modularity and the Flexibility of
Thought.* Oxford: Oxford University Press.

Carston, R. (1996). The Architecture of the Mind: Modularity and Modularization. In
D. Green (ed.), *Cognitive Science: An Introduction.* Oxford: Blackwell.

Chiappe, D. (2000). Metaphor, modularity, and the evolution of conceptual integration.
Metaphor and Symbol 15: 137–58.

Chiappe, D., and MacDonald, K. B. (2005). The evolution of domain-general mechanisms
in intelligence and learning. *Journal of General Psychology* 132: 5–40.

Cherniak, C. (1986). *Minimal Rationality.* Cambridge, MA: MIT Press.

Collins, J. (2005). On the input problem for massive modularity. *Minds and Machines* 15(1):
1–22.

Cooper, R. P., and Shallice, T. (2006). Hierarchical schemas and goals in the control of
sequential behavior. *Psychological Review* 113 : 887–931.

Copeland, J. (1993). *Artificial Intelligence: A Philosophical Introduction.* Oxford: Blackwell.

Corkill, D. (2003). Collaborating Software: Blackboard and Multi-Agent Systems & the
Future. In *Proceedings of the International Lisp Conference,* New York, New York,
October 2003.

Cosmides, L., and Tooby, J. (1994). Origins of domain specificity: The evolution of
functional organization. In L. Hirschfeld and S. Gelman (eds.), *Mapping the Mind:
Domain Specificity in Cognition and Culture,* 85–116. Cambridge: Cambridge University
Press.

Cosmides, L., and Tooby, J. (2000). The Cognitive Neuroscience of Social Reasoning. In M. S.
Gazzaniga (ed.), *The New Cognitive Neurosciences,* 2nd ed. Cambridge, MA: MIT Press.

Dennett, D. C. (1987). *The Intentional Stance.* Cambridge, MA: MIT Press.

Fodor, J. (1983). *The Modularity of Mind.* Cambridge, MA: MIT Press.

———. (2000). *The Mind Doesn't Work That Way: The Scope and Limits of Computational
Psychology.* Cambridge, MA: MIT Press.

———. (2005). Reply to Steven Pinker "So How *Does* the Mind Work?". *Mind & Language*
20(1): 25–32.

———. (2008). *LOT 2: The Language of Thought Revisited.* Oxford: Oxford University Press.

Ford, K. M., and Pylyshyn, Z. W. (eds.) (1996). *The Robot's Dilemma Revisited: The Frame Problem in Artificial Intelligence*. Norwood, NJ: Ablex.

Garfield, J. (ed.). (1987). *Modularity in Knowledge Representation and Natural-Language Understanding*. Cambridge, MA: MIT Press.

Gigerenzer, G., Todd, P. M., and the ABC Research Group. (1999). *Simple Heuristics That Make Us Smart*. New York: Oxford University Press.

Glymour, C. (1985). Comment: Fodor's holism. *Behavioral and Brain Sciences* 8: 15–16.

———. (1987). Android epistemology: Comments on "Cognitive wheels." In Z. W. Pylyshyn (ed.), *The Robot's Dilemma*. Norwood, NJ: Ablex, 65–76.

Gopnik, A., and Meltzoff, A. (1997). *Words, Thoughts and Theories*. Cambridge, MA: MIT Press.

Guha, R. V., and Levy, A. (1990). A Relevance Based Meta Level. *MCC Technical Report No. ACT-CYC-040-90*. Austin, TX: MCC Corp.

Hayes, P. (1987). What the Frame Problem Is and Isn't. In Z. Pylyshyn (ed.), *The Robot's Dilemma*. Norwood, NJ: Ablex.

Hayes-Roth, B. (1985). A blackboard architecture for control. *Artificial Intelligence* 26: 251–321.

Horgan, T., and Tienson, J. (1996). *Connectionism and the Philosophy of Psychology*. Cambridge, MA: MIT Press.

Hurley S. (2001). Perception and action: Alternative views. *Synthese* 129(1): 3–40.

James, W. (1890). *The Principles of Psychology*, vol. 1. Cambridge, MA: Harvard University Press.

Jackendoff, R. (1992). Is There a Faculty of Social Cognition? In R. Jackendoff, *Languages of the Mind*. Cambridge, MA: MIT Press, 69–81.

———. (1992). *Languages of the Mind*. Cambridge, MA: MIT Press.

Kirsh, D. (1991). Today the earwig, tomorrow man? *Artificial Intelligence* 47: 161–84. Reprinted in M. Boden (ed.), *The Philosophy of Artificial Life*. New York: Oxford University Press, 1996.

Lormand, Eric. (1994). The Holorobophobe's Dilemma. In K. Ford and Z. Pylylshyn (eds.), *The Robot's Dilemma Revisited*. Norwood, NJ: Ablex.

Ludwig, K., and Schneider, S. (2007). Fodor's challenge to the classical computational theory of mind. *Mind & Language* 23(1): 123–43.

Lyons, Jack C. (2001). Carving the mind at its (not necessarily modular) joints. *British Journal for the Philosophy of Science* 52(2): 277–302.

Machery, E., and Barrett, H. C. (2006). Debunking *Adapting Minds*. *Philosophy of Science* 73: 232–46.

Miller, G., Galanter, E., and Pribram, K. (1960). *Plans and the Structure of Behavior*. New York: Henry Holt.

Newell, A. (1990). *Unified Theories of Cognition*. Cambridge, MA: Harvard University Press.

Newell, A., and Simon, H. A. (1972). *Human Problem Solving*. Englewood Cliffs, NJ.: Prentice-Hall.

Pinker, S. (1994). *The Language Instinct*. New York: William Morrow.

———. (1997). *How the Mind Works*. New York: W.W. Norton.

———. (2005). So how does the mind work? *Mind & Language* 20(1): 1–24.

Pohl, T. (2005). *Cognitive Illusions: A Handbook on Fallacies and Biases in Thinking, Judgement and Memory*. Hove, UK: Psychology Press.

Prinz, J. J. (2006). Is the Mind Really Modular? In R. Stainton (ed.), *Contemporary Debates in Cognitive Science*. Oxford: Blackwell, 22–36.

Pylyshyn, Z. W. (1984). Computation and cognition. Cambridge, MA: MIT/Bradford.

————. (1987). *The Robot's Dilemma: The Frame Problem in Artificial Intelligence (Theoretical Issues in Cognitive Science)*. Norwood, NJ: Ablex.

Richerson, P., and Boyd, R. (2006). *Not by Genes Alone: How Culture Transformed Human Evolution*. Chicago: University of Chicago Press.

Russell, S., and Norvig, P. (2003). *Artificial Intelligence, A Modern Approach*. 2nd ed. Upple Saddle River, NJ: Prentice Hall.

Samuels R. (1998). Evolutionary psychology and the massive modularity hypothesis. *British Journal for the Philosophy of Science* 49: 575–602.

————. (2000). Massively Modular Minds: Evolutionary Psychology and Cognitive Architecture. In P. Carruthers and A. Chamberlain (eds.), *Evolution and the Human Mind*. Cambridge, UK: Cambridge University Press, 13–46.

————. (2005). The Complexity of Cognition: Tractability Arguments for Massive Modularity. In P. Carruthers, S. Laurence, and S. Stich (eds.), *The Innate Mind: Structure and Contents*. Oxford: Oxford University Press.

————. (2006). Is the Mind Massively Modular? In R. Stainton (ed.), *Contemporary Debates in Cognitive Science*. Oxford: Blackwell, 37–56.

————. (2010). Classical computationalism and the many problems of cognitive relevance. *Studies in History and Philosophy of Science Part A* 41(3): 280–93.

Segal, G. (1996). The Modularity of Theory of Mind. In P. Carruthers and P. Smith (eds.), *Theories of Theory of Mind*. Cambridge: Cambridge University Press.

Shanahan, M. P. (1997). *Solving the Frame Problem: A Mathematical Investigation of the Common Sense Law of Inertia*. Cambridge, MA: MIT Press.

Simon, H. (1962). The architecture of complexity. *Proceedings of the American Philosophical Society* 106: 467–82.

Sperber, D. (1994). The Modularity of Thought and the Epidemiology of Representations. In L. A. Hirschfeld and S. A. Gelman (eds.), *Mapping the Mind*. Cambridge: Cambridge University Press, 39–67).

————. (2002). In Defense of Massive Modularity. In I. Dupoux (ed.), *Language, Brain, and Cognitive Development*. Cambridge, MA: MIT Press, 47–57.

————. (2005). Modularity and Relevance: How Can a Massively Modular Mind Be Flexible and Context-Sensitive?. In P. Carruthers, S. Laurence, and S. Stich (eds.), *The Innate Mind: Structure and Content*. Oxford: Oxford University Press.

Sperber, D., and Wilson, D. (1996). Fodor's frame problem and relevance theory. *Behavioral and Brain Sciences* 19(3): 530–32.

Sperber, D., and Hirschfeld, L. (2007). Culture and Modularity. In P. Carruthers, S. Laurence, and S. Stich (eds.), *The Innate Mind: Culture and Cognition*. Oxford: Oxford University Press, 149–64.

Stanovich, K. (2004). *The Robot's Rebellion: Finding Meaning in the Age of Darwin*. Chicago: University of Chicago Press.

Tooby, J., and Cosmides, L. (1992). The Psychological Foundations of Culture. In J. Barkow, L. Cosmides, and J. Tooby (eds.), *The Adapted Mind*. New York: Oxford University Press.

Tooby, J., and Cosmides, L. (1995). Foreword. In S. Baron-Cohen, *Mindblindness: An Essay on Autism and Theory of Mind*. Cambridge, MA: MIT Press.

Traverso, P., Ghallab, M., and Nau, D. (2004). *Automated Planning: Theory & Practice*. San Francisco: Elsevier.

Whiten, A., Horner, V., and Marshall-Pescini, S. (2003). Cultural panthropology. *Evolutionary Psychology* 12: 92–105.

CHAPTER 5

PERCEPTION AND MULTIMODALITY

CASEY O'CALLAGHAN

PHILOSOPHERS and cognitive scientists of perception by custom have investigated individual sense modalities in relative isolation from each other. However, perceiving is, in a number of respects, multimodal. The traditional sense modalities should not be treated as explanatorily independent. Attention to the multimodal aspects of perception challenges common assumptions about the content and phenomenology of perception, and about the individuation and psychological nature of sense modalities. Multimodal perception thus presents a valuable opportunity for a case study in mature interdisciplinary cognitive science. This chapter aims to raise these issues against the background of unimodal approaches in the study of perception. It presents some of the central empirical findings concerning multimodality, and it explains the philosophical implications of these findings. Foremost, it aims to encourage and open avenues for future research.

1. UNIMODAL APPROACHES TO THE STUDY OF PERCEPTION

Sometimes you turn to look at the source of a *clunk*. Sometimes you see a pan on the stove and grasp the hot handle. Sometimes you smell the pepper you taste, or find a stinky dead mouse under the sink. In each case, you perceive using more than one sense. You hear the sound and see its source. You feel and see the handle. You smell and taste the pepper. You see and smell the mouse. By itself, this is no surprise: we perceive visually, auditorily, tactually, olfactorily, and gustatorily.

And yet, most work on perception by philosophers and cognitive scientists has focused on human vision. Empirical psychologists catalog visual illusions and deficits because they want to discover perceptual principles and mechanisms. Philosophers analyze visual experience to explain what we perceive and how we perceive it. This focus makes sense since vision is central to how most humans experience, understand, and navigate their surroundings.

Focusing exclusively on vision is risky. Nothing guarantees that what holds of seeing holds generally of perceiving. What is essential to seeing might not be essential to perceiving, and certain requirements on perceiving might lack salience in the visual case. Impressive diversity among the varieties of perceiving also might go underappreciated. Attention to the contrasts among smelling a scallion's odor, hearing the sounds of a colleague's footsteps, and seeing a building's collapse tells us things about perception that attention to vision alone cannot. Focusing upon modalities other than vision is good methodology in the study of perception.

Researchers in recent years increasingly have turned to nonvisual modalities for insights about perception and perceptual processes. Audition has attracted interest among scientists and philosophers, who have contrasted the objects and the spatial and temporal aspects of audition and vision (see, e.g., Bregman 1990; Handel 1993; McAdams and Bigand 1993; Blauert 1997; Casati and Dokic 1994; O'Callaghan 2007, 2008a). Others have explored smell and taste in order to test the claim that perceptual phenomenology is a variety of representational content (see Lycan 2000; Batty 2010). Martin (1992) contrasts tactile spatial experience with visual spatial experience and expresses skepticism whether any general account of perception holds across modalities. Noë (2004) even argues that visual awareness essentially involves the kind of exploratory activity characteristic to tactile and proprioceptive perception (see also O'Regan and Noë 2001; see Clark 2006 for a challenge from audition). Promising interdisciplinary research on perceptual modalities other than vision now is beginning to thrive. Such work raises new puzzles and invites novel questions, and it represents the kind of multimodal approach necessary to avoid a parochial, vision-based take on perception.

This approach nonetheless risks remaining *unimodal* in one striking way. It involves investigating each modality in isolation from the others. It is easy to find examples from empirical and philosophical literature. Sensory scientists study the transduction and transmission of information in a single sense or sensory pathway, or study the responses of cells or networks of cells to unimodal stimuli such as flashes, beeps, odors, or pinpricks. Philosophers of perception write about unimodal qualia or sensible qualities such as hue, pitch, and taste. When the topic is not unimodal characteristics, it frequently is unimodal experiences, such as the visual experience of seeing a shape, the auditory experience of hearing a crash, or the olfactory experience of smelling a corpse flower. As a result, theorizing about perception often takes place like this: we think about individual modalities, compare and contrast them, and compile the stories into an overall account of what it is to perceive.

This approach reveals a presupposition that is not innocent. The methodology suggests that we are thinking of each sense as *explanatorily independent* from the others. This assumption of independence is evident at physiological, functional, and experiential levels of explanation. At one extreme, the senses in this way of thinking involve different organs and pathways, perform discrete functions, are causally and informationally encapsulated, and constitute entirely distinctive modes of awareness. Thus, the modalities of sense perception and their associated sense experiences in principle are separable from each other. It is tempting to assume when examining just one sense that it is possible to get a complete account of perceiving with that sense modality that is explanatorily independent from the others. If so, the senses constitute independent domains of philosophical and scientific inquiry. One might even hold that assembling the sense-specific stories about seeing, hearing, touching, smelling, and tasting exhausts theorizing about perception.

What supports this approach? The senses sometimes come apart. You could cease entirely to see without losing touch, or become deaf without ceasing to see. However, you also could lose one or more visual capacities (color vision, the ability to see faces) without losing others (spatial vision), you could lose visual and tactual capacities simultaneously, and losing vision could impact or impair your touch or audition. So, I suspect the approach is sustained by the conviction that the modalities of perception and perceptual awareness *differ* dramatically. Distinctive sense organs respond to different forms of energy. These organs function in quite different ways, and each seems dedicated to performing a specialized kind of task. Distinct pathways of activation and areas of the brain are dedicated to the different senses. Experiences associated with different modalities also seem to differ remarkably. Seeing a pileated woodpecker differs in qualitatively dramatic ways from hearing it vocalize. Smelling unfamiliar odors and hearing musical sounds make it intuitive to think of each sense as experientially distinctive. Furthermore, the modalities differ in their effects. Each grounds different sorts of beliefs and actions. When riding your bike, thanks to vision you come to believe a pedestrian is crossing the street ahead; thanks to hearing you get out of the way of the car approaching from behind.

This perspective, and the methodology, would be viable if experiences associated with the different senses were entirely experientially discrete or encapsulated, or if they were exhausted by distinctive or sense-specific phenomenology. It would be defensible if the senses had disjointed functions, or if they were entirely causally independent or physiologically isolated.

The evidence, however, suggests that we should abandon this perspective. Perceiving is *richly* multimodal. Perceptual processes involve extensive interaction among sensory systems. The patterns of interaction demonstrate significant functional cooperation among senses. Adequately accounting for multimodal processes—those that involve interaction or coordination among more than one sense—suggests that explaining what happens in one modality requires appeal to others. The senses thus are not causally, functionally, or explanatorily independent units of study. We may see and hear, but what we see depends upon what we hear. This extends even to perceptual experience, where the products of such cooperation are evident in multimodal perceptual content and phenomenology. Any adequate philosophical understanding of perception and perceptual content should explain the respects in which perception is multimodal.

2. Some Varieties of Multimodality

We should from the start distinguish several ways in which perception could be multimodal. Though not exhaustive, this will help to highlight the targets.

Humans perceive thanks to *several* senses. We see, hear, touch, taste, and smell. The commonsense distinctions among sense modalities might correspond to folk psychological categories used to explain different ways of acquiring empirical beliefs or different patterns of responding to some surroundings. Seeing makes you believe that the car across the street is purple, and hearing makes you turn your head toward the collision. These categories perhaps are etiologically grounded in differences to perceptual experience that are recognizable from the first person—nobody would mistake a current experience of seeing a purple car for hearing. A great deal of science also has taught us in fascinating detail about the physiological differences among the eyes, ears, skin, tongue, and nose. It has taught us about nerves and brain activity, and about the evolution and function of sensory organs and pathways. It is not trivial to say that we perceive with different sense modalities, but it is a weak claim that in this respect alone perception and perceptual experience are multimodal.

Despite their differences, the senses work in concert. Perceiving frequently occurs in a multimodal context, among sources that collectively or individually stimulate more than one sense. Sensory organs and pathways are not activated entirely in isolation.

Activity in the different senses is responsible for your overall perceptual *experience* at any given time. Right now, you might hear some music and feel a breeze, see a sandwich and taste a sour candy. But, although perceptual experience comprises experiences drawn from the different senses, it does not seem fragmented or disjointed. There is a strong sense in which one's visual experience, auditory experience, tactile experience, and so on make up a single composite experience with discernible visual, auditory, tactile, etc. aspects or components. In this respect, perceptual experience is multimodally *unified* (see, e.g., Tye 2003; Bayne and Chalmers 2003). Multimodal unification, described in experiential terms, might be a matter of mere co-consciousness. But, since sensory experiences could be unified in a respect that leaves out other co-conscious experiences (such as headaches), it might instead be a matter of sharing a common framework or structure. A natural candidate for the structural respect in which *perceptual* experiences are unified is *spatial*, since many, even if not all, perceptual experiences include experiences of a common spatial framework.

Certain experiences, moreover, seem to belong together in a stronger sense, as when you both hold and see a fuzzy yellow tennis ball, or when you reach up to a location in space to grab a tennis ball you see approaching. You might also taste some item you smell, or see and feel a given shape. In these cases, sets of experiences associated with different sense modalities are *integrated* in a way that other unified experiences, such as concurrent experiences of tasting some wine and seeing some flowers, are not. They are not merely co-conscious and do not simply constitute a spatially unified experience. While experiences with quite different subject matter might be co-consciously or spatially unified, integration of this sort involves experiences that concern some common perceptible object or feature. Integration of this sort is unification of experiences associated with different modalities that nonetheless are the same with respect to what is perceived, rather than just unification among experiences that concern different things or features. Integration thus is closely related to what Bayne and Chalmers (2003) call *objectual unity*; integration, however, might hold with respect to features or properties rather than just individuals.

A final variety of multimodally unified or integrated experience concerns features whose perception *requires* the operation of distinct modalities. Gustation presents nice candidates, since much of what we regard as taste, and many of the tastes we attribute to foods and drinks, require both the tongue and nose. The flavors of wine, for instance, might include unified features of taste and smell that are not decomposable without remainder. Losing your tongue or your nose would render such features inaccessible to you.

The dramatic differences among experiences associated with different senses naturally raise some questions. How could experiences associated with different senses be unified in any manner other than being co-conscious? How could visual, auditory, and olfactory experiences, for example, reveal the same qualities, features, or individuals? Given these kinds of concerns, empiricists traditionally held both that sensory experiences are merely co-consciously unified and that the identification of common intentional objects of sensory experience requires extra-perceptual cognition, such as association or inference. In part for these reasons, empiricists traditionally answered Molyneux's question negatively. A traditional empiricist thus would explain spatial unity and integration in terms of processes that outstrip sense perception.

One lesson of this chapter is that we should abandon the traditional view. It is true that much sensory processing occurs in distinctive sensory pathways, and that many experiences, such as color experiences, are associated with a single sense. However, information from multiple modalities commonly is assembled and knit together to produce unified or integrated perceptual experiences. Explaining the perceptual *mechanisms* responsible for unification and integration would be a great advance in the cognitive science of perception and would impact a number of philosophical debates about perception.

We can make progress by considering an intriguing class of cases that suggests an important sense in which perceptual *processes* are multimodal. These cases reinforce the cooperative character of perceptual modalities, and ultimately help to explain multimodal unification and integration.

First, consider *adaptation*. When stimulation in a given modality is disrupted in order to introduce a discrepancy between experiences in two modalities, subjects adapt over time. For example, despite being disoriented when fitted with prism goggles that shift the optical scene, subjects later adapt and manage to get around normally (Helmholtz 1866/1925; Held 1965). On a natural understanding, adaptation involves adjustments to perceptual experience over time. Such reordering demonstrates a tendency to *calibrate* distinct modalities to each other, so adaptation illustrates a concern for the *relationships* among sense modalities. Adaptive *processes* thus are multimodal in that they reorganize experience in light of information drawn from multiple senses. However, someone might say that adaptive processes are not strictly speaking perceptual, but instead merely help to deal with reorganized perceptual information for the purposes of further cognition and action. This response denies that adaptation involves changes to perceptual experience over time, and suggests that it is just a matter of coping with disrupted patterns of sensory stimulation. This suggestion is compatible with persisting experiential discrepancy.

A more dramatic class of cases demonstrates that the multimodality of sense perception runs deeper. These involve *interaction* among sensory systems. Frequently, stimulation to one sensory system impacts processes and alters experiences that are commonly associated with another. Interactions of this sort are rampant, and they regularly lead to perceptual *illusions*. To be clear, activity in one sensory system causes perceptual illusions associated with another modality. Seeing can make you have illusory tactile feelings, and hearing can cause visual illusions. This is not just adaptation. Adaptation results from the ability of perception or action systems to adjust over time; it may have no immediate, apparent effect, and it persists after the intervention. Intermodal interactions and illusions happen immediately, and their effects disappear as soon as the stimulus does. The effects of adaptation and interaction might even compete, as when intermodal interaction resolves a discrepancy that otherwise would lead to adaptation (Welch and Warren 1980).

Crossmodal illusions are the most striking cases of multimodal interaction. As in the study of perception more generally, such illusions are evidence of the processes involved in normal functioning. They help to reveal perceptual organization and perceptual principles that otherwise are disguised. Indeed, explaining crossmodal illusions contributes to explaining multimodal unification and integration, and it draws attention to the most philosophically fruitful respects in which perception is multimodal. Interaction among the senses shows that perceiving involves not just the independent operation of discrete sense modalities. Understanding perception thus requires more than assembling accounts of the separate modalities considered in isolation from each other—it requires recognizing a unified multimodal perspective.

3. CROSSMODAL ILLUSIONS

What you perceive with one modality can impact what you perceive with another. Stated in such a general way, this is uncontroversial. If you see an object, you might now reach out to touch it and later taste it. When you hear a sound, you might turn and see its source. What is surprising is how directly one sense can impact another. For instance, just as a cue presented on one side of your visual field can, by attracting selective attention, enhance response time and accuracy for visual targets presented on that side (Posner 1988), an *auditory* cue presented to one side enhances response time and accuracy for *visual* targets presented on that side by attracting selective attention *across modalities*. Crossmodal cuing also operates across other pairs of modalities (see Spence et al. 2004 for review).

More strikingly, stimulation to one sense modality can cause an illusory experience in another. Vision, for instance, affects *spatial* aspects of experience in other modalities and frequently leads to illusions. First, it commonly causes illusory experiences of spatial *location* in other senses. Ventriloquism is an illusory experience of spatial location in audition that is caused by the visible location of an apparent sound source. Though best known from the case of hearing a puppeteer "throw" a voice, the ventriloquist illusion does not require speech—a flash can impact where you hear a concurrent beep. Ventriloquism is an illusory auditory experience caused by something visible (see Bertelson 1999). (Many are surprised to learn that ventriloquism does not involve throwing sound; it is an auditory perceptual illusion as of a sound coming from a place where its apparent source is seen.) Vision also captures proprioceptive location. For example, seeing a displaced image of your hand, or seeing a rubber hand, illusorily shifts where you *feel* your hand to be (Hay et al. 1965; Pick et al. 1969). Second, however, vision also causes spatial illusions of *shape*, *size*, and *orientation* in experiences associated with other modalities (see, e.g., Rock and Victor 1964). (With crossmodal effects, as with other perceptual illusions, subjects differ in susceptibility and in the extent of their illusions.)

Each of the cases above is compatible with the thesis that *vision wins*, or dominates another modality whenever a conflict exists. Deference to vision reinforces the impression that it is the primary perceptual modality, and could vindicate visuocentric theorizing about perception in philosophy and cognitive science.

Crossmodal effects on perceptual experience, however, are not limited to vision's impact upon other modalities, and they do not concern just spatial features. Audition, for example, leads to visual recalibrations. The start of a sound can alter when a light seems to switch on, so that the light's onset seems synchronous with the sound's. A sound's duration can alter the perceived duration of a visual stimulus. A quick beep can make a moving visible target appear to freeze (Vroomen and de Gelder 2000). Sound can alter visibly perceived rate, and even temporal order.

One might think that crossmodal effects are limited in the following way. Vision impacts other modalities concerning spatial features, and audition impacts other modalities for temporal features. But, other modalities, including touch, can impact the experience of temporal properties in vision, and audition even interacts to some extent with proprioception for spatial features (proprioception tends to win, but not always).

Moreover, other kinds of perceptible features drive crossmodal interactions and illusions. Qualitative characteristics can generate illusions across modalities, as when audition alters tactile experience of texture (Jousmäki and Hari 1998; Guest et al. 2002) or a smell alters taste. The McGurk effect is a more striking example that involves speech perception (McGurk and Macdonald 1976). When presented with the sound of the bilabial /ba/ (pronounced with the lips together) along with incongruent video of a speaker articulating the velar /ga/ (pronounced with the tongue at the back of the palette), many listeners report experiencing clearly the sound of the alveolar /da/ (pronounced with the tongue near the front of the palette), a kind of average or compromise. The effect quickly stops when you look away from the mouth.

One modality also can influence causal impressions in another, as the *motion-bounce* effect shows. Two disks on a screen traversing crossing paths can look either to *stream* past one another or, sometimes, to *bounce* and rebound from one another. A sound played when the disks coincide helps resolve the ambiguity and leads to a far higher portion of bounce percepts (Sekuler et al. 1997).

Finally, a fascinating crossmodal illusion discovered by Shams et al. (2000, 2002) involves audition's impact upon vision even when stimulation to each is unambiguous and even though there is no *obvious* conflict. In the sound-induced flash illusion, a single flash accompanied by two brief beeps causes many subjects to experience not just two beeps but *two* flashes. A sound impacts the number of events visually experienced. The effect continues for three and sometimes four beeps. But, a visual stimulus is required—the sound alone does not generate a visual experience. The illusion is asymmetric: vision does not impact audition in this way. Shams et al. report that this is a visual perceptual illusion caused by a sound, and that their results cannot be explained by extra-perceptual cognition or by subjects adopting a strategy to help respond to ambiguous or conflicting experiences.

4. Explaining Crossmodal Illusions

4.1. Multimodal Organizing Principles

Crossmodal illusions take place when stimulation of one sense impacts perceptual experience associated with another. Sometimes, then, a process connected with one sensory system impacts a process connected with another. That means information from one sense can change how another responds.

The sound-induced flash illusion and a battery of other cases show that the predisposition towards intermodal perceptual recalibration and reconciliation is quite strong. That crossmodal effects are so prevalent indicates that crossmodal biases and illusions are neither aberrations nor mere quirks. This suggests that, if they do not just stem from accidents or miscuing, crossmodal illusions do result from processes governed by *intermodal* perceptual organizing principles. It also suggests that such intermodal organizing principles apply quite generally. For one, they constrain perception even under ordinary conditions when ambiguity and conflict are absent from sensory information and no recalibration occurs. For another, they apply across different sets of sense modalities. Perhaps the principles governing interactions among modalities even all share some general function or rationale. Crossmodal effects in that case reveal that inescapably multimodal processes are pervasive in perception.

It is helpful to contrast these multimodal perceptual effects with *synaesthesia*, another case in which stimulation in one sensory system impacts experience ordinarily associated with another. In synaesthesia, subjects have, for example, color experiences in response to sounds, sound experiences in response to colors, texture experiences in response to tastes, or tactile experiences in response to sounds (see, e.g., Baron-Cohen and Harrison 1997; Cytowic 2002). In contrast to crossmodal perceptual illusions, synaesthesia is relatively rare. Synaesthesia that involves qualitative perceptual phenomenology occurs in roughly one in two thousand persons. More to the point, synaesthesia is a kind of experiential quirk that results from highly contingent facts about a person's sensory wiring, such as the proximity of otherwise functionally distinct pathways or brain regions. Synaesthesia differs in another important respect from crossmodal illusions. The processes responsible for synaesthesia, in contrast to those responsible for crossmodal illusions, nearly always produce illusions. Synaesthetes do not literally hear colors or taste roughness. There is no regular connection between the colors of things and the colors synaesthetes experience as a result of hearing sounds. The items synaesthetes perceive usually just lack the determinate qualities synaesthetically experienced. Synaesthesia is robustly illusory, and any benefits it confers, such as enhanced memory, are accidental.

In contrast to synaesthesia, crossmodal illusions result from intermodal biases and recalibrations that are not mere accidents. First, they concern types of features that are present and perceptually available, while synaesthesia involves outright illusion. For instance, hands commonly have locations we can see and proprioceptively feel them to have, but sound makers generally lack their synaesthetic hues. Second, crossmodal recalibrations in fact help to improve the accuracy of our perceptual responses given information from multiple sensory sources. Vision's impact on other modalities concerning spatial features often resolves ambiguities and corrects or minimizes perceptual conflicts and errors. Such ambiguities, conflicts, and errors might stem from differences to the resolving power, accuracy, or noisiness of sensory stimulation. That audition overrides vision for certain features or under certain conditions enhances our capacity to perceive temporal characteristics. Intermodal processes commonly even help to avoid illusion, as when vision corrects a front-back confusion in auditory localization. Moreover, such processes help to deal with simple physical facts, such as that the visual signal from some event arrives earlier than the auditory signal. So, although they sometimes lead to illusion, the principles that govern intermodal processes nevertheless are intelligible as advantageous. Synaesthesia, in contrast, does not stem from processes that in general enhance the capacity to perceive. Synaesthesia, unlike ordinary intermodal biases and recalibrations, is not in this way intelligible as adaptive.

4.2. Crossmodal Triggers and Mechanisms

What, then, are the *mechanisms* by which intermodal bias is exercised and recalibration takes place, and what are the *principles* that govern multimodal interaction and conflict resolution? Do any *general* principles exist that govern all crossmodal effects (see, e.g., Handel 2006)? Detailed answers to these questions currently are scarce. However, some philosophically relevant patterns are emerging.

Three primary factors are known to trigger or influence multimodal interaction (for review, details, and discussion, see, e.g., Welch and Warren, 1980; Stein and Meredith 1993; Spence and Driver 2004; Calvert et al. 2004; Handel 2006). *Spatial* and *temporal* information is most important. Being in the same place at the same time is the strongest trigger to intermodal coordination. Spatiotemporal proximity commonly cues recalibration when different senses disagree, whether the disagreement concerns space, time, or some other attribute. Perhaps this stems from two more basic principles: that two distinct things cannot occupy just the same place at the same time (cf. Bedford 2004), and that a single thing cannot have conflicting features. Crossmodal interactions and illusions weaken predictably once spatial and temporal information from different senses begins to diverge. Increasing disagreement beyond some threshold leads to perceptual separation with biasing toward the dominant modality, followed by complete dissociation without biasing. Spatiotemporal information figures among bottom-up influences on multimodal processes. One interesting upshot of the spatial parameters on crossmodal interactions concerns the representation of space in different modalities. Austen Clark (2010) argues that to explain crossmodal cuing of spatial selective attention, a common spatial framework must be represented as such across modalities. Perhaps a similar argument could be constructed for temporal features.

Both active and passive *attention* also impact the strength of crossmodal interactions and illusions. Whether directed intentionally or not, attention enhances multimodal processing and strengthens the inclination to reconcile discrepant information from distinct sensory sources. Attention facilitates feature binding, so perhaps some multimodal processes depend upon feature binding. Attention counts among top-down perceptual influences on multimodal processes.

Finally, *compellingness* might play a role in multimodal perception. For instance, hearing the sound of a teakettle while seeing spatially offset steam causes a stronger crossmodal spatial illusion than hearing the sound of a bell while seeing a spatially offset bell that does not vibrate in the expected way (Jackson 1953; cited in Handel 2006, 407). Top-down cognitive factors might therefore modulate the impact of more basic features, such as spatial and temporal characteristics, on crossmodal biases and illusions. These cases also show that care should be taken to distinguish the effects of automatic perceptual processes from those of response strategies adopted to deal with ecologically suspicious experimental tasks (see de Gelder and Bertelson 2003).

Physiologically, the mechanisms that drive crossmodal effects include connections between processes formerly thought to be modality-specific. Areas of the brain previously believed to be dedicated entirely to unimodal tasks now are known to be activated by extramodal stimuli or to respond to multimodal input. Such effects can occur quite early. For instance, V1 in the early visual system receives auditory information during the sound-induced flash experiment (Watkins et al. 2006). Furthermore, there are neurons dedicated to receiving multimodal input, while certain areas of the brain, such as the superior colliculus, play a critical role in consolidating multimodal information (see Stein and Meredith 1993). Top-down influences, such as processes associated with selective attention activated by one modality, also impact multiple modalities (see, e.g., Driver and Spence 2000). Multimodal interactions thus occur in various ways at different stages in perceptual processing, notably including those formerly thought unimodal (the essays in Calvert et al. 2004 provide a comprehensive review).

4.3. Why Multimodal Interaction?

Given these general triggers, what governs or determines how a multimodal interaction, recalibration, or illusion unfolds? For instance, why does vision sometimes win, while audition at other times impacts visual experience? What determines the strength of an illusion? Among the candidates for explaining particular patterns of bias, recalibration, and illusion are considerations such as which modality is more appropriate for the kind of information or feature in question, which modality is more accurate, and what perceptual outcome enhances the likelihood of correctness. Consider them in turn.

According to the *modality appropriateness hypothesis*, the modality that is most *inherently appropriate* for a particular kind of feature biases or dominates another when disagreement exists over that feature (see, e.g., Welch and Warren 1980). For instance, vision dominates audition when disagreement exists about space, and audition wins when the conflict concerns temporal features. Inherent appropriateness might appeal to physiological characteristics that make vision capable of fine spatial resolution and audition capable of fine temporal resolution. Sometimes, however, the modality that intuitively is not most appropriate for a given feature biases the one that is. Sound can impact felt texture, and vision biases speech perception, as in the McGurk effect. Furthermore, biasing effects can be reversed. Blurring vision or adding visual distractors increases auditory dominance for spatial features by decreasing visual reliability. Altering the reliability of a modality for a given feature changes dominance patterns. If appropriateness depends upon reliability, appropriateness is not an inherent characteristic of a sense modality.

In light of this, perhaps *deference to the most reliable modality* governs crossmodal recalibrations. If so, audition wins when it most reliably indicates a given feature, and vision wins when it does, independent from which is most inherently appropriate for a given feature. But this is not quite right, either, since crossmodal recalibrations come in degrees. In fact, four patterns are common when information about some feature such as spatial location or temporal onset disagrees. Consider, for example, visual and auditory information about spatial location. First, when the information differs wildly, we simply see one thing in one location and hear another thing in another location, with no attempt to resolve the disagreement. Second, when information differs significantly but not wildly, we might see something in one location and hear another thing in another location. But, in this case, one or both locations might be shifted or biased to some extent toward the other. Third, when the difference is less drastic, we might see and hear something in a common spatial location that is an average or compromise of the information stemming from the two modalities. Finally, we might see and hear something in a common spatial location that coincides with the actual location of the stimulus to one modality or the other. The first pattern involves no crossmodal bias, while the second involves bias without convergence. The third pattern involves convergence with mutual bias, while the fourth involves complete deference or dominance. The lesson is that total dominance in favor of one modality or the other in situations of conflict is not the rule. Ventriloquism, visual capture of proprioception, and McGurk effects each involve some measure of compromise.

A promising reply is that crossmodal biases and recalibrations *enhance* the reliability of perception in the face of noisy, fallible, or imprecise sensory stimulation. What does this involve? In ordinary cases when different senses agree, experience in each modality simply becomes more salient and detection improves. But when information conflicts and the senses do not agree, complete resolution in favor of one modality is a limit case. Frequently, averaging or weighting or some form of compromise occurs. To explain this, distinguish the reliability of a modality for a given kind of feature from the reliability of perception overall for that feature. Deferring completely to the modality that is most reliable for a given feature need not always lead to the optimal or most reliable perceptual result. Rather, the likelihood of some worldly situation in the face of the given visual and auditory stimulation, for example, is determined by the reliability of each modality relative to the other for that feature. Enhancing the reliability of the perceptual results thus should involve a weighting of evidence from the different sources. In fact, intermodal biasing and recalibration frequently conforms to a *weighting function* that incorporates the relative reliability of multiple modalities in order to determine the strength of the contribution from a given modality to the perceptual compromise. Thus, for instance, visual cues might to a great extent impact the auditory experience of spatial location, while auditory cues make a significantly smaller contribution than visual cues to the visual experience of spatial location. When the reliability of each modality matches, the perceptual result in each involves averaging. Weighting functions should predict the degree to which biasing takes place, and they should explain perceptual errors caused by multimodal recalibration. In addition to the spatial and temporal crossmodal illusions, for instance, weighing both visual and auditory contributions nicely captures why a sound biases an ambiguous visual display from mostly streaming to mostly bouncing percepts, or why visual cues cause the auditory McGurk illusion even though the auditory stimulus is unequivocal. In the latter case, deference to the auditory cue arguably would make more sense given audition's strength at resolving spectral and temporal information about speech sounds. However, the compromise that occurs when a visual cue accompanies the auditory signal enhances the reliability of the resulting percept, given that each modality bears information about spoken phonemes. Similar explanations have been applied to other crossmodal illusions, including the sound-induced flash illusion (see Shams et al. 2005). Weighting functions that enhance perceptual reliability thus show promise in explaining multimodal adjustments across a range of cases (see Handel 2006, ch. 9). A critical but open question is just how closely crossmodal recalibrations conform to what is statistically optimal.

5. Conflict, Content, and Phenomenology

Commonly, sensory stimulation and processes associated with one perceptual modality alter processes associated with another modality. This results from perceptual principles that govern even ordinary cases in which no illusions occur: multimodal triggering cues might be absent; or the triggers might be present, but if no conflict exists, there is no need for reconciliation; or multimodal interactions might correct sensory noise or inaccuracies that would have caused perceptual error. Taking advantage of several modalities enhances the reliability of perception.

Explaining multimodal effects and principles impacts a number of other topics in the philosophy and cognitive science of perception. Here let us shift from explaining multimodal interactions to explaining their impact on how we understand perception. The rest of this section argues that multimodality has consequences concerning perceptual content and phenomenology.

5.1. Why Conflict Matters

Different senses sometimes bear conflicting information about some feature. Consider what happens in cases of intermodal conflict. The telling fact is that divergent information from different senses so frequently is *reconciled*: conflicts are resolved. It is striking that biasing, reconciliation, and convergence take place at all when different senses bear different information. Why not simply settle for different perceptions with different senses?

Explaining patterns of multimodal interaction that alter and reconcile information from different senses as conflict resolution means recognizing that information from different senses, which may be encoded in different sense-specific ways, is information about something common. Resolving crossmodal conflicts means that perceptual systems in effect treat information from different senses as information about the same things, features, or subject matter. Since conflicting information is treated as information about something available to more than one sense modality, multimodal reconciliation exhibits a perceptual concern for common objects across the senses.

This has been reflected in what researchers have called *unity assumptions* that govern crossmodal interactions. Unity assumptions embody the criteria according to which perceptual systems determine whether to treat stimulation from distinct senses as belonging together, where belonging together means being a possible subject for multimodal recalibrations.

The importance of this point has been underestimated. Multimodal interaction is not just random or accidental, since it resolves or minimizes differences among modalities. It reconciles information from different senses by recalibrating responses from different sensory systems in a principled way. Reconciliation, however, requires a conflict in information, and genuine conflict requires disagreement. Disagreement requires a common subject matter, or it is merely apparent. Even apparent disagreement requires the presupposition of a common subject matter. So, to explain what perceptual processes do in multimodal contexts as recalibrating in order to reconcile conflicting stimuli or divergent information from different senses requires recognizing that perceptual systems treat information drawn from different senses as in this way commensurable.

Understanding crossmodal illusions and intermodal interactions thus promises to shed light upon ordinary perception because similar considerations might help to explain multimodal unity and integration even when no recalibrations or illusions occur. Conflict resolution requires conflict detection, which requires treating information from different sources as commensurable. Since they require conflict detection, the mechanisms that trigger intermodal recalibrations depend upon treating perceptual information from different senses as commensurable. But it is plausible that commensurability figures prominently in determining what ought to be unified or integrated. Perhaps the capacity to detect and resolve conflicts is a de facto prerequisite to unity and integration because it demands recognizing and thus solving the problem of commensurability. If the multimodal processes responsible for determining when information from different sensory sources is commensurable are among those that determine the content and phenomenology of perceptual experiences, then they provide a glimpse at part of what is responsible, at least causally, for the perceptual sense that experiences associated with several modalities belong together or are unified.

5.2. Common Content

Reconciling information from different senses demonstrates a perceptual concern for items or features that are common to different modalities. But, it is not enough that different senses happen to bear information about a shared subject matter. Remarkably, amid so much divergent information across the senses, certain information streams are treated as possibly conflicting or agreeing. Since distinct sensory systems bear information that differs both in content and in format, reconciliation also requires treating certain information that different senses bear *as commensurable*. Conflict resolution thus demonstrates that perceptual systems sometimes treat information from different senses *as* information concerning the same thing. The upshot of this is that the patterns of multimodal activity involved in recalibrating and reconciling information drawn from different senses amount to the exercise of a kind of perceptual grasp upon sources of sensory stimulation that are common across modalities. What is noteworthy is that such a perceptual grasp cannot be explained in terms that are entirely unique or specific to a given modality, since it requires, for instance, treating auditory information as commensurable with or as sharing a source or subject matter with visual information. So, this grasp cannot be characterized entirely in modality-specific terms. Rather, given that multimodal principles are deployed in response to information drawn from several modalities, and given that they involve treating such sensory information as commensurable, multimodal processes reveal either a multimodal or modality-independent (amodal) *way* of grasping the common sources of sensory stimulation. Crossmodal illusions are evidence of this grasp.

If multimodal processes exercise or implement a perceptual grasp upon common sources of sensory stimulation that is not modality-specific, this supports the claim that there exists a variety of perceptual content that cannot be captured in modality-specific terms. Some theories of content require a causal connection. Representations that result from multimodal recalibrations are causally connected to things or features accessible to more than one sense, and in a way that runs through more than one sensory system. Some theories require reliable or counterfactual supporting connections. Multimodal processes reliably identify and track items and features accessible to multiple modalities—audible location, for instance, frequently responds accurately to changes in visual information about location. And they do so in a way that is transferable across modalities—you do not lose track of a basketball when you simultaneously grab it and stop looking at it.

Furthermore, theories of content entail satisfaction or veridicality conditions. This provides an argument that perceptual states possess multimodal or amodal content, which can, in addition, be shared by perceptual states associated with different modalities. This is because multimodal processes, including biasing and recalibration, ground perceptual states with correctness conditions that cannot be characterized exhaustively in terms that are specific to a given modality. For instance, perception that involves multimodal recalibration is veridical only if a single feature instance (which cannot intelligibly be ascribed distinct determinate values) or individual (which cannot intelligibly be ascribed conflicting characteristics) is responsible for the stimulation to distinct senses. It is not enough in such cases that an *auditory-location* and a *visual-location*, or that a *visual-object* and a *tactual-object*, are perceptually represented. The auditory-location and visual-location must be identified; the visual-object and tactual-object must be identified. Thus, many crossmodal illusions are not just misperceptions through a given modality of some feature; they also involve mistakenly perceptually *identifying* what is perceived through different modalities. For instance, visual capture of the felt location of your hand is not just a proprioceptive illusion of spatial location as it also involves mistakenly identifying what you see with what you feel. Ventriloquism is not just an auditory spatial illusion; it is perceptually misidentifying the seen puppet as the sound source. Ventriloquism would involve an illusion even if it literally involved "throwing" a sound to the puppet's location. Visual capture and ventriloquism each involve illusions of identity.

What sorts of things might figure among such contents? The features and items in question likely belong to two groups. First, broad structural features such as spatial and temporal ones serve as basic triggers that drive intermodal interactions. They frequently are the respects in which senses disagree, so representations of spatial and temporal features must be commensurable across modalities. This suggests that spatial and temporal features figure in perceptual contents common across modalities. Spatial and temporal information alone is not enough, however. That vision represents one place and audition represents another does not explain reconciliation. Representation of all of the locations, perhaps populated with different sense-specific features, does not yet explain why conflicts across modalities are a problem. Explaining reconciliation requires recognizing that spatial or temporal information *about something* conflicts. Spatial and temporal conflict likely involves items or individuals to which spatial and temporal features are attributed. What are the candidates for items that might figure in contents ascribed to different modalities? Individuals, such as *objects* and *events*, play a critical role in multimodal perception. They may have apparent spatial or temporal locations, and features such as shape, rate, or duration. They also may be attributed special features such as color, taste, or loudness that are accessible to a unique modality. Individuals are good candidates among the multimodal targets of attention since, plausibly, they are tracked across modalities, as when you turn to see the source of a sound. These perceptible individuals thus, plausibly, also anchor perceptual demonstratives and are available as common subjects of empirical thought. If explaining crossmodal recalibration requires positing a perceptual grasp upon such individuals, then since they must be identified across modalities, there must be a common way of grasping or representing them that applies to perceptual states associated with multiple modalities. If so, perceptual states that represent individuals accessible to multiple modalities possess a kind of content that cannot exhaustively be characterized in modality-specific terms.

5.3. Shared Phenomenology

I have been discussing perceptual processing, explanation, and content. What about perceptual experience? In particular, what about "what it's like," or the phenomenal character of perceptual experience?

So far, the argument has been that we perceive common features (individuals or properties) that are perceptually grasped or represented as common across multiple modalities. This is required to explain conflict and reconciliation. But, prima facie, common content is compatible with distinctive phenomenal character for every pair of perceptual experiences that occur in distinct modalities. Crossmodal illusions might stem from lower-level causal and informational processes that nevertheless lead to phenomenally distinctive experiences associated with different modalities. The point is that even if causal interaction takes place between subpersonal perceptual processes that lead to experiences in different senses, and even if explaining those interactions requires positing grasping or representing as of common items at some (perhaps merely subpersonal) level, this need not be reflected as such in phenomenology. Causal influence does not guarantee constitutively dependent experiences. On a widespread traditional picture, each sense contributes entirely *distinctive* phenomenal characteristics, and sense-specific contributions *exhaust* overall perceptual experience. Perhaps the traditional picture is safe.

It is useful to consider this issue in the context of the contemporary debate over intentionalism about the phenomenal character of perceptual experience. Some *intentionalist* theories hold that the phenomenal features of perceptual experiences supervene upon perceptual content, or representational features (see, e.g., Harman 1990; Dretske 1995; Tye 2000; Byrne 2001). Thus, if multimodal perception shows that perceptual experiences associated with different senses can share content, intentionalism implies that they also must share phenomenal character. This consequence has been used to argue against intentionalism (see, e.g., Lopes 2000; O'Dea 2006). It is intuitive to think that we perceptually experience things and features in ways that are *distinctive* to each modality. Even if we can see and touch the same thing, and even if something common must figure in the contents of sight and touch, it is natural to suspect there is a distinctively visual way of experiencing it and a different, recognizably tactual way of experiencing it. So, even if common sensibles figure in common contents, there are still different ways of perceptually experiencing or entertaining them with different senses, and thus distinct, modality-specific phenomenal characters.

Some respond with an *intramodal* restriction to the intentionalist thesis: within a perceptual modality, phenomenal character supervenes on representational content. There are at least two ways to spell out this claim. Consciously perceiving something to be blue, to be illuminated on its left side, or to occlude part of a surface surely involves a distinctively visual experience. Perceiving something as high-pitched or as loud involves a distinctively auditory experience. At any time, in everyday conditions, one's visual or auditory experiences arguably all involve at least some modality-specific phenomenal features, since they represent at least some properties, such as colors or pitches, that are proper sensibles. Thus, the overall or total phenomenal character of a visual or an auditory experience arguably *is* distinctive.

Nevertheless, it would be a mistake to concede the stronger point implicit in the initial challenge to intentionalism, which is that the representation of any given feature by any given modality is associated with a distinctive, modality-specific phenomenal character. This stronger view implies that perceptual phenomenality is exhausted by modality-specific phenomenality; that is, that phenomenal character

is exhausted by modality-specific phenomenal features. For the intramodal intentionalist, accepting this means endorsing the claim that a difference in modality
suffices for a difference in phenomenal character, even holding fixed the features
represented.

But we should reject this. For ordinary human perceivers, in whom perceptual
principles govern multimodal bias, interaction, and reconciliation, perceptual experiences include phenomenal characteristics that must be shared by or be common
to experiences associated with different senses. Evidence exists that multimodal
sharing is reflected even in perceptual experience and that perceptual phenomenality is not exhausted by modality-specific characteristics.

Consider cases in which *intermodal* binding takes place. Just as distinct features
from a single modality, such as color and shape, are *intramodally* bound in visual
experience—a single perceptible thing seems to have both color and shape, and to
be distinct from other things that have color and shape—features drawn from different senses, such as visible shape and felt texture, or visible motion and audible
location, sometimes perceptibly belong to a single item thanks to intermodal binding (see, e.g., Pourtois et al. 2000). The very same thing perceptually seems to have
features drawn from more than one sense modality. That which is experienced to
have a visible feature is experienced also to have a tactile feature (or an audible feature). That which bears the visible feature is identified with that which bears the
tactile feature, so a single thing is experienced as having both visible and tactile
features. To illustrate the point, imagine the contrast between watching and listening to a movie under ordinary conditions and watching and listening either to a
poorly dubbed foreign film or to one in which the soundtrack is temporally offset
from the visual cues. In each of the latter two cases, intermodal binding breaks
down. In order to capture the phenomenal character of experiences in which intermodal binding does take place, some aspect of perceptual phenomenality must
account for the experiential sense in which the item seen is the very same item
touched or heard. So, some aspect of perceptual phenomenal character is common
to the experiences of seeing and touching or hearing (alternatively, the multimodal
experience involves a brute phenomenal feeling of sameness). This aspect of perceptual phenomenality might be multimodal in nature, or it might have an entirely
amodal character. Either way, distinctive sense-specific characteristics do not jointly
exhaust the phenomenal features of perceptual experience. Either experiences ordinarily associated with different senses share at least some aspects of phenomenal
character, or perceptual experiences are not exhausted by distinctive phenomenal
characteristics that are specific to individual modalities (see O'Callaghan 2008b).

This, however, implies that considering visual examples and invoking characteristics that are proprietary to vision does not suffice to characterize the phenomenology even of what we ordinarily count as visual perceptual awareness. That
approach abstracts from the context of other modalities and thus ignores their contributions to integrated multimodal perceptual experience. Not only do we assess
vision and visual experience from the perspective of multimodal awareness,
but ordinary examples of visual experience themselves cannot exhaustively be

characterized, understood, or explained without comprehending vision's role in integrated multimodal perceptual experience. A complete account of vision and visual experience requires recognizing not just causal connections with other modalities but also constitutive relationships among the contents and the phenomenology of experiences commonly associated with different modalities.

6. THE SENSES

Multimodal perception, especially crossmodal interactions and illusions, bears on how we understand the nature of the senses.

6.1. Are the Senses Modular?

One might be tempted to think the senses are *modular* (for instance, in the sense of Fodor 1983, 2000). The senses often are mentioned as examples of psychological modules. Suppose that perception in general is modular and thus impenetrable by beliefs and other forms of cognition. And suppose that the processes associated with each sense are specific or dedicated to that sensory domain and that they operate mostly independently from other senses. Then one might hold that audition, for instance, is informationally encapsulated and impenetrable by information from other sensory domains. One might count audition among the modules.

However, this view is far too strong. Multimodal processes involve causal interaction and information transmission among sensory systems. Perceptual processing in vision is shaped and changed by auditory information. Connections between sensory systems exist at numerous physiological locations and at different functional or computational stages. Shimojo and Shams (2001) argue explicitly against what they call the historically prevalent view that the sense modalities are distinct modules on the grounds of multimodal plasticity and interaction.

However, it is difficult to find work that advocates such a strong version of the modularity thesis for the senses. Some hold that only an early part of the visual system is modular (e.g., Pylyshyn 1999). Some hold that even early vision's modularity depends upon a number of distinct systems working in concert that detect, for example, color, motion, edges, and form. In that case, early vision itself employs a number of modules. Perhaps it is no surprise that complete sensory systems fail to act as strict modules. Multimodal effects make trouble only if one believes that all perceptual modules are modality-specific. But, on a weakened modularity view, why maintain this? Perhaps, while vision, audition, smell, taste, touch, and proprioception are not strictly modular, functional modules responsive to common multimodal or amodal features and individuals, such as space, time, objects, and events (and perhaps even spoken language), might exist. This would seem necessary to uphold even a less-than-massively modular view of the mind in which perceptual systems are distinguished from higher cognition by their modularity (see, e.g., Fodor 2000). The important upshot is that modules specific to vision, audition, and the rest do not exhaust perceptual capacities. Certain critical perceptual processes require input from several sensory systems and impact experience associated with multiple modalities. In addition to any sense-specific functions, perception involves multisensory tasks.

6.2. Individuating the Senses

Does the multimodality of perception challenge even the common understanding that we have five (or so) different senses? We have assumed that we can individuate the senses in a way that corresponds to folk notions of seeing, hearing, feeling, tasting, and smelling. But multimodal perception casts doubt on several competing accounts of the senses.

A number of proposals for individuating the senses have been advanced. It is useful to group them into four broad kinds. Different accounts, respectively, individuate the senses in terms of their *objects*, in terms of characteristics of *experiences*, in terms of some aspect of perceptual *processes* (such as responsiveness by sense organs to different kinds of *energy*, *physiological* channels, or *function*), and by mere *convention*. How do these accounts fare given the results discussed in this chapter?

Can we individuate the senses in terms of their intentional *objects*, or *sensibles*, (a view commonly attributed to Aristotle; cf. Grice 1962)? Since some features and individuals, such as things we both see and hear, figure among the objects of more than one modality, citing objects alone does not suffice to distinguish the senses. This forces us to appeal to proper sensibles or objects—things perceived only with one sense. But knowing whether something is available to one sense or two senses assumes we can count the senses. Furthermore, each sense has several proper sensibles, such as pitch and loudness for hearing, or hue and brightness for vision. Thus, in light of multimodal considerations, individuating senses in terms of their objects results either in violations of common sense distinctions, such as seeing and hearing, or in too many senses.

Perhaps, in light of the problem of common sensibles, we can appeal to distinctive characteristics of *experiences* associated with each sense to individuate the modalities, as Grice (1962) suggests. If the phenomenological conclusion of Section 5.3 is correct, then this account faces similar problems. If there are aspects of phenomenology common among experiences usually associated with different modalities, such as seeing and hearing, then neither the experience of seeing nor of hearing is exhausted by entirely distinctive phenomenal characteristics. Therefore, it is plausible that no phenomenal feature is distinctive to all and only visual experiences. Thus, we cannot individuate the senses experientially or phenomenologically. However, suppose we admit some additional perceptual modality, or supra-modality, beyond the five traditional senses. This additional supra-modality then could accommodate, for instance, the distinctive amodal or multimodal experience of intermodal binding, while distinctive modality-specific experiences remained for vision, audition, and the rest. Such an account faces at least two obstacles. It must explain the experiential sense that the very item seen to have color and shape also is felt to have texture and form. It thereby trades one "horizontal" intermodal sharing problem for another "vertical" one. Moreover, it owes an account of what is experientially distinctive about supra-modally experiencing an object or feature (in contrast, for instance, to seeing it). Some also might find it odd to say that a supra-modal, amodal, or multimodal *modality* exists in addition to the traditional senses.

Heil (1983) suggests we individuate the senses in terms of the physical energy to which they respond, while Keeley (2002) suggests we individuate the senses in terms of organs evolutionarily dedicated to picking up information of a certain kind. Neither of these accounts strictly conflicts with the multimodal results discussed above, but once we take them to apply to *perceptual* systems, multimodal processes become a problem. The end sense organs are not hooked up to functionally discrete systems that are dedicated to responding to unique information or features in the way that the eye responds to light, the ears to mechanical pressure waves, and the nose to chemicals. Multimodal perceptual processes discern and respond to constancies and commonalities in stimulation across sense organs. Individuating senses either by end organs or by physical energy types therefore may not suffice to individuate corresponding perceptual systems. Since the senses on such accounts are not *perceptual* modalities, their theoretical interest to psychology and philosophy is minimized. Seeing interests philosophers and cognitive scientists not just because it involves having an organ that responds to light, but because seeing is a way of perceiving.

Nudds (2003) argues that, given trouble individuating the senses, we should say that the senses are conventional categories we treat as different ways of perceiving. Nevertheless, according to Nudds, knowing that someone is seeing is informative because it suggests to us the kinds of things that person is likely to perceive, such as colors, shapes, and objects. Knowing someone is hearing makes it more likely that the person perceives pitches, sounds, and events. But we should resist thinking the senses correspond to physiologically or psychologically real kinds. The senses, on this view, are no more than folk psychological concepts with limited value to empirical science. Though unsettling, and though Nudds's argument from the failure of extant accounts does not rule out deeper similarities and differences that ground conventional distinctions among the senses, multimodal aspects of perception might lend support to this line of thought.

In light of this, perhaps we should restrict talk of the senses to low-level systems such as the eye and very early visual processes, which can be individuated as Heil and Keeley suggest. But these senses are not *perceptual* modalities, and they do not correspond neatly to categories of experience. Seeing, hearing, and the rest are ways of perceiving whose individuation might be messy in the way Nudds suggests. Perceiving, therefore, might be deeply multimodal in that it involves principles and processes that span multiple sensory systems.

7. MULTIMODALITY IN PERCEPTION

In summary, crossmodal illusions demonstrate that perception involves interactions among processes associated with different modalities. Patterns of crossmodal bias and recalibration reveal the organization of multimodal perceptual processes. Multimodal interactions obey intelligible principles; they resolve conflicts; they enhance the reliability of perception; they are not mere quirks or accidents. Multimodal processes also demonstrate a concern across the senses for common features and individuals, for the following reason: the intermodal biasing and recalibration responsible for crossmodal illusions requires that information from sensory stimulation associated with different senses is taken to be commensurable. Since conflict resolution requires a common subject matter, commensurable information from different senses shares or traces to a common source. Crossmodal processes thus amount to the exercise of a principled perceptual grasp upon the common sources of sensory stimulation across modalities.

Further philosophical work is needed to characterize the varieties of multimodality and their bearing on traditional theories of perception, perceptual experience, and the senses. The implications for research methodology are clear: studying the various modalities in isolation reveals just the surface of the story about perception. Philosophical and empirical work thus should not proceed on the assumption that it is possible to understand perception and perceptual experience in terms of a single modality entirely in isolation from the others. For one, it is not possible simply to extrapolate or translate claims about vision into claims about other modalities. Moreover, vision itself may resist an exhaustive understanding that does not appeal to nonvisual modalities. Without recognizing perception's deep multimodality, we overlook the role of the several senses in perception.

What does this mean for how we understand the nature of perception? Handel (2006) says perceiving is about solving *correspondence problems* through the detection of contrast and change. A correspondence problem is one of identifying an individual or feature, either at a time or over time, given sensory information that varies from location to location and from moment to moment. (The original correspondence problem concerns how to reconcile two retinal images to yield information about depth.) Perceiving, in this view, is about using contrasts and changes in noisy and variable sensory stimulation to detect in a scene sensible individuals that bear relatively stable characteristics.

One important lesson of multimodal effects is that an analog of the correspondence problem within a modality holds between modalities. Perceiving involves determining that what you hear is what you see, that the object you feel to be in your hand is the one you see, or that what tastes bitter is what you feel on your tongue. An empirical account of perception should explain how this is accomplished. Philosophers and cognitive scientists should ask what makes the accomplishment significant.

Here is a start. One upshot deals with the relationships among experiences commonly associated with different modalities. Just as perceiving colors and visual objects involves detecting constancies in hue and shape despite variations and changes in appearance due to lighting and viewing angle, perceiving ordinary objects and events implicates constancies in features detected with different senses, across changes to the modality with which one perceives. Without appeal to such constancies, shifting sense modalities would perceptually seem to result in altogether different objects of experience. If different modalities shared no intentional objects or features, perceptual experience would in one noteworthy way seem fragmented. We would perceive no relationships among things experienced through different modalities. Our sense of the cohesiveness of the world as we perceive it through different senses therefore depends upon our perceptually keeping track of common items and features across different modalities. Critical aspects of perceptual experience thus stem from multimodal functioning. Spatio-temporal unity, objectual unity, and integration are tied to our capacity to detect constancies and solve correspondence problems across modalities.

Another upshot deals with the nature of experiences associated with a given modality. We perceptually identify and keep track of things despite contrasts in sensory information and in presentation across modalities. This, it was argued, grounds varieties of perceptual content and phenomenal character that are common to perceptual experiences associated with different modalities. It follows that characterizing perceptual content and perceptual phenomenology requires appeal to terms beyond those that are proprietary to a given modality. Characterizing an ordinary occasion of perceiving by means of audition therefore requires appeal to terms that are not inherently auditory. Talk of sounds, pitch, timbre, loudness, and audible-location, whose perceptual significance is purely auditory, needs to be supplemented with talk of sound sources—objects that make sounds, or events of sound production—and locations shared with vision. A similar claim holds for vision. We might therefore be justified in saying that we can visually perceive something *as* the sort of thing that could have auditory, tactual, olfactory, or gustatory qualities rather than as something with visual significance alone. Likewise, you might hear something *as* the sort of thing that could be seen or touched. So, the experience of perceiving with one modality could embody *perceptual* expectations that implicate another. For instance, seeing a convincing hologram invites expectations that are violated when you learn you can put your hand through it without resistance. Such visual expectations are perceptual rather than just results of extra-perceptual cognition or association. The scope of experiences associated with a given modality, as a result, might be far greater than traditional views imagine.

Solving crossmodal correspondence problems, on the approach discussed, requires a common amodal or multimodal code that is shared among modalities. The proprietary or distinctive aspects of sense perception and experience thus may distract from what is most noteworthy about perceiving. Perhaps it is in grasping multiple sensory perspectives as perspectives upon a common source that *the* world—a single, unified world of things and happenings—becomes available as a subject for attention, empirical belief, and action. If so, perceiving is a thoroughly multimodal affair.

ACKNOWLEDGMENTS

Thanks are due, for great questions and helpful feedback, to audience members at the Harvard Perception Workshop, Southern Methodist University, and Bates College, where I presented some of this material. Many thanks, also, to Tim Bayne, Alex Byrne, David Chalmers, Philippe Chuard, John Doris, Justin Fisher, Fiona Macpherson, Mohan Matthen, Mark Okrent, Diana Raffman, Susanna Siegel, Barry Smith, Jeff Speaks, Wayne Wu, and the editors, for enjoyable conversations and generous comments on this work. I hope these ideas will encourage others to contribute to this emerging topic.

REFERENCES

Aristotle. (1984). De Anima. In *The Complete Works of Aristotle: The Revised Oxford Translation*. Oxford: Oxford University Press.

Baron-Cohen, S., and Harrison, J. E. (1997). *Synaesthesia: Classic and Contemporary Readings*. Malden, MA: Blackwell.

Batty, C. (2010). A representational account of olfactory experience. *Canadian Journal of Philosophy* 40: 511–38.

Bayne, T. J., and Chalmers, D. J. (2003). What Is The Unity of Consciousness? In A. Cleeremans (ed.), *The Unity of Consciousness: Binding, Integration, and Dissociation*. Oxford: Oxford University Press.

Bedford, F. L. (2004). Analysis of a constraint on perception, cognition, and development: One object, one place, one time. *Journal of Experimental Psychology: Human Perception and Performance* 30: 907–12.

Bertelson, P. (1999). Ventriloquism: A Case of Cross-Modal Perceptual Grouping. In G. Aschersleben, T. Bachmann, and J. Müsseler (eds.), *Cognitive Contributions to the Perception of Spatial and Temporal Events*. Amsterdam: Elsevier.

Blauert, J. (1997). *Spatial Hearing: The Psychophysics of Human Sound Localization*. Cambridge, MA: MIT Press.

Bregman, A. S. (1990). *Auditory Scene Analysis: The Perceptual Organization of Sound*. Cambridge, MA: MIT Press.

Byrne, A. (2001). Intentionalism defended. *Philosophical Review*, 110: 199–240.

Calvert, G., Spence, C., and Stein, B. E. (2004). *The Handbook of Multisensory Processes*. Cambridge, MA: MIT Press.

Casati, R., and Dokic, J. (1994). *La Philosopie du Son*. Nimes, France: Chambon.

Clark, A. (2006). That lonesome whistle: A puzzle for the sensorimotor model of perceptual experience. *Analysis* 66: 22–25.

Clark, A. (2010). Cross modal cuing and selective attention. In F. Macpherson (ed.), *The Senses: Classic and Contemporary Philosophical Perspectives*. Oxford: Oxford University Press.

Cytowic, R. E. (2002). *Synesthesia: A Union of the Senses*. Cambridge, MA: MIT Press.

de Gelder, B., and Bertelson, P. (2003). Multisensory integration, perception and ecological validity. *Trends in Cognitive Sciences* 7: 460–67.

Dretske, F. (1995). *Naturalizing the Mind*. Cambridge, MA: MIT Press.

Driver, J., and Spence, C. (2000). Multisensory perception: beyond modularity and convergence. *Current Biology* 10: R731–35.

Fodor, J. (1983). *The Modularity of Mind*. Cambridge, MA: MIT Press.

———. (2000). *The Mind Doesn't Work That Way: The Scope and Limits of Computational Psychology*. Cambridge, MA: MIT Press.

Grice, H. P. (1962). Some Remarks About the Senses. In R. J. Butler (ed.), *Analytical Philosophy, Series 1*. Oxford: Blackwell.

Guest, S., Catmur, C., Lloyd, D., and Spence, C. (2002). Audiotactile interactions in roughness perception. *Experimental Brain Research* 146: 161–71.

Handel, S. (1993). *Listening: An Introduction to the Perception of Auditory Events*. Cambridge, MA: MIT Press.

Handel, S. (2006). *Perceptual Coherence*. Oxford: Oxford University Press.

Harman, G. (1990). The Intrinsic Quality of Experience. In J. Tomberlin (ed.), *Philosophical Perspectives*, vol. 4. Atascadero, CA: Ridgeview.

Hay, J. C., Pick, H. L., and Ikeda, K. (1965). Visual capture produced by prism spectacles. *Psychonomic Science* 2: 215–16.

Heil, J. (1983). *Perception and Cognition*. Berkeley: University of California Press.

Held, R. (1965). Plasticity in sensory-motor systems. *Scientific American* 213: 84–94.

Helmholtz, H. (1866/1925). *Treatise on Physiological Optics*, vol. 3. New York: Optical Society of America.

Jackson, C. V. (1953). Visual factors in auditory localization. *Quarterly Journal of Experimental Psychology* 5: 52–65.

Jousmäki, V., and Hari, R. (1998). Parchment-skin illusion: Sound-biased touch. *Current Biology* 8: R190.

Keeley, B. L. (2002). Making sense of the senses: Individuating modalities in humans and other animals. *The Journal of Philosophy* 99: 5–28.

Lopes, D. M. M. (2000). What is it like to see with your ears. *Philosophy and Phenomenological Research* 60: 439–53.

Lycan, W. (2000). The Slighting of Smell. In N. Bhushan and S. Rosenfeld (eds.), *Of Minds and Molecules: New Philosophical Perspectives on Chemistry*. Oxford: Oxford University Press.

Martin, M. (1992). Sight and Touch. In T. Crane (ed.), *The Contents of Experience*. Cambridge: Cambridge University Press.

Matthen, M. (2005). *Seeing, Doing, and Knowing: A Philosophical Theory of Sense Perception*. Oxford: Oxford University Press.

McAdams, S., and Bigand, E. (1993). *Thinking in Sound: The Cognitive Psychology of Human Audition*. New York: Oxford University Press.

McGurk, H., and MacDonald, J. (1976). Hearing lips and seeing voices. *Nature* 264: 746–48.

Noë, A. (2004). *Action in Perception*. Cambridge, MA: MIT Press.

Nudds, M. (2003). The significance of the senses. *Proceedings of the Aristotelian Society* 104: 31–51.

O'Callaghan, C. (2007). *Sounds: A Philosophical Theory*. Oxford: Oxford University Press.

———. (2008a). Object perception: Vision and audition. *Philosophy Compass* 3:803–29.

———. (2008b). Seeing what you hear: Crossmodal illusions and perception. *Philosophical Issues* 18: 316–38.

O'Dea, J. (2006). Representationalism, supervenience, and the cross-modal problem. *Philosophical Studies* 130: 285–95.

O'Regan, J. K., and Noë, A. (2001). A sensorimotor account of vision and visual consciousness. *Behavioral and Brain Sciences* 24: 939–1031.

Pick, H. L., Warren, D. H., and Hay, J. C. (1969). Sensory conflict in judgments of spatial direction. *Perception and Psychophysics* 6: 203–05.

Posner, M. I. (1988). Structures and Functions of Selective Attention. In T. Boll and B. Bryant (eds.), *Master Lectures in Clinical Neuropsychology*. Washington, DC: American Psychological Association.

Pourtois, G., de Gelder, B., Vroomen, J., Rossion, B., and Crommelinck, M. (2000). The time-course of intermodal binding between seeing and hearing affective information. *Neuroreport* 11: 1329–333.

Pylyshyn, Z. W. (1999). Is vision continuous with cognition? The case for cognitive impenetrability of visual perception. *Behavioral and Brain Sciences* 22: 341–423.

Rock, I. and Victor, J. (1964). Vision and touch: An experimentally created conflict between the two senses. *Science* 143: 594–96.

Sekuler, R., Sekuler, A. B., and Lau, R. (1997). Sound alters visual motion perception. *Nature* 385: 308.

Shams, L., Kamitani, Y., and Shimojo, S. (2000). What you see is what you hear. *Nature*, 408: 788.

———. (2002). Visual illusion induced by sound. *Cognitive Brain Research*, 14: 147–152.

Shams, L., Ma, W. J., and Beierholm, U. (2005). Sound-induced flash illusion as an optimal percept. *NeuroReport* 16: 1923–927.

Shimojo, S. and Shams, L. (2001). Sensory modalities are not separate modalities: Plasticity and interactions. *Current Opinion in Neurobiology* 11: 505–9.

Spence, C. and Driver, J. (eds.). (2004). *Crossmodal Space and Crossmodal Attention*. Oxford: Oxford University Press.

Spence, C., McDonald, J., and Driver, J. (2004). Exogenous Spatial-Cuing Studies of Human Crossmodal Attention and Multisensory Integration. In Spence, C. and Driver, J. (eds.), *Crossmodal Space and Crossmodal Attention*. Oxford: Oxford University Press.

Stein, B. E., and Meredith, M. A. (1993). *The Merging of the Senses*. Cambridge, MA: MIT Press.

Tye, M. (2000). *Consciousness, Color, and Content*. Cambridge, MA: MIT Press.

———. (2003). *Consciousness and Persons: Unity and Identity*. Cambridge, MA: MIT Press.

Vroomen, J. and de Gelder, B. (2000). Sound enhances visual perception: Cross-modal effects of auditory organization on vision. *Journal of Experimental Psychology: Human Perception and Performance*, 26: 1583–90.

Watkins, S., Shams, L., Tanaka, S., Haynes, J. D., and Rees, G. (2006). Sound alters activity in human V1 in association with illusory visual perception. *NeuroImage*, 31: 1247–56.

Welch, R. B. and Warren, D. H. (1980). Immediate perceptual response to intersensory discrepancy. *Psychological Bulletin*, 88: 638–67.

CHAPTER 6

....

EMBODIED COGNITION*

....

LAWRENCE A. SHAPIRO

EMBODIED cognition is one of several burgeoning research programs, including enactive cognition and extended cognition, that variously seeks to replace, revise, or at least upset the reigning cognitivist conception of mind according to which cognitive processes involve computations over symbolic representations. This chapter provides an introduction to embodied cognition and its close kin, focusing on historical roots, core concepts, methodological practices, and future prospects. Worth emphasizing is that embodied cognition is a research program in its infancy. This is evident in the fact that there is as yet no firm consensus among practitioners about how to understand crucial concepts such as "embodiment." Nor is there unanimity about the goals of embodied cognition, or whether and to what extent ideas and methods from traditional cognitive science should have a place in embodied cognition research. Also like new research programs, embodied cognition has drawn heavy fire from members of the "old guard," who see embodied cognition as incoherent, full of false promises, or tacitly relying on the tried-and-true ideas of old-school cognitivism to explain its successes.

1. TRADITIONAL COGNITIVISM

....

Traditional cognitivism, or more simply cognitivism, refers to the theory of mind and accompanying methods for its study that emerged from research in the latter half of the twentieth century in the fields of cybernetics (Weiner, von Neumann), computer

* I am grateful to Fred Adams, Ken Aizawa, and Gary Hatfield for helpful comments on earlier drafts. I completed much of the work for this chapter while a fellow at the Centre for the Foundations of Science, University of Sydney, and am especially grateful to Mark Colyvan for making that possible. Many of the ideas in this chapter have received fuller treatment in *Embodied Cognition* (Shapiro 2011).

science (Turing, Newell, Simon, Minksy, McCarthy), neuroscience (McCulloch), psychology (Neisser, Sternberg, Rock, Pylyshyn), linguistics (Chomsky), and philosophy (Putnam, Dennett, Fodor). Although it would be incorrect to attribute to these various disciplines a common goal to reconceive the nature of mind and outline new methods for its investigation, it is certainly true that individuals in each of these fields had an awareness of the goings-on in the other fields. Indeed, in many cases, these individuals were making significant contributions in several fields.

Ignoring niceties, these various fields supported a computational description of the mind. In this view, mental processes (e.g., problem solving, remembering, perceiving, comprehending, producing language and so on) are computational processes. Mental objects (beliefs, memories, perceptions, etc.) are symbolic structures with representational content. As Von Eckardt (1995) has noted, this computational characterization of the mind is intended as much more than a metaphor, as the common but unfortunate epithet "computer metaphor" explicitly suggests. According to cognitivism, mental processes literally are computational, and mental objects literally are symbolic structures over which computational processes operate.

Newell and Simon gave early voice to this idea. In the late 1950s, Newell and Simon (and Shaw) developed a computer program that they called General Problem Solver (GPS). The goal of GPS was to solve problems in the same manner that human beings do. But, because GPS solved problems by first representing them symbolically and then searching for solutions through rule-guided operations over symbolic objects, the suggestion that human cognitive processes were *like* the processes in which GPS engaged assumed that human cognition is also computational and symbolic. Thus, it is no surprise to see Newell and Simon claiming that "the processes going on inside the subject's skin—involving sensory organs, neural tissue, and muscular movements controlled by neural signals—are also symbol manipulating processes; that is, patterns in various encodings can be detected, recorded, transmitted, stored, copied, and so on, by the mechanisms of this system" (1961, 2014).

The notion that cognition is symbol manipulation also receives vivid expression in computational theories of perception. These theories of perception might usefully be seen as modernized versions of the inference-based theory of perception that Helmholtz articulated in the late nineteenth century. Almost all computational theories of vision, for instance, assume that the task of vision is to construct 3-D representations of objects from 2-D representations on the retinal surface. Perception requires inference, because 2-D retinal projections must be amplified. On their own, they are inadequate to specify the structure of the objects that cause them. Accordingly, computational vision theorists investigate algorithms that can "recover" 3-D structure from 2-D representations (e.g., Marr 1982). As Rosenfeld, a computer vision scientist, describes it: "A variety of techniques have been developed for inferring surface orientation information from an image; they are known as 'shape from X; techniques, because they infer the shape of the surface…from various clues…that can be extracted from the image" (1988, 865). Among these techniques are those that compute shape from shading, shape from motion, or shape from texture, and so on. All of these techniques begin with a 2-D retinal "image" from which are constructed representations of particular features (e.g., edges, blobs, contours, patches) to which are applied computational operations with the goal of deriving 3-D shape.

The preceding examples illustrate the dedication to computationalism that is central to the cognitivist conception of mind, but they also suggest something about the tasks that cognitivists have taken as paradigmatic of intelligence, as well as the methodology for understanding the mind to which cognitivists are often committed. Regarding the first point, cognitivists such as Newell and Simon took cognition to be most clearly in evidence when solving mathematical equations, playing chess or checkers, and working out problems such as the Tower of Hanoi puzzle. Obviously, these are precisely the kinds of activities for which operations over symbolic structures have the greatest chance of success.

As regards methodology, cognitivism treats cognition, including perception, as a constructive process in which computational operations transform a static representation into a goal state. Because cognition begins with an input representation, the psychological subject can be conceived as a passive receptor of information. This is especially clear in perception research, where the subject is often sitting in a chair, with head stable in a chin rest, eyes focused straight ahead toward a monitor on which stimuli are presented. Because the cognitivist's primary concern is the discovery of algorithms by which inputs (such as those representing shading) are transformed into outputs (such as those representing shape), experimental methods should provide an environment that isolates just those stimuli that will be relevant to an investigation of the mental process of interest. Furthermore, because the psychologist's goal is to understand which computations over which representations suffice to realize a given psychological capacity, there is no need to consider features of the subject's body or environment. Psychological processing starts at the moment that stimuli are encoded in computationally tractable representations. Indeed, a number of philosophers (Fodor 1980; Stich 1983; Segal 1989; Egan 1991) have endorsed (in various ways) the idea that cognitivist investigations of the mind be methodologically solipsistic, in the sense that they should proceed without regard to the existence of anything but the brain of the subject in which, presumably, all computational processing takes place. As far as many cognitivists are concerned, psychological subjects are literally brains in vats.

2. Gibsonian Foundations of Embodied Cognition

Although a number of researchers in embodied cognition give credit to Maurice Merleau-Ponty for inspiration (Varela, Thompson, and Rosch 1991; Clark 1997; Noë 2004), the phenomenological framework in which Merleau-Ponty developed his ideas, pregnant with the thought of Brentano, Husserl, and Heidegger, is an unlikely foundation on which to rest a contemporary science of the mind that brings with it a tremendous amount of technological sophistication as well as aspirations to the same standards of objectivity that mark other scientific endeavors (but see Varela, Thompson, and Rosch 1991). Better to look at J. J. Gibson's rigorous, groundbreaking studies of visual perception and the school of ecological psychology that he fathered (1950, 1966, 1979). Of special interest in the present context are Gibson's claims that: (i) there is information present in the environment that is (ii) available to an active agent and that (iii) specifies to this agent a world of affordances.

The information about which Gibson speaks is contained in the optic array. Converging on every point in an illuminated environment, including the sense organs of a perceiver, is light reflected from surrounding surfaces. Preserved in this array of light is information about surfaces' shapes, textures, and colors. This is because the intensity and wavelength of light from one surface will differ from the intensity and wavelength of light reflected from other surfaces, in effect causing a projection onto an organism's sense organs of a group of solid visual angles, the edges of which mark the boundaries of surfaces in the environment. As the organism moves, the elements of the optic array undergo lawful transformations. Some visual angles expand, others contract, new ones appear, and so on. However, within this changing pattern of visual angles will be a variety of constant features. These invariants, Gibson argued, specify, or carry information, about structure in the environment. Perception, according to Gibson, is the process by which this information is detected.

An example of an invariant is the nondisruption of edges in covering surfaces (Gibson 1979, 76). An observer's motion relative to one surface that is in front of another will cause the edges of the distal surface to change while the edges of the proximal surface remain constant. This invariant specifies the relative distances of the objects to the observer. Similarly, objects at different distances from an observer will, if the same size, be cut by the horizon in the same proportion (178). Thus, the fact that two trees have their top thirds above the horizon specifies to the observer that they are the same height, regardless of their relative distances from the observer (and, of course, regardless of whether the observer has conscious access to this invariant).

Before explaining Gibson's account of affordances, it is worth pausing to contrast Gibson's view of perception with the cognitivist's. For Gibson, perception is strongly coupled with action, because most of the invariants within the optic array do not appear except against a background of change. Perceivers actively scan, move within, or otherwise manipulate the environment. For the cognitivist, however, perception is a process that begins and builds from static images that flutter across the retina. The vision scientist Richard Gregory neatly encapsulates the antithesis of Gibson's view: "Perceptions are constructed, by complex brain processes, from fleeting fragmentary scraps of data signaled by the senses...Current sensory data (or stimuli) are simply not adequate directly to control behavior in familiar situations"

(1972, 707).[1] Gregory's characterization of the perceptual process motivates typical cognitivist accounts of, for instance, the perception of object size: the size of an object must be computed from a representation of the amount of visual angle it subtends together with a representation of its distance from the observer. In contrast, Gibson denies the need for computations over representations to determine object size. Object size is already specified by invariants in the optic array such as horizon-cuts and the amount of graded texture an object's base obscures. These invariants render the need for computation otiose.[2]

We come finally to the most controversial (and maligned[3]) element of Gibson's theory of perception: affordances. Gibson believed that information in the optic array sufficed to specify opportunities for action, thus providing observers with an ability to perceive e.g., the chair as something that can be sat upon, the surface as something that can be walked on, the food as nourishing, and so on. Because the present interest in Gibson is as an inspiration for embodied cognition, this is not the place to consider whether Gibson could mount an adequate defense of this view. The significance of affordances for current purposes concerns their agent-relativity. Whether a chair can be sat upon, and thus whether an observer correctly perceives it as something that can be sat upon, depends on the properties of the observer as well as those of the chair. The perception of a given chair as being something on which one can sit is thus a perception that is tied to a particular observer. Similarly, to use another of Gibson's examples, a branch affords a resting place for a bird, but not a person. Accordingly, a bird can perceive the branch as affording a resting place for itself, but a person cannot. The perception of affordances, in short, is an observer-relative form of perception.[4] The very same chair might be perceived as "sittable upon" by one observer but not another.[5] Gibson believed that how the world is perceived, and what it is perceived to afford, depends on the kinds of interactions an observer of a particular sort—having a body of a particular kind—can make with its environment.

To summarize, the ecological school of psychology Gibson developed brought with it a conception of psychology quite distinct from the cognitivist school that within Gibson's lifetime came to predominate. For cognitivists, cognition consists in computational operations on informationally meager symbolic representations in order to produce new representations that serve cognitive functions such as perception, language comprehension, problem solving, and so on. For Gibson, percep-

---

[1] Mace (2005) quotes this passage as well. See Mace (1974) for more passages in the same vein.

[2] See Goldstein (1981) and Hatfield (1988, 1991) for suggestions about how to reconcile features of Gibson's staunchly anti-representationalist views with conceptions of representation that Gibson apparently failed to consider.

[3] Fodor and Pylyshyn (1981).

[4] However, Gibson (1979) insisted that affordances are "objective"—therefore not in the eye of the beholder.

[5] Relevant here are Warren's studies of observers' perceptions of steps as climbable, and passages as passable (Warren 1984; Warren and Whang 1987).

tion is the detection of information that, with no further embellishment, suffices to specify features of an observer's world. According to Gibson, the active observer could, by collecting and sampling the wealth of information contained within the optic array, know its world in terms relative to its needs. We will see these Gibsonian themes recurring time and again as we now turn toward a discussion of embodied cognition.

3. Some Conceptions of Embodiment

As mentioned previously, there is no consensus about what it means to say that cognition is embodied. Unlike cognitivists, who can agree on a fairly specific understanding of what it means to say that cognition is computational, embodied cognition researchers have no such rallying point. Some attempts to define embodied cognition have been made (Riegler 2002), and lists of criteria for embodiment have been drawn (Anderson 2007). But at this point in the development of embodied cognition, such efforts are just as likely to straitjacket research as they are to promote it. More profitable are discussions of embodied cognition that seek thematic connections between the disparate research programs that identify themselves as embodied cognition.[6] This said, it is worth considering some prominent views about the nature of embodiment.

Varela, Thompson, and Rosch, whose *The Embodied Mind* is widely seen as a foundational text in the embodied cognition movement, say that from the embodied perspective "cognition is no longer seen as problem solving on the basis of representations; instead, cognition in its most encompassing sense consists in the enactment or bringing forth of a world by a viable history of structural coupling" (1991, 205). Varela et al. are here expressing an idea close to that which Gibson sought to capture with the concept of affordances. An organism's conception of the world is a function of its history—both phylogenetic and ontogenetic—of interactions with properties in its environment. Because properties of an organism's body determine the form these interactions must take, organisms with relevantly different bodies will conceive the world differently (i.e., will conceive *different* worlds).

Ester Thelen, another revered figure in embodied cognition, along with colleagues suggests a similar understanding of embodiment:

> [t]o say that cognition is embodied means that it arises from bodily interactions with the world. From this point of view, cognition depends on the kinds of experiences that come from having a body with particular perceptual and motor capabilities that are inseparably linked and that together form the matrix within which reasoning, memory, emotion, language, and all other aspects of mental life are meshed. (2001, 1)

---

[6] Wilson (2002) has an excellent discussion in this vein.

Thelen's characterization of embodiment thus echoes Varela et al.'s. Embodied cognition emphasizes the special form cognition takes as a result of the idiosyncratic properties of the cognizer's body.

Randall Beer, whose support for embodied cognition arises from his painstaking and insightful studies of simulated robot behavior, says that "embodiment emphasizes the role of an agent's own body in its cognition...An embodied nervous system can utilize the natural biomechanics of its body, the geometry of its sensory surfaces, and its ability to actively control these surfaces to simplify many problems" (2003, 211). Beer's research shares the dynamical systems perspective that Thelen also champions, according to which cognitive behavior is the product of interactions between brain, body, and world that become intelligible through the lens of dynamical systems theory. In this conception of embodiment, behavior that the cognitivist would have assumed to be the product of purely computational processes taking place within the brain becomes instead the output of a dynamical system consisting of brain, body, and environment, whose changes over time can be modeled with differential equations. The salient point is that the tools of dynamical systems theory can often (or, at least, sometimes) explain, and often (or, at least, sometimes) explain *better* how cognition arises than would a traditional cognitivist account.

Andy Clark, who among philosophers has doubtless done most to advance and refine conceptions of embodied cognition, emphasizes two aspects of embodied cognition that he draws from converging fields of research (1997).[7] First is the idea that embodied cognition eschews "excessive world modeling" (1997, 23). This idea, although distinct from Varela et al.'s of "world making," also has a Gibsonian ancestor. As we have seen, Gibson believed that the presence of information in the environment obviated a need for inferential operations on impoverished representations. An object's size does not need to be calculated from representations of visual angle and distance because information about object size is already specified within the optic array. Embodied cognition theorists have developed this Gibsonian idea, seeking to discover how various cognitive activities that cognitivists have sought to explain in representationally rich terms might be better or as well explained using representationally scanty resources.

Clark's second point about embodied cognition reflects another Gibsonian theme: researchers in embodied cognition, because they reject or at least harbor misgivings about the need for world modeling, prefer to investigate the organism-environment "fit" that facilitates an organism's ability to access information external to it. Different species of organism, owing to differences in the properties of their bodies (including their sensory organs) will fit the environment in distinct ways. Differences between the bodies of a koala and a quoll account for differences in how each kind of organism interacts with a tree. As these interactions differ, so

---

[7] These two points are actually culled from three that Clark makes in the first chapter of *Being There*. The third point is that the second point applies to human beings.

too the variety of information that becomes available to each organism differs.[8] But also, because koalas and quolls use trees for different ends, the information to which each is tuned to respond may not be the same.

These two claims of Clark's, that a cognizer's need to build models of the world might often (or on occasion) be satisfied instead by access to information already present in the environment, and that organisms' access to information will differ depending on how particular organisms are able to interact with their environment and which interactions with the environment are important for them, are part of what embodied cognition theorists mean when they claim that cognition is *situated*. The picture of the situated cognizer differs considerably from that of the symbol processor. The former is interacting with its environment in a manner designed (or intended) to extract information that is relevant to its needs. The latter uses its sense organs to, in effect, take a picture of the world, which it then submits for processing, awaiting the final result: a new and more refined picture, suitable for guiding action. The situated cognizer is the bicyclist who crashes to the ground if he stops moving; the symbol processor is the archer who holds steady, closes her eyes, and visualizes the target.

4. Support for Embodied Cognition

The above characterizations of embodied cognition, which are admittedly vague, suffice to motivate a variety of research projects within a variety of distinct fields. At the risk of oversimplification, seeing these projects as motivated by one or more of the following questions is convenient:

> CONCEPTUALIZATION: How might the body limit or constrain an organism's conception of the world?
> CONSTITUTION: How might the body play a constitutive role in cognitive processing?
> REPLACEMENT: How can an organism's interaction with the environment replace representational processes thought to have been at the core of cognition?

These questions are not intended to be distinct. Answering one may partly answer others, and indeed research directed toward one of these questions will often contribute answers to the others. The questions, it should be obvious, have their Gibsonian forebears. CONCEPTUALIZATION asks exactly how properties of a body

---

[8] Note that this is not necessarily true. Different kinds of organism may be able to pick up the same information in different ways, or despite differences in their body types. However, when organisms differ in body type, especially in certain ways, this suggests that each has access to information that is unavailable to the other.

influence the conception an organism can build of its world, and is thus concerned with many of the same issues Gibson confronted in his belief that perceivers see the world in terms of what its objects afford them. CONSTITUTION asks how properties of an organism's body might be integrated into the processes of information collection and information processing—a question that Gibson studied in his attempt to understand how a perceiver's actions make accessible the invariants in the optic array. REPLACEMENT advances the Gibsonian idea that the processes of perception, and cognition more generally, are not serial stages of representation construction, but consist instead in an active engagement with an informationally rich environment.

Another significant feature of the research programs that address these questions is the explicit effort they make to distance their goals and methods from those that traditional cognitivists pursue. Nearly universal among research articles falling within the embodied cognition domain is an introductory section in which cognitivism is glossed for the purpose of painting a background against which the author's goals and methods can stand in stark relief. Given embodied cognition's identity as an alternative to cognitivism, the extent to which this contrast retains its clear definition becomes a matter of pressing importance. We shall return to this issue below. Now is the time to consider the various lines of research that have sought answers to the three questions above.[9]

4.1. Conceptualization

The question of conceptualization might be clarified by analogy to the more familiar thesis of linguistic determinism, regardless of the current standing of this thesis.[10] According to linguistic determinism, the structure of the language one learns influences and limits one's understanding of the world. Thus, people from different linguistic communities will conceive of the world in different ways.[11] Similarly, some researchers in embodied cognition argue that the structural properties of an organism's body influence and limit its conceptualization of the world. Lakoff and Johnson's studies of metaphor are the most recognized line of work in this area (1980, 1999). Lakoff and Johnson first show that the use of metaphor for understanding and developing concepts is commonplace. Thus, for instance, the concept of an ar-

---

[9] This tripartite division differs from one previously suggested (Shapiro 2004, 2007), but is one developed much more extensively in Shapiro (2011).

[10] Boroditsky, although not endorsing full blown linguistic determinism, has done much in recent years to revive interest in the idea that the language one speaks influences one's capacity to think in certain ways. (see, e.g., Boroditsky 2001).

[11] For example, Boas and Whorf identified somewhere between four and seven "Eskimo" words for snow, and it has become common lore that this shows that the Inuit conception of snow must differ from that possessed by members of linguistic communities where there are fewer words to name snow. Pullum (1991) is widely viewed as having debunked Boas and Whorfs' claims, and, a fortiori, those gross exaggerations of them.

gument might be understood by the metaphor "argument is war." This metaphor structures the concept of argument insofar as it forces a picture of argument as a violent engagement in which there are winners and losers, in which ground can be gained or lost, in which the parties involved must be on guard for attacks from un-expected quarters, and so on. One who understands argument as being like war has a different conception of it than one who understands argument as being like solv-ing a puzzle, where parties may work together to offer competing ideas in an effort to come to a final and satisfying conclusion.

Lakoff and Johnson are aware that this account of concept learning risks charges of circularity. A metaphor connecting *argument* to *war* can structure one's under-standing of *argument* only if one already has an understanding of *war*. Similarly, one's understanding of *war*, if Lakoff and Johnson are right, has been shaped by metaphoric connections to other concepts. This circle of concepts must drop anchor at some point, and in fact does so at the level of *basic* concepts, that is, concepts that a human being understands without recourse to metaphoric connections to other concepts.

Lakoff and Johnson argue that many of the basic concepts are spatial: a reflec-tion of those properties of the human body that determine motion through space. Human concepts of *up* and *down*, for instance, emerge from the vertical orientation that human bodies typically assume in wakeful moments. Similarly, the concepts *front* and *back* have their origin in the fact that sensory organs for sight, smell, and taste are on the same plane of the human body. The ascent from these basic con-cepts to more abstract ones follows a complex route, but the details are not impor-tant in this context. An organism's understanding of its world is embodied, Lakoff and Johnson contend, in the sense that the concepts through which it views the world are grounded in properties of its body.

A counterfactual helps to illuminate this point. If human beings did not have erect bodies and sensory systems facing in the same direction, then their basic con-cepts, and a fortiori, concepts built on these, would differ. Their understanding of the world would differ from the understanding of the world that actual human beings possess. "Imagine," Lakoff and Johnson suggest, "a spherical being living out-side of any gravitational field, with no knowledge or imagination of any other kind of experience. What could UP possibly mean to such a being?" (1999, 76). And, with no concept (or a very distinct concept) of *up*, these organisms would no longer share with human beings those concepts that have been structured in terms of *up* (for instance, these organisms may not conceive happiness as a *boosting* of spirits, or as a feeling of *buoyancy*).

In a series of clever experiments, Boroditsky and Ramscar (2002) show an effect that might be taken as evidence for the kind of embodied conceptualization that Lakoff and Johnson describe. Boroditsky and Ramscar exposed subjects to the am-biguous sentence "Next Wednesday's meeting has been moved forward two days." The sentence is ambiguous because, depending how one conceives of "moving for-ward," the new meeting day may be either Monday or Friday. Subjects were then asked "What day is the meeting now that it has been rescheduled?" Boroditsky and

Ramscar were interested in whether primes that suggested to subjects motion through space would influence the subjects' interpretation of the ambiguous sentence. The data show that exposure to primes suggesting thoughts of forward motion would make "Friday" the more probable response; when primes suggested thoughts of something moving toward the subject, the subject would more likely respond "Monday." In explaining these data, Boroditsky and Ramscar argue that subjects conceive of time in spatial terms. When already prompted to think about forward motion, subjects are inclined to see the rescheduling of the meeting as moving away (toward Friday), and when subjects are compelled to reflect on objects moving toward them, they tend to understand the meeting as coming toward them—moving from Wednesday to Monday.

On the assumption that the kind of motion one can take through space is a function of the kind of body one has, Boroditsky and Ramscar's research can be seen as suggesting an answer to CONCEPTUALIZATION. One way in which the body constrains or limits an organism's conception of the world is by the freedom of motion it permits. Human motion is generally in two dimensions and usually forward. These limits are a consequence of the means for motility available to human beings. Had human beings wings, or were they to live in aquatic or gravity-free environments, perhaps they would not conceive of time in terms of motion through space. Or, perhaps time's arrow would not be conceived as pointing ahead.[12]

Focus on CONCEPTUALIZATION marks a departure from traditional cognitivism insofar as it takes very seriously the body's role in concept acquisition. Whereas cognitivists have tended toward a view of mind that allows its investigation to proceed without detailed investigation of the properties of the body that forms a shell around it, work on answering the question of CONCEPTUALIZATION shows the danger in this strategy. A more complete understanding of how organisms conceive their world, and why they develop the conceptions of the world that they do, demands careful attention to the nature of interactions with the world that an organism's body permits. Because cognition takes place in a body, one should not be surprised to find that the properties of the body have a significant impact on the products of cognition. It is recognition of this point that, some in embodied cognition believe, is lost in traditional cognitivism.

Notice, however, that recognition of this point does not require that cognitivists give up their central doctrine, that is that cognition is computation over symbolic

---

[12] Liberties are being taken with Boroditsky and Ramscar's research for the purpose of illustrating a line of research in embodied cognition. Boroditsky and Ramscar (2002) in fact deny that their research supports what they call *strong* embodiment because it is enough to influence subjects' conception of time that they be *thinking* about forward or backward motion rather than, as some in embodied cognition might insist, that they be actually moving themselves forward or backward. As we will see below, embodied cognition theorists who emphasize the importance of sensorimotor processes in cognition do not always require that an organism actually be engaged in motor activity for cognition to depend on sensorimotor processes. Boroditsky and Ramscar also emphasize that language has a role in how human beings conceive time, as Boroditsky's work on Mandarin-speaking subjects seeks to show (Boroditsky 2001).

representations. Lakoff and Johnson's conclusions, for instance, are consistent with the possibility that basic concepts are encoded symbolically, and that more abstract concepts evolve in virtue of computational operations on the representations of concepts such as *up, down, back*, and *front*. Thus, research on CONCEPTUALIZATION might better be viewed not as a challenge to cognitivism, but as a splash of cold water in the faces of cognitivists who have resisted the need to consider how an organism's body might shape and limit the representations that are available to computational processes.

4.2. Constitution

CONSTITUTION asks about the integration of the body within cognitive processing. The issue of integration is a sticky one. Cognitivists of course accept that cognitive processing takes place within a body—how could it not given the location of the brain? Moreover, cognitivists understand that the body's motions and other activities will determine what the sensory systems can make available to the brain. Thus, the answer to CONSTITUTION that embodied cognition researchers seek, if it is truly to distinguish embodied cognition from cognitivism, must go beyond the unremarkable assertion that an organism's body plays an important causal role in cognitive processing. At the very least, embodied cognition researchers must explain how the causal role the body plays is not one to which cognitivists can readily assent, or is not one that cognitivism can, with little trouble, accommodate. More striking would be evidence that the body is actually a constituent in cognitive processing, where the difference between a cause and a constituent is like the difference between the report of the starting gun, which causes the race to begin, and the individuals participating in the race, who are constituents of the race itself.[13] A satisfying answer to CONSTITUTION will describe the body as a piece of the cognitive process itself rather than as link in a causal chain that extends, further upstream, to cognition.

For a clearer sense of this point, consider Churchland, Ramachandran, and Sejnowski's (1994) discussion under the heading "Movement (of Eye, Head, Body) Makes Many Visual Computations Simpler" (1994, 51). The movements Churchland et al. mention include head motions, which provide for motion parallax—an indication of relative depth of objects in the visual field. Although head movements such as these, it is true, are rarely mentioned in the studies of "pure" or "computational" vision that Churchland et al. are criticizing, one must take care not to conclude that cognitivists must reject head movements as an important element in perception. The cognitivist might very well incorporate head movements into a computational description of vision, constructing an algorithm that combines representations of these movements with representations of other portions of the visual scene.[14] Thus, answers to CONSTITUTION must seek to establish something

---

[13] See Block (2005) for further discussion.
[14] In fairness, Churchland, Ramachandran, and Sejnowski are *not* denying this possibility.

more: they must show, as, for instance, Noë believes, that cognitive processes such as perception are "in part *constituted* by our possession and exercise of bodily skills" (2004, 25, emphasis added).[15]

Bearing on the issue of Constitution are studies of sensorimotor processes involved in cognition as well as research that models cognition as the unfolding activity of a dynamical system. Because dynamical systems' approaches to cognition figure heavily in the next section, this section will focus on sensorimotor accounts of cognition. At the core of these accounts is the conception of cognition as constituted in part by motor activity or, at any rate, by a kind of knowledge about the motor activity that would be involved (if not actually involved) in acquiring information about the world. The theory of perception that O'Regan and Noë have been developing falls squarely within this category.

O'Regan and Noë (2001) argue that an organism's ability to perceive is in part constituted by its tacit knowledge of the myriad sensorimotor contingencies that distinguish one form of perception from another. One sort of contingency arises from the nature of the sensory apparatus itself.[16] Thus, because the retina is a concave surface, a perceived straight line will appear on the retina as a great arc. As the perceiver shifts his focus to a point above the line, the arc shifts higher on the retina, becomes shorter, and changes shape. The laws that define these regular transformations resulting from contingencies of the sensory apparatuses serve to define what it is to be having a visual perceptual experience in contrast to, say, an auditory perceptual experience, which would be individuated by its own set of sensorimotor contingencies.

Successful perceivers, according to O'Regan and Noë, will have mastery, or tacit knowledge, of the sensorimotor contingencies particular to their own sensory apparatuses. But it is not just motion of the eye that reveals sensorimotor contingencies. Often these contingencies show themselves only when the entire organism is in motion, moving around objects, picking them up, seeing them against a background of other objects, etc. The actions an organism takes in its efforts to perceive display a kind of skill—a skill that in turn reflects familiarity with the sensorimotor contingencies that impose order on patterns of changing stimulation that would otherwise appear as an uncrackable cipher—a code waiting to be broken.

Moreover, because sensorimotor contingencies will differ depending on the idiosyncratic nature of sensory apparatuses and skills necessary for their use, perceptual experience is likely to differ for different kinds of perceivers. "If perception is in part constituted by our possession and exercise of bodily skills...then it may also depend on our possession of the sort of bodies that can encompass those skills, for only a creature with such a body could have those skills. To perceive like us, it fol-

---

[15] For further elaboration of this point, see Aizawa (2007) and Adams and Aizawa (2008, 168–69).

[16] O'Regan and Noë (2001) also discuss sensorimotor contingencies that are the product of objects in the world rather than the apparatuses used to perceive these objects.

lows, you must have a body like ours" (Noë 2004, 25).[17] The body relativity of perception, Noë thinks, requires a method of investigation distinct from that to which cognitivism is committed. Even were one to retain the strategy of explaining perception as the achievement of algorithmic processes, the body must enter this explanation in ways cognitivists never foresaw. "The point is not that algorithms are constrained by their implementation, although that is true. The point, rather, is that the algorithms are actually, at least in part, formulated *in terms of* items at the implementational level. You might actually need to mention hands and eyes in the algorithms!"[18] (Noë 2004, 25, emphasis in original).

The idea that the active body is an important part, and perhaps constituent, of cognition is also apparent in Glenberg's (1997) embodied theory of memory. Glenberg proposes that the function of memory is, in contrast to most cognitivist views, not simply to remember events, but to encode past events in terms of bodily activity. In doing so, the function of memory becomes more ecologically situated than a view of memory as simple recall would suggest. Memory helps to guide an organism in its interactions with the environment by revealing how previous actions can "fit" with present circumstances in order to produce new and appropriate actions.

Making this account embodied, Glenberg argues, is the content of the encodings involved in memory. Features of the environment are encoded not from a third person perspective, but in terms of how an organism with a body like *so* can interact with them. Memories, in turn, consist in encodings of how one's body did in fact interact with features of the environment. Because one's current perception of the environment and one's memory of the environment are encoded in the same way—in terms relative to action—it is possible to combine (or, in Glenberg's terminology, *mesh*) the present with the past, thereby creating representations of various potential actions.

Thus, to find one's way home from the Sydney Opera House, one retrieves not a mental map of Sydney, but instead combines memories encoded in terms of previous actions—walk down these steps, follow this sidewalk, step onto the ferry here—with perceptions of present opportunities for action. Memory guides action, in effect, by bringing past action to bear on present situations for action, and it does this by means of a common body-based code for action.

In addition to his work on memory, Glenberg, as well as others, have accumulated results suggesting a tight connection between motor activity and cognition. For instance, Kaschak and Glenberg (2000) and Glenberg and Kaschak (2002) have found that when subjects must pull a lever toward themselves to indicate comprehension of a sentence, they are slower to react when the sentence is about an action involving a transfer *away* from the subject. Tucker and Ellis (1998) exposed subjects

---

[17] Note the support this kind of claim might give to various answers to CONCEPTUALIZATION.

[18] This claim might seem to fall prey to the criticism offered earlier of a certain interpretation of Churchland et al.'s (1994) point about head movements. It is one thing for heads and eyes to be mentioned or represented in an algorithm, but quite another for heads and eyes actually to be part of an algorithm. This latter claim has the oddness of a category mistake.

to images of objects with handles on either their right or left sides, and oriented either right side up or upside down. The task was to identify the direction of the object's orientation. Subjects were quicker to respond when the hand used in the response was the same hand that would be used to grasp the handle of the object. These experiments are designed to show, at least, that there is a link between action and cognition in the sense that the cognitive activity appears to draw on motoric resources. Cognition is embodied, accordingly, to the extent that motor programs are integrated with other processes that result in, for instance, language comprehension or object identification.[19]

Proponents of sensorimotor theories of cognition see further support for their views in recent investigations of canonical and mirror neurons. These neurons have been found in region F5 of the macaque's premotor cortex as well as in area PF of the macaque's parietal lobe and perhaps the superior temporal sulcus.[20] In human brains, the growing consensus is that these neurons are present in the superior temporal sulcus (Rizzolatti and Craighero 2004) and in the inferior frontal gyrus (Broca's area), which is thought to be a homologue of the macaque's F5 (Gallese et al. 1996). Canonical neurons and mirror neurons are bimodal, discharging when observing a stimulus but also when acting. Canonical neurons respond selectively to objects; that is, a canonical neuron will fire in response to the observation of a cup and also in preparation for grasping. The actions during which canonical neurons discharge are appropriate to the objects observed. Thus, the canonical neuron that fires in response to the observation of a small object is also the neuron that fires when the monkey makes a precise prehension grip, and the neuron that fires when the monkey is viewing a large object is the same one that fires in the course of making a whole-hand prehension (Gabarini and Adenzato 2004). Mirror neurons, in contrast, respond to the observation of actions rather than objects. So, the same mirror neuron that is involved in a grasping action will fire when the monkey observes another monkey (or human being) grasping an object. Similarly, some mirror neurons respond selectively to the observation of tool use (Ferrari, Rozzi, and Fogassi 2005).

Although the function of canonical and mirror neurons remains a subject of intense controversy, their existence lends support to the view that cognition includes a motor constituent. If the same neurons that are involved in observing objects and actions are also involved in guiding action, a case might be made that objects and actions are understood in terms of the actions the observer must take to use that object, or to replicate that action. Because actions are actions of a body, the idea gains strength that the body is indeed a constituent in cognitive processing.

[19] Adams (2010) charges Glenberg with drawing a strong *constitution* explanation for his data when they can be explained as well with a weaker *causal influence* conclusion.

[20] Gallese et al. (1996) distinguish the mirror neurons in F5 from neurons in STS, but other researchers do not.

4.3. Replacement

If CONCEPTUALIZATION asks how the body affects one's capacity to conceive the world, and CONSTITUTION asks how the body becomes part of one's cognitive processing, REPLACEMENT asks how the body and its interactions produce cognitive abilities that do not require for their explanation any of the traditional resources of cognitivism.

Answers to REPLACEMENT generally take two forms, each of which draws on evidence from both psychology and robotics. First, a number of psychologists and roboticists have argued that organisms are able to exploit features of their environments in a manner that does away with the need to represent the environment.[21] Second, some psychologists and roboticists have proposed that cognition be understood as emerging from the dynamic interactions of the body, environment, and brain, in the course of which there is simply no time, nor need, for the construction and manipulation of representations. We shall examine both these approaches to REPLACEMENT in turn.

4.3.1. Using the World as Its Own Model

A natural rationale for believing that cognition depends on representation is the idea that the world out there must somehow get "in here." How else to see the Harbor Bridge but to form a representation of it (certainly, the bridge itself would not fit inside one's head)? How else to find one's way home but to form a representation of the path that will take one there? Although these questions are offered rhetorically, some researchers have taken them quite seriously, providing a "here's how else" response.

An influential response to this question about what, if not representations, permits access to the external world is Ballard, Hayhoe, Pook, and Rao's (1997) theory of deictic codes. "Deictic" refers to an act of pointing. A deictic code binds an object in the world to a cognitive process by, in effect, instructing the process where to look to find the object, thus freeing the process from the need to construct a representation of the object. By analogy, rather than having to remember the word for rattlesnake, one can communicate the immediate danger of a rattlesnake bite to a friend by shouting "Watch out for that!" while pointing to the snake.

The pointing Ballard et al. studied was not done by fingers, but by eye saccades. Subjects were shown a model (on a computer monitor) consisting of blocks of different colors, and their task was to reconstruct the model from a collection of randomly scattered colored blocks (also on the computer monitor). A representationally rich strategy would require that subjects memorize the colors and locations of blocks in the model and then consult this representation in memory to reconstruct the model in the workspace. Ballard et al. found instead that subjects would saccade to the model, then back to the block resources, where they would pick up (with a

---

[21] See Shapiro (2008b).

mouse) a block of the right color, then back to the model, and then to the work space, where they would drop the block in its correct position. The pattern of saccades suggests that subjects never bothered to memorize the color and location of blocks in the model. The first saccade to the model was for the purpose of choosing from the resources a block of the correct color. The second saccade, after the block had been picked up, was for the purpose of deciding where to position the block in the work space. Rather than encoding facts about the color and position of blocks in the model, subjects appear to have been encoding information about *where* information about color and location could be found.

Of course, even if Ballard et al. are correct that deictic pointers can replace the need to build certain kinds of representations for the purpose of reconstructing these block scenes, they have not shown that subjects who rely on deictic pointers do not also make use of representations of some sort or other. Perhaps subjects retain a representation of the colors and positions of blocks in working memory as they shift their focus from the model to the work space. Indeed, how else could a deictic pointer suffice for the experimental task if information about color and position were not represented in some fashion? In short, perhaps deictic pointers merely *attenuate* the representations on which subjects must rely when reconstructing the block scene, leaving open the possibility (and, it would seem, requirement) that representations of color and position are present in some stages of model reconstruction.[22]

Using their Hollywood Squares paradigm, Richardson and Spivey (2000) collected data that seem to support Ballard's idea that human beings use deictic codes, at least sometimes, to do the work typically attributed to more robust representations. Subjects viewed a computer monitor divided into quadrants. In sequence, a head would appear in one quadrant, state a fact (e.g., "Shakespeare's first plays were historical dramas; his last was *The Tempest.*"), and then disappear. After the fourth head disappeared, the screen would remain blank except for the lines separating the quadrants, and in a neutral voice the computer would make a statement (e.g., "Shakespeare's first play was *The Tempest.*") that the subject had to evaluate as true or false. Despite the fact that the heads delivered the information auditorily, and that subjects could see that the heads were no longer present, subjects would twice as often fixate on the quadrant that contained the head which imparted the relevant information than they would any other quadrant.

Richardson and Spivey take their results to show that subjects use deictic pointers to supplement or ease the burden on internal working memory. Rather than (or in addition to) storing representations of the facts the heads had uttered, subjects encoded the facts in terms of where to "look" to find them again. Presumably, storing a fact such as "look here for Shakespeare information" is less costly than storing the information about Shakespeare. Using deictic codes allows us "to employ the body's environment as [a] sort of notice board of 'virtual post-it notes' that complement our internal memory" (Richardson and Spivey 2000, 173).

---

[22] Thanks to Gary Hatfield for bringing this point to the author's attention.

Mindful of the challenge raised above against Ballard et al.'s conclusions, Richardson and Spivey's more temperate claim that deictic pointers can *decrease* the representational burden on working memory is no doubt for the better. Insofar as subjects must remember the location of the head that uttered the sentence of interest, and must remember some details about what was said if they are to look to the quadrant from which the utterance was spoken, the experimental task apparently requires more than just deictic pointers for its successful completion. Indeed, one might go a step farther and argue, as we saw earlier in the discussion of parallax, that motions of the body (e.g., turning toward the correct quadrant) might well cue or facilitate a cognitive ability such as perception or, in this case, memory, but that the presence of such cues is something that cognitivists can easily accommodate within a computational description of the stages involved in the cognitive process.

The idea that task-specific information might be stored (or left) in the environment rather than represented has also been prominent in Brooks's approach to the design of autonomous mobile robots. Indeed, it was Brooks who coined the expression that labels this subsection (Brooks 1991). From the perspective of traditional AI, the design for an autonomous mobile robot (i.e., a robot that is capable of moving about an environment, avoiding obstacles, setting courses, pursuing goals, and so on) requires perceptual modules that produce representations of the environment, a central processing module comprising submodules for learning, planning, knowledge representation, and other tasks, and action modules that convert the deliverances of the central processes into behavior. Brooks claims that the construction of robots on the basis of this sense-think-act architecture has produced few successes. As a practical matter, testing particular modules is very difficult because the performance of each module depends on the performance of others. Thus, one cannot test a single module before having designed every other module (143).

Rather than decomposing the capacities of mobile robots along the lines of sense-think-act, Brooks instead developed an architecture consisting of layers, each of which directly connects sensing with acting. The layers are defined in terms of the actions they perform. For instance, an object avoidance layer will move the robot forward until a sensor detects an obstacle. The activation of the sensor causes, automatically, avoidance behavior. Running parallel to the object avoidance layer is a "visit distant object" layer. This layer causes the robot to move toward a distant object. The two layers, despite running in parallel, assist each other. As the robot heads toward the object determined by the second layer, the first layer insures that the robot avoids obstacles in its path. If the first layer causes the robot to change course to avoid an obstacle, the second layer then steers the robot back on course toward the distant object. Adding layer upon layer, each of which connects sensing directly to acting, the behavior of the robots becomes quite surprising. Brooks's *Creatures* (as he calls them) are able to move about a cluttered environment, collect soda cans, and deposit them in a recycling bin.

Crucially, the versatility of a Creature does not depend on the existence of a central processing module—there is none. Brooks describes a Creature as "a collection of competing behaviors. Out of the local chaos of their interactions there emerges, in the eye of an observer, a coherent pattern of behavior. There is no central purposeful locus of control" (1991, 144). There are, Brooks insists, no representations of the environment, no representations of rules to guide action, no representations of goals. Each layer acts more like a reflex than it does a reasoner.[23] And just as reflexes simply connect a stimulus to an action without the need of intervening representations, "the state of the world determines the action of the Creature" (145). In short, Creatures do not need to represent features of their world: they need only to respond to these features. In this way, the world itself is responsible for guiding the behavior of Creatures.

In summary, deictic encoding and Brooks's layer-driven robots provide alternatives to the representation-heavy accounts of cognitive behavior that have long been on offer. At the heart of these alternatives is a reorientation toward letting the world do the work, or hold the information, that is required for the completion of a task. Although Brooks's claim that his Creatures do not employ representations of any sort remains controversial,[24] and although the use of deictic codes does not eliminate the need for some representation, both research projects might, in contrast to traditional cognitivist ones, be best described as *representation-lite*.

4.3.2. *Dynamical Systems Approaches to Cognition*

A dynamical system is any system whose changes of state over time can be modeled with a rule, which usually takes the form of a differential equation, but need not. A common example of a dynamical system is a pendulum. A pendulum's state space consists of all the possible angles and angular velocities the pendulum can have. At any instant, the pendulum will be in a particular state—at a particular angle with a particular angular velocity. Newton's law F = ma describes the evolving behavior of the pendulum over time. Thus, given Newton's law and a specification of the angle and angular velocity of a pendulum at time t1, it is possible to calculate the pendulum's angle and angular velocity at t2.

The mathematical tools necessary to describe dynamical systems have been applied to cognitive behavior with some surprising results. Moreover, advocates of the dynamical systems approach claim that dynamical analyses of cognition often do not avail themselves of representations, and so are an alternative to the computational explanations traditional cognitivists favor. As an example, Thelen, Schöner, Scheier, and Smith (2001) used the tools of dynamical systems theory to understand the classic "A-not-B" error that Piaget (1955) identified.[25] Between the ages of seven

[23] The reflex analogy is the author's, not Brooks's, but it seems to capture his intent.
[24] In fact, it is hard to see how they could not. For instance, insofar as creatures can keep a heading toward a distant point, they would seem to need, at a minimum, a representation of this heading. See Wilson (2004, 177) for discussion.
[25] Replication of this error takes quite a bit of precision, and there remains controversy over the robustness of this error.

and twelve months, infants who watch an object hidden repeatedly in spot A, and are then shown the object hidden in spot B, will reach to spot A to retrieve the object. Piaget argued that infants exhibit this behavior because they have entered a stage of development in which objects are in part identified with their location. Infants reach to spot A because this spot, is, in a sense, a constituent of the object that was there. They do not understand that the same object may exist in a different location.

Thelen et al. dispense with explanations in terms of what infants represent or know about objects and their relations to them. Instead they model the infant's behavior as the evolution of a dynamical system whose state space is defined by variables representing (i) features of the task environment (parameterized in terms of quantities such as the distance of location A from location B), (ii) the specific cue the experimenter uses to elicit reaching behavior, and (iii) the infant's history of prior reaching toward spot A. A tracing of the dynamics of this system, Thelen et al. argue, explains why infants exhibit the "A-not-B" error, and does so without the "need to posit such individual and separate mechanisms such as egocentric or allocentric coding or memory or response inhibition deficits or incomplete object knowledge" (2001, 4). Indeed, the model they develop shows that there is nothing more mysterious about the infant's behavior than there is, for example, the behavior of an American who, when driving in Australia, constantly turns on the windshield wipers when trying to engage the turn signal. Most significantly, the model shows the behavior to emerge from complex interdependencies between the activities of several variables over time. Computations over symbolic representations never enter this model.

Beer (2003) has used the tools of dynamical systems theory to build simulations of robots that exhibit object discrimination behavior. The robot's task in these experiments is to catch circles and avoid diamonds as these objects fall straight down from above. The robot travels back and forth horizontally; it must center itself beneath a falling circle in order to catch it, and must move away from the path of a falling diamond to avoid it. The robot has an eye that "sees" along seven rays that are equally distributed over an angular range of $\pi/6$. When a falling object intersects one of these rays, a corresponding sensory neuron receives an input inversely proportional in strength to the distance of the falling object from the robot (so that the closer the object is to the robot, the stronger the input). Discrimination between circles and diamonds requires that the robot move back and forth, causing the falling shape to interrupt its rays of vision. Circles and diamonds will break these rays in different ways depending on the horizontal position of the robot relative to them.[26]

The robot's "nervous system" comprises a three-layer connectionist network with seven input nodes (corresponding to the robot's seven rays of vision), five hidden nodes, and two output nodes. The dynamical system Beer models consists

---

[26] However, because both circles and diamonds are bilaterally symmetrical, if the robot is directly beneath a diamond as it starts to fall, it is unable to distinguish it from a circle (Beer 2003, 215).

of this nervous system, the falling objects, and the robot's horizontal motion. These components of the system are always in flux—the objects are falling, the levels of activation the input nodes receive are changing, and the robot's body is moving back and forth. The interdependencies between elements of the system and the changes in robot behavior they produce over time make the system a natural target for a dynamical systems analysis. Beer models the system's behavior with sixteen differential equations that provide a complete description of what the state of the system will be at any future time given a description of the system's state at the present time. The analysis also enables Beer to understand how the system eventually succeeds in making the discriminations between circles and diamonds that it does.

As with Thelen et al's account of the "A-not-B" error, Beer's explanation of object categorization focuses exclusively on the state of a system of interdependent parts as it evolves through time. There is no reason to describe the system's behavior as the product of computations over symbolic representations. Indeed, the robot's representations of the world are at best attenuated: they consist only in responses to visual rays as these are "broken" by a falling shape. Beer observes that when one views cognition as an achievement growing from the interactions of a brain, a body, and an environment:

> the focus shifts from accurately representing an environment to continuously en-
> gaging that environment so as to stabilize coordinated patterns of behavior that
> are adaptive for the agent...Indeed, a dynamical approach raises important ques-
> tions about the very necessity of notions of representation and computation in
> cognitive theorizing (2003, 210).

Beer's sentiment is popular among researchers who have turned to dynamical systems theory as a tool for understanding cognition (see Port and Van Gelder 1995). For this reason, modeling cognition as a dynamical system, like using the world as its own model, is seen as an answer to REPLACEMENT.

Clark (1997) raises a cautionary note regarding the excitement that dynamical systems research has generated. He observes that dynamical systems accounts of cognition seem to be mainly descriptive. Thelen et al., for instance, describe how the environmental setting, cueing paradigm, and past reaching behavior interact to produce the "A-not-B" effect. Thelen et al. do not, however, identify the underlying mechanisms that are responsible for the behavior that causes the "A-not-B" effect. As Clark makes the point, "commanding a good pure dynamical characterization of the system falls too far short of possessing a recipe for building a system that would exhibit the behaviors concerned" (1997, 120). In effect, Clark is challenging dynamical systems researchers to move beyond the neo-behaviorism implicit in their approach. It is one thing to use equations to describe how a system behaves, and quite another to uncover the causes of this behavior.

Some who offer dynamical systems explanations of cognition appreciate the force of Clark's challenge and have taken steps toward meeting it. Beer, especially, is concerned with not just describing the behavior of his robots, but with understanding the causes of the behavior. Indeed, he seems to have Clark's point in mind when he says, "it could be argued that this characterization is primarily descriptive in nature, detailing what the structure is without explaining how it arises" (2003, 226). In response, he takes the bull by its horns, presenting an account of "how this structure emerges from the interaction between the agent and its environment" (226). If we overlook the details, the essence of this account describes how within the dynamics of the agent, its horizontal motions, and the falling shape, various attractors appear that slowly draw the agent toward a particular action. Although the complete story is full of complexity, the significant point in the present context is that some people working within the dynamical systems framework are sensitive to Clark's worries about explanatory inadequacy—and are directing their efforts accordingly.

5. EXTENDED COGNITION

Embodied cognition and extended cognition are sometimes not distinguished, or are treated as complementary theses (Wilson 2002; Anderson 2007). This is probably a mistake. The questions of CONCEPTUALIZATION, CONSTITUTION, and REPLACEMENT, while doubtless philosophically rich, depend at least as much for their resolution on the direction empirical results take as they do on conceptual matters. In contrast, extended cognition seems to rest more centrally on philosophical issues concerning the nature of the mind. One such issue that arises immediately is whether the mind is where cognitive processes are. If not, then one might readily assent to the idea that cognitive processes may include parts of the environment external to an organism, because doing so requires no commitment to the more dubious claim that an organism's mind is at least partly outside its body.

Given the largely philosophical nature of the extended cognition thesis, the prominent place of philosophers involved in this debate is unsurprising. Thus, Clark and Chalmers (1998)'s study is the *locus classicus* for an articulation and defense of extended cognition, Rowlands (1999) is perhaps the first book-length argument for extended cognition, and Wilson (2004) devotes great effort to the development of a framework for extended cognition. Adams and Aizawa (2008)'s study offers the first sustained criticism of the many facets of extended cognition. Rupert (2009) followed with another book-length critique. Claims for and against extended cognition, however, do not arise *only* from within philosophy departments: psychologists too have offered defenses (Spivey, Richardson, and Fitneva 2004) and criticisms (Wilson 2002) of the thesis. The psychologist Merlin Donald (1991) inspired many contemporary discussions of extended cognition when he argued for a stage in the evolution of cognition in which external props come to perform the function of memory storage. With his careful study of the steps involved in navigational tasks aboard a large naval vessel, the psychologist and anthropologist Edwin Hutchins (1995) has drawn attention to the possibility that cognition arises from groups of individuals in interaction with each other and with specially designed tools.[27]

At the center of the extended cognition thesis is the recognition that many cognitive activities require for their completion the exploitation of features in the environment. Of vital importance is that these features are exploited for the information they contribute toward performing a cognitive task, such as remembering the location of an address (Clark and Chalmers 1998), computing the product of large numbers (Wilson 2004), or solving a logic problem (Rowlands 1999). In many cases, proponents of extended cognition note that agents often store or "offload" information in the environment so as to reduce the load on their internal memory (e.g., by writing down directions to the party rather than trying to remember them). Moreover, agents structure their environments to make the information they contain more accessible (e.g., by alphabetizing their libraries).

The truth of these points is not in dispute. The question, of course, is how facts such as these bear on the question of the mind's boundaries. Do parts of the environment become parts of the mind by virtue of their role in cognitive processes? Advocates of extended cognition respond affirmatively to this question on the condition that the role the environment plays is of the right sort. In general, philosophers have identified two prospects for what this right sort of role might be.

First, a case for extended cognition might rest on the claim that external resources play a functionally identical role to internal resources, and because psychological kinds are identified by their functional roles, there can be no justification for denying that some external resources are parts of the mind. Thus, Clark and Chalmers (1998) imagine an Alzheimer's victim who uses a notebook to store information that an unafflicted person would simply memorize. If the notebook entries are indeed functionally identical to items in memory, then this is sufficient reason to conclude that the notebook entries are part of the Alzheimer's victim's mind. To deny this would require an appeal to nonfunctional considerations, but this would then require the development of new criteria for taxonomizing psychological states.[28]

Defenders of extended cognition might also appeal to the role the environment plays in a dynamical systems analysis of cognition. According to these analyses, nervous system, body, and environment are *coupled*. Although "coupling" is a term often waved about in discussions of extended cognition as if it were a magic wand, the term has a precise meaning within dynamical systems theory. Nervous system, body, and environment are coupled when each are represented by variables within a differential equation that describes the evolving state that results from their interaction. Because the coupling of these components produces cognition, cognition comprises all of these components. "Where, then, is the mind?" Clark asks (1997,

---

[27] For a critical review of Hutchins (1995), see Wilson (1998).

[28] See Shapiro (2008a) for discussion.

68). "Is it indeed 'in the head,' or has mind now spread itself, somewhat profligately, out into the world.... In a sense, then, human reasoners are truly *distributed* cognitive engines" (68).

Both the functional parity defense of extended cognition and the coupling defense have been challenged. Aizawa and Adams (2001, 2008) and Rupert (2004) have argued that features of the environment typically cannot play the role of genuine psychological states because they are not integrated in the correct way with other items and processes involved in cognition. Rupert, for instance, discusses the phenomenon of negative transfer. Subjects learn a list of paired associations, for instance the names of husband and wife pairs. After the learning period, subjects are tested until they succeed to some criterion in recalling the name of the wife after exposure to the name of the husband. They are then told that a number of couples have divorced and remarried and that they must learn a new list of associations. Memorizing this new list is much more difficult for subjects because the list they had previously memorized interferes with the learning process. Old knowledge interferes with the acquisition of new knowledge.

Although negative transfer is a property of human memory, there is no reason to expect the Alzheimer's victim's memory to exhibit the same property. In learning a new list of paired associations, the Alzheimer's victim has merely to erase her old list and write down the new one. But negative transfer is just one instance of a psychological tendency that distinguishes natural human memory from the externalized memory that the Alzheimer's victim has manufactured. Furthermore, memory is just one cognitive trait among many that will show tendencies that cannot be reproduced by systems that include external components.

Against the coupling defense for extended cognition, Wilson has argued that even if "the forces that drive cognitive activity do not reside solely inside the head" (2002, 630), one need not conclude that cognition is itself distributed. Although this is correct, Wilson might be accused of missing the point about coupling. There is no doubt that interacting parts of a dynamical system can be studied in isolation. For instance, the behavior of the dynamical system consisting of one pendulum hanging from the bob of another has been the subject of intense investigation, but of course it is also possible to study the behavior of single pendulums in isolation from each other. It is not possible, however, to study the behavior of the linked pendulums by studying the behavior of the pendulums in isolation from each other. Similarly, although it is possible to study nervous systems, bodies, and environments apart from each other in order to understand the dynamics that result from their interaction, one cannot investigate them independently. The point Wilson seems to have missed is that cognition, according to proponents of extension, is not a component that can be studied apart from other components, but is the *outcome* of dynamically interacting components.

Adams and Aizawa (2008) offer another line of criticism that might apply to both the functional parity and coupling justifications for extended cognition. They argue that any attempt to extend the boundaries of the mind beyond the head must stake a claim on the "mark of the cognitive." The point is clear in other contexts. Either defending or denying the claim that viruses are alive requires that one have in hand a mark of the living. But if it is reasonable to ask for the criteria of life in order to decide whether something is alive, it should be reasonable as well to ask for criteria of cognition in order to assess claims that cognitive processes may extend into the environment.[29] Adams and Aizawa argue that the criteria that proponents of extended cognition offer are too extravagant, with the result that thermostats or plants end up possessing cognition. In contrast, they defend a more austere conception of cognition that would make extended cognition untenable. Insofar as Adams and Aizawa are right that the case for extended cognition must rest on an adequate definition of cognition, one sees again why the controversy over extended cognition has drawn so much philosophical attention.

6. Conclusion

The health and prospects of the embodied cognition research program are certainly not as glorious as some of its champions would suggest. Uncontestable, however, is the vigor and range of research that falls within the scope of embodied cognition. The questions that CONCEPTUALIZATION, CONSTITUTION, and REPLACEMENT ask are well-formed in the sense that continuing philosophical and empirical investigation can steer the way toward their resolution. Especially impressive is the sense of unity that researchers in embodied cognition display. As the foregoing has shown, there may be no single goal toward which embodied cognition researchers strive. However, there is a common theme—that traditional cognitive science has somewhere gone wrong, and that a fix requires attending more carefully to how an organism's body and interactions with the environment contribute to cognition.

Traditional cognitive scientists should welcome the challenge. Presumably, cognition is not computational by definition. Certainly there is no longer reason to adopt a Cartesian confidence in the complete distinctness of mind and body. If it appears that cognition *must* be computational, and that its operations can or should be understood apart from consideration of the happenings in the body and the environment, this is probably the result of enculturation in the reigning cognitivist orthodoxy.[30] This is not to say that the cognitivist paradigm is wrong—perhaps it is correct after all. The point is simply that cognitive science is not *necessarily* a special case of computer science. Perhaps there are other and better perspectives from

---

[29] Adams and Aizawa's request for a mark of the cognitive also bears on the prospects of REPLACEMENT, for those seeking to replace standard cognitive explanations with nonstandard ones owe an account of why their *explananda* are nonetheless cognitive.

[30] Rowlands (1999) too makes this point.

which to understand the mind. Insofar as the claims of embodied cognition are coherent and not wildly implausible, they deserve serious reflection.

It is worth noting that cognitivism has already faced a challenge from connectionism that is similar in spirit to that which embodied cognition poses. Indeed, some describe embodied cognition as the next stage in the battle that began with the tossing of the connectionist gauntlet (Varela et al. 1991; Franklin 1995; Clark 1997). Like connectionists, embodied cognition researchers conceive of themselves as offering a new framework for studying the mind. Although this is not the place to evaluate the current standing of the dispute between connectionists and cognitivists, those familiar with the history of this dispute might do well to think of the embodied cognition challenge to cognitivism in its terms. Drawing on this history suggests that embodied cognition might do well to remain open to the following possibilities.

First, the explanatory methods and tools of embodied cognition might turn out to be very well-suited to some cognitive *explananda* but not others. Already apparent is that much of embodied cognition research seems to focus on the perceptual end of the cognitive spectrum, or offers models for behavior that in some cases are of questionable cognitive significance (e.g., Thelen's 1995 work on the development of stepping behavior in infants). Recognizing that their research often focuses on tasks that are far removed from the problem solving and reasoning tasks that cognitive science originally targeted, some embodied cognition researchers have expressed confidence that their theoretical approach can "scale up." But, although efforts to see how far "up" the cognitive spectrum embodied cognition might travel are at this point well worth taking, reasonable also would be to question whether the spectrum is indeed a spectrum after all. Perhaps categorization tasks and chess playing do not differ only in cognitive degree. Perhaps the variety of tasks labeled as cognitive requires different conceptual tools for their explanation. If this is right, embodied cognition may be wrong to paint itself as an *alternative* to cognitivism, just as cognitivism would be wrong to present itself as an overarching theory of cognition. In this view, embodied cognition and cognitivism can live happily side by side, having divided the cognitive bounty along lines that suit their individual talents.

A second possibility is that embodied cognition and cognitivism might find ways to merge. Suggestive in this regard is Hatfield's development of a "'noncognitive psychological' approach to perceptual theory" (1988, 178), according to which perceptual processes are understood to be representational, in contrast to Gibson, but not symbolic, in contrast to cognitivism (see also Hatfield 1991). Hatfield's noncognitive psychological stance thus integrates important insights from Gibson's ecological psychology with a conception of perception as representational. The reason to mention Hatfield's noncognitive psychological approach in the present context is simply to note the possibility of a middle ground between areas thought to be unbridgeable.[31] We have seen already that embodied cognition is not defined

---

[31] See Norman (2002) for a different style of reconciliation between ecological and cognitivist psychologies.

by its antagonism toward representation, although some within the field have urged nonrepresentational treatments of cognition. Moreover, cognitivism is not committed to the unimportance of the body in cognitive processes. There is room here for reconciliation, and one might expect that a combination of resources can only benefit ongoing efforts to understand the mind.

REFERENCES

Adams, F. (2010). Embodied Cognition. *Phenomenology and the Cognitive Sciences* 9: 619–628.
Adams, F., and Aizawa, K. (2001). The bounds of cognition. *Philosophical Psychology* 14: 43–64
———. (2008). *The Bounds of Cognition*. (Malden, MA: Blackwell Publishing).
Aizawa, K. (2007). Understanding the embodiment of perception. *Journal of Philosophy* 104: 5–25.
Anderson, M. (2007). How to Study the Mind: An Introduction to Embodied Cognition. In F. Santoianni and C. Sabatano (eds.), *Brain Development in Learning Environments: Embodied and Perceptual Advancements*. Cambridge: Cambridge Scholars Press.
Ballard, D., Hayhoe, M., Pook, P., and Rao, R. (1997). Deictic codes for the embodiment of cognition. *Behavioral and Brain Sciences* 20: 723–67.
Beer, R. (2003). The dynamics of active categorical perception in an evolved model agent. *Adaptive Behavior* 11: 209–43.
Block, N. (2005). Review of Alva Noë's *Action in Perception*. *Journal of Philosophy* 102: 259–72.
Boroditsky, L. (2001). Does language shape thought?: Mandarin and English speakers' conceptions of time. *Cognitive Psychology* 43: 1–22.
Boroditsky, L., and Ramscar, M. (2002). The roles of body and mind in abstract thought. *Psychological Science* 13: 185–89.
Brooks, R. (1991). Intelligence without representation, *Artificial Intelligence* 47: 139–59.
Churchland, P. S., Ramachandran, V., and Sejnowski, T. (1994). A Critique of Pure Vision. In C. Koch and J. Davis (eds.), *Large-Scale Neuronal Theories of the Brain* (Cambridge, MA: MIT Press).
Clark, A. (1997). *Being There: Putting Brain, Body and World Together Again* (Cambridge, MA: MIT Press).
Clark, A., and Chalmers, D. (1998). The extended mind. *Analysis* 58: 7–19.
Donald, M. (1991). *Origins of the Modern Mind* (Cambridge, MA: Harvard University Press).
Egan, F. (1991). Must psychology be individualistic?, *The Philosophical Review* 100: 179–203.
Ferrari, P., Rozzi, S., and Fogassi, L. (2005). Mirror neurons responding to observation of actions made with tools in monkey ventral premotor cortex. *Journal of Cognitive Neuroscience* 17: 212–26.
Fodor, J. (1980). Methodological solipsism as a research strategy in cognitive psychology. *Behavioral and Brain Sciences* 3: 63–73.
Fodor, J., and Pylyshyn, Z. (1981). How direct is visual perception?: Some reflections on Gibson's "ecological approach." *Cognition* 9: 139–96.
Franklin, S. (1995). *Artificial Minds*. Cambridge, MA: MIT Press.
Gabarini, F., and Adenzato, M. (2004). At the root of embodied cognition: Cognitive science meets neurophysiology. *Brain and Cognition* 56: 100–6.

Gallese, V., Fadiga, L., Fogassi, L., and Rizzolatti, G. (1996). Action recognition in the premotor cortex. *Brain* 11: 593–609.

Gibson, J. J. (1950). *The Perception of the Visual World* (Boston: Houghton Mifflin).

———. (1966). *The Senses Considered as Perceptual Systems* (Prospect Heights, IL: Waveland Press).

———. (1979). *The Ecological Approach to Visual Perception* (Boston: Houghton-Mifflin).

Glenberg, A. (1997). What memory is for. *Behavioral and Brain Sciences* 20: 1–55.

Glenberg, A., and Kaschak, M. (2002). Grounding language in action. *Psychonomic Bulletin & Review* 9: 558–65.

Goldstein, E. (1981). The ecology of J. J. Gibson's perception. *Leonardo* 14: 191–195.

Gregory, R. (1972). Seeing as thinking: An active theory of perception. *London Times Literary Supplement*, June 23.

Hatfield, G. (1988). Representation and content in some (actual) theories of perception. *Studies in History and Philosophy of Science* 19: 175–214.

———. (1991). Representation in Perception and Cognition: Connectionist Affordances. In W. Ramsey, S. Stich, and D. Rumelhart (eds.), *Philosophy and Connectionist Theory* (Hillsdale, NJ: Lawrence Erlbaum Associates).

Hutchins, E. (1995). *Cognition in the Wild* (Cambridge, MA: MIT Press).

Kaschak, M., and Glenberg, A. (2000). Constructing meaning: The role of affordances and grammatical constructions in sentence comprehension. *Journal of Memory and Language* 43: 508–29.

Lakoff, G., and Johnson, M. (1980). *Metaphors We Live By* (Chicago: University of Chicago Press).

———. (1999). *Philosophy in the Flesh: The Embodied Mind and Its Challenge to Western Thought* (New York: Basic Books).

Mace, W. (1974). Ecologically Stimulating Cognitive Psychology: Gibsonian Perspectives. In W. Weimer and D. Palermo (eds.), *Cognition and the Symbolic Processes I* (Hillsdale, NJ: Lawrence Erlbaum).

———. (2005). James J. Gibson's ecological approach: Perceiving what exists. *Ethics and the Environment* 10: 195–208.

Marr, D. (1982). *Vision*. (San Francisco: Freeman).

Newell, A., and Simon, H. (1961). Computer simulation of human thinking. *Science* 134: 2011–17.

Noë, A. (2004). *Action in Perception* (Cambridge, MA: MIT Press).

Norman, J. (2002). Two visual systems and two theories of perception: An attempt to reconcile the constructivist and ecological approaches. *Behavioral and Brain Sciences* 25: 73–144.

O'Regan, J., and Noë, A. (2001). A sensorimotor account of vision and visual consciousness. *Behavioral and Brain Sciences* 24: 939–1031.

Piaget, J. (1955). *The Child's Construction of Reality* (London: Routledge and Kegan Paul).

Port, R., and Van Gelder, T. (eds.). (1995). *Mind as Motion: Explorations in the Dynamics of Cognition*. Cambridge, MA: MIT Press.

Pullum, G. (1991). *The Great Eskimo Hoax and Other Irreverent Essays on the Study of Language*. Chicago: University of Chicago Press.

Richardson, D., and Spivey, M. (2000). Representation, space, and Hollywood Squares: Looking at things that aren't there anymore. *Cognition* 76: 269–95.

Riegler, A. (2002). When is a cognitive system embodied?. *Cognitive Systems Research* 3: 339–48.

Rizzolatti, G., and Craighero, L. (2004). The mirror-neuron system. *Annual Review of Neuroscience* 27: 169–92.

Rosenfeld, A. (1988). Computer Vision: Basic Principles. *Proceedings of the IEEE* 76: 863–68.

Rowlands, M. (1999). *The Body in Mind: Understanding Cognitive Processes.* Cambridge: Cambridge University Press.

Rupert, R. (2004). Challenges to the hypothesis of extended cognition. *The Journal of Philosophy* 101: 1–40.

———. (2009). *Cognitive Systems and the Extended Mind.* Oxford: Oxford University Press.

Segal, G. (1989). Seeing what is not there. *The Philosophical Review* 98: 189–214.

Shapiro, L. (2004). *The Mind Incarnate.* Cambridge, MA: MIT Press.

———. (2007). The embodied cognition research programme. *Philosophy Compass* 2: 338–46.

———. (2008a). Functionalism and mental boundaries. *Cognitive Systems Research* 9: 5–14.

———. (2008b). Symbolism, Embodiment, and the Broader Debate. In M. de Vega, A. Glenberg, and A. Graesser (eds,), *Symbols and Embodiment: Debates on Meaning and Cognition.* Oxford: Oxford University Press.

———. (2011). *Embodied Cognition.* New York: Routledge.

Spivey, M., Richardson, D., and Fitneva, S. (2004). Thinking Outside the Brain: Spatial Indices to Visual and Linguistic Information. In J. Henderson and F. Ferreira (eds.), *The Interface of Language, Vision, and Action: Eye Movements and the Visual World.* New York: Psychology Press.

Stich, S. (1983). *From Folk Psychology to Cognitive Science.* Cambridge, MA: MIT Press.

Thelen, E. (1995). Motor development: A new synthesis. *American Psychologist* 50: 79–95.

Thelen, E., Schöner, G., Scheier, C., and Smith, L. (2001). The dynamics of embodiment: A field theory of infant perseverative reaching. *Behavioral and Brain Sciences* 24: 1–86.

Tucker, M., and Ellis, R. (1998). On the relations of seen objects and components of potential actions. *Journal of Experimental Psychology: Human Perception and Performance* 24: 830–46.

Varela, F., Thompson, E., and Rosch, E. (1991). *The Embodied Mind: Cognitive Science and Human Experience.* Cambridge, MA: MIT Press.

Von Eckardt, B. (1995). *What is Cognitive Science?.* Cambridge, MA: MIT Press.

Warren, W. (1984). Perceiving affordances: Visual guidance of stair climbing. *Journal of Experimental Psychology: Human Perception and Performance* 10: 683–703.

Warren, W., and Whang, S. (1987). Visual guidance of walking through aperatures: Body-scaled information for affordances. *Journal of Experimental Psychology: Human Perception and Performance* 13: 371–83.

Wilson, M. (2002). Six views of embodied cognition. *Psychological Bulletin and Review* 9: 625–36.

Wilson, R. (1998). Joint Review of Edwin Hutchins' *Cognition in the Wild* and Ron McClamrock's *Existential Cognition: Computational Minds in the World. Mind* 107: 486–92.

———. (2004). *Boundaries of the Mind. The Individual in the Fragile Sciences: Cognition.* Cambridge: Cambridge University Press.

CHAPTER 7

....

ARTIFICIAL INTELLIGENCE

....

DIANE PROUDFOOT AND
B. JACK COPELAND

ALAN Turing wrote, "One way of setting about our task of building a 'thinking machine' would be to take a man as a whole and to try to replace all the parts of him by machinery" (1948, 420).[1] "It is customary," he said, "to offer a grain of comfort, in the form of a statement that some particularly human characteristic could never be imitated by a machine....I cannot offer any such comfort, for I believe that no such bounds can be set" (1951, 486). Indeed Turing thought it probable that once "machine thinking...started, it would not take long to outstrip our feeble powers" (c.1951, 475). Here he anticipated the scenario—sometimes called "runaway AI"— that disturbs many modern technological futurists (see Section 5; also Turing 1951, 485–86).

Pursuing human-level AI has a particular virtue, in Turing's view; he said, "I believe that the attempt to make a thinking machine will help us greatly in finding out how we think ourselves" (1951, 486). Human-level intelligent systems have been called the "Holy Grail" of AI (e.g. Moor 2003, 211) and "the Big AI Dream" (Lenat 2008a, 12; see also Brooks 1996), and the quest to build such machines dominated AI's early years. This quest came to be severely criticized. For example:

> AI research and experiment has paid far too much attention to the development
> of machinery and programs which seek directly or indirectly to imitate human
> performance....This focus...does not seem to have been very productive...[and]

---

[1] For more on Turing and the history of AI, see Copeland (2004a); Copeland and Proudfoot (1996, 2005, 2006).

is more likely to produce interesting curiosities such as ELIZA than working AI applications. (Whitby 1996, 56)

AI largely changed direction in the 1980s and 1990s, concentrating on building domain-specific systems and on sub-goals such as self-organization, self-repair, and reliability. Computer scientists aimed to construct "intelligence amplifiers"[2] for human beings, rather than imitation humans (this is sometimes described as a contrast between human-computer interaction and artificial intelligence—see, e.g., Winograd 2007). Some AI professionals call this conception of the field "narrow AI," and several lament it. Douglas Lenat says, for example, "I was heartbroken to see the whole field in retreat...from the 'strong AI' goal to...the 'raisin bread' notion of AI:...little tiny flickers of intelligence here and there in applications like Roombas [the iRobot vacuum cleaner], where the AI is analogous to the raisins in raisin bread" (Lenat 2008a, 20–21).[3] However, even talk of "flickers of intelligence" in such contexts seems dangerously reminiscent of the hyperbole that damaged AI's reputation last century (see Section 7).

In the twenty-first century, several AI researchers are once more arguing for, and working toward, human-level AI—or "artificial general intelligence" (AGI).[4] Conferences and workshops have once again focused on human-level AI and AGI.[5] John McCarthy says that he would "be inclined to bet on this 21st century" as the time when human-level AI will be achieved (2007, 1174). There is still, however, opposition: for example, Jordan Pollack—who exhorts AI to focus instead on "mindless intelligence"—recently berated the field's "preoccupation with mimicking human-level intelligence." AI, he claims, "behaves as if human intelligence is next to godliness" (2006, 51).[6]

In this chapter we discuss central philosophical issues concerning human-level AI. Section 1 describes and defends Turing's "imitation game" (now known simply as the Turing test), his proposed benchmark for success in the quest to develop human-level AI. Section 2 rebuts what is possibly the most famous objection to the thesis that thinking or intelligence (especially human-level intelligence) can be duplicated in a computer, John Searle's "Chinese Room argument." In Section 3 we turn to one of AI's fundamental assumptions: the universal digital computer—or some other Turing-machine equivalent system, such as a Turing-type neural net—is the *right kind of machine* to think. We discuss and reject an argument that this assumption *must* be true; and we argue that it is an open question whether there may be more to human-level intelligence than human-level AI as traditionally conceived ("Turing-machine AI") can capture. In Sections 4 and 5 we describe and criticize

---

[2] Lenat (2008b, 281) uses this term.

[3] The Roomba domestic vacuum cleaner is based on the robotics research carried out by Rodney Brooks at MIT. See irobot.com for information on the Roomba.

[4] For discussion of ambiguities in the notion of "human-level" AI, see, e.g., McDermott (2007); Goertzel (2007); Sloman (2008).

[5] See, e.g., Cassimatis, Mueller, and Winston (2006); see Feigenbaum (2003); Lenat (2008a); Goertzel (2007); McCarthy (2007, 2010); Minsky, Singh, and Sloman (2004); Zadeh (2008).

[6] This dispute is discussed in Proudfoot (2011).

the recent claims by "technological futurists" that human-level—and superhuman-level—AI will be achieved in the near future, and that human beings will themselves become artificial intelligences. Section 6 highlights the tendency to make-believe and misplaced anthropomorphism in AI.

1. THE TURING TEST

How are we to tell when we reach human-level AI (assuming this is possible)? The Turing test is one putative solution, first outlined in Turing's "Intelligent Machinery" (1948, 431).

1.1. The Imitation Game

Turing based his test on a computer-imitates-human game, describing three versions of this game in 1948, 1950, and 1952. The famous version appears in a 1950 article in *Mind*, "Computing Machinery and Intelligence" (Turing 1950). Here Turing describes an "imitation game" involving an interrogator and two contestants, one male (A) and one female (B). The interrogator communicates with A and B from a separate room (nowadays this would probably be by means of a keyboard and screen); the interrogator's task is to find out, by asking questions, which contestant is the man, and the man's goal is to make the interrogator choose wrongly. Turing then asked, "What will happen when a machine takes the part of A in this game?" (1950, 441). In this new game, a computer imitates a human being (man or woman).[7] The interrogator's task is to discover which contestant is the computer; to do so the interrogator is permitted to ask any question, or put any point, on any topic. The computer performs satisfactorily if it does no worse in the computer-imitates-human game than the man in the man-imitates-woman game.

Turing claimed, notoriously, that the question *Can machines think?* is "too meaningless to deserve discussion" and proposed replacing it with the question *Are there imaginable digital computers which would do well in the imitation game?* (1950, 449, 441). In a 1952 radio broadcast, "Can Automatic Calculating Machines Be Said to Think?," he said:[8]

> Well, that's my test. Of course I am not saying at present either that machines really could pass the test, or that they couldn't. My suggestion is just that this is

---

[7] Some theorists argue that the machine's task is to impersonate a *woman*, or a man imitating a woman (e.g., Genova 1994; Sterrett 2003; Traiger 2003; Lenat 2008a, 2008b). On this, see Copeland (2003a).

[8] This broadcast contains the third version of Turing's computer-imitates-human game. Unlike the 1948 and 1950 versions, this is not a 3-player game. See Copeland (2003a).

the question we should discuss. It's not the same as "Do machines think," but it seems near enough for our present purpose...(Turing et al. 1952, 495)

Turing believed that a machine would pass his test (1953, 569), but only after "at least 100 years"—that is to say, not before 2052 (Turing et al. 1952, 495).[9]

1.2. An "Emotional Concept"

The orthodox interpretation of Turing's test is that it provides an *operational definition* of intelligence (or thinking) in machines, in terms of *behavior* (or behavioral dispositions or capacities). (Examples of the orthodox interpretation include Hodges 1992; French 2000; Cohen 2005; Harnad 2008.)

In his 1952 broadcast, however, Turing denied that he was proposing a definition (Copeland 1999, 2003a). He said:

> I don't want to give a definition of thinking, but if I had to I should probably be unable to say anything more about it than that it was a sort of buzzing that went on inside my head. But I don't really see that we need to agree on a definition at all. (Turing et al. 1952, 494–95)

Moreover, in "Intelligent Machinery" Turing made the philosophical motivation for his test explicit (Proudfoot 2005, 2011) This report for the National Physical Laboratory demonstrates that, contrary to the orthodox interpretation, Turing was no behaviorist. In a section entitled "Intelligence as an emotional concept," before describing the earliest version of the computer-imitates-human game, he said:

> The extent to which we regard something as behaving in an intelligent manner is determined as much by our own state of mind and training as by the properties of the object under consideration. If we are able to explain and predict its behaviour or if there seems to be little underlying plan, we have little temptation to imagine intelligence. (1948, 431; see also Turing et al. 1952, 500)

The imitation game is an experiment to see if the interrogator will "imagine intelligence" in the machine; the interrogator, Turing said, "must be taken in by the [machine's] pretence" (Turing et al. 1952, 495). "Intelligent Machinery" sets out the thesis that whether an entity is intelligent is determined in part by our responses to the entity's behavior. For Turing, the concept of intelligence is an "emotional"—that is to say, *response-dependent*—concept.[10] For this reason, the orthodox behaviorist interpretation is wrongheaded, whether it takes Turing to be offering a definition of intelligence or merely a logically sufficient condition (examples of the latter interpretation include Block 1995; Eberbach, Goldin, and Wegner 2004). Moor's (1976,

---

[9] In his 1950 paper, Turing predicted that "in about fifty years' time" an average interrogator would not have a greater than seventy percent chance of detecting the machine (after five minutes of playing the game) (1950, 449).

[10] For an analysis of Turing's approach to the concept of intelligence, see Proudfoot (forthcoming).

1987) "inductive test" reading of the Turing test also overlooks the importance for Turing of the interrogator's response, as does Shieber's (2007) "interactive proof" interpretation.

1.3. Objections to the Test

Scientists have raised several objections to Turing's test (see Proudfoot 2011). These include: the test is too *difficult*, and fails to provide a practical way forward for AI (Cohen 2005); the research and development required to construct a machine capable of passing the test are too *expensive* (Waltz 2006); engineering resources are better used to build AI aids for humans (Lenat 2008b); and trying to build *human-like* systems will not help us to understand (the generic concept of) intelligence, and diverts attention from AI's real achievements (Hayes and Ford 1995; Pollack 2006; Lenat 2008a).[11] Critics claim that "work directed at success in the Turing test is neither genuine nor useful AI research" (Whitby 1996, 55) and that the influence of the test has been "a tragedy for AI" (Hayes and Ford 1995, 975). Opponents of Turing's imitation game conclude that it should be "relegated to the history of science" (Ford and Hayes 1998, 79).

However, these objections typically misconstrue Turing's test—which was not offered as AI's only goal. Frequently they involve highly speculative empirical claims (Patrick Hayes and Kenneth Ford, for example, assert that a "truly human-like program would be nearly useless" (1995, 975)). In addition, both scientists and philosophers sometimes take criticisms that properly apply to Hugh Loebner's annual Prize Contest in Artificial Intelligence (which is based on the Turing test) and transfer them to Turing's actual test. For example, Block (1995) argues, based on the success of primitive and plainly unintelligent programs in Loebner's test, that the Turing test is too easy.

The philosophical objections to the test include the following: the test is not justified by any *theory* of intelligence (Churchland 1996); *stupid* machines can pass the test, in part due to gullible interrogators (Block 1981, 1995); and the imitation game does not test for *mind* (Searle 2002—see Section 2 below). In our view, the standard philosophical objections fail (see, e.g., Copeland 2003a, 2004c, 2004e; Copeland and Proudfoot 2008). Also, proposed alternative tests of intelligence in machines fail to improve on Turing's test (see Proudfoot 2006). (Recent examples of alternative tests include Nilsson's (2005) "employment test" and Feigenbaum's (2003) "Feigenbaum Test.")

We discuss two influential philosophical objections in greater detail.

1.4. The Shannon-McCarthy/Block Objection

In 1956 Claude Shannon and John McCarthy said:

---

[11] Engineers do employ "reverse Turing tests" (e.g., Rui and Liu 2004; Ponec 2006).

[I]t is possible, in principle, to design a machine with a complete set of arbitrarily chosen responses to all possible input stimuli…Such a machine, in a sense, for any given input situation…merely looks up in a "dictionary" the appropriate response. With a suitable dictionary such a machine would surely satisfy Turing's definition [of thinking] but does not reflect our usual intuitive concept of thinking. (1956, v–vi)

This objection has occurred to several theorists but nowadays is usually credited to Ned Block (1981, 1995). A "blockhead" is a (hypothetical) program able to play the imitation game successfully, for any fixed length of time, by virtue of including a look-up table. This large, but finite, table contains all the exchanges between program and interrogator that could (in principle) occur during the length of time in question. Such a program, so the objection goes, does not think.

The formal point on which the Shannon-McCarthy objection rests—the encapsulability in a look-up table of all the relevant behavior—would have been obvious to Turing (see Copeland 2003a). In his 1950 paper, Turing described a "mimicking digital computer" (a machine that uses a finite look-up table to mimic the behavior of any "discrete state machine" that has a finite number of possible states) (1950, 448). Well aware (even in those pioneering days of electronic computing) that storage capacity and speed are crucial to performance, Turing stressed that the mimicking machine "must have an adequate storage capacity as well as working sufficiently fast" (Turing 1950, 448), and in his 1952 broadcast he said, "To my mind this time factor is the one question which will involve all the real technical difficulty" (Turing et al. 1952, 503). These remarks suggest a reply to the Shannon-McCarthy objection (Copeland 2004c). Given practical limitations on storage capacity, their hypothetical "machine with a complete set of arbitrarily chosen responses to all possible input stimuli" simply cannot be built. And even if it could be built, it would not succeed in the imitation game—it would take too long to answer the questions.

If, as Shannon and McCarthy believed, Turing intended his test as a definition of "thinking" (or as a logically sufficient condition), the mere possibility of their machine would be enough to defeat the test. However, there is no textual evidence to suggest that this was Turing's intention; indeed, his denial that he was offering a "definition of thinking" (see Section 1.2) and his emphasis on *real-world* machines suggests otherwise (see further Copeland 2003a, 2004c).[12]

1.5. Fiendish Experts

Fiendish expert objections to the Turing test are of the form: "An expert could unmask the computer by asking it…". An imitation-game interrogator with expert knowledge in psychology could use recent discoveries of characteristic weaknesses in human reasoning to detect any computer not specifically programmed to reproduce them (Lenat 2008b). Likewise, an interrogator who has studied humans' response times to certain sorts of questions (especially in experiments using priming) could employ this knowledge to identify the machine (French 1990). A computer

[12] Also Proudfoot (forthcoming).

may also be identified by its "superarticulacy," unless the computer is specially (and perversely) programmed so as to avoid this (Michie 1993). In the annual Loebner Contests, interrogators use knowledge of typical weaknesses in AI programming to identify the machine (e.g., inputting nonsense utterances and probing the contestants' common-sense knowledge).

Turing anticipated this type of objection, implying in his *Mind* paper that a computer might be identified by its speed and accuracy at arithmetic (1950, 442). Although the interrogator is permitted to ask any question (or put any point) she likes, not just *any* interrogator is permitted. The interrogator in the 1948 version of the test (which is limited to chess-playing) is to be "rather poor" at chess, and in the 1952 version "*should not be expert about machines*" (1948, 431; Turing et al. 1952, 495 (italics added); see further Copeland 2004e).

In our view, the Turing test is central to AI. According to Edward Feigenbaum, "Computational Intelligence *is* the manifest destiny of computer science, the goal, the destination, the final frontier" (2003, 39). But how are we to tell when we have reached our destination? Turing's test is the only game in town.

2. THE CHINESE ROOM ARGUMENT

John Searle's Chinese Room argument (CRA, for short) aims to show that cognition *cannot* consist in computation. Searle contrasts what he calls "strong AI," which aims to *duplicate* intelligence, thought, understanding, and other cognitive abilities in a computational system, with "weak AI," which aims only to *simulate* intelligent behavior. Searle's principal target is strong AI, but his argument strikes also at the Turing test and at computational theories of mind.

2.1. The Argument

In its original form (Searle 1980a), the CRA addresses the case of a human clerk—call him or her Clerk—who "handworks" a GOFAI ("Good Old-Fashioned AI"[13]) program. (Section 2.4 discusses Searle's application of the argument to connectionist AI.) Clerk, a monolingual English speaker, works alone with paper and pencil inside a room equipped with slots labeled "input" and "output." The program is given to Clerk in the form of a set of rulebooks (in English), which occupy many shelves inside the room.[14] People outside the room pass sheets of paper inscribed with Chinese characters through the input slot; after much computation Clerk

---

[13] Haugeland (1985).

[14] One not uncommon misunderstanding takes Searle to be asserting that the rulebooks contain a *look-up table* that pairs possible inputs directly with ready-made outputs. So misinterpreted, Searle's argument is weakened. In fact, Searle's intention is that the rulebooks may contain any GOFAI program that is claimed by its creators (or others) to understand Chinese—or indeed to have any "cognitive states" whatsoever (Searle 1980a, 417).

eventually passes back more sheets through the output slot. Unknown to Clerk, the input consists of a story followed by questions about the story, and the output consists of responses to the questions (all in Chinese). Clerk in effect carries out (on paper) all the processing steps that a digital computing machine would carry out if given the same program and the same input.

The symbols processed by the program need not be Chinese characters, which merely form a vivid example. The symbols could be expressions of any natural or computer language, and the program that Clerk handworks could be any AI program whatsoever, present or future. Searle intends his argument to be completely general. The incoming symbols could even be digital representations of the world that are generated by the artificial sense organs of a humanoid robot and the program be the robot's (so-called) "perception and understanding" software.

To the programmers outside, the verbal behavior of the Chinese Room—that is, the system that includes the rulebooks, Clerk, Clerk's pencils, rubbers, and worksheets, the input and output provisions, and any clock, random-number generator, or other equipment that Clerk may need in order to execute the precise program in question—is indistinguishable from that of a native Chinese speaker. Indeed, the Room may successfully pass a Turing test conducted in Chinese. But does the Room *understand* Chinese?

Here is Searle's argument:

> [Clerk] do[es] not understand a word of the Chinese... [Clerk] ha[s] inputs and outputs that are indistinguishable from those of the native Chinese speaker, and [Clerk] can have any formal program you like, but [Clerk] still understand[s] nothing. For the same reasons, [the] computer understands nothing.

In sum:

> [W]hatever purely formal principles you put into the computer, they will not be sufficient for understanding, since a human will be able to follow the formal principles without understanding. (1980a, 418)

Responses to the CRA within AI have been diverse. For example, according to Ray Kurzweil, an influential technological futurist (see Section 5), Searle has a "biology-centric view of consciousness," a "bias that computers are inherently incapable of 'mental life,'" and "a basic lack of understanding of technology" (2002, 131, 164, 170). Others in AI are too ready to accept Searle's argument. For example, Pollack remarks:

> John Searle's "Chinese Room" argument is hateful because, in fact, he's correct. Neither the room nor the guy in it pushing symbols "understands" Chinese. But this isn't really a problem, because nobody actually "understands" Chinese! We only think we understand it. (2006, 51)

2.2. Why the CRA Does Not Work

The fundamental problem should be obvious: the Chinese Room argument is *not logically valid*. The proposition that the formal symbol-manipulation carried out by

Clerk does not enable *Clerk* to understand the Chinese story by no means entails the quite different proposition that the formal symbol manipulation carried out by Clerk does not enable the *Room* to understand the Chinese story. Searle's argument is no more valid than this one: *Clerk has no taxable assets in Japan* entails *The organization of which Clerk is a part has no taxable assets in Japan*. The CRA fails to show that the Room—that is, the complete computing machine—does not understand Chinese.

It is important to distinguish this, the *logical reply* to the argument (Copeland 1993a, 1993b, 2002a) from what Searle calls the *systems reply*. The systems reply is the following claim:

> While it is true that the individual person who is locked in the room does not understand the story, the fact is that he is merely part of a whole system and the system does understand the story. (Searle 1980a, 419)

As Searle points out, the systems reply is worthless, since it "simply begs the question by insisting without argument that the system must understand Chinese" (Searle 1980a, 419). The logical reply, on the other hand, is a point about entailment. The logical reply does not beg the question, since it involves no claim about the truth or falsity of the statement that the Room can understand Chinese.

2.3. The Part-Of Principle

In the course of his discussion of the systems reply, Searle formulates a fresh version of the CRA. He retells the story in such a way that the system is *part of Clerk*:

> My response to the systems theory is quite simple: Let the individual…
> memoriz[e] the rules in the ledger and the data banks of Chinese symbols, and
> [do] all the calculations in his head. The individual then incorporates the entire
> system.…We can even get rid of the room and suppose he works outdoors. All
> the same, he understands nothing of the Chinese, and a fortiori neither does the
> system…If he doesn't understand, then there is no way the system could under-
> stand, because the system is just a part of him. (1980a, 419)

This version of the argument turns on a thesis that we call Searle's "Part-Of" principle (Copeland 1993b, 175): *if Clerk does not understand the Chinese story, then no part of Clerk understands the Chinese story*. Searle does not explain why he thinks the Part-Of principle is true. Yet the principle is certainly not self-evident (a number of potential counterexamples are discussed in Copeland 2002a and Copeland and Proudfoot 2006). Might not a module within Clerk's brain be able to understand Chinese? Perhaps various modules in Clerk's brain perform other tasks (for example, producing solutions to sets of tensor equations) without Clerk him- or herself being able to carry out these tasks—and even if Clerk strenuously denies being able to do so. (Such an equation-solving module may be responsible for our ability to catch cricket balls and other moving objects (McLeod and Dienes 1993)).

One might seek to uphold the Part-Of principle by maintaining that, since a part of Clerk is solving tensor equations (understanding Chinese, etc.), *so is Clerk*. Clerk's denial cuts no ice, any more than the blindsighted subject's sincere denial that he can see the light spot shows that he cannot see it. However, this response is not available to Searle. It is a cornerstone of Searle's overall case that Clerk's sincere report "I don't speak a word of Chinese" (1980a, 418) suffices for the claim that Clerk does not understand the Chinese inputs (Copeland 1993a, 1993b, 2002a; Proudfoot 2002). This thesis, like the Part-Of principle itself, is left totally unsupported by Searle. In the absence of a successful defense of these two key assumptions, the revised version of the CRA stalls.

2.4. The Chinese Gym

According to Searle, connectionist AI, like traditional symbol-processing AI, succumbs to the CRA:

> Imagine that instead of a Chinese room, I have a Chinese gym: a hall containing many monolingual English-speaking men. These men would carry out the same operations as the nodes and synapses in a connectionist architecture…and the outcome would be the same as having one man manipulate symbols according to a rule book. No one in the gym speaks a word of Chinese…Yet with appropriate adjustments, the system could give the correct answers to Chinese questions. (1990, 22)

Once again the logical reply is sufficient to refute the argument. One could agree with Searle that participation in this simulation does not suffice to provide any of the participating human clerks with an understanding of Chinese. But there is no entailment from this proposition to the claim that the simulation *as a whole* cannot ("with appropriate adjustments") come to understand Chinese. The fallacy involved in moving from part to whole is even more glaring here than in the initial version of the argument.

There are additional difficulties for Searle. First, not every connectionist network can be simulated by a team of human clerks (see Section 3.5)—Searle simply assumes, incorrectly, that all connectionist architectures fall prey to the Chinese gym scenario (Searle 1990, 22). Second, the gymnasium version of the CRA contains an additional logical fallacy. The gymnasium version consists of two inferences:

(1) No individual clerk in the gym understands Chinese.
Therefore, (2) The simulation as a whole (call it *G*) does not understand Chinese.
Therefore, (3) The network being simulated (call it *N*) does not understand Chinese.

As already pointed out, the inference from (1) to (2) involves a part-whole fallacy. The inference from (2) to (3) involves a different fallacy, which we call the *simulation fallacy* (Copeland 2002a). To commit the simulation fallacy is to argue as follows:

> *x* is a simulation of *y*; it is not the case that *x* has property Φ ∴ it is not the case that *y* has property Φ.

In effect, Searle himself has drawn attention repeatedly to the simulation fallacy, in the course of emphasizing the difference between a computer simulation of a phenomenon (e.g., mentation) and the phenomenon itself.[15] Paraphrasing one of Searle's colorful illustrations:

> Barring miracles, you could not run your car by doing a computer simulation of the burning of gasoline, but no-one in his right mind would infer from this that you could not run your car by burning gasoline. (Compare Searle 1990, 23; 1989, 38.)

Notwithstanding Searle's own warnings, the gymnasium version of the CRA commits this very fallacy, by arguing as follows:

> G is a simulation of N; it is not the case that G understands Chinese ∴ it is not the case that N understands Chinese.

2.5. Wittgenstein and the CRA

Wittgenstein frequently employed the idea of a human being acting "like a reliable machine" (1965, 119). A "living reading-machine" (1972, § 157) is a human being or other creature who is given (as input) written signs, for example Chinese characters, arithmetical symbols, logical symbols, or musical notation, and who produces (as output) text spoken aloud, solutions to arithmetical problems, proofs of logical theorems, notes played on a piano, and suchlike. In Wittgenstein's view, an entity that manipulates symbols *genuinely* reads (calculates, and so on) only if he or she has a particular *history*, involving learning and training, and participates in a social *environment* that includes normative constraints and further uses of the symbols (1989b, 180; 1983, 425, 257; see Proudfoot 2004a, 2004b). Clerk, the inhabitant of Searle's Chinese room, is a *mere* reading machine, lacking the history and environment necessary for understanding (or meaning anything by) Chinese characters. In this respect Wittgenstein agreed with Searle's stance; he said, for example, "If a man makes Chinese noises we don't say he talks Chinese, unless he has done other things first and can do other things afterwards" (1989b, 55). However, Wittgenstein's arguments also generate objections to the CRA (see Proudfoot 2002).

First, to Searle's conviction that Clerk does not understand the Chinese symbols. In his original statement of the CRA, Searle himself is Clerk, and he says, "[I]t seems to me quite obvious in the example that I do not understand a word of the Chinese stories" (1980a, 418). But what is to justify this claim? Searle states that human brains create "awareness" or "mental life" (1980b, 454, 452), and also that "[o]nly a being that could have conscious intentional states could have intentional states at all" (1992, 132). Perhaps, then, what Searle finds "quite obvious" is that (in the Chinese room) he lacks the conscious phenomena correlated with speaking or

---

[15] Usually Searle's examples highlight what logicians would call a *contraposed* form of the fallacy: *x* is a simulation of *y*; *y* has property Φ; therefore *x* has property Φ.

reading Chinese. Using Wittgenstein's vocabulary, Searle (or Clerk) is *meaning-blind*—he lacks the "impressions" which "are found by experience to accompany sentences" (1988, § 175*ff*.; 1974, 45; see Proudfoot 2009). However, Wittgenstein used a reader-memorizer thought experiment to argue that such phenomena are neither necessary nor sufficient for meaning or understanding (1972, §§ 156–160). He argued that, instead, it is *use* that matters (1988, § 184). Searle offers no reason to think that Clerk is more than merely meaning-blind, or that meaning-blindness *is* an obstacle to understanding. Without a justification of the claim that Clerk does not understand the Chinese symbols, the CRA cannot get off the ground.

Wittgenstein's arguments tell, second, against Searle's account of the difference between Clerk (or the computer) and a genuine Chinese speaker. As we have seen, Searle claims that Clerk lacks awareness and mental life; the computer, he says, "attaches no meaning, interpretation, or content" to the symbols it manipulates (1982, 4; see also Searle 1987, 423). It is these, Searle claims, that the genuine Chinese speaker (and the human user of the computer) possess. In short, Searle appears to endorse the conception of the mind that Wittgenstein called "the 'accompanying' picture":

> We think of the way a Chinese sentence is a mere series of sounds for us, which just means that we don't understand it and we say this is because we don't have any thoughts in connection with the Chinese sentence (e.g. the Chinese word for "red" doesn't call up any image in us). (Wittgenstein 1989b, 286; see also 1974, 152)

Wittgenstein argued that this conception fails to explain meaning and understanding—for example, because it "simply duplicates language with something else of the same kind" (1974, 152; see Proudfoot 1997, 2009). In particular he attacked the hypothesis that a genuine speaker possesses an "interpretation" that is to determine (or constitute) the meaning of a symbol. For example, if a Chinese character were to "call up" an interpretation in Clerk, *how* would this enable Clerk to understand the symbol? If the "interpretation" is itself a symbol, then ex hypothesi it does not suffice for understanding. If it is not a symbol, *what is it?* Wittgenstein said, "The mental act seems to perform in a miraculous way what could not be performed by any act of manipulating symbols" (1965, 42). For the conception of the mind underlying the CRA to be persuasive, Searle owes us an account of the notions of "awareness" and "interpretation" that is not vulnerable to Wittgenstein's arguments.

Enough for now about the Chinese Room argument.

3. Hypercomputation and "Wide" Mechanism

In Section 2 we criticized an a priori argument against strong AI. In this section, we explain and criticize an a priori argument *for* strong AI. We call this argument the *apodeictic* ("it must be so") argument for human-level AI.

The philosophical assumption underlying strong human-level AI, as the field is traditionally conceived, is this: *everything about the human mind that matters to cognition can be duplicated in a Turing machine* (and in the real world can be duplicated in some fast, memory-rich, possibly robot-embodied computer that is logically equivalent to a Turing machine). The apodeictic argument attempts to justify this assumption, and runs as follows. Given that the mind is scientifically explicable rather than a mystery, the mind is a mechanism, an information-processing machine; since the set of possible operations that can be carried out by information-processing machines is identical to the set of operations that can be carried out by the universal Turing machine (or UTM, the most abstract and general form of information-processing machine), the mind must ultimately be explicable in terms of the computational properties of the UTM.[16] In this section we raise an objection to this argument, and outline a challenge to the philosophical assumption underlying (traditional) human-level AI.

3.1. Computers and Computers

In 1936, when Turing thought up the UTM, a *computer* was not a machine at all, but a human being, a mathematical assistant (Copeland 1997a, 2000; Copeland and Proudfoot 2010). The human computer calculated by rote, in accordance with some effective method supplied by an overseer prior to the calculation.[17] Turing used the term *computer* and its cognates in this sense in his 1936 paper, saying for example, "Computing is normally done by writing certain symbols on paper" (1936, 75) and "The behaviour of the computer at any moment is determined by the symbols which he is observing, and his 'state of mind' at that moment" (1936, 75). The Turing machine (or, as Turing called it, "computing machine") is an idealization of the human computer:

> We may compare a man in the process of computing a real number to a machine which is only capable of a finite number of conditions…The machine is supplied with a "tape". (Turing 1936, 59)

A *hypercomputer* is any information-processing machine (notional or real) that is able to achieve more than a human rote worker can in principle achieve (Copeland and Proudfoot 1999).[18] Hypercomputers compute—in a broad sense of "compute"—functions or numbers, or more generally solve problems or carry out information-processing tasks that lie beyond the reach of the universal Turing machine (see Copeland 2002–3)). When we introduced the term "hypercomputation" in 1999 we were naming an emerging field with a substantial history: speculation

---

[16] See, e.g., Boden (1988) at page 5 and page 259 ("If a psychological science is possible at all, it must be capable of being expressed in computational terms"), and the discussion on pages 177–79 of Boden (2006). There are many examples of this form of argument in the literature; several are discussed in Copeland (2000).

[17] The human computer is, in Wittgenstein's sense, a living reading machine—see Wittgenstein's 1939 discussion with Turing (Wittgenstein 1989a, 36–37).

[18] See also Copeland (1997b, 1998a, 1998b, 1998c); Copeland and Sylvan (1999).

that there may be physical processes—and so, potentially, machine operations—whose behavior falls beyond the scope of the UTM stretches back over several decades (a historical survey is given in Copeland 2002b).[19] The literature contains numerous descriptions of idealized hypercomputers. To mention only a few examples, a hypercomputer is in principle realizable by certain classical electrodynamical systems (Scarpellini 1963), the physical process of equilibration (Doyle 1982), an automaton traveling through relativistic space-time (Pitowsky 1990, Hogarth 1994, Shagrir and Pitowsky 2003), a quantum mechanical computer (Kieu 2002, 2003),[20] an interneural connection (Siegelmann and Sontag 1994, Siegelmann 2003), and a temporally evolving sequence of Turing machines, representing for example a learning mind (Copeland 2002b, 2004f).[21] It remains unknown, however, whether hypercomputation is permitted or excluded by real-world physics.

The hypothesis that the mind is a form of hypercomputer is yet to be fully explored. As philosopher and physicist Mario Bunge remarked, the traditional computational approach "involves a frightful impoverishment of psychology, by depriving it of nonrecursive [non Turing-machine computable] functions" (Bunge and Ardila 1987, 109). Might hypercomputational aspects of the mind be involved in (what Gödel[22] and Turing[23] called) "mathematical intuition" (see Penrose 1994)?[24] Might hypercomputation be involved in other aspects of cognition? The possibility that the mind is hypercomputational raises a challenge to the philosophical assumption that everything that matters to human cognition can be replicated in a Turing machine.

3.2. Narrow versus Wide Mechanism

In this section and the next we explain how the mere possibility of hypercomputation undercuts the apodeictic argument for human-level AI.

A *narrow mechanist* claims that the Turing machine provides a maximally general notion of mechanism; as Allen Newell puts it, the very concept of a determinate physical mechanism is *formalized* by the concept of the Turing machine (1980, 150). So, according to the narrow mechanist, the mind, being a machine, *must* be logically reducible to—or, as Hilary Putnam says, "perspicuously representable" as—a Turing

---

[19] See, e.g., Da Costa and Doria (1991); Doyle (1982); Geroch and Hartle (1986); Komar (1964); Kreisel (1967, 1974); Penrose (1989, 1994); Pour-El and Richards (1979, 1981); Scarpellini (1963); Stannett (1990, 2003); Vergis et al. (1986).

[20] See further Hagar and Korolev (2006).

[21] See also Bringsjord (2003); Syropoulos (2008).

[22] See, e.g., Gödel's remarks reported by Wang (1996, 184ff).

[23] Turing (1939); see also Copeland (2004b, 2004d, 2006, 2011).

[24] Penrose's notorious "Gödel argument," which we believe does not succeed, is criticized in Copeland (1998a).

machine.[25] A *wide mechanist*, on the other hand, maintains that some information-processing machines are *not* perspicuously representable as Turing machines (Copeland 2000). Thus the wide mechanist claims that the general concept of an (information-processing) mechanism *cannot* be formalized by the concept of the Turing machine.

This claim is demonstrated, the wide mechanist argues, by the existence of examples of notional hypercomputers, such as those mentioned in Section 3.1. These are mechanisms, in the relevant sense of "mechanism"—they are machines, even though the processes they carry out go beyond "mechanical" processes, in the technical sense of "mechanical" used in mathematical logic (see further Copeland 2000)—and by their nature cannot be perspicuously represented as Turing machines.

To give a simple illustration of how an information-processing machine can *fail* to be logically reducible to a Turing machine: certain analogue-type machines making use of real-valued quantities are not reducible to Turing machines. Essentially this is because not all real numbers are Turing-machine computable. An example of such a machine is the hypercomputational *extended Turing machine* (ETM), due to Fred Abramson (1971). The ETM cannot be logically reduced to (or perspicuously represented as) a standard Turing machine. This is because any arbitrary real number (e.g., a computable real such as π or an uncomputable real such as τ[26]) can be stored on a single square of the ETM's (notional) tape, whereas in the standard Turing machine each (non-blank) square stores only a single *discrete* symbol (e.g., "0" or "1"). The result is that the ETM is able to compute functions that are not computable by any standard Turing machine (see further Copeland and Sylvan 1999; Copeland 2002b).

The wide mechanist sees no reason to accept the common view that the claim that the mind is scientifically explicable amounts to little more than (or entails) the claim that the mind is reducible to a Turing machine. The wide mechanist maintains that the mind is a machine and therefore scientifically explicable, but argues that, since not all information-processing machines are logically reducible to Turing machines, it is an open question whether the mind is so reducible (see further Copeland 2004f).

3.3. Narrow Mechanism and the Foundations of AI

In discussions of the foundations of AI, the thesis of narrow mechanism appears to be an uncritical but firmly held assumption. The mind (and any other intelligent

---

[25] The term *perspicuous representation* is used by Putnam (1992, 6). He says, "it does not seem that there is any principled reason why we must be perspicuously representable as Turing machines, *even assuming the truth of materialism*...[o]r any reason why we must be representable in this way at all—even non-perspicuously—under the idealization that we live forever and have potentially infinite external memories" (1992, 7).

[26] τ is $0.h_1 h_2 h_3 \ldots$, where each h_i is the ith value (0 or 1) of the halting function.

system), if a machine, must be logically reducible to a Turing machine, and so a Turing-equivalent computer *must* be capable, in principle, of rivaling human intelligence. This is the apodeictic argument for human-level AI—and it is altogether too swift.

To focus on one example of this form of reasoning (by one of AI's most influential thinkers), Newell argues from his narrow mechanist claim cited above—that the concept of a determinate physical mechanism is formalized by the Turing-machine concept—to the conclusion that (what he calls) a physical symbol system "can become a generally intelligent system," that is, can become a system with "the same scope of intelligence as we see in human action" (Newell 1980, 170; Newell and Simon 1976, 116). A "physical symbol system" is a physical instantiation of a symbol-processing system closely related to LISP (a programming language used in AI). Newell argues as follows: a physical symbol system is in essence a UTM, and so (given the thesis of narrow mechanism) "contains the potential for being any other system, if so instructed"—and therefore contains the potential for becoming a generally intelligent system (1980, 170).

If successful, this a priori argument would be of much value, since it would demonstrate the Newell-Simon "Physical Symbol System Hypothesis," which holds that the processing of structures of symbols by a (Turing-equivalent) digital computer is sufficient to produce human-level intelligence (Newell and Simon 1976; Newell 1980). However, Newell offers no appropriate justification for the crucial narrow mechanist assumption upon which his argument rests.[27] If the wide mechanist is correct, it is false that the UTM contains the potential for becoming "any other system."

The apodeictic argument begs the central (empirical) question of whether every process involved in cognition is logically reducible to a Turing-machine process. All the hard work lies in settling this question.

3.4. Hypercomputation and "Pure" Simulation

As noted in Section 3.2, the extended Turing machine is not logically reducible to the Turing machine. Nevertheless, the ETM and other hypercomputers (whether analogue or digital) are in a certain sense *simulable* by the Turing machine—simulable in the sense that, given any *finite* number N of outputs produced by an ETM or another hypercomputer, some Turing machine can write out in order these N outputs.[28] This is a particular instance of the general fact that, no matter what finite sequence of discrete symbols one might select, a Turing machine can be designed to write out the selected sequence (by employing a technique equivalent to a look-up

---

[27] The justification that Newell does offer (which is a priori) is an example of what we call the "equivalence fallacy"; see Copeland (2000, 2003b).

[28] To any required (finite) degree of precision, if the outputs are nondiscrete.

table). Moreover, if those N outputs are all the outputs that the hypercomputer in question ever actually produces (because, say, it wears out), then the mechanism's *entire* output-behavior is simulated by the Turing machine. In short: any mechanism that, in the course of its lifetime, produces only a finite number of outputs *must* be simulable by the UTM.

An upholder of the apodeictic argument for human-level AI might turn to these considerations for support (since human beings certainly wear out and perish). The apodeictic argument can be recast as follows: it must be possible to *simulate* the human mind by a Turing machine, therefore *weak* human-level AI (at least) must be achievable in principle. This is the *simulation version* of the apodeictic argument. James Culbertson presents a form of the argument in which the simulation is effected by a finite neural network that is equivalent to a Turing machine with a finite tape. The simulation will have "behavioral properties just like John Jones or Henry Smith," Culbertson says, and will be able to "ingeniously solve problems, compose symphonies, create works of art and literature and engineering, and pursue any goals" (Culbertson 1956, 100, 114). Here, of course, Culbertson moves from weak to strong AI, first claiming that John Jones's behavior is simulated by the network (weak AI), and then claiming that the network itself ingeniously solves problems and so forth (strong AI).

Simulation in the sense at issue (i.e., the sense in which it is true that every mechanism producing only a finite number of outputs is simulable by Turing machine) involves no requirement that the simulating machine be able to calculate what the mechanism's N+1st output *would* be if (counterfactually) it were to produce one. Simulations that do not satisfy this requirement are "pure"—or "mere," or "surface-level"—simulations. A pure simulation has no explanatory value.[29] The simulating machine can use any convenient method, such a finite look-up table or a hash function, to produce the outputs (a blockhead is a "pure" simulation of a human conversationalist—see Section 1.4). Pure simulation tells us little or nothing about the nature of the mechanism being simulated, and so, if the goal is the explanation of mind, pure simulation falls far short.

Pure simulation does have a legitimate role in weak AI (why attempt to write a program that will serve as an explanatory model if a look-up table will do the job in hand?). However, the prospects for achieving weak human-level AI *in the actual world* by means of pure simulation are extraordinarily poor. A blockhead involves a look-up table that is, Block says, "too vast to exist" (Block 1995, 381). In an *imaginary* world, a look-up table that is too vast to exist in the actual world might enable a computer to simulate the behavior of a human being (and to pass

---

[29] Putnam has emphasized that although a Turing machine is able to simulate "*every* physical system whose behavior we want to know only up to some specified level of accuracy and whose 'lifetime' is finite," it is a quite different matter to claim that "such a simulation is in any sense a *perspicuous representation* of the behavior of the system" (1992, 6).

the Turing test), but the computer might require more storage registers than there are fundamental particles in the actual universe, and its clock speed might require signals to travel faster than the actual speed of light. Culbertson says that his neural network simulation of John Jones (itself essentially a look-up table) may "need more neurons than there are atoms in the whole universe" (Culbertson 1956, 101).

The problem with the simulation version of the apodeictic argument, then, is that it does *not* inform us that human-level AI is achievable (in principle) *in the actual world*. The simulation version of the argument is of little more value than an assurance that human-level AI is achievable in fairyland.

3.5. Hyper-Nets and the Chinese Gym

We return briefly to the Chinese Room argument. In extending the CRA to connectionism (Section 2.4 above), Searle assumes that a team of human clerks can (in principle) replicate the working of any connectionist network. But this assumption is false.

Paul Smolensky was one of the first to speculate that connectionist networks may challenge "the claim that the class of well-defined computations is exhausted by those of Turing machines" (1988, 3). Subsequently, Hava Siegelmann described a type of connectionist network (involving real-valued synaptic weights) that, like the ETM, exceeds the computational bounds of the Turing machine (Siegelmann and Sontag 1994; Siegelmann 2003). We call any neural network that cannot be simulated by a Turing machine (except in the pure sense) a *hyper-net*.

A team of clerks can always be replaced by a single clerk (as Searle himself allows (1990, 22)). Given this and the *Church-Turing thesis*, it follows that a hyper-net *cannot* be replicated by a party of clerks—even a party of clerks who labor forever. The Church-Turing thesis states that any mathematical task that can be carried out by a human clerk "working to fixed rules, and without understanding" (Turing 1945, 386) can equally be carried out by a Turing machine (for further discussion, see Copeland 1997a). If a party of clerks can replicate a hyper-net, then a single clerk can replicate a hyper-net, and so (by the Church-Turing thesis) a Turing machine can replicate a hyper-net. Since this is a contradiction, it is to be concluded that a party of clerks cannot replicate the working of a hyper-net.

Hyper-nets simply fall outside the scope of the gymnasium version of the CRA.[30] Searle's claim that the CRA is applicable to all "parallel machines"—systems having "many computational elements that operate in parallel and interact with one another" (1990, 22)—is mistaken.

---

[30] For additional discussion, focusing on the idea of an *accelerating* Turing machine, see Bringsjord, Bello, and Ferrucci (2001); Copeland (2002c); Copeland and Shagrir (2011).

3.6. Wide Mechanism and the Finiteness of the Universe

In this section we briefly consider an influential objection to wide mechanism. (The concept of hypercomputation is itself controversial and frequently misunderstood. We have responded to numerous objections to hypercomputation and wide mechanism in our previous publications and will not repeat those discussions here; see Copeland and Proudfoot 2004; section 3 of Copeland 2002b sets out and answers seventeen objections; see also Copeland, 2000, 2004f, 2006.[31]) The objection we will consider is based on the supposed *finiteness of the universe*.

The argument runs: since, in a finite and bounded universe, *any* set of outputs of *any* mechanism is necessarily finite in number, every mechanism can be simulated by a Turing machine, and therefore hypercomputation is ruled out. We discussed the issue of simulation in Section 3.4: any mechanism that produces only a finite set of outputs is simulable (in the pure sense) by a Turing machine, but this implies nothing about the nature of the mechanism. Machines that cannot be perspicuously represented as Turing machines (such as the ETM, for example) may contingently produce only a finite number of outputs. Pure simulation is one thing, explanation quite another. The argument from the finiteness of the universe cannot show that wide mechanism is false. Nor, indeed, does the guarantee, in a finite and bounded universe, of the existence of a pure Turing-machine simulation of any hypercomputational mechanism show that the hypercomputational mechanism is in effect *superfluous*. Running the hypercomputational mechanism might, after all, be the only feasible means of ascertaining the outputs in question.[32]

We know of no persuasive argument for the claim that the mind, if a machine, must be a Turing machine. The central issues are *empirical*, and open. Is the physical universe in any respect hypercomputational—and if so, is this respect relevant to the functioning of the mind? Since the answers to these questions are unknown, it is premature to insist that the mind can be explained in terms of the computational properties of the universal Turing machine. Similarly, a priori arguments that Turing-equivalent computers, if fast enough and supplied with enough memory, *must* be capable of human-level cognition are to be regarded with suspicion. Even if a positive answer to the metaphysical question "Can machines think?" is assumed, it remains an open, empirical question whether Turing-equivalent computers are capable of rivaling the intellectual capacities of human beings.

---

[31] One common line of objection stems from a well-known paper by Robin Gandy arguing that *whatever can be calculated by any discrete, deterministic, mechanical device is Turing-machine computable* (Gandy 1980); if Gandy is right, there can be no such thing as a discrete, deterministic, mechanical hypercomputer. Gandy's conclusion is challenged in Copeland and Shagrir 2007, where some hidden, and questionable, assumptions in Gandy's argument are exposed.

[32] See further Copeland (1997b, 2002b, 2004f).

4. "Moore's Law" and Human-Level AI

For several influential "technological futurists," the question *Can machines think?* is settled by the remarkable speed of computer innovation.

4.1. The Futurists

Recently futurists have predicted that intelligent computers will emerge in the near future. For example:

> How soon could…an intelligent robot be built? The coming advances in computing power seem to make it possible by 2030. (Joy 2000)
>   [T]oday's very biggest supercomputers are within a factor of a hundred of having the power to mimic a human mind. Their successors a decade hence will be more than powerful enough. (Moravec 1998)

Futurists claim that philosophical naysayers simply ignore advances in AI; Searle, according to Kurzweil, "clearly expects the twenty-first century to be much like the twentieth century, and considers any significant deviation[s] from present norms to be absurd on their face" (2002, 132).

  Since futurists regard the answer to the traditional philosophical question *Can machines think?* as settled (in the affirmative), they focus on other philosophical questions. These include: Will intelligent machines be conscious? Will they have freewill, or rights? How should they be treated? Futurists have influenced popular discussion, much of which centers on the question whether intelligent (and super-intelligent) machines will be predictable by humans, and if they are to be welcomed or feared.

4.2. "Moore's Law"

Futurists such as Bill Joy and Hans Moravec (who expects "fully intelligent machines in a few decades" (1998)) typically base their optimistic predictions directly on increases in *computing power*. "Moore's Law" has been widely cited as a guarantee of intelligent machines in the near future. But what did Gordon Moore actually say?

  In 1965 Moore foresaw "a proliferation of electronics," and in 1995 a "revolution in society" as a result of information technology (1965, 114; 1995, 17). These remarks say nothing about *intelligence*. His detailed projections were restricted to computational resources at the chip level. In 1965 he predicted (on the basis of cost analyses) that, at least until 1975, the number of components in an integrated circuit would roughly double each year (1965, 115). In 1975 he said that (after another five years) the rate of increase might approximate a doubling every two years (1975, 13). In 1995 he remarked that "[t]he definition of 'Moore's Law' has come to refer to almost anything related to the semiconductor industry that when plotted on semi-log paper approximates a straight line" (1995, 2).

There is a simple fallacy in any attempt to use Moore's Law to guarantee, or even (by itself) render more likely, the existence of intelligent computers. Even if the futurists' optimistic predictions about processing power can properly be derived from Moore's projections concerning the number of components in an integrated circuit (and even if we ignore the expectation that his growth curve will flatten out fairly soon), their predictions have *no* implications for either the (metaphysical) question "Can a machine think?" or the question "If machines can think, are (Turing-equivalent) computers the right sort of machines to think?". Thus the futurists' predictions about processing power do not (and cannot) demonstrate the *technological* possibility of AI. The same is true, of course, of any other forecast of artificial intelligence that is based solely on predicted increases in computing power.

Some futurists appeal to Moore's Law specifically to justify optimistic claims about increases in the *speed* of (artificial) thought. For example, Nick Bostrom (Director of the Future of Humanity Institute at the University of Oxford) claims:

> If Moore's law continues to hold...the speed of artificial intelligences will double at least every two years. Within fourteen years after human-level artificial intelligence is reached, there could be machines that think more than a hundred times more rapidly than humans do. (Bostrom 2003a, 763; see also Bostrom 2006).

Likewise, the Singularity Institute for Artificial Intelligence (which Kurzweil cofounded) claims that, on the basis of hypothesized exponential increases in chip speed, "[a]t the very least it should be physically possible to achieve a million-to-one speedup in thinking."[33] But again the appeal to Moore's (or other similar) projections is fallacious. These provide no reason to think that relative increases in *computer* speed will be matched by increases in speed of *thought* (let alone exactly matched, as Bostrom claims).

When, as here, engineering issues give way to metaphysical questions, futurists typically hand-wave toward materialism—in particular, computationalism. For example, Bostrom states:

> It would suffice for the generation of subjective experiences that the computational processes of a human brain are structurally replicated in suitably fine-grained detail...This attenuated version of substrate-independence is quite widely accepted. (2003b, 244)

So much for the "hard problem" of consciousness!

5. Singularitarianism and the Future of AI

Several futurists claim, not only that human-level and superhuman-level AI will soon be achieved, but also that *human beings* will become artificial intelligences—by means of "nanobot" implants and ultimately by becoming "software-based." Some

---

[33] http://singinst.org/overview/whatisthesingularity. (Accessed February 8, 2010).

futurists, perhaps most famously Ray Kurzweil, go further and predict a "singularity"—a dramatic technological change with unpredictable consequences.[34]

5.1. Kurzweil's Law

The "Law of Accelerating Returns" (or "Kurzweil's Law") holds that the rate of progress of any evolutionary process (along with the "returns" of this process) increases exponentially over time, and the rate of exponential growth itself grows exponentially (Kurzweil 2001). This, Kurzweil claims, applies to AI:

> Within a few decades, machine intelligence will surpass human intelligence, leading to The Singularity—technological change so rapid and profound it represents a rupture in the fabric of human history. The implications include the merger of biological and nonbiological intelligence, immortal software-based humans, and ultra-high levels of intelligence that expand outward in the universe at the speed of light. (2001)

The Singularity will arrive by "2045, give or take" (Kurzweil in Else 2009).

Kurzweil's specific predictions include:[35] reverse engineering all regions of the human brain (yielding "the software of intelligence"); humans "enhanced" through biotechnology and nanotechnology ("By the 2030s, we will be more non-biological than biological" (2007a, 19)); human-level AI (in 2001 Kurzweil predicted hardware matching the computational capacity of the brain by 2010, and software enabling a machine to pass the Turing test by 2029);[36] and superhuman-level AI.

[34] Modern Singularitarians claim a history of like-minded thought, frequently citing von Neumann, Good, or Vinge. In his 1958 obituary of von Neumann, the mathematician Stanislaw Ulam referred to a conversation (with von Neumann) centering on "the ever accelerating progress of technology and changes in the mode of human life, which gives the appearance of approaching some essential singularity in the history of the race beyond which human affairs, as we know them, could not continue" (1958, 5). In 1965, Jack Good—a leading codebreaker at Bletchley Park, working with Turing on both Enigma and Tunny (and programming Colossus, the world's first large-scale electronic digital computer), and with Max Newman at the Manchester Computing Laboratory, home of the first stored-program electronic computer—predicted that "ultraintelligent" machines would be built in the twentieth century (1965, 78). Once such machines existed, he said, "there would...unquestionably be an 'intelligence explosion'...Thus the first ultraintelligent machine is the last invention that man need ever make" (Good 1958, 33). In 1993, Vernor Vinge predicted a "singularity"—an "exponential runaway" of technology (Vinge 1993; see also his comments in Wolens 2009). Vinge predicts superhuman intelligent AI by 2030 (Vinge 1993).

[35] Except where noted, all claims and quotations in this (and the following) paragraph are from Kurzweil (2001).

[36] Kurzweil, it seems, believes that exponential improvements in *software* (in addition to exponential growth in computing power) *will* suffice to bring about human-level AI. However, many theorists have argued that human-level AI requires more, including: *embodiment* (on Brooks's "physical grounding hypothesis," see, e.g. Brooks 1999), *situatedness* (on Dreyfus's arguments for situated AI and "Heideggerian AI," see Dreyfus 1992, 2007; Dreyfus and Dreyfus, 1986), and *history*, which may include training (Dreyfus 1992; Hawkins 2008; on Wittgenstein's arguments, see Proudfoot 2004a, 2004b).

Kurzweil also predicts that we will be able to "download" the human brain: an individual's "personal mind file" will be "reinstantiated" in a nonbiological medium (see, e.g., Moravec 1988 and Goertzel 2007 for the same claim). The "newly emergent person" may choose life "out on the web" or in an artificial body—longevity, even immortality, will be ensured by proper data storage (technological enhancements to the brain will prevent boredom (Kurzweil 2004)). This "freeing of the human mind from its severe physical limitations of scope and duration [is] the necessary next step in evolution," Kurzweil claims.

5.2. Objections to Kurzweil's Singularitarianism

Kurzweil's Law predicts exponential growth in technological development, techno-logical innovation, and in fact in "every" aspect of information technology (Kurz-weil 2007b). Predictions of such great scope require strong evidence. However, the methodology of Kurzweil's extrapolations from past technological progress, in par-ticular his assumption that exponential growth will continue long enough to reach the "Singularity," has been roundly criticized (e.g., Ayres 2006; Hawkins 2008). His forecasting methods have been described as "para-science" and "naïve" (Modis 2006, 105–6); he has been accused of being "[i]nfatuated with statistics and seduced by the power of extrapolation" (Nordmann 2008), and of using "the argument from repetition" (Devezas 2006, 113). His claims about the history of computing (which are one of the bases for his forecasts) have also been criticized (Proudfoot 1999a, 1999b).

In addition, Kurzweil's claims about the nature of intelligence have been at-tacked (e.g., Hawkins 2008; McDermott 2006); likewise his claims about the human brain—with regard to computational power, "software," and the prospects of re-verse engineering (e.g., Horgan 2008; Zorpette 2008). Even optimists within AI dis-pute Kurzweil's time frame and method for achieving artificial intelligence (e.g., Hofstadter 2008); progress in building task-specific devices does not, it is pointed out, justify his forecast for human-level AI (McDermott 2006, 1231). Moore himself (see Section 4) says that the Singularity will never occur (2008).

Given these powerful criticisms, what explains the popularity and influence of "Singularitarianism"? Critics suggest that it is a quasi-religious response to the fear of death (Zorpette 2008; Coates 2006, 122)—the Singularity is "the rapture of the geeks" and Kurzweil's predictions "the creed of the Singularitarians" (Zorpette 2008; see also Modis 2006, 111; Horgan 2008, 41; Sharkey 2009, 28). Kurzweil's claims have been called "escapist, pseudoscientific fantasies" (Horgan 2008, 41) that promise "transhumanist bliss" (Jones 2008)—a way of "enjoying the benefits of religion without [religion's] ontological burdens" (Dembski 2002, 99).

Kurzweil insists that the critics who challenge his extrapolations mistakenly take an "intuitive linear view" of technological progress. In his view, it seems, evi-dence of trends in technological development that run counter to his predictions are always based on recent experience, which is too brief to exhibit the underlying exponential growth (Kurzweil 2001; see also 2006c, 10–14; 2008, 10). But this is merely to beg the question against his critics. By disposing summarily of counter-evidence, Kurzweil risks making Singularitarianism unfalsifiable—and indeed more akin to religious faith than science.

5.3. "Uploading the Mind"

The (Kurzweillian) Singularitarian's answer to the question *What makes a human being A at one time and an entity B (an uploaded file) at a future time the one person?* is, in effect, that *A*'s brain and *B* are "functionally equivalent."[37] This theory of persistence conditions for persons is fraught with familiar metaphysical difficulties. If personhood implies embodiment, or the capacity to have moral rights, or phenomenal consciousness, an electronic file may not even be a candidate for being a person. (On whether a person (at time *t1*) can be identical to an entity (at time *t2*) that is not a person, see, e.g., Olson 1997.) There are also powerful considerations in favor of the necessity of bodily continuity to personal identity over time (see, e.g., Williams 1973a, 1973c).

The theory faces, in particular, the notorious *duplication problem* (see Williams 1973b, 1973d). If *A* at *t1* and *B* at *t2* are one person, in virtue of the functional equivalence of *A*'s brain and *B*, then by the same token *A* at *t1* and *C* (another uploaded file, also functionally equivalent to *A*'s brain) at *t2* are also one person. By transitivity, *B* and *C* are one person—although in different locations at the same time and subject to different inputs (e.g., *B* in a robot body and *C* living "out on the web"). This consequence is prima facie absurd.

One response to this thought-experiment (paralleling a response to split-brain and "fission" thought-experiments[38]) is: allow *no more than one* instantiation (biological or nonbiological) of *A*'s "mind file" at any one time. However, this solution is, inappropriately, a merely practical response to a logical problem. (Even as such, it is problematic: given the prevalence of computer bugs and viruses, it is risky to run *B* only after *A* dies—what if the untested file fails to work?) In considering the thought-experiment, the only conclusions seem to be: either *A* does *not* survive death or *A* "survives as" *both* *B* and *C*.[39] However, the latter option does too severe damage to the concepts of personhood and survival for us to say that *A* persists.

Kurzweil does ask: if *B* is running while *A* is still alive, is one person in two places at once? His answer is *No*: if his own mind file is reinstantiated while he is alive, that person is *not* Ray Kurzweil (2001; see also 2006c, 383–84). However, he offers no justification for this answer. Since Kurzweil claims (strikingly like Descartes) that the body and brain are merely the "house" of the individual (2006b, 40),

---

[37] In places the "mind file" is solely the downloaded brain, in others Kurzweil adds the "nervous system, endocrine system, and other structures that our mind file comprises" (2006c, 325). He does not define "functional equivalence," but it is plainly a qualitative identity.

[38] On split-brain and fission thought experiments, see, e.g., Wiggins (1967) and Parfit (1987).

[39] On the notion of "survives as," see Parfit (1975).

it is unclear what justification he *could* have. And in any case, if A and B are not identical when A is alive, why are they identical when A is *dead*?[40]

Kurzweil underestimates the metaphysical difficulties for his vision of "immortal software-based humans" because he runs together two different hypotheses. He claims that reinstantiating A's mind file (and terminating A) is "essentially identical to" the gradual replacement of the components of A's biological brain with nonbiological equivalents (which process itself, he claims, is "not altogether different from" normal biological turnover) (2001; see also 2006c, 385). However, these scenarios are importantly different, and in the second bodily continuity is (at least, arguably) preserved. Contrary to Kurzweil's claims, Singularitarianism does *not* offer immortality.

In general Singularitarianism pays little attention to philosophical problems. Kurzweil allows that his vision of the future gives rise to "polite philosophical debates" (2001), but appears to believe that these debates are peripheral to the *science* (he contrasts a "philosophical" with a "scientific" position (Brooks et al. 2006)). However, it is plain that the question whether "uploading the mind" is possible—or whether machines can think or computationalism is true—is fundamental, not peripheral, to Singularitarianism. Even if the futurists could justify their forecasts of technological progress, they must also solve the philosophical problems that beset their speculations about the future of AI and humanity.

6. ARTIFICIAL INTELLIGENCE OR MAKE-BELIEVE INTELLIGENCE?

Modern technological futurists are not alone in making overly optimistic claims. Misplaced anthropomorphism and make-believe are common both inside and outside AI, leading to the impression that aspects of human-level AI have already been achieved.

Human beings readily treat robots and face agents in virtual worlds as animate objects with human characteristics; an illustrative example is an observer's spontaneously *apologizing* to Kismet (a robotic "head" with exaggerated "facial" features) for moving too close to it (Breazeal and Fitzpatrick 2000). Some researchers in AI deliberately exploit the tendency to anthropomorphize machines: they build anthropomorphic robots in order to facilitate human-computer interaction and to give their "autonomous" devices multiple opportunities to acquire new behaviors

---

[40] In *The Singularity Is Near*, Kurzweil concedes that a person A and (the electronic file functionally equivalent to A's brain) B are *not* identical when A is dead, just because they are not identical when A is alive (2006c, 384). This concession falsifies his theory of personal identity over time. However, Kurzweil holds on to his theory, saying merely "Despite these dilemmas my personal philosophy remains based on patternism—I am principally a pattern that persists in time" (2006c, 386).

(e.g., Breazeal and Fitzpatrick 2000). Anthropomorphic robots are also used to test models of human social and psychological development (e.g., Scassellati 2000, 1998; Mataric 1994). However, AI researchers also succumb to unwitting anthropomorphism and make-believe (see Drew McDermott 1976 on one sort of anthropomorphism in AI, "wishful mnemonics"; see also Proudfoot 2011 on the "forensic problem" of anthropomorphism).[41]

For example, Valentino Braitenberg described his (very simple) imaginary robot vehicles as possessing "egotism," "free will," and "the a priori concept of 2-dimensional space" (1984, 81, 68, 41). The engineers who built Braitenberg's vehicles out of modified LEGO bricks said that one (named "Frantic") is "a philosophical creature," which "does nothing but think" (Hogg, Martin, and Resnick 1991, 8). Masaki Yamamoto said that Sozzy, a robot vacuum cleaner, possesses "joy, desperation, fatigue, sadness" (1993, 213). Cog, Rodney Brooks's famous upper-torso humanoid robot, is to "delight in learning, abhor error, strive for novelty, recognize progress," and "exhibit a sense of humour" (Dennett 1994, 140–41). Kurzweil claims that nanobots will have "arms and hands" and "little fingers," and that twenty-first century computers will go to houses of worship for meditation and prayer, in order "to connect with their spiritual dimension" (1999, 139, 153).

Even when AI researchers aim to avoid anthropomorphism (for example, by explicitly denying that anthropomorphic robots such as Kismet have *genuine* thoughts or emotions), they may unwittingly anthropomorphize their machines by attributing *expressive behaviors* to the machines without qualification. Thus Kismet's creator says that the robot *smiles*, *grins*, and *frowns* (Breazeal 2001, 585; 2003, 147).

The tendency to anthropomorphize machines biases us to find minds in the oddest places. The examples above show that much anthropomorphism is simply misplaced: Frantic, the Lego-brick robot vehicle, is *not* "philosophical" and does not "think"—and Sozzy, the robot vacuum cleaner, is not sad. The traditional goal of AI is to build "*machines with minds*, in the full and literal sense" (Haugeland 1985, 2). If so, then the burden of proof rests on the AI theorist to demonstrate, of any candidate artificial intelligence, that optimistic claims about the device's cognitive abilities are not merely the product of misplaced anthropomorphism or make-believe. How is this to be demonstrated? AI must address the problem.

7. Conclusion

Notoriously, last century AI attracted much criticism for incautious attributions of psychological properties to computers. To pick just two examples of overly confident claims: in 1958 Simon and Newell said, "[T]here are now in the world machines

---

[41] On anthropomorphism in AI, see also Proudfoot (1999a, 2004a, 2004b). Some computer scientists do challenge the make-believe in AI; Sharkey, for example, refers to "the trickster role" of AI devices (2009, 29).

that think, that learn, and that create" (1958, 8), and in 1982 Roger Schank's company Cognitive Systems, Inc. proclaimed, "Our programs understand English…We give our computer programs the same kind of knowledge that people use, so our programs understand a sentence just the way a person does."[42] Our discussions of futurism, Singularitarianism, and anthropomorphism show that, alongside AI's numerous remarkable achievements, wild and unphilosophical optimism remains alive and well.

In our view, neither optimism nor pessimism can be justified a priori. In this chapter we have shown, for example, that Searle's Chinese Room argument, an a priori attack on strong AI, is ineffective; and we have criticized the attempt to argue a priori that universal computers *must* be capable of matching the human intellect. Even if it is assumed that the answer to the metaphysical question "Can machines think?" is *Yes*, the claim that a Turing-equivalent computer is capable of rivaling the intellectual capacities of humans remains an empirical hypothesis, and one whose truth or falsity is presently unknown. Only trying will tell.[43]

REFERENCES

Abramson, F. G. (1971). Effective computation over the real numbers. *Twelfth Annual Symposium on Switching and Automata Theory*. Northridge, CA: Institute of Electrical and Electronics Engineers.

Ayres, R. U. (2006). Review [of *The Singularity is Near*]. *Technological Forecasting & Social Change* 73: 95–100.

Block, N. (1981). Psychologism and behaviorism. *Philosophical Review* 90(1): 5–43.

———. (1995). The Mind as the Software of the Brain. In E.E. Smith and D.N. Osherson (eds.), *An Introduction to Cognitive Science*, vol. 3, 2nd ed. Cambridge, MA: MIT Press.

Boden, M. A. (1988). *Computer Models of Mind*. Cambridge: Cambridge University Press.

———. (2006). *Mind as Machine: A History of Cognitive Science*, vol. 1. Oxford: Oxford University Press.

Bostrom, N. (2003a). When machines outsmart humans. *Futures* 35: 759–64.

———. (2003b). Are we living in a computer simulation? *Philosophical Quarterly* 53(211): 243–55.

———. (2006). How long before superintelligence? *Linguistic and Philosophical Investigations* 5(1): 11–30.

Braitenberg, V. (1984). *Vehicles: Experiments in Synthetic Psychology*. Cambridge, MA: MIT Press.

Breazeal, C. (2001). Affective interaction between humans and robots. In J. Kelemen and P. Sosik (eds.), *ECAL 2001, LNAI 2159*. Berlin: Springer-Verlag.

---

[42] From an advertising leaflet distributed by Cognitive Systems, Inc., quoted in Winograd and Flores (1986, 128).

[43] This chapter draws on research supported in part by Royal Society of New Zealand Marsden grant no. UOC905.

————. (2003). Toward sociable robots. *Robotics and Autonomous Systems* 42: 167–175.

Breazeal, C., Fitzpatrick, P. (2000). That certain look: Social amplification of animate vision. *Proceedings of the AAAI Fall Symposium, Socially Intelligent Agents—The Human in the Loop*, http://people.csail.mit.edu/paulfitz/pub/AAAIFS00.pdf.

Bringsjord, S., Bello, P., and Ferrucci, D. (2001). Creativity, the Turing test, and the (better) Lovelace test. *Minds and Machines* 11: 3–27.

Bringsjord, S., and Zenzen, M. (2003) *Superminds: People Harness Hypercomputation, and More*. Dordrecht: Kluwer.

Brooks, R. A. (1996). Prospects for human level intelligence for humanoid robots. *Proceedings of the First International Symposium on Humanoid Robots (HURO-96)*, Tokyo, Japan, October 1996, http://people.csail.mit.edu/brooks/publications.html

————. (1999). *Cambrian Intelligence: The Early History of the New AI*. Cambridge, MA: MIT Press.

Brooks, R. A., Kurzweil, R., and Gelernter, D. (2006). Gelernter, Kurzweil debate machine consciousness. KurzweilAI.net, 6 December 2006, http://www.kurzweilai.net/articles/art0688.html?printable=1.

Brooks, R. A., and Stein, L. A. (1994). Building brains for bodies. *Autonomous Robots* 1: 7–25.

Brunette, E. S., Flemmer, R. C., and Flemmer, C. L. (2009). A review of artificial intelligence. *ICARA 2009—Proceedings of the Fourth International Conference on Autonomous Robots and Agents*. Wellington: IEEE.

Bunge, M., and Ardila, R. (1987). *Philosophy of Psychology*. New York: Springer-Verlag.

Cassimatis, N., Mueller, E. T., and Winston, P. H. (eds.) (2006). Achieving human-level intelligence through integrated systems and research: Introduction to this Special Issue. *AI Magazine* 27(2): 12–14.

Churchland, P. M. (1996). Learning and conceptual change: The view from the neurons. In A. Clark and P. Millican (eds.), *Connectionism, Concepts, and Folk Psychology: The Legacy of Alan Turing*, vol. II. Oxford: Oxford University Press.

Coates, J. F. (2006). Discussion [of *The Singularity is Near*]. *Technological Forecasting & Social Change* 73: 121–27.

Cohen, P. R. (2005). If not Turing's test, then what? *AI Magazine* 26(4): 61–67.

Copeland, B. J. (1993a). *Artificial Intelligence: A Philosophical Introduction*. Oxford: Blackwell.

————. (1993b). The curious case of the Chinese gym. *Synthese* 95: 173–86.

————. (1997a). The Church-Turing thesis. In E. Zalta (ed.), *Stanford Encyclopedia of Philosophy*, http://plato.stanford.edu (updated in 2006).

————. (1997b). The broad conception of computation. *American Behavioral Scientist* 40: 690–716.

————. (1998a). Turing's o-machines, Penrose, Searle, and the brain. *Analysis* 58: 128–38.

————. (1998b). Even Turing machines can compute uncomputable functions. In C. Calude, J. Casti, and M. Dinneen (eds.), *Unconventional Models of Computation*. London: Springer-Verlag.

————. (1998c). Super Turing-machines. *Complexity* 4: 30–32.

————. (1999). A Lecture and Two Radio Broadcasts on Machine Intelligence by Alan Turing. In K. Furukawa, D. Michie, and S. Muggleton (eds.), *Machine Intelligence 15*. Oxford: Oxford University Press.

————. (2000). Narrow versus wide mechanism. *Journal of Philosophy* 96: 5–32.

————. (2002a). The Chinese Room from a logical point of view. In J. Preston and M. Bishop (eds.), *Views into the Chinese Room: New Essays on Searle and Artificial Intelligence*. Oxford: Oxford University Press.

———. (2002b). Hypercomputation. In B. J. Copeland (ed.), (2002–3). *Hypercomputation*. Special issue of *Minds and Machines* 12(4): 461–502.

———. (2002c). Accelerating Turing machines. *Minds and Machines* 12: 281–301.

———. (ed.) (2002–3). *Hypercomputation*. Special issue of *Minds and Machines* 12(4), 13(1).

———. (2003a). The Turing test. In J. H. Moor (ed.), *The Turing Test: The Elusive Standard of Artificial Intelligence*. Dordrecht: Kluwer.

———. (2003b). Computation. In L. Floridi (ed.), *The Blackwell Guide to the Philosophy of Computing and Information*. Oxford: Blackwell.

———. (ed.) (2004a). *The Essential Turing*. Oxford: Oxford University Press.

———. (2004b). Introduction to Turing (1939). In B. J. Copeland (ed.), *The Essential Turing*. Oxford: Oxford University Press.

———. (2004c). Introduction to Turing (1950). In B. J. Copeland (ed.), *The Essential Turing*. Oxford: Oxford University Press.

———. (2004d). Introduction to Turing (c.1951). In B. J. Copeland (ed.), *The Essential Turing*. Oxford: Oxford University Press.

———. (2004e). Introduction to Turing et al. (1952). In B. J. Copeland (ed.), *The Essential Turing*. Oxford: Oxford University Press.

———. (2004f). Hypercomputation: Philosophical issues. *Theoretical Computer Science* 317: 251–67.

———. (ed.) (2005). *Alan Turing's Automatic Computing Engine: The Master Codebreaker's Struggle to Build the Modern Computer*. Oxford: Oxford University Press.

———. (2006). Turing's thesis. In A. Olszewski, J. Wolenski, and R. Janusz (eds.), *Church's Thesis after 70 Years*. Frankfurt: Ontos Verlag.

———. (2011). From the *Entscheidungsproblem* to the personal computer. In M. Baaz, C. Papadimitriou, D. S. Scott, H. Putnam, and C. Harper (eds). *Kurt Gödel and the Foundations of Mathematics*. Cambridge: Cambridge University Press.

Copeland, B. J., and Proudfoot, D. (1996). On Alan Turing's anticipation of connectionism. *Synthese* 108: 361–77.

———. (1999). Alan Turing's forgotten ideas in computer science. *Scientific American* 280(4): 99–103.

———. (2004). The computer, Artificial Intelligence, and the Turing test. In C. Teuscher (ed.), *Alan Turing: Life and Legacy of a Great Thinker*. Berlin: Springer Verlag.

———. (2005). Turing and the computer. In B. J. Copeland (ed.), *Alan Turing's Automatic Computing Engine: The Master Codebreaker's Struggle to Build the Modern Computer*. Oxford: Oxford University Press.

———. (2006). Artificial Intelligence: History, Foundations, and Philosophical Issues. In P. Thagard (ed.), *Handbook of the Philosophy of Science: Philosophy of Psychology and Cognitive Science*. Amsterdam: Elsevier.

———. (2008). Turing's test: A Philosophical and Historical Guide. In R. Epstein, G. Roberts, and G. Beber (eds.), *Parsing the Turing Test: Philosophical and Methodological Issues*. Berlin: Springer.

———. (2010). Deviant encodings and Turing's analysis of computability. *Studies in History and Philosophy of Science, Part A* 41(3): 247–52.

Copeland, B. J., and Shagrir, O. (2007). Physical computation: How general are Gandy's principles for mechanisms? *Minds and Machines* 17: 217–31.

———. (2011). Do accelerating Turing machines compute the uncomputable? *Minds and Machines: Special Issue on the Philosophy of Computer Science* 21: 221–39.

Copeland, B. J., and Sylvan, R. (1999). Beyond the universal Turing machine. *Australasian Journal of Philosophy* 77: 46–66.

Culbertson, J. T. (1956). Some uneconomical robots. In C. E. Shannon and J. McCarthy (eds.), *Automata Studies*. Princeton, NJ: Princeton University Press.

da Costa, N.C.A., and Doria, F.A. (1991). Classical physics and Penrose's thesis. *Foundations of Physics Letters* 4: 363–73.

Dembski, W. (2002). Kurzweil's impoverished spirituality. In J. W. Richards, (ed.), *Are We Spiritual Machines? Ray Kurzweil vs. the Critics of Strong A.I.* Seattle: Discovery Institute.

Dennett, D. C. (1994). The practical requirements for making a conscious robot. *Philosophical Transactions of the Royal Society of London* (Series A) 349: 133–46.

Devezas, T. C. (2006). Discussion [of *The Singularity Is Near*]. *Technological Forecasting & Social Change* 73: 112–21.

Doyle, J. (1982). What is Church's thesis? Laboratory for Computer Science, MIT. In B. J. Copeland (ed.), (2002–3). *Hypercomputation*. Special issue of *Minds and Machines* 12(4): 519–20.

Dreyfus, H. L. (1992). *What Computers Still Can't Do*. Cambridge, MA: MIT Press.

———. (2007). Why Heideggerian AI failed and how fixing it would require making it more Heideggerian. *Artificial Intelligence* 171: 1137–60.

Dreyfus, H. L., and Dreyfus, S. E. (1986). *Mind over Machine*. Oxford: Blackwell.

Duffy, B. R. (2003). Anthropomorphism and the social robot. *Robotics and Autonomous Systems* 42: 177–90.

Eberbach, E., Goldin, D., and Wegner, P. (2004). Turing's ideas and models. In C. Teuscher (ed.), *Alan Turing: Life and Legacy of a Great Thinker*. Berlin: Springer-Verlag.

Else, L. (2009). Ray Kurzweil: A singular view of the future. *New Scientist Opinion* 2707, May 6, 2009, http://new.scientist.com.

Feigenbaum, E. A. (2003). Some challenges and grand challenges for computational intelligence. *Journal of the ACM* 50(1): 32–40.

Ford, K., and Hayes, P. (1998). On conceptual wings: Rethinking the goals of artificial intelligence. *Scientific American Presents* 9(4): 78–83.

Franklin, S. P. (1995). *Artificial Minds*. Cambridge, MA: Bradford Books.

French, R. (1990). Subcognition and the limits of the Turing test. *Mind* 99: 53–65.

———. (2000). The Turing test: The first 50 years. *Trends in Cognitive Science* 4: 115–22.

Gandy, R. (1980). Church's thesis and principles for mechanisms. In J. Barwise, H. J. Keisler, and K. Kunen, (eds.), *The Kleene Symposium*. Amsterdam: North-Holland.

Genova, J. (1994). Turing's sexual guessing game. *Social Epistemology* 8(4): 313–26.

Geroch, R., and Hartle, J. B. (1986). Computability and physical theories. *Foundations of Physics* 16: 533–50.

Goertzel, B. (2007). Human-level artificial general intelligence and the possibility of a technological singularity. A reaction to Ray Kurzweil's *The Singularity is Near*, and McDermott's critique of Kurzweil. *Artificial Intelligence* 171: 1161–173.

Good, I. J. (1965). Speculations Concerning the First Ultraintelligent Machine. In F. L. Alt and M. Rubinoff, (eds.), *Advances in Computers*. New York: Academic Press.

Hagar A., and Korolev A. (2006). Quantum hypercomputability? *Minds and Machines* 16: 87–93.

Harnad, S. (2008). Running commentary on "Computing Machinery and Intelligence." In R. Epstein, G. Roberts, and G. Beber (eds.), *Passing the Turing Test: Philosophical and Methodological Issues in the Quest for the Thinking Computer*. Berlin: Springer.

Haugeland, J. (1985). *Artificial Intelligence: The Very Idea*. Cambridge, MA: MIT Press.

Hawkins, J. (2008). [interviewed in] Tech Luminaries Address Singularity, [and in] Expert View. *IEEE Spectrum*, June 2008, http://spectrum.ieee.org/computing/hardware/tech-luminaries-address-singularity.

Hayes, P., and Ford, K. (1995). Turing test considered harmful. *IJCAI-95 Proceedings of the Fourteenth International Joint Conference on Artificial Intelligence*, August 20–25, vol. 1. San Francisco: Morgan Kaufmann.

Hodges, A. (1992). *Alan Turing: The Enigma*. London: Vintage.

Hofstadter, D. (2008). [interviewed in] Tech Luminaries Address Singularity. *IEEE Spectrum*, June 2008, http://spectrum.ieee.org/computing/hardware/tech-luminaries-address-singularity.

Hogarth, M. L. (1994). Non-Turing computers and non-Turing computability. *Proceedings of the Biennial Meeting of the Philosophy of Science Association* 1: 126–38.

Hogg, D. W., Martin, F., and Resnick, M. (1991). Braitenberg creatures. Epistemology and Learning Memo #13, MIT Media Laboratory. Cambridge, MA.

Horgan, J. (2008). The consciousness conundrum. *IEEE Spectrum*, June: 36–41.

Jones, R. A. L. (2008). Rupturing the nanotech rapture. *IEEE Spectrum*, June: 64–67.

Johnson-Laird, P. (1987). How Could Consciousness Arise from the Computations of the Brain? In C. Blakemore and S. Greenfield (eds.), *Mindwaves*. Oxford: Basil Blackwell.

Joy, W. (2000). Why the future doesn't need us. *Wired Magazine* 8(4), http://www.wired.com/wired/archive/8.04/joy.pr.html.

Kieu, T. D. (2002). Quantum hypercomputation. In B. J. Copeland (ed.), (2002–3). *Hypercomputation*. Special Issue of *Minds and Machines* 12(4): 541–61.

———. (2003). Computing the noncomputable. *Contemporary Physics* 44: 51–77.

Komar, A. (1964). Undecidability of macroscopically distinguishable states in quantum field theory. *Physical Review* (2nd series) 133B: 542–44.

Kreisel, G. (1967). Mathematical Logic: What Has It Done for the Philosophy of Mathematics? In R. Schoenman (ed.), *Bertrand Russell: Philosopher of the Century*. London: George Allen and Unwin.

———. (1974). A notion of mechanistic theory. *Synthese* 29: 11–26.

Kurzweil, R. (1999). *The Age of Spiritual Machines: When Computers Exceed Human Intelligence*. New York: Viking Press.

———. (2001). The law of accelerating returns. KurzweilAI.net, March 7, 2001, http://www.kurzweilai.net/articles/art0134.html?printable=1.

———. (2002). Locked in His Chinese Room: Response to John Searle. In J. W. Richards (ed.), *Are We Spiritual Machines? Ray Kurzweil vs. the Critics of Strong A.I.* Seattle: Discovery Institute.

———. (2004). A dialogue on reincarnation. KurzweilAI.net, January 6, 2004, http://www.kurzweilai.net/articles/art0609.html?printable=1.

———. (2006a). Why we can be confident of Turing test capability within a quarter century. KurzweilAI.net, July 13, 2006, http://kurzweilai.net/meme/frame.html?main=/articles/art0683.html.

———. (2006b). Reinventing humanity: The future of human-machine intelligence. *The Futurist* 40(2): 39–46.

———. (2006c). *The Singularity Is Near: When Humans Transcend Biology*. New York: Penguin Books.

———. (2007a). Let's not go back to nature. *New Scientist* 2593: 19.

———. (2007b). Foreword to *The Intelligent Universe*. KurzweilAI.net, February 2, 2007, http://www.kurzweilai.net/articles/art0691.html?printable=1.

———. (2008). The Singularity: The last word. *IEEE Spectrum*, October: 10.

Lenat, D. B. (2008a). The voice of the turtle: Whatever happened to AI? *AI Magazine* 29(2): 11–22.

————. (2008b). Building a Machine Smart Enough to Pass the Turing Test: Could We, Should We, Will We? In R. Epstein, G. Roberts, and G. Beber (eds.), *Parsing the Turing Test: Philosophical and Methodological Issues in the Quest for the Thinking Computer*. Berlin: Springer.

Mataric, M. J. (1994). Learning to Behave Socially. In D. Cliff, P. Husbands, J-A. Meyer, and S. Wilson (eds), *From Animals to Animats: Proceedings of the Third International Conference on Similarities of Adaptive Behavior (SAB-94)*. Cambridge, MA: MIT Press.

McCarthy, J. (2007). From here to human-level AI. *Artificial Intelligence* 171: 1174–82.

————. (2010). Formalizing common sense knowledge in mathematical logic. *The Rutherford Journal: The New Zealand Journal for the History and Philosophy of Science and Technology* 3, http://rutherfordjournal.org.

McDermott, D. (1976). Artificial Intelligence Meets Natural Stupidity. *SIGART Newsletter* 57: 4–9.

————. (2006). Kurzweil's argument for the success of AI. *Artificial Intelligence* 170: 1227–33.

————. (2007). Level-headed. *Artificial Intelligence* 171: 1183–86.

McLeod, P., and Dienes, Z. (1993). Running to catch the ball. *Nature* 362: 23.

Michie, D. (1993). Turing's test and conscious thought. *Artificial Intelligence* 60: 1–22.

Minsky, M., Singh, P., and Sloman, A. (2004). The St. Thomas common sense symposium: designing architectures for human-level intelligence. *AI Magazine* 25(2): 113–24.

Modis, T. (2006). Discussion [of *The Singularity Is Near*]. *Technological Forecasting & Social Change* 73: 104–12.

Moor, J. H. (1976). An analysis of the Turing test. *Philosophical Studies* 30: 249–57.

————. (1987). Turing Test. In S. C. Shapiro (ed.), *Encyclopedia of Artificial Intelligence*, vol. 2. New York: Wiley.

————. (2003). The Status and Future of the Turing Test. In J. H. Moor (ed.), *The Turing Test: The Elusive Standard of Artificial Intelligence*. Dordrecht: Kluwer.

Moore, G. E. (1965). Cramming more components onto integrated circuits. *Electronics* 38(8): 114–17 (The article is reprinted in *Proceedings of the IEEE*, 86(1), 1998.)

————. (1975). Progress in digital integrated electronics. *Technical Digest*, IEEE International Electron Devices Meeting, 21 (1975): 11–13.

————. (1995). Lithography and the future of Moore's law. *Proceedings of SPIE—The International Society for Optical Engineering* 2438: 2–17.

————. (2008). [interviewed in] Tech Luminaries Address Singularity. *IEEE Spectrum*, June. http://spectrum.ieee.org/computing/hardware/tech-luminaries-address-singularity

Moravec, H. (1988). *Mind Children: The Future of Robot and Human Intelligence*. Cambridge, MA: Harvard University Press.

————. (1998). When will computer hardware match the human brain? *Journal of Evolution and Technology* 1, http://www.jetpress.org/volume1/moravec.htm.

Newell, A. (1980). Physical symbol systems. *Cognitive Science* 4: 135–83.

Newell, A., and Simon, H. A. (1976). Computer science as empirical inquiry: Symbols and search. *Communications of the Association for Computing Machinery* 19: 113–26.

Nilsson, N. J. (2005). Human-level artificial intelligence? Be serious! *AI Magazine* 26(4): 68–75.

Nordmann, A. (2008). Singular simplicity. *IEEE Spectrum*, June: 60–63.

Olson, E. T. (1997). *The Human Animal: Personal Identity without Psychology*. Oxford: Oxford University Press.

Parfit, D. (1975). Personal Identity. In J. Perry (ed.), *Personal Identity.* Berkeley and Los Angeles: University of California Press.

———. (1987). *Reasons and Persons.* Oxford: Oxford University Press.

Penrose, R. (1989). *The Emperor's New Mind: Concerning Computers, Minds, and the Laws of Physics.* Oxford: Oxford University Press.

———. (1994). *Shadows of the Mind: A Search for the Missing Science of Consciousness.* Oxford: Oxford University Press.

Pitowsky, I. (1990). The physical Church thesis and physical computational complexity. *Iyyun* 39: 81–99.

Pollack, J. B. (2006). Mindless intelligence. *IEEE Intelligent Systems* 21(3): 50–56.

Ponec, M. (2006). Visual reverse Turing tests: A false sense of security. *Proceedings of the 2006 IEEE Workshop on Information Assurance.* United States Military Academy, West Point, N.Y.

Pour-El, M. B., and Richards, I. (1979). A computable ordinary differential equation which possesses no computable solution. *Annals of Mathematical Logic* 17: 61–90.

Pour-El, M. B., and Richards, I. (1981). The wave equation with computable initial data such that its unique solution is not computable. *Advances in Mathematics* 39.

Proudfoot, D. (1997). On Wittgenstein on cognitive science. *Philosophy* 72: 189–217.

———. (1999a). How human can they get? *Science* 284(5415): 745.

———. (1999b). Facts about Artificial Intelligence. *Science,* 285(5429): 835.

———. (2002). Wittgenstein's Anticipation of the Chinese Room. In J. Preston and M. Bishop (eds.), *Views into the Chinese Room: New Essays on Searle and Artificial Intelligence.* Oxford: Oxford University Press.

———. (2004a). Robots and Rule-Following. In C. Teuscher (ed.), *Alan Turing: Life and Legacy of a Great Thinker.* Berlin: Springer-Verlag.

———. (2004b). The implications of an externalist theory of rule-following for robot cognition. *Minds and Machines* 14(3): 283–308.

———. (2005). A new interpretation of the Turing test. *The Rutherford Journal: The New Zealand Journal for the History and Philosophy of Science and Technology* 1, http://rutherfordjournal.org.

———. (2006). Review of James Moor (ed.), The Turing Test: The Elusive Standard of Artificial Intelligence. *Philosophical Psychology* 19(2): 261–65.

———. (2009). Meaning and mind: Wittgenstein's relevance for the "Does Language Shape Thought?" debate. *New Ideas in Psychology* 27: 163–83.

———. (2011). Anthropomorphism and AI: Turing's much misunderstood imitation game. *Artificial Intelligence* 175: 950–57.

———. (forthcoming). Rethinking Turing's test.

Putnam, H. (1992). *Renewing Philosophy.* Cambridge, MA: Harvard University Press.

Richards, J. W. (ed.) (2002). *Are We Spiritual Machines? Ray Kurzweil vs. the Critics of Strong A.I.* Seattle: Discovery Institute.

Rui, Y., and Liu, Z. (2004). ARTiFACIAL: Automated reverse Turing test using FACIAL features. *Multimedia Systems* 9(6): 493–502.

Scarpellini, B. (1963). Zwei unentscheitbare Probleme der Analysis. *Zeitschrift für mathematische Logik und Grundlagen der Mathematik* 9. Published in translation as "Two undecidable problems of analysis" (with a new commentary by Scarpellini). In B. J. Copeland (ed.), (2002–3). *Hypercomputation.* Special Issue of *Minds and Machines* 13(1): 49–77.

Scassellati, B. (1998). Imitation and Mechanisms of Shared Attention: A Developmental Structure for Building Social Skills. Presented at the Autonomous Agents 1988 workshop *Agents in Interaction: Acquiring Competence through Imitation,*

Minneapolis/St. Paul, May 1998, http://www.ai.mit.edu/projects/cog/Publications/scaz-imitation.pdf.

Scassellati, B. (2000). Investigating Models of Social Development Using a Humanoid Robot. In B. Webb and T. Consi (eds.) *Biorobotics*. Cambridge, MA: MIT Press.

Searle, J. R. (1980a). Minds, brains and programs. *Behavioral and Brain Sciences* 3: 417–24.

———. (1980b). Intrinsic intentionality: Reply to criticism of "Minds, brains and programs." *Behavioral and Brain Sciences* 3: 450–56.

———. (1982). The myth of the computer. *The New York Review of Books* 29(7) (April 29): 3–6.

———. (1987). Minds and Brains without Programs. In C. Blakemore and S. Greenfield (eds.), *Mindwaves*. Oxford: Basil Blackwell.

———. (1989). *Minds, Brains and Science*. London: Penguin.

———. (1990). Is the brain's mind a computer program? *Scientific American* 262(1): 20–25.

———. (1992). *The Rediscovery of the Mind*. Cambridge, MA: MIT Press.

———. (2002). Twenty-One Years in the Chinese Room. In J. Preston and M. Bishop (eds.), *Views into the Chinese Room: New Essays on Searle and Artificial Intelligence*. Oxford: Oxford University Press.

Shagrir, O., and Pitowsky, I. (2003). Physical hypercomputation and the Church-Turing thesis. In B. J. Copeland (ed.), (2002–3). *Hypercomputation*. Special Issue of *Minds and Machines* 13(1): 87–101.

Shannon, C. E., and McCarthy, J. (eds.) (1956). *Automata Studies*. Princeton, NJ: Princeton University Press.

Sharkey, N. (2009). [Interviewed by Nic Fleming] The revolution will not be roboticised. *New Scientist*, 203(2723): 28.

Shieber, S. (2007). The Turing test as interactive proof. *Noûs* 41(4): 686–713.

Siegelmann, H. T. (2003). Neural and super-Turing computing. In B. J. Copeland (ed.), (2002–3). *Hypercomputation*. Special Issue of *Minds and Machines* 13(1): 103–14.

Siegelmann, H. T., and Sontag, E. D. (1994). Analog computation via neural networks. *Theoretical Computer Science* 131: 331–60.

Simon, H. A., and Newell, A. (1958). Heuristic problem solving: The next advance in operations research. *Operations Research* 6: 1–10.

Sloman, A. (2008). The well-designed young mathematician. *Artificial Intelligence* 172: 2015–34.

Smolensky, P. (1988). On the proper treatment of connectionism. *Behavioral and Brain Sciences* 11: 1–23.

Stannett, M. (1990). X-machines and the halting problem: Building a super-Turing machine. *Formal Aspects of Computing* 2: 331–41.

———. (2003). Computation and hypercomputation. In B. J. Copeland (ed.), (2002–3). *Hypercomputation*. Special Issue of *Minds and Machines* 13(1): 115–53.

Sterrett, S. G. (2003). Turing's Two Tests for Intelligence. In J. H. Moor (ed.), *The Turing Test: The Elusive Standard of Artificial Intelligence*. Dordrecht: Kluwer.

Syropoulos, A. (2008). *Hypercomputation: Computing beyond the Church-Turing Barrier*. New York: Springer.

Traiger, S. (2003). Making the Right Identification in the Turing Test. In J. H. Moor (ed.), *The Turing Test: The Elusive Standard of Artificial Intelligence*. Dordrecht: Kluwer.

Turing, A. M. (1936). On computable numbers, with an application to the Entscheidungsproblem. In B. J. Copeland (ed.), *The Essential Turing*. Oxford: Oxford University Press.

———. (1939). Systems of logic based on ordinals. In B. J. Copeland (ed.), *The Essential Turing*. Oxford: Oxford University Press.

———. (1945). Proposed electronic calculator. In B. J. Copeland (ed.), *Alan Turing's Automatic Computing Engine: The Master Codebreaker's Struggle to Build the Modern Computer*. Oxford: Oxford University Press.

———. (1948). Intelligent machinery. In B. J. Copeland (ed.), *The Essential Turing*. Oxford: Oxford University Press.

———. (1950). Computing machinery and intelligence. In B. J. Copeland (ed.), *The Essential Turing*. Oxford: Oxford University Press.

———. (1951). Can digital computers think? In B. J. Copeland (ed.), *The Essential Turing*. Oxford: Oxford University Press.

———. (c.1951). Intelligent machinery, a heretical theory. In B. J. Copeland (ed.), *The Essential Turing*. Oxford: Oxford University Press.

———. (1953). Chess. In B. J. Copeland (ed.), *The Essential Turing*. Oxford: Oxford University Press.

Turing, A. M., Braithwaite, R., Jefferson, G., and Newman, M. (1952). Can automatic calculating machines be said to think? In B. J. Copeland (ed.), *The Essential Turing*. Oxford: Oxford University Press.

Ulam, S. (1958). John von Neumann, 1903–1957. *American Mathematical Society Bulletin* 64(3): 1–49.

Vergis, A., Steiglitz, K., and Dickinson, B. (1986). The complexity of analog computation. *Mathematics and Computers in Simulation* 28: 91–113.

Vinge, V. (1993). The coming technological singularity: How to survive in the post-human era. NASA Technical Reports Server, Accession No. N94-27359, http://ntrs.nasa.gov.

———. (2008). Signs of the singularity. *IEEE Spectrum*, June: 76–82.

Waltz, D. L. (2006). Evolution, sociobiology, and the future of artificial intelligence. *IEEE Intelligent Systems*, 21(3): 66–69.

Wang, H. (1996). *A Logical Journey: From Gödel to Philosophy*. Cambridge, MA: MIT Press.

Whitby, B. (1996). The Turing Test: AI's Biggest Blind Alley? In P. Millican and A. Clark (eds.), *Machines and Thought: The Legacy of Alan Turing*, vol. I. Oxford: Oxford University Press.

Wiggins, D. (1967). *Identity and Spatio-Temporal Continuity*. Oxford: Blackwell.

Williams, B. (1973a). Personal Identity and Individuation. In B. Williams, *Problems of the Self: Philosophical Papers 1956–1972*. Cambridge: Cambridge University Press.

———. (1973b). Bodily Continuity and Personal Identity. In B. Williams, *Problems of the Self: Philosophical Papers 1956–1972*. Cambridge: Cambridge University Press.

———. (1973c). The Self and the Future. In B. Williams, *Problems of the Self: Philosophical Papers 1956–1972*. Cambridge: Cambridge University Press.

———. (1973d). Are Persons Bodies? In B. Williams, *Problems of the Self: Philosophical Papers 1956–1972*. Cambridge: Cambridge University Press.

Winograd, T. (2007). Shifting viewpoints: Artificial intelligence and human-computer interaction. *Artificial Intelligence* 170: 1256–58.

Winograd, T. A., and Flores, F. (1986). *Understanding Computers and Cognition*. Norwood, NJ: Ablex.

Wittgenstein, L. (1965). *The Blue and Brown Books*. New York: Harper.

———. (1972). *Philosophical Investigations*, 2nd ed. Edited by G. E. M. Anscombe and R. Rhees; translated by G. E. M. Anscombe. Oxford: Basil Blackwell.

———. (1974). *Philosophical Grammar*. Edited by R. Rhees; translated by A. J. P. Kenny. Oxford: Basil Blackwell.

———. (1983). *Remarks on the Foundations of Mathematics*, rev. ed. Edited by G. H. von Wright, R. Rhees, and G. E. M. Anscombe; translated by G. E. M. Anscombe. Cambridge: MIT Press.

———. (1988). *Remarks on the Philosophy of Psychology*, vol. I. Edited by G. E. M. Anscombe and G.H. von Wright; translated by G. E. M. Anscombe. Chicago: University of Chicago Press.

———. (1989a). *Wittgenstein's Lectures on the Foundations of Mathematics, Cambridge 1939: From the Notes of R. G. Bosanquet, Norman Malcolm, Rush Rhees & Yorick Smythies*. Edited by C. Diamond. Chicago: University of Chicago Press.

———. (1989b). *Wittgenstein's Lectures on Philosophical Psychology 1946–47: Notes by P. T. Geach, K. J. Shah, and A. C. Jackson*. Edited by P. T. Geach. Chicago: University of Chicago Press.

Wolens, D. (2009). Singularity 101 with Vernor Vinge. *h+ magazine* Spring 2009, http://hplusmagazine.com/articles/ai/singularity-101-vernor-vinge.

Yamamoto, M. (1993). "SOZZY": A hormone-driven autonomous vacuum cleaner. *Proceedings of the International Society for Optical Engineering* 2058: 211–22.

Zadeh, L. A. (2008). Toward human level machine intelligence—is it achievable? The need for a paradigm shift. *IEEE Computational Intelligence Magazine* 3(3): 11–22.

Zorpette, G. (2008). Waiting for the Rapture. *IEEE Spectrum*, June: 32–35.

CHAPTER 8

EMOTIONS: HOW MANY ARE THERE?

JESSE J. PRINZ

THERE are estimated to be about two thousand emotion words in English, though most of these words are rarely used, some are synonyms, and many languages have much smaller vocabularies (Russell 1991). Still, the size of the emotion lexicon makes one wonder, how many emotions are there? Might there be two thousand? More? Is there any limit on the number? Or is the stock of emotions actually quite small? Perhaps we have a limited set of primitives that can be extended into a significant variety. But what are the primitives, and how do they get extended? These are the questions this chapter grapples with. Counting emotions is an interesting project in itself, but it is also a way to raise more general questions about the nature of emotions. First of all, we need to have some general handle on what emotions are before any tabulation can begin. Second, in surveying the emotions we may find evidence that supports certain theories of the emotions and counts against others. This chapter does not propose to give an actual estimate on the number of emotions, except to say that we start with a few and end up with a lot. That may sound unsurprising, but on some leading theories of emotion we start and end with many, and on others we start and end with few. So the vague answer to the how-many question will require some assessment of where those theories go wrong. And in presenting this answer, this chapter will defend a particular theory of the emotions called the somatic appraisal theory, which does a better job explaining the range of emotions than its competitors.

1. THEORIES OF EMOTION AND EMOTION CARDINALITY

To count emotions we need to know what we are counting—we need to know what emotions are. Unfortunately, there are almost as many theories of emotion as there are emotion theorists. If these were surveyed, we would never get to the task of counting. Fortunately, the competing theories of emotion can be grouped into several broad categories. A brief discussion of these will illustrate one of the factors that has made it so difficult to settle on how many emotions there are.

First consider cognitive theories. These come in many varieties, but the basic idea is that every emotion necessarily involves a cognitive state. For example, some psychologists endorse "appraisal theories" according to which an emotion is a state that follows from a (perhaps automatic and unconscious) assessment of how one is faring in relation to the environment (e.g., Frijda 1986; Lazarus 1991; Scherer 1993). Philosophers appeal to a variety of different cognitive states: Armon-Jones (1989) defines emotions as construals of one's situations, Solomon (1976) equates emotions with judgments, and Greenspan (1988) says they are thoughts. For psychologists, the cognitive states in question are usually regarded as antecedents of the emotions, whereas philosophers tend to define them as components. Either way, they are part of the identity conditions that can be used to distinguish one emotion from another. Sadness, for example is presumed to involve the idea of loss, and fear is presumed to involve the idea of danger. If you do not have the requisite cognitive state, you do not have the emotion.

On the face of it, cognitive theories may appear to proliferate the number of emotions beyond plausibility. It might appear that every thought or construal can qualify as an emotion, and that would be absurd. To avoid unconstrained emotion proliferation, cognitive theorists usually restrict the category to certain kinds of cognitive states. The psychologists' preferred term *appraisal* captures a common thread: an appraisal can be defined as a cognitive state that represents a relationship between an organism and some set of circumstances that bears on well-being. Thoughts about loss qualify as appraisals, because a loss is a circumstance in which something that an organism cares about has been removed or eliminated; thoughts about danger qualify as appraisals, because danger is a circumstance in which an organism is under threat of being harmed. It is helpful to restrict the class of emotions to the class of appraisals, but even this restriction may be too permissive. Intuitively, not every appraisal is an emotion. You can recognize that the cigarette you are smoking poses a threat to your health without being afraid of it. Cognitive theorists need to distinguish genuine emotions from the appraisals that we make dispassionate. To use a popular metaphor, cognitive theorists typically insist that an appraisal can qualify as an emotion if and only if is "hot." Cool recognition of danger and loss will not do. Theories differ on what hotness consists in. One possibility is that a hot judgment is one that triggers or is accompanied by or a state that motivates behavior, such as an action tendency or a felt change in physiological arousal.

Cognitive theories can vary on the question of how many emotions there are. There could be a relative small set of emotions, corresponding to a set of biologically prepared hot appraisals. Or it could turn out that the range of possible emotions is open-ended. That would require some process of acquiring new hot appraisals. Unfortunately, cognitive theorists tend not to say much about whether and how new emotions can be acquired. Presumably, there are three ways in which cognitive theorists could accommodate acquisition of new emotions. First, there might be a way to combine appraisals, so that, for example, one can construe an event as both dangerous and as a loss at the same time (imagine what a child experiences when her parents leave her at school for the first time). Second, there could be a process by which biologically prepared appraisals get modified or elaborated (for example, the emotion of xenophobia might emerge when the fear appraisal is combined with judgments about foreignness) (see Oatley and Johnson-Laird 1997). Third, there might be a mechanism for generating entirely new hot appraisals—that is, a mechanism that can transform cool appraisals into hot appraisals (perhaps the Japanese emotion *amae* emerges when people are enculturated to have strong feelings associated with the otherwise cool appraisal that one has a dependency relationship on another person or institution). Admittedly, this third process is a little dubious, because *ex nihilo, nihil fit*. It is not clear how an appraisal could become hot if it does not build on a previously existing hot appraisal. Even culturally distinctive emotions, such as *amae*, may derive from a common core, such as the species-typical emotion known as "attachment" that binds children to their parents.

The "hot" metaphor is a double-edged sword. It helps the defender of cognitive theories escape the unwanted consequence that every appraisal judgment is an emotion, but it renders the approach vulnerable to a fatal objection. Once it is admitted that appraisals are not sufficient for emotions, it is easy to see that they are also not necessary. The added element of hotness that is supposed to convert a dispassionate appraisal into an emotion can qualify as an emotion in its own right. Take the judgment that smoking is dangerous, which does not qualify as an emotion in and of itself. Then imagine that we add something to it that makes it hot—an anxious feeling or avoidance tendency, for example. Now imagine that we had this hot component on its own: we feel anxious or initiate an avoidance response. That hot component is a good candidate for being an emotion on its own. Suppose you never judge cigarettes to be dangerous, but you have an anxious feeling in their presence: your heart races, and you exhibit avoidance behavior, such as reeling away when you see someone lighting up. You may insist that cigarette smoking is safe, but your phenomenology and behavior suggest that you are actually afraid of cigarettes. You would report the anxious feeling as fear, and others would describe your behavior as a sign of fear.

This theoretical point can be bolstered empirically. There is good evidence that we can have states that are recognizably emotional in the absence of judgments. For example, states that feel and function like emotions can be caused by spinal cord pathologies, by chemicals that act on the autonomic nervous system, and by making certain facial expressions (see Zajonc 1984 for a review). There is also evidence that a state that functions and feels like an emotion can be initiated via direct pathways from perceptual systems of the brain to ancient brain nuclei that orchestrated autonomic and behavioral responses (LeDoux 1996). Thus, cognitive states do not seem to be necessary for emotions (for more argument, see Prinz 2004a).

In terms of the present discussion, the point can also be put as follows: If we are interested in counting the emotions, we might think the trick is to count them by the accompanying thoughts. But this method of counting will not do. First, there may be emotions with no accompanying thoughts (for reasons just suggested), and, in addition, the thoughts might prove misleading. Suppose you form the thought that cigarettes are dangerous while feeling anxious on some occasions, and while feeling exhilarated on others. If the thought is what counts, both states qualify as what one might call cigarette-fear. But that does not seem plausible. On occasions when you have a positive feeling of exhilaration adjoining the thought that cigarettes are dangerous, it would be better to say that you are feeling fearless. For thrill-seekers, danger can be a source of pleasure. So it seems a mistake to count emotions by cognitions.

Let us turn, then, away from cognitive theories of emotion, to noncognitive theories. Here there are several options. Consider first the so-called circumplex theory (Russell 1980). In this approach, emotions are states that can be characterized along two dimensions: valence and arousal. The state we call fear is a negative state with high arousal, and sadness is a negative state with low arousal. Emotions such as contentment are low arousal and positive, and emotions such as jubilation are high arousal and positive. Some circumplex models suggest that positive and negative valence are independent dimensions, so that emotion space must actually have three dimensions (Watson and Tellegen 1985). Pleasant emotions are defined, in this approach, as high in positive valence and low in negative valence, and conversely for unpleasant emotions. An emotion such as relief might be low in negative valence, but not high in positive valence, so it would not be especially pleasant. Boredom might be low is positive valence but not especially negative.

Defenders of the cimcumplex model tend to construe the dimensions as continuous. That raises interesting questions for emotion counting. If each emotion is a point in a continuous space, then it would seem to follow that there are infinitely many emotions. It would also seem to follow that there are no categorical boundaries between emotions as there is no principled way to mark where one emotion begins and another ends. Both of these implications reveal how much the circumplex approach departs from folk psychology, especially when considered together. Folk psychology divides emotions into discrete categories; we have words for individual emotions, and we do not assume that every difference in arousal levels corresponds to a different emotion. In response, defenders of the circumplex model might propose that emotion space is not equally well traversed throughout; emotional states may tend cluster together in subregions of that space, and our emotion labels may correspond to these clusters. In this approach, there is a sense in which each point in emotion space is a unique emotion, but adjacent clusters of points are phenomenologically and functionally similar enough to be treated as alike. Emotion categories emerge as attractors in arousal/valence space.

This move will not do justice to folk psychology, however, because there are emotions that are similar in valence and arousal but distinct according to folk psychology. Jealousy and anger, for example, may both be unpleasant and high in arousal. Indeed, intense fear, revulsion, and despair might fall into this region of arousal/valence space as well. Two or three simple dimensions cannot do justice to the fact that emotions seem to have very distinct identities. Jealousy and anger do not differ on a continuous dimension; they differ in that one involves romantic infidelity and the other involves insults. Emotions also seem to be associated with distinctive behavioral and physiological response. Fear, revulsion, and despair are all high in arousal, but they are differentiated by fight behavior, nausea, and tears. We tend to take such differences into consideration when counting emotion categories. We treat identical points in arousal/valence space as distinct emotions, and we do not do so arbitrarily: differences in functional role and physiology are implicated. We also treat different points in arousal/valence space as alike. For example, mild and intense disgust differ in arousal, but they are easy to group together because their underlying physiology (wrinkled nose and intestinal rejection) are recognizably similar, and both arise in similar contexts (e.g., decay, pollution, bodily fluids, deformation, and animality). The total physiological profile and elicitation conditions of an emotion matter more than any simple dimension, such as arousal. Circumplex theory cannot ignore these factors without departing from folk classification. If the theory departs from folk classification, then it can be accused of missing its target. The circumplex model is not a good account of what we mean by "emotions"; at best, such models capture some of the dimensions that are constitutive of emotions while leaving others out. If we rely on just arousal and valence, we will undercount and overcount: we will treat different points in arousal/valence space as distinct when they should be co-classified, and we will erroneously assume that the same point always corresponds to the same emotion.

What we need is a theory of emotions that is noncognitive but also richer in resources than circumplex theories. In the remainder of this section, we will consider two possibilities. The first is that approach associated with Paul Ekman. According to Ekman (1994), all emotions have the following features:

1. Automatic appraisals
2. Distinct elicitors
3. Presence in other primates
4. Quick onset
5. Brief duration
6. Unbidden occurrence
7. Distinctive physiology

The first item on Ekman's list may give the impression that he is defending a cognitive theory, but that would be misleading. Ekman distinguishes automatic appraisals from what he calls "extended appraisals," which are slow, conscious, and often deliberative. He does not say much about what goes into an automatic appraisal, but his characterization suggests that it is often little more that a perceptual representation of the input that elicits an emotion. Each emotion has a characteristic set of elicitors, and an appraisal is a process by which we recognize that something falls into this set. A crawling insect, loud sudden noise, or loss of support might be among the fear elicitors, and the automatic appraisal be regarded as the process by which we recognize that something belongs to this class; once that recognition takes place, patterns of physiological, expressive, and behavioral changes also take place. Some of these changes, along with the recognition mechanisms that trigger them are constitutive of an emotion in Ekman's view.

Griffiths (1997) calls Ekman's theory the "affect program" approach, because each emotion can be regarded as a mini-program in the mind, with distinctive inputs, outputs, and intervening computations. In this chapter, we will refer to it as the "Darwinean modules" approach to draw attention to two important features. First, Ekman's characterization of emotions implies that he regards each emotion as a module, in Fodor's (1983) sense of the term. According to Fodor, modules are characterized by the following features:

1. Informationally encapsulated
2. Domain specific
3. Innate
4. Operate on shallow inputs
5. Fast
6. Automatic
7. Distinctive neurophysiology

The features have been numbered this way to draw attention to parallels between Ekman's account and Fodor's. Automatic appraisals are, in Ekman's view, isolated from information in central cognition. Distinctive elicitors entail that each emotion is domain specific (fear deals with dangers, and sadness with losses). The fact that emotions are found in other primates suggests that they are innate. The fact that they are quick and brief is consistent with the idea that they are fast and operate on shallow (relatively unprocessed) inputs. The fact that they are unbidden implies that they are automatic. And the fact that they have distinctive physiology in consistent with the assumption that they may be dissociable in the central nervous system. So, as Griffiths points out, Ekman has a modular theory of the emotions. It is also a Darwinian theory, in that he thinks emotions are products of evolution. He built his career on providing evidence for universal facial expressions of emotion, and, as we will see, he has downplayed the cultural contributions to emotions.

To determine the number of emotions, Ekman thinks we need to determine the number of Darwinian modules that exist. His early research focused on six emotions that seem to have widely recognized facial expressions, but he has expanded the list in recent years to include about fifteen: amusement, anger, contempt, contentment, disgust, embarrassment, excitement, fear, guilt, pride in achievement, relief, sadness/distress, satisfaction, sensory pleasure, and shame (Ekman 1999).

Ekman calls these "basic emotions," by which he means they are adaptations to fundamental life tasks, but he does not think that, in addition to these, there are some nonbasic emotions. Ekman's basic emotions are not building blocks from which other emotions are derived; he thinks these are the only emotions. The list just presented comprises his considered speculations about the sum total of existing emotions.

This may sound a bit odd on the face of it given the large number of emotion words we have in English and the active research program that exists on cultural variation in emotions. But Ekman has a story to tell about why there seem to be more than fifteen emotions. He says that in addition to the components inside a Darwinean emotion module, there are other factors that can influence how we count the emotions. Ekman mentions three ways in which learning can influence our emotions. First, we can learn different strategies for coping with our emotions and different attitudes toward them. One example is what Ekman and Friesen (1969) call display rules: we learn when it is appropriate to express certain emotions, and that can lead us to suppress expressions under certain circumstances. Second, we can expand on the set of things that elicit our emotions by learning that certain things are, say, dangerous or threatening (Ekman 2003, ch. 1). Presumably the expansions are outside the Darwiniean modules, and therefore, they do not lead to the acquisition of new emotions, but only what Ekman calls variations on a theme. Third, we may come to interpret previously existing emotions in new ways. Earlier, we mentioned Ekman's distinction between automatic appraisals, which are components of emotions, and extended appraisals, which are not. Extended appraisals are slow cognitive processes that analyze the meaning of a current affective state. Once again, this does not lead to new emotions, but only to new interpretations of old emotions. Presumably, the new interpretations sometimes lead to new words. For example, when we interpret anger in moralistic terms, we may call it indignation. This is not a different emotion from anger, but rather a reinterpretation of the basic biological state. Ekman (1999) sometimes invites us to think of anger as an emotion family, which implies that there are several different kinds of anger, each of which is a distinct emotion. But this terminology is a little misleading given his Darwinean module view. If the differences between two instances of anger depend on processes outside the module that constitutes anger, then two instances are not different species of the same emotion genus, but rather the same species with different, contextually appropriate labels. Elaborated appraisals, variations on a theme, and display rules have in impact on our emotions, but none provide us with a means of acquiring something genuinely new.

Ekman's account is not as minimalist as the circumplex approach: he believes in discrete emotions, and each of these has distinctive physiology and a distinctive class of elicitors. But like the circumplex approach, Ekman's approach is parsimonious: he thinks there are just over a dozen emotions. If he is right, there is a nicely circumscribed range of phenomena for emotion researchers to investigate. Because emotions are modules, they can be studied without studying central cognition or complex, symbolically mediated forms of social and cultural learning. More accurately, there may be interesting ways in which emotions influence cognition and culture, but the influence of cognition and culture on the emotions is very limited, if Ekman is right. This makes emotions easy to count and comparatively easy to study.

Before assessing the Darwinian module approach, let us introduce one final theory for comparison. It is the somatic appraisal theory defended elsewhere (Prinz 2004a). The theory builds on William James's (1884) account according to which emotions are perceptions of patterned changes in the body (a theory brought back into vogue by Damasio 1994). When we perceive or reflect on certain objects and events, our bodies respond in systematic ways: some events make the body prepare for flight, some for fighting, some cause tears, and so on. One might think that these changes are the *results* of emotions: the loss of a loved one made you sad, so you cried. But James argues that this is backward, as the emotion is not the cause of the bodily change, but the perception thereof. Sadness is the perception of tearing eyes, a closing throat, a limp posture, a wrenching gut, and other physiological changes that tend to occur under conditions of loss. The details may differ, but the overall global pattern is what we experience as sadness. In the years since James's seminal discussion, a number of authors have defended his theory, citing evidence that the neural correlates of emotion are also involved in bodily perception (Damasio et al. 2000), changes in bodily states affect our emotions (Strack et al. 1988), and deficits in bodily perception diminish emotions and emotional intensity (Hohmann 1966; though see also Chwalisz et al. 1988).

As it stands, however, James's theory suffers from a fatal flaw. It does not capture the idea that emotions are meaningful. If an emotion were nothing but a perception of bodily change, then it would not make sense to say an emotion is about anything. And, if emotions have no meaning or content, then it becomes impossible to distinguish them from other perceptions of changes in the body, such as twitches, tingles, muscle aches, perspiration, and so on. In terms of the present topic, James's theory risks proliferating emotions beyond comprehension, so that every perceived bodily change corresponds to an emotional state.

The somatic appraisal theory is designed to compensate for this flaw in James's formulation. Emotions are perceptions of changes in the body, as James suggested, but they also carry information about circumstances that bear on well-being. Suppose that a loss of a loved one causes you to cry and enter a bodily state that you would experience as a feeling of sadness. The perception of that bodily state registers as the relevant bodily changes, but it also does something else: it carries the information that there has been a loss. By "carries the information," the term is used in the kind of way that has become familiar in recent naturalistic semantic theories (Dretske 1981; Fodor 1990). This particular bodily change and the experience thereof are under the law-like control of losses. Losses cause these changes to occur, and when they do, there is a high probability that there has been a loss. Moreover, the bodily changes that occur and the perception thereof *have the function* of carrying information about loss. They were set up (by evolution) as responses to loss. The downtrodden state of the body is a form of disengagement and withdrawal from the world that allows an organism to regroup and build up new resources before facing the world in the absence of an object that it once relied on. Tears may signal distress to conspecifics and the need for help. So, when an organism enters this bodily pattern, it both carries information that a loss has occurred and it has the function of doing so. On popular semantic theories, these two elements (carrying information and having a teleologically function) are regarded as individually or jointly sufficient for having meaning (see, e.g., Dretske 1988). Thus, our perception of tears and other bodily symptoms of sadness signify loss: the perception registers the bodily changes, but represents the loss. Somatic perceptions have semantic content.

If this view is right, then the meaning of a bodily perception is very much like the meaning of an appraisal judgment. Both represent a relationship between an organism and its circumstances that bear on well-being. The difference is that appraisal judgments are cognitive states, and as such they require possession of concepts. If cognitive theories are right, we need a concept of loss in order to be sad. On the neo-Jamesian view that is being presented, these concepts are unnecessary. One can become a detector of losses without a concept of loss. For example, imagine the visual experience that an infant has when a caregiver moves out of sight. That visual experience reliably detects the loss (or disappearance) of a loved one, even though it does not contain the concept loss. The infant's mind may contain a biologically prepared link between such visual experiences and the somatic pattern associated with sadness. The disappearance of a caregiver may cause tears to flow, and the perception of those tears represents loss. Thus, where cognitive theories say that appraisals are judgments, the view being presented says that the perception of the bodily state is semantically akin to an appraisal: it is a somatic appraisal. As on a cognitive appraisal theory, somatic appraisals can be individuated by their meaning. Fear and sadness are different, in part, because they represent different circumstances. But they also differ in how they feel, because they are constituted by perceptions of different bodily states. So emotions are individuated by both their somatic and their semantic content.

The somatic appraisal theory faces an immediate question: if emotions do not contain judgments, how do they come to be reliably caused by things such as losses, dangers, and threats? The answer to this question was already hinted at in the example of the infant crying. For a small class of emotions, there may be a biologically prepared set of perceptual elicitors. The sadness response is elicited by perceptual experiences of caregivers leaving; the fear response is elicited by perceptions of loud noises, darkness, sudden loss of support, and other experiences that correspond to dangers; the anger response is elicited when a baby's limbs are aggressively constrained, when food is forced in the mouth, and so on. This list is not supposed to be complete or accurate as it is only meant to illustrate the main idea that our initial state (at birth or through maturation early in life) equips us with a few bodily patterns and perceptual elicitors that trigger those patterns. One can think of the elicitors for each pattern as comprising a mental file (called a triggering file). The items in the triggering file determine the conditions under which the emotion arises, and that helps to ensure that emotions have meaning; they are not arbitrary twinges.

So far, the somatic appraisal theory has much in common with Ekman's Darwinean modules. Ekman is not a Jamesian, but he does say that each emotion is associated with a physiological pattern. Ekman says these patterns are evolved adaptations, and that is also true in the somatic appraisal theory. He also says that emotions exploit automatic appraisals, which, in the interpretation above, are a lot like triggering files. Ekman says these appraisals are components of emotions, while the somatic appraisal theory says they are causes, rather than components, but the difference is not especially important for present purposes. Within this context, the crucial question is whether the accounts differ in a way that would lead to a different assessment of how many emotions there are. So far, we have said nothing that would suggest any difference on that question. We have said we have a small number of biologically prepared somatic appraisals, but have not said anything about whether that set can expand.

As it happens, however, the set of emotions can expand. The somatic appraisal theory is compatible with three ways of acquiring new emotions. Emotions are individuated by their semantic content and their somatic profile (the pattern of bodily changes the perception of which constitutes the emotion). A change in semantic content could lead to the creation of a new emotion, and the introduction of new bodily patterns could as well. In addition, two existing emotions could be mixed together to create a third, with a compound semantic content and a hybrid bodily pattern. In principle, Ekman's account could be modified to allow changes of all three kinds. The next section will describe each kind of change in more detail, and will introduce examples to suggest that such changes do in fact occur and should be regarded as cases of learning new emotions.

2. Acquiring New Emotions

As just indicated, the somatic appraisal theory, which the author endorses, offers three possible ways of acquiring new emotions: new semantic content, new bodily patterns, and new blends. Let us consider these in turn.

As noted above, Ekman thinks that the initial set of elicitors for our biologically prepared emotions can be expanded through learning. But he says that such additions are restricted to things that closely relate to the initial set (Ekman 2003, 24). For example, he says people in New Guinea learn to fear wild pigs, but he notes that, like the initial elicitors, wild pigs are dangerous. Emotions get elaborated by learning that novel objects belong to the same category as the ones that are already disposed to arouse us. When this happens, there is no new emotion—just an old emotion applied to new cases. In contrast, elaborations of an emotion can lead to changes in what emotions represent, and these semantic changes can change the identity of our emotions.

To see how this works, recall the terminology introduced above. Emotions occur under the control of triggering files, which can change in two ways. First, new items can be added. This can happen when a new item is similar to an old item, falls into a category represented in the file, gets associated with an old item through ordinary conditioning, or gets associated through social learning. To illustrate the last of these options, consider the wild pig example. It is plausible that our initial triggering file for fear includes a mechanism that allows us to extrapolate new dangers from the facial expressions of conspecifics. When children in New Guinea see their elders expressing fear of wild pigs, that may cause perceptual representations of those animals to be added to the fear file. This kind of change does not necessarily change the semantic content of fear, because the emotion continues to be a detector of dangerous things. But, in principle, such additions could have a semantic impact. Suppose you live in a group that displays fear responses to many benign things. At a certain point, after expanding the triggering file, the bodily response will no longer carry information about danger. If information has a bearing on semantic content (which the author believes it does), then the emotion may change its meaning.

Second, new triggering files can be acquired. We can speculate that once components have been added to a triggering file, they can splinter off. For example, among the elicitors for anger are cases of social injustice (e.g., inequities that stem from social institutions or abuses of power). Elicitors pertaining to injustice may tend to collect together in the anger file and form an independent functional unit. Over time, elicitors pertaining to injustice may become more closely linked to each other than to other items in the anger file (such as glaring eyes, or rude drivers on the freeway). Once a collection of representations pertaining to injustice gains this kind of autonomy, it may begin to function as its own triggering file. The somatic state associated with the new file is just like anger, but it has a narrow meaning: when this somatic state arises by means of the new triggering file, it represents injustice. It carries information that an unjust event has occurred, and it has (through learning) the function of tracking such events. We call this emotion *indignation*. It is an offshoot of anger, but deserves to be considered an emotion in its own right because it has a more restricted functional role. The anger caused by, say, a computer crash would not be called indignation, nor would self-directed anger, as when we berate ourselves for doing something especially stupid. A similar story might be told for schadenfreude, moral disgust, *amae* (the Japanese emotion of dependency), and *accidie* (the medieval Christian term for the torpor associated with the drudgery of religious practice), among others.

The cases that we have just been discussing are ones in which the somatic profile of an emotion remains constant but the meaning changes, giving rise to a new category. Emotions can also be transformed and acquired by learning new kinds of bodily responses. Some of our bodily responses are biologically prepared, but the body is not immune to the influences of learning and culture. Earlier was mentioned Ekman's notion of culturally learned display rules, which influence the way and extent to which people express emotions. Ekman and others have noticed that in Japan, strong expression of negative emotions in public settings is discouraged. One consequence of this is that emotional expressions become subtler, and, to some extent, involve different facial actions. A recent study showed that Americans tend to focus on the mouth when interpreting emotions, and Japanese tend to focus on the eyes (Yuki et al. 2007). Even the emoticons used in electronic communication reflect this difference. Americans typically represent a happy face with the emoticon :) comprising neutral eyes and a smile, whereas Japanese prefer to use this emoticon ^_^ comprising happy eyes and a neutral mouth. These preferences and recognition strategies reflect the fact that people in Japan and North America acquire different habits for expressing happiness. Interestingly, such differences are not restricted to cultures such as Japan, where repression of expressions is encouraged. Dacher Keltner has even observed differences in how Americans and English people smile, suggesting that the English smile reflects greater sensitivity to social status hierarchies, which are salient in that culture (see Max 2005). These findings suggest that common facial expressions can vary in subtle ways across cultures. That is born out by the fact that facial recognition varies across cultures even for expressions that have been described by Ekman as universal (Russell 1994). In addition, there may be some facial expressions that are highly recognizable in some cultures, but not in others. For example, in classic Indian dance, performers are trained to make distinctive and exaggerated expressions for emotions such as wonder, tranquility, and erotic love. There is also evidence that in some parts of India, women express shame by biting their tongues (Menon and Shweder 1994). Ekman would say that such differences in display rules do not amount to differences in the underlying emotions. But this would be a mistake: if emotions are experiences of bodily changes, different bodily habits can change the nature of an emotion. We should not assume, for example, that Japanese, English, and Americans experience happiness in the same way.

In addition to facial expressions, people can acquire more encompassing patterns of bodily change. Consider the practice of lowering the head or upper body to signal respect. Once this becomes a habit, as it seems to be in some cultures, respect takes on a new somatic profile. Or consider the Christian expression of reverence depicted in many Renaissance paintings: the head tilts upward to the side, the eyes look up, and the hands are clasped in prayer or placed gently on the chest. It is not inconceivable that this configuration became habitual for some Christian worshipers and transformed the somatic profile of their feelings of devotion. There are also various culture-bound psychiatric disorders that have distinctive bodily profiles, such as *latah* (a hyperstartle disorder in Southeast Asia) and hysteria (which includes symptoms such as hysterical paralysis). Or consider *amok*, the Malay word for a species of anger that involves frenzied behavior. When people acquire a culturally transmitted disposition to run amok, they acquire an emotion that differs somatically from what we are used to in the West.

Often, when we acquire new somatic profiles the semantic content of an emotion remains constant. This may be true in the case of Japanese and American smiles. But that does not mean the emotion remains constant. If emotions are perceptions of changes in the body then, trivially, the emotions are different. We sometimes use different words to name emotions that are somatically distinct but semantically alike. This is most obvious in the case of emotions that differ in intensity, such as fear and terror, anger and rage, joy and euphoria. Arguably, fear and terror are semantically alike, but differ in physiological arousal. Sometimes the somatic difference is more extreme. For example, in Malay, the word *amok* refers to frenzied anger and *marah* refers to a more sullen kind of anger, close to what we call brooding. Both may occur in response to insults and offenses, but they feel very different. Vocabulary can be used to distinguish different somatic profiles. Labels can also be introduced at different levels of abstraction so that we can co-classify emotions that are somatically different, but semantically alike. We can talk about different forms of happiness or different forms of anger, and when we do so, we are often referring to somatically distinct states that are similar in meaning.

So far, we have been talking about two ways to acquire new emotions on the somatic appraisal theory: we can acquire new triggering files and new bodily habits. There is also a third way to acquire new emotions: we can blend existing emotions together. Blending has not been widely investigated in emotion theory, but it deserves more attention (see Plutchik 1962). Ekman (1999) expresses some doubts about whether emotions can blend, and this may owe to his modular view. However, blending occurs quite regularly, and can be a source of new emotions.

It is useful to distinguish blended emotions from mixed emotions. Mixed emotions occur when we have two concurrent feelings about the same object or event. For example, if someone gives a gift you do not like, you might feel grateful and disappointed at the same time. When this happens, there is no pressure to say you have acquired a new emotion, sad-gratitude. But sometimes co-occurring emotions blend together in a more integrated way. Blending involves two conditions that make it different from mixing. First, with mixed emotions, it is usually easy to specify different aspects of an eliciting event toward which the co-occurring emotions are directed: you are grateful that someone gave you a present, but sad about the present itself. With blends, the co-occurring emotions are harder to tease apart in this way. Second, blends are not just one-shot occurrences; they get stored as such, and can acquire their own triggering files. They may begin as mixed emotions, but they become integrated and autonomous.

Consider contempt. Contempt blends together anger and disgust and perhaps also a feeling of dominance or superiority. The facial expression of contempt tends to include a lowered brow, like anger, and a curled lip, like disgust. It may originate when a single object elicits both of these emotions. Imagine hearing about a corrupt politician. You are angry about the violation of public trust, but also disgusted, because corruption is construed as a kind of contamination: the corrupt politician sullies the good name of the office he or she holds. When this combination of emotions first arises, the anger and disgust may be directed at slightly different properties, but there is also a sense in which they clearly have a common object: the politician, in this example. Moreover, the combination of anger and disgust is likely to occur again and again. Over time, the combination may acquire its own class of elicitors: corruption, hypocrisy, disrespect, and so on. These initially reside in the intersection of the triggering files for anger and disgust, but they can form a cluster of their own and splinter off. Thus contempt becomes an enduring emotion. Other blended emotions may include envy (desire and resentment), thrills (happiness and fear), nostalgia (happiness and sadness), and horror (fear and disgust). In each case, the emotions combine previously existing elements and acquire proprietary triggering files that determine the semantic content of the blend. Nostalgia, for example, is not just any blend of happiness and sadness, but happiness and sadness caused by ruminating about the past. Other blends of happiness and sadness may also occur, tuned to different sets of elicitors.

We have just described three ways in which new emotions can be generated if the somatic appraisal theory is correct. We can change the semantic content of an emotion by altering triggering files, we can change the somatic profile of an emotion by learning new bodily habits, and we can create new hybrids by blending emotions together. The last of these processes may implicate the first two as well; blends typically modify semantic content by introducing new triggering files, and the somatic profile of a blend may take on emergent properties that arise when the body integrates somatic states from two distinct emotions.

Ekman's Darwinean module view could be adapted to accommodate these methods of acquiring distinct emotions, but, as he presents the view, the number of emotions remains constant. He rejects the idea of new semantic contents, saying that the themes of our emotions remain constant as we learn new elicitors. He rejects changes in bodily profile because he thinks emotional physiology is biologically fixed and universal. He rejects blends, because he has doubts about whether two emotion modules can run their affect programs concurrently. The examples presented in this section provide some reasons for thinking that Ekman is wrong. There seem to be genuine cases of new emotions as our rich emotion vocabulary would suggest.

3. How to Count Emotions

Let us conclude by drawing out some implications that the somatic appraisal theory has for counting the emotions. If cognitive appraisal theories are right, we count emotions by looking for hot judgments; if circumplex theories are right, we could by looking for attractors in arousal-valence space; and if the Darwinian module theory is right, we look for dissociable affect programs that have homologues in other species. Things work a bit differently on the somatic appraisal theory. Emotion counting can proceed in the following three broad steps, which can be pursued concurrently.

First, bodily patterns must be identified. The body is a continuous system, and there are infinitely many bodily states, but, within this variation, there are also recognizable patterns. For example, among the endlessly varied configurations of the face, some are co-classified and recognized as emotion. Work on expression recognition has been very helpful in emotion research. There are also related research programs on recognizing posture, respiration, and action tendencies. One expectation of the somatic appraisal theory is that emotions involve widespread bodily changes rather than highly isolated somatic events. So we should look for facial expressions that occur with changes in respiration, posture, and action. We should also expect that these co-occurrences will hang together, so that the occurrence of one tends to promote the occurrence of others. This prediction is borne out by empirical findings. For example, Levenson et al. (1990) found that when people make emotionally interpretable facial expressions, systematic physiological changes occur throughout the body. These coordinated bodily responses must have a source in the nervous system. Therefore, research on bodily patterns should include efforts to identify underlying neural circuits for body regulation, as well as the circuits that are used to recognize when bodily changes occur. If emotions are perceptions of bodily changes, these latter circuits will also be the neural correlates of emotion.

Second, there must be systematic investigation of emotion elicitors. We should determine what things trigger our emotions. When we do that, we can try to distinguish direct and indirect elicitors. A direct elicitor is something that is mentally represented in a way that can initiate an emotion without being analyzed by further representations. An indirect elicitor requires more processing before it initiates an emotion. Only the former qualify as triggers. A trigger is something that brings about a psychological response causally rather than evidentially. A trigger does not need to categorized or deliberated upon to have an emotional effect. Once we identify direct elicitors (or triggers), we should try to determine which ones are associated together in triggering files. That task is tricky, but progress can be made. One strategy for determining what elicitors belong together is to determine what elicitors trigger that same pattern of bodily response. It must be noted, however, that in the somatic appraisal theory, the same bodily response can operate under the control of different triggering files. So the fact that two elicitors cause the same response does not entail that they correspond to the same emotion. To overcome this hurdle, we should seek evidence pertaining to the degree to which classes of elicitors are associated together in semantic memory. It is not entirely clear at the outset what form that evidence should take. We could look for priming effects to see if two elicitors are closely linked, we could do multidimensional scaling studies on descriptions of elicitors to look for semantic similarities, and we could look for dissociations. Once we have some idea of what elicitors belong together in the same triggering files, we can also try to determine what semantic content those files impart on the emotions they trigger. The collection of items in a triggering file collectively determines when an emotion will occur, and that contributes to the significance of the emotion. Emotions do not represent each individual elicitor, but rather the more general organism/environment relationship that a class of elicitors collectively instantiate.

Once we have identified all the bodily patterns and the meaning-conferring triggering files that cause those patterns to occur, we will have, in effect, identified the set of emotions. We can count up those file-body pairs and estimate how many emotions there are. However, we may want to do a third thing before counting: we may want to distinguish the emotions that are biologically basic from those that owe more to learning and enculturation. That way, we can count these two categories separately, and give a more complete answer to questions about how many emotions there are. The notion of biological basicness is related to the notion of innateness, which has proven difficult to define. For present purposes we can think of biologically basic emotions as ones that are typical of the species and not attributable to any general purpose learning mechanism. There is no algorithm for determining which emotions are biologically basic, and, in the adult state, basic emotions may be indistinguishable from nonbasic emotions. A nonbasic emotion has no more parts than a basic emotion; it is just an offshoot of a basic emotion acquired through one of the methods discussed above (new triggering files, learned bodily habits, or blending). One useful approach can be found in Ekman's research: we can look for universally recognized facial expressions and universal elicitors. The more evidence we find for universality, the more evidence we will have for the claim that an emotion is basic. However, one complication in using this method is that the emotions that seem to be the best candidates are never *completely* universal. As noted above, the expression of happiness, which is the most recognized expression that has ever been studied, is not exactly the same cross-culturally. Elicitors of this expression presumably vary quite a bit as well, and in some cultures it is even customary to make what we regard as a happy expression when expressing despair (laughing at a funeral, for example). Consequently, when we look at adult behavior the emotions we find are rarely, if ever, truly universal (Prinz 2004b). Thus, we should supplement research on adults with developmental and comparative research. By looking for the bodily patterns and elicitors in young children and nonhuman primates, we might do a better job determining which of our emotions are truly basic.

A list of basic emotions will not be offered here (see Prinz 2004a; 2004b). As the basic emotions may not correspond to emotions that are exactly like the ones we have names for in adult psychology, any list would be potentially misleading. Ekman may be right that the number of basic emotions is relatively small, and the items on his list are helpful insofar as they reflect decades of research on this topic. Emotions akin to many of the items on his list are likely to be basic. It is considerably harder to count nonbasic emotions for at least two reasons. First, as suggested above, it is difficult to individuate triggering files, and in some cases, there may be no principled way to determine whether two closely related sets of elicitors constitute one file or two. There can be files that pass some tests for independence and not others. Second, the number of nonbasic emotions is potentially open-ended, as we can continue acquiring new emotions. To count nonbasic emotions, do we consider all cultures that have ever existed, all that currently exist, all future cultures? Each option presents insuperable difficulties because we do not have access to all the people we need in the past, present, or (most obviously!) future. So we cannot expect an accurate count of nonbasic emotions. We can, however, try to determine how many nonbasic emotions are highly prevalent in a given culture at a given time. The foregoing methods of counting could be applied to arrive at an estimate. We may also use emotion vocabulary as a clue, though there are undoubtedly unnamed emotions, and emotion words that are synonymous. We also need to be mindful that emotions can be counted at different levels of abstraction. Some words may refer to large categories, and some to emotion terms that refer to states that differ only in intensity. When we encounter cases of the latter kind, do we say there is one emotion or two? Do we say there is a different emotion for every level of intensity? To some extent, the answers to such questions depend on the theoretician and the explanatory goals of the theory. Emotions that differ in intensity are thereby psychologically different, but the difference may not matter for most explanatory purposes.

These complications underscore that there is no settled answer to the question in the title of this chapter. We start out with a few emotions and end up potentially with many. How many? Depends on how we count. But this response should not lead to the conclusion that the exercise of counting emotions is a waste of time. On the contrary, it is a good way of testing theories of emotions and arriving at defensible principles of individuation. Cognitive theories seem to undercount, because they leave out emotions that occur in the absence of cognition. Circumplex theories undercount too, because they cannot distinguish between emotions that are alike in valence and arousal, but different in meaning or bodily pattern. They also overcount by presenting emotions in a continuous space, and by distinguishing emotions that are semantically alike, but different in arousal. The Darwinian modules theory undercounts, because it restricts the class of emotions to a small set of evolved responses, and neglects the evidence for emotional learning. The theory that does best justice to our folk ways of classifying emotions as well as the empirical evidence for emotional diversity is the somatic appraisal theory. If that theory is right, then we start out with a handful of emotions but then acquire new ones, and there may be no limit on the number of possible emotions we can possess.

REFERENCES

Armon-Jones, C. (1989). *Varieties of Affect*. Toronto: University of Toronto Press.

Chwalisz, K., Diener, E., and Gallagher, D. (1988). Autonomic arousal feedback and emotional experience: Evidence from the spinal cord injured. *Journal of Personality and Social Psychology* 54: 820–28.

Damasio, A. R. (1994). *Descartes' Error: Emotion, Reason, and the Human Brain*. New York: Gossett/Putnam.

Damasio, A. R., Grabowski, T. J., Bechara, A., Damasio, H., Ponto, L. L. B.; Parvizi, J., and Hichwa, R. D. (2000). Subcortical and cortical brain activity during the feeling of self-generated emotions. *Nature Neuroscience* 3: 1049–56.

Dretske, F. (1981). *Knowledge and the Flow of Information*. Cambridge, MA: MIT Press.

———. (1988). *Explaining Behavior*. Cambridge, MA: MIT Press.

Ekman, P. (1994). All emotions are basic. In P. Ekman and R. J. Davidson (eds.), *The Nature of Emotion: Fundamental Questions*. New York: Oxford University Press.

Ekman, P. (1999). Basic emotions. In T. Dalgleish and T. Power (eds.), *The Handbook Of Cognition And Emotion*. Sussex, UK: John Wiley.

———. (2003). *Emotions Revealed*. New York: Times Books.

Ekman, P., and Friesen, W. V. (1969). The repertoire of nonverbal behavior: Categories, origins, usage, and coding. *Semiotica* 1: 49–98.

Fodor, J. A. (1990). *A Theory of Content and Other Essays*. Cambridge, MA: MIT Press.

———. (1983). *The Modularity of Mind*. Cambridge, MA: MIT Press.

Frijda, N. H. (1986). *The Emotions*. Cambridge: Cambridge University Press.

Greenspan, P. (1988). *Emotions and Reasons*. New York, NY: Routledge.

Griffiths, P. (1997). *What Emotions Really Are: The Problem of Psychological Categories*. Chicago: University of Chicago Press.

James, W. (1884). What is an emotion? *Mind* 9: 188–205.

Hohmann, G. W. (1966). Some effects of spinal cord lesions on experienced emotional feelings. *Psychophysiology* 3: 143–56.

Lazarus, R. S. (1991). *Emotion and Adaptation*. New York: Oxford University Press.

LeDoux J. E. (1996). *The Emotional Brain*. New York: Simon and Schuster.

Levenson, R. W., Ekman, P., and Friesen, W. V. (1990). Voluntary facial action generates emotion-specific autonomic nervous system activity. *Psychophysiology* 27: 363–84.

Max, D. T. (2005). National smiles: Do Americans smile differently than the English?. *New York Times Magazine*, December 11.

Menon, U., and Shweder, R. (1994). Kali's tongue: Cultural psychology and the power of "shame" in Orissa, India. In S. Kitayama and H. Markus (eds.), *Emotion and Culture: Empirical Studies of Mutual Influence*. Washington, DC.: American Psychological Association.

Oatley, K., and Johnson-Laird, P. N. (1987). Towards a cognitive theory of emotions. *Cognition and Emotion* 1: 29–50.

Plutchik, R. (1962). *The Emotions: Facts, Theories and a New Model*. New York: Random House.

Prinz, J. J. (2004a). *Gut Reactions: A Perceptual Theory of Emotion*. New York: Oxford University Press.

———. (2004b). Which Emotions Are Basic?. In D. Evans and P. Cruse (eds.), *Emotion, Evolution and Rationality*. Oxford: Oxford University Press.

Russell, J. A. (1980). A circumplex model of affect. *Journal of Personality and Social Psychology* 39: 1161–78.

———. (1991). Culture and the categorization of emotions. *Psychological Bulletin* 110: 426–50.

———. (1994). Is there universal recognition of emotion from facial expression? A review of the cross-cultural studies. *Psychological Bulletin* 115: 102–41.

Scherer, K. R. (1993). Studying the emotion antecedent appraisal process: An expert system approach. *Cognition and Emotion* 7: 325–55.

Solomon, R. (1976). *The Passions: Emotions and the Meaning of Life*. Indianapolis: Hackett.

Strack, F., Martin, L. L., and Stepper, S. (1988). Inhibiting and facilitating conditions of facial expressions: A nonobtrusive test of the facial feedback hypothesis. *Journal of Personality and Social Psychology* 54: 768–77.

Watson, D., and Tellegen, A. (1985). Towards a consensual structure of mood. *Psychological Bulletin* 98: 219–35.

Yuki, M., Maddux, W. W., and Masuda, T. (2007). Are the windows to the soul the same in the East and West? Cultural differences in using the eyes and mouth as cues to recognize emotions in Japan and the United States. *Journal of Experimental Social Psychology* 43: 303–11.

Zajonc, R. B. (1984). On the primacy of affect. *American Psychologist*, 39: 117–23.

CHAPTER 9

....

ATTENTION

....

CHRISTOPHER MOLE

THOSE psychological phenomena that get counted as the *cognitive* ones are, more or less without exception, those that relate in some way to attention. The diversity among cognitive phenomena is therefore matched by diversity among the explanatory contexts in which attention figures. And this diversity is one of the reasons cognitive scientists have not settled on any standard definition of *attention*. In place of a definition they typically preface their discussions of attention by citing a remark from Chapter XI of William James's *The Principles of Psychology*:

> Everyone knows what attention is. It is the taking possession by the mind, in clear and vivid form, of one out of what seem several simultaneously possible objects or trains of thought. (James 1890, 381)

The authors who cite this passage from James often quote only the first of these two sentences, their point being that attention is so familiar that there is no need for us to have a definition in order to establish what it is that we are talking about. For those who do want to say something more substantive about what attention is, the second part of James's remark serves to characterize the distinctive explanatory role that claims about attention play. There are, James suggests, several trains of thought that might be followed by minds like ours, given the various cognitive opportunities afforded by the sort of environment that we find ourselves in. It is not simply chance that explains our ending up with the trains of thought that we actually have. Instead there are some things that catch our attention, and some things to which we deliberately attend. Attention may be defined by reference to this explanatory role: it is that phenomenon which explains the selective directedness of our mental lives.

The cognitive scientists who study attention frequently do so in the hope of illuminating such philosophically puzzling phenomena as consciousness, free will, and the interface between perception and cognition. Despite the philosophical relevance of this work, there has been relatively little discussion by philosophers of the topic of attention. The reasons for philosophy's neglect of the topic can be traced back to widespread suspicions about the reality of such inner acts as attendings, or, at least, to suspicions about their explanatory importance. These suspicions originate with the later work of Wittgenstein. In *Philosophical Investigations* § 412, for example, Wittgenstein says of "turning my attention onto my own consciousness" that "This surely is the queerest thing there could be!" (Wittgenstein 1953). Other mid-twentieth century philosophers shared Wittgenstein's reluctance to cast attention in any explanatorily weight-carrying role (see, e.g., Geach 1957, 64).

Some historical factors have also contributed to attention's status as a proprietary explanandum for the more purely empirical branches of cognitive science. Attention enjoys a central place on cognitive science's explanatory agenda (and, especially, on the agenda of cognitive *psychology*) partly on account of the role that twentieth-century theories of attention played in laying the philosophical foundations that needed to be in place in order for cognitive psychology to establish itself as a scientifically respectable enterprise. The mid-twentieth century attempts to explain attention were among the first research programs to establish the scientific credentials of cognitive science's distinctively computational explanatory approach.

This establishment of cognitive science's scientific credentials was a philosophical achievement as well as a scientific one, requiring, as it did, the overthrow of the then-influential logical positivist idea that the criteria of meaningfulness for psychological theories obliged those psychologists who were proposing and testing such theories to trade only in claims about publicly observable entities, such as distal stimuli and behavioral responses. At the end of the 1950s that positivistic idea was discredited; it was research on attention, together with research on language, that most thoroughly discredited it. Noam Chomsky's 1959 review of B. F. Skinner's 1957 book, *Verbal Behavior,* convinced many of those studying the mind that the most complex and most distinctively human aspects of cognition could not be explained by theories that restrict themselves to mentioning only publicly observable patterns of association between stimuli and responses (Chomsky 1959). In place of that behaviorist approach, a more computation-based approach was needed, in which the explanation of mental phenomena was achieved via a description of the rules governing the transformation of representations by the unseen information-handling processes taking place within the brain. In the case of language, this new computation-based approach had already been exemplified in Chomsky's own 1957 book, *Syntactic Structures.* Outside of the linguistic domain, the information-handling approach that came to be distinctive of the then-emerging field of cognitive science had been most successfully deployed as a tool for understanding human cognition by Donald Broadbent in his 1958 book, *Perception and Communication.* The concluding chapter of Broadbent's book gives what was, at the time, the most sustained and authoritative attack on the positivistic approach that had led psychologists to be "chary of postulating events within the organism" (Broadbent 1958, 302).

Broadbent's approach to attention, in *Perception and Communication* and subsequently (e.g., Broadbent 1971), was hugely influential. In many ways it is that approach that continues to set the terms for discussions of attention taking place today. It is now

universally agreed, however—and has been since around the beginning of the 1990s—
that the debates prompted by Broadbent's work were in some way misguided, that
their eventual collapse without a satisfactory resolution was inevitable, and that we
need to avoid falling into the same mistakes when framing our current theories.

The first half of this chapter will consider Broadbent's approach to the explanation of attention. It will consider the lessons that we have learned from its failure,
while identifying certain ways in which that approach continues to be influential.
To some readers this will be a familiar story. But the familiarity of Broadbent's explanatory project and the history of its eventual collapse should not be taken as a
sign that cognitive science has come to any clear understanding of how its current
debates can avoid facing a similar fate. On the contrary, several of our current theories share some of Broadbent's assumptions. The lack of consensus about the reasons for the failure of Broadbent's project means that it is unclear whether the
persistence of these assumptions is problematic. An understanding of the strengths
and weaknesses of the groundwork that Broadbent laid therefore remains essential
to our understanding of where our theorizing about attention currently stands.

The collapse of Broadbent's project marked the beginning of a rather extended
period in which the paradigm within which attention was studied and theorized got
renegotiated. Although it had become clear by the early 1990s that nothing quite like
Broadbent's theory could be made to work, it is only in the last ten years or so that
any consensus has emerged as to what, if anything, should replace it. In the period
immediately following the abandonment of Broadbent's project, cognitive scientists ceased to address the topic of attention all at once, or to address it very directly.
Attention labs did, of course, continue to carry out research, and a variety of attention-related phenomena continued to be studied and explained, but there was no
expectation that there would be a single unified theory of attention to which all of
these various explanations contributed. More recently cognitive scientists working
on attention have begun to converge on a new unifying paradigm for its explanation. This paradigm, based on what is sometimes known as *the biased-competition
model* (and sometimes as *the integrated competition model*), remains controversial.
But as a working hypothesis it has proved its usefulness, and, although there are
disagreements as to how it should be interpreted, it has no real rivals in the current
literature. The last part of this chapter will try to explain why this model may prove
to be of some philosophical importance.

One of the biased-competition model's features that is of particular interest to
those cognitive scientists whose agenda is a philosophical one is that there are some
interpretations of the model that suggest a departure from our previous picture of
the way in which attention is related to the processes that underpin it in the brain.
It has traditionally been natural to assume that there are some particular brain processes, implemented at identifiable cortical loci, the operation of which accounts for
the fact that a person is paying attention to one thing rather than another. According to some interpretations of the biased-competition model, this assumption is
mistaken. In these interpretations the facts about attention emerge, not from the
operation of any cortical processes in particular, but from broadly distributed

patterns of activation that take place throughout the cortex, and perhaps through-out the organism as a whole (see Allport 2011 for suggestions along these lines). The methodological and explanatory consequences of this shift in our understanding of attention's metaphysical relation to its basis in the brain, if such a shift does turn out to be required, would be considerable. They represent some of the clearest points at which purely philosophical work has a contribution to make to the cognitive scien-tist's explanatory project.

1. BROADBENT

Broadbent's 1958 book presented the case for using communication-theoretic con-ceptual resources when explaining perceptual phenomena. By *communication theory* Broadbent meant not only the abstract theory of computation, of the sort that was then being built on the foundations laid by Turing and Church. He also meant to include the applied theory that was being used in such academic projects as the design and construction of Newell and Simon's General Problem Solver (Newell, Shaw, and Simon 1957) and also, more conspicuously, in the design and construction of such high-profile information technologies as the automated tele-phone exchange and the microchip (both of which were at crucial stages in their development at the time when Broadbent was writing).

Broadbent was particularly concerned about showing that ideas from commu-nication theory can be put to work in giving a theory that identifies certain bottle-necks in the processing capacity of the information channels that are responsible for perception. He was himself reluctant to claim that such a theory of perceptual capacity bottlenecks provides the explanation of attention, preferring to speak of his work as providing us with a theory of "vigilance" or of "the selective uptake of information," (1982) but these qualms were not shared by many of Broadbent's con-temporaries, nor by those who attempted to produce theories of attention in the decades that followed. From the late 1950s until the early 1990s most of cognitive psychology's discussions of attention were conducted in the communication-theo-retic terms that *Perception and Communication* introduced. This was not because that book's account of the bottlenecks in perceptual processing was universally agreed upon; in fact, almost everyone took it that Broadbent's theory needed to be recast, augmented, or modified in some way. But they also held that to be starting with Broadbent's theory, and with the idea that attention-related phenomena can be understood as resulting from bottlenecks in perceptual processing capacity, was to be starting in the right place.

It is only partially true to say that this consensus is a thing of the past. While nobody now thinks that attention should be explained by a theory that looks quite like Broadbent's, it remains true that the guiding idea that Broadbent introduced to the literature retains its influence. Broadbent's idea that attention's role in cognition is the management of bottlenecks in information processing capacity is now so widely endorsed as to be treated by many psychologists as if it were uncontentious, or simply analytic. For this reason it is commonplace to find attention and "capacity-limitation management" being treated as synonymous. One clear example of such a treatment (although there are many more that could be given) is found in a 2002 article by Emanuela Bricolo and her collaborators. They write: "In general, set-size effects are taken to indicate that processing of the array of elements depends on limited capacity resources, that is, it involves attention." (Bricolo et al. 2002, 980).

Even in the most sophisticated and ambitious of our current theories of attention (which we shall see more of later) this link between attention and capacity limitation is presented as a more or less axiomatic starting point. On the first page of the introduction to their 2008 book, *Principles of Visual Attention*, Claus Bundesen and Thomas Habekost set out their central claim as being that:

> Attention, at least in the visual domain, is the working of a few specific mechanisms that follow a unified set of mathematical equations. The purpose of these mechanisms is *selectivity* [...] The importance of selectivity becomes clear when one considers the severe capacity limitations that characterize humans. (Bundesen and Habekost 2008, 3)

The same Broadbentian idea—that attention is primarily a matter of capacity-limitation management—can be seen in the work of John Reynolds and Robert Desimone. When setting the stage for their presentation of the biased competition model of attention they write that:

> The visual system is limited in its capacity to process information. However, it is equipped to overcome this constraint because it can direct this limited capacity channel to locations or objects of interest. (Reynolds and Desimone 2000, 233)

Such quotations suggest that even these innovative recent theorists endorse Broadbent's idea that the explanation of attention is to be given by producing an account of perceptual capacity limitations, and of their cognitive handling. But, although all of these theorists endorse Broadbent's capacity-limitation idea, they, like everyone else since at least the early 1990s, think that something was fundamentally misconceived in Broadbent's own treatment of that idea. They, like everyone else, agree that the debates prompted by Broadbent's capacity-bottleneck view eventually proved to be fruitless, and, like everyone else, they take this fruitlessness to be a sign that the theoretical framework of those debates suffered from some sort of confusion. The question of *where* those debates went wrong is, however, one to which several different answers have been given, and on which no consensus has been reached.

2. The Legacy of Broadbentian Thinking

The body of data that Broadbent was seeking to account for when he proposed his theory of capacity bottlenecks in perceptual processing was drawn from experiments that were primarily concerned with gauging the extent to which we are ignorant of the things to which we are not attending. Those experiments took various forms, but much the most popular of the methods for studying attention in Broadbent's time involved auditory stimuli. One particular method involved presenting subjects with two streams of speech, played simultaneously through headphones, with one stream being played to each ear. These experiments were known as "dichotic listening experiments." An early finding from such experiments is that, when paying attention to the auditory stream being played into one ear, subjects find it relatively easy to ignore the stream being played to the other: Streams that are physically separated in this way tend not to interfere with one another.

The dichotic listening result that provided the primary motivation for Broadbent's theory was the discovery that when subjects attend to the speech played to one ear, they may still detect simple physical changes to the stream of speech played in the other ear—they may, for example, notice that the speech has become louder, or that its pitch has increased—but those same subjects remain completely oblivious to changes in the unattended stream's *semantic* properties. Subjects fail to notice if the unattended stream shifts from meaningful speech to the meaningless but speech-like nonsense that can be produced by playing the tape of a speech backwards. They also fail when given tests in which they are asked to recognize which of the words on a subsequently presented list were played to the unattended ear, even in conditions where those words had been played to that ear repeatedly.

Broadbent took this evidence to show that one's information-processing resources have sufficient capacity to encode the simple physical properties of all the stimuli that one is presented with, but have only a limited capacity for the encoding of the semantic properties of those stimuli. The resulting model depicts perceptual processing as proceeding in two stages. In the first stage a large capacity "sensory system" processes the physical features of all stimuli in parallel. On the basis of these physical features a subset of the representations generated by this large capacity system are selected to be passed on to a second "perceptual system," which has a smaller processing capacity, and which has the job of processing the stimuli's semantic properties.

The basic architecture of Broadbent's model—that of two systems with a bottleneck between them—was shared by the models that were proposed by Broadbent's rivals. Those who took issue with Broadbent's picture did so, not by disputing his claims about there being two separate systems with different capacities, but by disputing his claims about which properties were detected by the pre-bottleneck system and which by the post-bottleneck system. The now-familiar *cocktail party effect*, studied by Neville Moray (1959), gives a clear illustration of the way in which these disputes were conducted.

The cocktail party effect is a rather more commonplace phenomenon than its chic name might suggest. It is the phenomenon that occurs when one participates in one conversation while filtering out the others that are going on in the same room. This kind of attentive focusing is relatively easy, and is relatively easy to sustain, but only so long as the conversations that one is ignoring are not conversations pertaining to oneself: If one's own name is mentioned in one of the other conversations that is taking place in the room, then one tends to get distracted by it. This is a problem for Broadbent because the two-systems model, as Broadbent initially proposed it, entails that the semantic properties of stimuli are not explicitly represented in the nervous system unless and until those stimuli have been attended. The explicit representation of semantic properties is, according to this model, the work of the small capacity, post-bottleneck system, and the stimuli that have been processed by that system have, ipso facto, been given the subject's attention. If this were right, then no semantic property of an unattended stimulus could have a psychological effect, and it would follow that the semantic properties of unattended stimuli could not have an effect on the subsequent direction of the subject's attention. Unattended words might attract attention to themselves on account of their pitch or volume, but not on account of their meaning. The cocktail party effect shows that this prediction of Broadbent's theory is not borne out: it shows that occurrences in unattended speech of at least one semantic property—"being an instance of one's own name"—*can* explain the fact that one's attention gets drawn to that speech.

Investigations of the cocktail party effect did not sink Broadbent's theory entirely. Instead they initiated a prolonged debate, conducted by both sides in more or less Broadbentian terms, about the extent and character of the pre-bottleneck processing. If "being an instance of one's own name" were the *only* semantic property to have effects on attention when it occurs in the unattended stream, then one could retain the spirit of Broadbent's theory by adding this as an explicit exception to the claim that all semantic processing is the work of the post-bottleneck system. It would be ad hoc, perhaps, but not implausible, to claim that one's name has a special psychological status, and so has special cognitive apparatus devoted to its detection. Modified in such a way as to allow for this, Broadbent's theory would now say that pre-bottleneck processing is responsible for the detection of simple physical features, and also for own-name detection. Adding explicit exceptions for special stimuli such as one's name could not, however, give the Broadbentian a satisfactory account of all the psychological effects that dichotic listening experiments gave rise to.

An effect that is particularly problematic for a Broadbentian conception of the pre-bottleneck system was demonstrated in a classic study conducted by Anne Treisman in 1960. Treisman's experiment used a dichotic listening task in which the two streams of speech unexpectedly switch ears. If the two streams are both spoken by the same speaker, and if an auditory shadowing task is being completed, requiring the subject to keep her attention on one stream while ignoring the other, then, if the contents of the two streams switch over so that the sentence that was being started in the right ear is completed over at the left, the result is that the subject will unwittingly make a brief switch to shadowing the stream presented over on the left (Treisman 1960). Although Treisman herself attempted to give an explanation of this effect in broadly Broadbentian terms, the effect seems fatal to Broadbent's understanding of where semantic processing takes place, and of how much of it is going on: In order to explain why it is that the subject's attention switches to the previously unattended ear when the message from the previously attended ear moves over there, it seems we must allow that the sounds presented to the unattended ear are being recognized as containing a good semantic continuation of the attended message. But this requires that arbitrary semantic properties of the words presented to the unattended ear are being detected. And that seems to refute Broadbent's claims about semantic processing being an exclusively post-bottleneck matter, except in special cases.

Findings such as Treisman's do not by themselves require us to reject the two-systems-and-a-bottleneck architecture that Broadbent proposed. They require only that we differ from Broadbent in our understanding of the way in which cognitive labor is divided between the two components of that architecture. The debates that were prompted by Broadbent's work were, as we have said, primarily concerned with this division of labor. It was only after the eventual collapse of those debates (around the end of the 1980s) that psychologists began to ask whether the basic outlines of the Broadbentian architecture might themselves constitute a mistake.

According to Broadbent's chief rivals in the first phase of these debates, as they took place in the 1960s and 1970s, the lesson from research such as Moray's and Treisman's is that the attentional bottleneck comes after the system in which semantic properties are detected. This theory—known subsequently as the "late selection" theory—says that the parallel, pre-bottleneck system does all of the psychological heavy lifting. The only processes that happen in the post-bottleneck low-capacity system are those that allow semantic properties to be consciously accessed and remembered.

Throughout the 1960s and 1970s, and well into the 1980s, most debates about attention were concerned with more or less nuanced versions of the issue that divided the defenders of early selectionist theories (such as Broadbent's) and their late selectionist rivals (built on the proposals made by Deutsch and Deutsch 1963, and later developed by Duncan 1980). Even when they were not explicitly asking whether attention occurs early or late in the processing stream, the debates in this period that were concerned with the role of attention in cognition were often addressed to the same underlying question: that of where in the perceptual processing stream the attentional bottleneck occurs. These debates continued to be understood within the Broadbentian framework. They were understood as being settled by the facts about whether the operations in question take place after the attentional bottleneck, and so "requiring attention," or before it, and so "automatically."

The assumed links between attention and conscious access (see, e.g., Norman 1968), and between pre-attentive processing and "automaticity" (see Shiffrin and Schneider 1977) led some psychologists in this period to take the debates about the locus of attention to be empirical versions of what were traditionally philosophical questions about free will, and about the degree to which conscious decision making plays a role in determining the course of our actions and thoughts. Some researchers took this incursion into philosophical territory as a reason to be wary of the debates about attention (see, e.g., Allport 1980, 113). Others were more enthusiastic about the philosophical ambitions that attention research had acquired. Michael

Posner and Steven Petersen, writing in 1990, were quite explicit about the philo-sophical ambitions of their work on attention. They claimed that:

> If there is hope of exploring causal control of brain systems by mental states, it
> must lie through an understanding of how voluntary control is exerted over more
> automatic brain systems. We argue that this can be approached through under-
> standing the human attentional system at the levels of both cognitive operations
> and neuronal activity. (Posner and Petersen 1990, 25)

Since the issues underlying the debates between early and late selection theories had acquired this somewhat assorted philosophical baggage, it should be no surprise that the eventual demise of those debates was not the result of any single discovery that settled or dissolved the underlying issues. Instead, as is the way with paradigm shifts, psychology's eventual retreat from the Broadbentian debates was the result of various accumulated anomalies, and of a general sense that those debates had run into a state of empirical and theoretical fruitlessness. The interest in the Broadben-tian framework dropped off, and by the beginning of the 1990s the way was open for Allport, a long-standing critic of that framework, to make the case for abandoning it altogether.

In an influential 1993 article, Allport claimed to identify eight different assump-tions on which the debate between early and late selection theories depended. He pointed out, for example, that those who expected there to be a clear winner in the debate between early bottleneck theories and late bottleneck theories needed to make the assumption that there was just one such bottleneck to be located; they needed to assume that that bottleneck's location was stable from one situation to the next; and they needed to assume, also, that the bottleneck occurs in a system with earlier and later parts, not in a massively parallel system without any prevailing direction of information flow (since in a system of the latter sort, it could only be arbitrary to label one part of the system as "early" and another as "late"). On the basis of various cognitive neuroscientific considerations, Allport claimed that each of these assumptions turns out to be problematic.

Allport also, as a somewhat separate point, claimed to identify several ambigui-ties and equivocations in the various ways in which the term *attention* had been used. About this last point he was somewhat more tentative:

> With rare exceptions, the controversies over early and late selection, and over
> controlled versus automatic processing have relied on behavioural human perfor-
> mance data based on normal adult human subjects. Consideration of a wider
> range of evidence [...] calls into question many of the traditional assumptions
> on which these controversies were predicated. Such a review may also bring into
> focus something of the *heterogeneity* of attentional functions, which preoccupa-
> tion with these old controversies has tended to obscure. (Allport 1993, 188)

With so many reasons to be dissatisfied, and with such a long history of debate winding down into fruitlessness, the psychologists of the 1990s turned away from the early/late selection debate more or less entirely. The work of attention labs shifted toward the more piecemeal project of attempting to explain various attention-related phenomena on their own terms, rather than as explananda for a single unified theory of attention. The attempt to locate the bottleneck of attention was given up. This did not mean, however, that the last vestiges of the Broadbentian picture were abandoned. On the contrary, the now fragmentary research that was directed at various attention-related phenomena retained several elements of Broadbent's approach.

One of the attentional phenomena that continued to be studied throughout the period when the Broadbentian project was running into fruitlessness (and after) was that of serial, attention-demanding visual search: search of the sort that is required when trying to find one's keys on a crowded table, or the one red O on a page filled with red Xs and blue Os.

Another of the phenomena that continued to be studied throughout this period was the so-called "psychological refractory period": the impairment, when responding to a stream of response-demanding stimuli, in the execution of responses to those stimuli that occur while previous stimuli are still being processed.

The theories that were proposed to account for these phenomena retain a recognizably Broadbentian flavor. Anne Treisman's enormously influential "Feature Integration Theory" was proposed as a theory that is intended to account for the first of these phenomena—the phenomenon of serial visual search. According to the Feature Integration Theory, there is a large processing capacity for the parallel encoding of the simple features of the stimuli with which we are presented, and there is a much smaller capacity—limited to one location at a time—for the encoding of the more complex properties that emerge from the ways in which these simple features are combined (Treisman and Gelade 1980; Treisman 1993). With its clear distinction between those simple properties processed prior to an attention-related bottleneck, and those complex properties that are encoded only for the stimuli that make it through the bottleneck, this theory retains a clearly Broadbentian flavor.

An equally Broadbentian note is struck by Harold Pashler's attempts to explain the effect known as the psychological refractory period. In psychological refractory period experiments, two different stimuli are presented in quick succession, both of them requiring the subject to produce a response. If the interval between the stimuli is long enough, then the two responses can be produced equally quickly. But, as that interval gets shorter, a point is reached at which the second response gets held up. In Pashler's account, this delaying of the second response occurs, not because of limitations in the subject's capacity for identifying the two stimuli in quick succession, but because the resources involved in selecting the response that each stimulus demands are not available for the task of selecting the response appropriate to the second stimulus until after they have finished with the business of selecting a response appropriate to the first. Pashler is explicit about his intention to retain Broadbentian elements in his theory, which gives elegant explanations for a wide range of empirical data (Pashler 1984, 1994). The elements of Broadbent's picture that are retained here are the ideas that the production of a response to a stimulus depends on a linear stream of processing, that the first stages of this stream have a large capacity for handling several stimuli simultaneously, and that there is a central selective bottleneck with the capacity to process only one stimulus at a time.

Pashler and Treisman both retain aspects of the Broadbentian picture in their theories, albeit on a smaller scale and on a different empirical footing. Among the Broadbentian claims that they retain there are claims—those about linearity and about localizability—that are among those that Allport's influential 1993 article had complained about. It is unclear, therefore, whether those who accept Allport's critique of the Broadbentian debates ought to be entirely unhappy about the persistence of these assumptions.

A similar residual Broadbentianism can be discerned in some other more recent theories that have been offered as explanations of attention-involving phenomena. Nilli Lavie's theory of the effects of *perceptual load* on the efficiency of attentional filtering is unusual in the extent to which she, like Pashler, is explicit about the intention to salvage Broadbentian ideas by retaining within her theory the idea that attentional selection arises from a single bottleneck (although on her account, a moving one), occurring in a system with earlier and later parts. In a 2004 paper, Lavie, Hirst, De Fockert, and Viding even suggest that in the light of such a theory we should regard the early 1990s' rejection of the Broadbentian framework as premature. They write:

> The existence of discrepant evidence even with the same task has led some to doubt that the early and late selection debate can ever be resolved (e.g. Allport, 1993)
> However, Lavie (Lavie, 1995, 2001; Lavie and Tsal, 1994) has recently suggested that a resolution to the early and late selection debate may be found if a hybrid model of attention that combines aspects from both is considered. (Lavie et al. 2004, 340)

Other writers, in contrast, have suggested that the fragmentation of theorizing that followed the early 1990s' rejection of the Broadbentian framework did not go far enough. In a review article published in 2001, Jon Driver writes:

> This account [Broadbent's] is still heavily influential today. The distinction between a parallel "preattentive" stage encoding simple physical properties vs. a serial "attentive" stage encoding more abstract properties remains common in the current literature. Indeed, a dichotomous preattentive/attentive split is often assumed as given, perhaps too readily [...] Indeed, some authors (e.g. Allport, 1980, 1987, 1993) have argued that Broadbent's ingenious ideas may if anything have been almost too influential; once exposed to them, it becomes hard to think about attentional issues in any other way! (Driver 2001, 56)

One way to make sense of this long-standing ambivalence about the critique of the Broadbentian paradigm is to realize that, to the extent that psychology has followed Allport's 1993 article in rejecting the elements that went into Broadbent's theory, it has not done so simply by following Allport's rejection of the assumptions on which those elements depended. It has, instead, taken to heart Allport's remarks about attention's heterogeneity, and has put a philosophical twist upon them. The

philosophical twist owes something to the eliminative materialism that Paul Churchland introduced to the philosophical literature at around the time when psychology's
shift away from Broadbentian thinking was taking place (Churchland 1992). With
this eliminativist thinking in mind, the heterogeneity of attentional phenomena is
taken as evidence that the word *attention* must be multiply ambiguous, and this ambiguity is taken as a sign that the word belongs to a defective folk theory, which psychologists ought to be concerned with replacing, rather than elaborating.

This piece of eliminativism is sometimes treated as absolutely orthodox. In a
1997 textbook on attention, and again in that book's 2006 edition, Elizabeth Styles
tells her readers that "attention is not a single concept," and asks them to "accept
that to try to define attention as a unitary concept is not possible and to do so would
be misleading" (Styles 1997, 9). It is unclear, however, that the case for elimination
here will stand up to philosophical scrutiny. The dubiousness of this eliminativist
thinking was noted by Raja Parasuraman in the introduction to his 1998 book, *The
Attentive Brain*. Parasuraman identifies a tendency among psychologists to wonder
whether, "when confronted with [...] a list of [attention's] putative functions," our
response should be to "question the very concept of attention":

> If attention participates in all those functions, is it separate from each or is it an
> integral part of them? Or is attention epiphenomenal? Alternatively, if attention is
> not a single entity with a single definition, is it not an ill-conceived concept?
> (Parasuraman 1998, 3)

As Parasuraman goes on to note, this would be much too hasty as an argument for
eliminativism. There is no reason to think that a phenomenon participating in diverse functions must be ontologically dubious or ill-conceived. Nor is there any
reason to think that it will not be susceptible to explanation by a single unified
theory. What does follow from the heterogeneity of attention's functions is that we
should not expect a unified theory of attention to be given at the same level of description as that at which these heterogeneous functions are described. But the same
lower-level mechanism might nonetheless be at work in all of them. The attempt to
uncover ubiquitous common mechanisms, identified at a relatively low level of description, is one of the distinctive features of the new approach to attention that has
been establishing itself since around the beginning of the present century.

3. BIASED COMPETITION

We have seen that in the period following the collapse of the Broadbentian project,
the consensus among psychologists was that there was no single explanation of attention to be given, and that, for much of the 1990s, unified theories of attention
were neither proposed nor thought to be needed. "Attention" was taken to be the
name of, at most, a broad topic with various aspects, each of which was in need of

its own explanatory treatment. Over approximately the last ten years this attitude towards the explanation of attention has become less and less popular. The prospects for a single unified theory of attention have increasingly come to be regarded as optimistic ones.

So it is that in the recent literature we find Claus Bundesen and Thomas Habekost introducing the received view by writing that "Many have claimed that attention is simply an ill-defined term referring to a broad class of so-called 'attentional phenomena.'" In this connection Bundesen and Habekost cite Allport's 1993 article as well as Edgar Rubin's *Die Nichtexistenz der Aufmerksamkeit* (Rubin 1965/1925). (They also cite a 2006 article by John Duncan, although this is somewhat less fair: Duncan's view, at least in that article, is not that attention should be eliminated from our ontology, but only that it is a family resemblance concept.) With this eliminativist thinking established as being the orthodoxy, Bundesen and Habekost are then able to present their own theory as a revolutionary one: "In this book we shall argue for a quite different position: Attention, at least in the visual domain, is the working of a few specific mechanisms that follow a unified set of mathematic equations." (Bundesen and Habekost 2008, 3)

In a similar vein, and with a similar theoretical framework guiding them, Pastukhov, Fisher, and Braun have also argued for a return to theories that attempt to explain attention by identifying a unified cognitive basis, at least on a modality by modality basis. In an article from 2009, they present a series of experiments comparing the attention demands of various visual discrimination tasks, with a view to examining any contrast that there might be between the effects of dividing attention between similar tasks and of dividing it between dissimilar ones. On the basis of their discovery that the concurrent performance of dissimilar tasks seems to compromise attentive performance just as much as does the concurrent performance of similar ones, Pastukhov, Fisher, and Braun conclude (very much contrary to the suggestions that dominated the immediately post-Broadbent era) "that selective visual attention is a single, undifferentiated 'specific resource'" (Pastukhov et al. 2009, 1172).

The model within whose broad framework both of these sets of authors take themselves to be working is that of the biased (or "integrated") competition model, proposed and developed by John Duncan (2006), in collaboration with Robert Desimone, John Reynolds, and others. The competition that gives this model its name is not competition for passage through any particular bottleneck in processing capacity. Instead it is a competition arising from mutual inhibition, of the sort that takes place between patterns of activation throughout the central nervous system. The case for thinking that such a competition takes place does not come solely, or even primarily, from an inference to the best explanation of behavioral data. It comes, instead, from our understanding of the functional anatomy of the visual cortex (and from evidence suggesting that the processing in other sensory modalities is analogous, although here, as elsewhere in the recent literature, it is the visual case that is the best understood and most discussed). In visual cortex, as throughout the brain, there are inhibitory connections between neurons within each level of the processing hierarchy. Because of these inhibitory connections, the establishment of a strong pattern of firing in one group of neurons will tend to suppress the establishment of similar patterns in other neurons at the same level. The functional consequence of this is that when various stimuli are presented, or when the stimuli that are presented are ambiguous, a competition takes place between the various possible representations that any given level of processing might produce in response to those stimuli. In more anterior parts of the visual cortex, where the size and complexity of the receptive fields is greater, the representations that participate in these competitions will be of larger and more complex objects. In the early visual cortex they will be simple: two points of light falling within the receptive field of a single cell will compete in a struggle to suppress one another's representation (Connor et al. 1996, Geng et al. 2006, Jack et al. 2006).

These anatomically based functional claims provide an explanation of the selectivity of attention if they are supplemented with two further claims: First, that a stimulus receives a subject's attention when and only when it is victorious in the distributed, multilevel competition that these inhibitory connections give rise to. Second, and crucially, that there are top-down signals, originating in a network of frontal and parietal regions, by which one's current task priorities and intentions exert a biasing influence on the way in which this competition turns out. In some accounts the way in which this biasing influence operates is by boosting the baseline from which certain stimuli start. According to other accounts, the biasing influence operates by selectively boosting contrast gains. In yet other accounts, its effect is to sharpen the tuning curve of certain cells, so that the firing of those cells is driven by a narrower range of stimuli than would otherwise have been the case. In the most sophisticated of the recent computational models (Reynolds and Heeger's 2009 "Normalization Model"), a way has been found in which all of these effects can be understood as emerging from different interactions of the same parameters (although in the current state of our theorizing, this last model is not yet pitched at a level that makes any specific proposals about the cellular mechanisms by which such interactions might be implemented).

The biased competition model owes a large part of its appeal to the simplicity of the explanatory resources with which it operates, and to the variety of phenomena that make sense when described in its terms. Such phenomena are uncovered by a broad spectrum of methodological approaches. At the personal, phenomenological end of this methodological spectrum there are studies depending on introspective reports of our experiences in conditions of binocular rivalry. In these conditions, two different but overlapping stimuli are presented, each in a different color. Thanks to the use of a stereoscope, or to the wearing of spectacles with differently colored lenses, one of these stimuli is received only by the left eye and the other only by the right. What we experience in these conditions is very naturally described in the biased competition model's terms: the experience is of one stimulus succeeding, at least for a while, in suppressing the other one, either because, after a certain amount of struggle, that stimulus happens to be the one that wins out, or else because that stimulus is stronger and more salient from the start, and so wins the competition easily, or, in the most interesting case (although also the case for which introspection is least reliable), because the subject can intentionally bias the competition against the losing stimulus by focusing attention on those parts of the other stimulus that are making a strong showing. A plausible interpretation of the distinctive pattern of experiences that is produced in these conditions of binocular rivalry is that the experience is a phenomenal manifestation of the later stages of the intercellular competition for cortical representation. What is a lot less clear, however, is whether the phenomenology of binocular rivalry shows that such a competition is being biased by a top-down attention signal. There is good evidence that deliberately paying attention to one stimulus can have *some* suppressing effect on the perception of the other, but the effect is not strong, and it seems not actually to increase the dominance of the attended stimulus, only to prolong the periods when neither stimulus is dominant (Meng and Tong 2004, experiment 2). The results here are consistent with the idea that attention plays some role in settling the competition between stimuli in conditions of binocular rivalry, but the effect of attention in these conditions is weak when compared with attention's role in settling on the interpretation of less ecologically peculiar stimuli, such as the Necker cube or Rubin vase (Meng and Tong, experiment 3).

The phenomenology of one's shifting awareness in conditions of binocular rivalry is naturally described as the manifestation of a competition, and perhaps of a biased competition. This same interpretation suggests itself when we look at the more objectively gathered data that we find when using fMRI and single-cell recording methods to gauge the rates of neuronal firing elicited by attended and unattended stimuli in simple lab tasks. In these cases the evidence for top-down biasing is clearer. In single cell recordings in the macaque, Reynolds and Chelazzi observe that pairs of stimuli that are presented in the receptive field of a single cell seem to suppress one another, pulling down the rate of that cell's firing when compared to the rate that a single stimulus produces when presented on its own. When attention is directed on one of these stimuli rather than another, a winner emerges in the competition, with the result that the firing rate returns to the level typically elicited when that stimulus is presented on its own (Reynolds, Chelazzi, and Desimone 1999). In a series of experiments examining a diverse range of brain areas, Sabine Kastner has observed an analogous effect in fMRI studies with humans. A cluster of stimuli, none of which is being attended, elicits a relatively weak BOLD signal when compared to the signal that those same stimuli elicit when attention is being paid to one of them (Kastner and Ungerleider 2000, 2001).

There are at least two ways in which this biased competition model breaks from the assumptions of its Broadbentian predecessors. First, the idea that attended stimuli are all and only those that have been victorious in a biased competition taking place *throughout* the cortex is, by itself, a break from the tradition of looking for localizable bottlenecks that have only attended stimuli downstream from them. This non-Broadbentian aspect of biased-competition thinking is widely acknowledged. More radical, and less frequently noticed, is the fact that the biased competition model enables us to break altogether from Broadbent's capacity limitation view of attention's function. We have seen above that Reynolds and Desimone set the stage for the biased competition model (and that Bundesen and Habekost did so for their own version of a competition-based model) by talking about the need for attention as arising from a need for capacity-limitation-management. But the intercellular competition on which this model is based need not be a competition for limited capacity resources in anything like Broadbent's sense. The intercellular competition that Reynolds and Desimone are concerned with would have just the same selective effects whether the capacity for processing were large or small. That competition would be selective whatever the capacity of the participating systems, just as a well-organized league selects a single champion, however skilled and however numerous the competitors. Instead of understanding attention as serving the function of managing "the severe capacity limitations that characterize humans" (in Bundesen and Habekost's phrase 2008, 3), the biased competition model can instead be seen as opening the door to an understanding of attention as functioning in the regulation and management of that selectivity that is necessary to our practical and epistemic agency. One such function for attention's selectivity, emphasized in John Campbell's philosophical treatments of attention and reference, is in enabling us to fix concepts onto the deliverances of perception in such a way as to enable the things perceived to be things about which we can exercise our rationality (see Campbell 2002, Smithies 2011). Another such function—one that was noted by Allport and by Odmar Neumann in the late 1980s (Allport 1987, Neumann 1987), and that has more recently been revived in philosophical work by Wayne Wu (2008)—is in enabling us to execute a coherent course of action, even when faced with an affordance-rich environment that obliges us to select one of the several objects in our environment as the target for action, and to select one of the several actions afforded by that object as the action that is to be performed.

4. CONFLICTING INTERPRETATIONS OF BIASED COMPETITION

One of the ways in which the biased competition model has proved its intellectual fecundity is by providing those who are looking for the top-down source of attentional control signals with an operationalizable account of the sorts of modulation for which they should be looking. Katherine Armstrong and Tirin Moore, for example, have had great success in their investigations of the contribution made to attention in the macaque by activity in the frontal eye fields. Their technique for examining the attention-related activity in these cortical areas involves probing the ways in which microstimulation of them has competition-biasing effects in the visual cortex (Armstrong et al. 2006; Moore and Armstrong 2003). The biased competition model can also claim success as a working hypothesis for the interpretation of noninvasive work with humans. It is by using the biased competition model in the interpretation of their fMRI results that Kastner and Ungerleider are able to locate a particular network of cortical areas that show signs of having a special role in creating the top-down competition-biasing signals that the model postulates (Kastner and Ungerleider 2000).

There is room for controversy, however, about how exactly these findings should be understood. According to one way of understanding the biased competition model, attention is implemented by a competition that takes place throughout the cortical processing stream, and the outcome of that competition is influenced by a number of interacting factors. These include simple perceptual salience. They also include such factors as congruence with the agent's intentions and desires: factors whose influence on the competition is mediated by "top-down" signals. Those top-down signals might be especially interesting, from a psychological point of view, and they might have a stronger-than-usual influence on the way in which the attention-constituting competition comes out, but they have no special status from the point of view of attention's metaphysics. It is, in this interpretation of the theory, the whole of the intercellular competition that is the neural mechanism by which attention is realized.

This interpretation of biased-competition contrasts with an alternative interpretation according to which attention is realized, not by the whole of the intercellular competition, but only by the processes responsible for the top-down signals that bias it.

The first of these interpretations is the one that Desimone and Duncan themselves recommend. They write that:

> The approach we take differs from the standard view of attention, in which attention functions as a mental spotlight enhancing the processing (and perhaps binding together the features) of the illuminated item. Instead the model we develop is that attention is an emergent property of many neural mechanisms working to resolve competition for visual processing and control of behavior. (Desimone and Duncan 1995, 194)

But the second interpretation is implied by some claims made by the authors whose research employs the biased competition model. The article in which Kastner and Ungerleider claim to identify "the neural basis of biased competition in human visual cortex" is a case in point. They remark:

> Functional brain imaging studies reveal that biasing signals due to selective attention can modulate neural activity in visual cortex [...]. [T]he source of top–down biasing signals likely derives from a distributed network of areas in frontal and parietal cortex. Attention-related activity in frontal and parietal areas does not reflect attentional modulation of visually evoked responses, but rather the attentional operations themselves. (Kastner and Ungerleider 2001, p. 1263)

This is not an explicit disagreement with the Desimone and Duncan interpretation of biased-competition, but it is clear that Kastner and Ungerleider must favor the alternative interpretation, since in the interpretation that Desimone and Duncan favor the things that Kastner and Ungerleider say would fail to make sense. There would be no sense in asking which of the factors influencing the intercellular competition originates from "the attentional operations themselves" if it is the whole competition that is the mechanism by which attention operates. Kastner and Ungerleider's claim that some of the signals that contribute to the outcome of the intercellular competition are "due to selective attention" while some of the sources of these signals "are the attentional operations themselves" must be taken as ruling out the first interpretation of the biased competition model. That interpretation would make their claim as strange as the claim that Tottenham's 2–1 victory over Liverpool consisted in (or, more bizarrely, that it *caused*) one of their two goals in particular.

This dispute between the two rival interpretations of the biased competition model has not been clearly articulated in the current literature.[1] This may be because the issue has been thought to be a merely verbal one, depending not on one's understanding of the mechanisms by which attention is brought about, but only on the question of which components of those mechanisms one wants to label as attention. There is an extent to which this is true. The two interpretations of the biased competition model are interpretations of the same model: they are not separated by their answers to any questions about what happens in the brain when attention is paid, only by their understanding of which of the things that happens is the realizer of attention. But this is not merely a verbal matter. There is a genuine disagreement here: one that is, perhaps, of primarily philosophical interest, but that might nonetheless have methodological consequences.

The first interpretation of the model sees attention as an emergent phenomenon, realizable by diverse processes in diverse contexts. The second sees attention as a particular process—the process of competition-biasing—having a particular (although perhaps only disjunctively specifiable) locus in the brain. If the latter view is right, then we should expect to be able to make good inductive inferences about the processes that determine whether attention is paid across a range of task conditions. If the first view is right, then there is no reason to expect the processes that lead to victory for the attended stimuli in particular lab tasks will be the processes that determine the outcome of that competition in general (see Mole 2011 for more on this).

The settling of this dispute will depend on our making progress with certain metaphysical questions concerning the nature of emergence, and concerning the causal and explanatory status of psychological phenomena that have no particular neural processes corresponding to them. This is therefore one of the places where the philosophically informed interdisciplinarity that is distinctive of cognitive science proves to be indispensable.

[1] One reason for this may that there is one use of "model" and "interpretation"—the logician's use—on which it makes no sense to ask for the interpretation of a model. In this usage a model *is* an interpretation—but in this sense, of course, the "biased competition model" is not really a model at all. In the logician's usage, the biased-competition model is a *theory*.

REFERENCES

Allport, D. A. (1980). Attention and Performance. In G. Claxton (ed.), *Cognitive Psychology: New Directions*. London: Routledge & Kegan Paul.

⸻. (1987). Selection for Action: Some Behavioural and Neurophysiological Considerations of Attention and Action. In H. Heuer and D. F. Saunders (eds.), *Perspectives on Perception and Action*. Hillsdale, NJ: Erlbaum.

⸻. (1993). Attention and Control: Have We Been Asking the Wrong Questions? A Critical Review of Twenty Five Years. In S. Kornblum and D, Mayer (eds.), *Attention and Performance 14: Synergies in Experimental Psychology, Artificial Intelligence and Cognitive Neuroscience*. Cambridge, MA: MIT Press.

⸻. (2011). Attention and Integration. In C. Mole, D. Smithies, and W. Wu (eds.), *Attention: Philosophical and Psychological Essays*. New York: Oxford University Press.

Armstrong, K. M., Fitzgerald, J. K., and Moore, T. (2006). Changes in visual receptive fields with microstimulation of frontal cortex. *Neuron* 50(5): 791–98.

Bricolo, E., Gianesini, T., Fanini, A., Bundesen, C., and Chelazzi, L. (2002). Serial attention mechanisms in visual search: A direct behavioral demonstration. *Journal of Cognitive Neuroscience* 14(7): 980–93.

Broadbent, D. E. (1958). *Perception and Communication*. Elmsford, NY: Pergamon Press.

⸻. (1971). *Decision and Stress*. London: Academic Press.

⸻. (1982). Task combination and selective intake of information. *Acta Psychologica* 50(3): 252–90.

Bundesen, C., and Habekost, T. (2008). *Principles of Visual Attention: Linking Mind and Brain*. Oxford: Oxford University Press.

Campbell, J. (2002). *Reference and Consciousness*. Oxford: Oxford University Press.

Chomsky, N. (1957). *Syntactic Structures*. Oxford: Mouton.

⸻. (1959). A review of B. F. Skinner's *Verbal Behavior*. *Language*, 35(1): 26–58.

Churchland, P. M. (1992). Activation vectors versus propositional attitudes: How the brain represents reality. *Philosophy and Phenomenological Research* 52(2): 419–424.

Connor, C. E., Gallant, J. L., Preddie, D. C. and Van Essen, D. C. (1996). Responses in area V4 depend on the spatial relationship between stimulus and attention. *Journal of Neurophysiology* 75(3): 1306–8.

Desimone, R., and Duncan, J. (1995). Neural mechanisms of selective visual attention. *Annual Review of Neuroscience* 18(1): 193–222.

Deutsch, J. A., and Deutsch, D. (1963). Attention: Some theoretical considerations. *Psychological Review* 70(1): 80–90.

Driver, J. (2001). A selective review of selective attention research from the past century. *British Journal of Psychology* 92(1): 53–78.

Duncan, J. (1980). The locus of interference in the perception of simultaneous stimuli. *Psychological Review* 87(3): 272–300.

⸻. (2006). Brain mechanisms of attention. *The Quarterly Journal of Experimental Psychology* 59(1): 2–27.

Geach, P. (1957). *Mental Acts: Their Content and Their Objects*. New York: The Humanities Press.

Geng, J. J., Eger, E., Ruff, C. C., Kristjánsson, A., Rotshtein, P., and Driver, J. (2006). On-line attentional selection from competing stimuli in opposite visual fields: Effects on human visual cortex and control processes. *Journal of Neurophysiology* 96(5): 2601–12.

Jack, A. I., Shulman, G. L., Snyder, A. Z., McAvoy, M., and Corbetta, M. (2006). Separate modulations of human V1 associated with spatial attention and task structure. *Neuron* 51(1): 135–47.

James, William (1890/1981). *The Principles of Psychology*. Cambridge MA: Harvard University Press.

Kastner, S., and Ungerleider, L. G. (2000). Mechanisms of visual attention in the human cortex. *Annual Review of Neuroscience* 23: 315–41.

——.(2001). The neural basis of biased competition in human visual cortex. *Neuropsychologia* 39(12): 1263–76.

Lavie, N. (1995). Perceptual load as a necessary condition for selective attention. *Journal of Experimental Psychology: Human Perception and Performance* 21(3): 451–68.

——. (2001). Capacity Limits in Selective Attention: Behavioural Evidence and Implications for Neural Activity. In J. Braun, C. Koch, and J. Davis (eds.), *Visual Attention and Cortical Circuits*. Cambridge, MA: MIT Press.

Lavie, N., and Tsal, Y. (1994). Perceptual load as a major determinant of the locus of selection in visual attention. *Perception & Psychophysics* 56(2): 183–97.

Lavie, N., Hirst, A., de Fockert, J. W., and Viding, E. (2004). Load theory of selective attention and cognitive control. *Journal of Experimental Psychology: General*, 133(3): 339–54.

Meng, M., and Tong, F. (2004). Can attention selectively bias bistable perception? Differences between binocular rivalry and ambiguous figures. *Journal of Vision*, 4(7): 539–51.

Mole, C. (2011). *Attention Is Cognitive Unison*. New York: Oxford University Press.

Moore, T., and Armstrong, K. M. (2003). Selective gating of visual signals by microstimulation of frontal cortex. *Nature* 421 (6921): 370–73.

Moray, N. (1959). Attention in dichotic listening: Affective cues and the influence of instructions. *The Quarterly Journal of Experimental Psychology* 11(1): 56–60.

Neumann, O. (1987). Beyond Capacity: A Functional View of Attention. In H. Heuer and A. F. Sanders (eds.), *Perspectives on Perception and Action*. Hillsdale NJ: Erlbaum.

Newell, A. Shaw, J. C., and Simon, H. A. (1957). Empirical exploration of the logic theory machine. In *Papers Presented at the February 1957, Western Joint Computer Conference*. New York: Association of Computing Machinery.

Norman, D. (1968). Toward a theory of memory and attention. *Psychological Review* 75(6): 522–36.

Parasuraman, R. (1998). *The Attentive Brain*. Cambridge, MA: MIT Press, 577.

Pashler, H. (1984). Processing stages in overlapping tasks: Evidence for a central bottleneck. *Journal of Experimental Psychology: Human Perception and Performance* 10(3): 358–77.

——. (1994). Dual-task interference in simple tasks: Data and theory. *Psychological Bulletin* 116(2): 220–44.

Pastukhov, A., Fischer, L., and Braun, J. (2009). Visual attention is a single, integrated resource. *Vision Research* 49(10): 1166–73.

Posner, M. I., and Petersen, S. E. (1990). The attention system of the human brain. *Annual Review of Neuroscience* 13: 25–42.

Reynolds, J., and Desimone, R. (2000). Competitive Mechanisms Subserve Selective Visual Attention. In A. Marantz, Y, Miyashita, and W. O'Neil (eds.), *Image, Language, Brain: Papers from the First Mind Articulation Project Symposium*. Cambridge, MA: MIT Press.

Reynolds, J., Chelazzi, L., and Desimone, R. (1999). Competitive mechanisms subserve attention in macaque areas V2 and V4. *Journal of Neuroscience* 19(5): 1736–53.

Reynolds, J. H., and Heeger, D. J. (2009). The normalization model of attention. *Neuron* 61(2): 168–85.

Rubin, E. (1965/1925). *Die Nichtexistenz Der Aufmerksamkeit* (Psykologiske Tekster, 1). Copenhagen: Akademisk Forlag.

Shiffrin, R. M., and Schneider, W. (1997). Controlled and automatic human information processing: I. Detection, search and attention. *Psychological Review* 84(1): 1–66.

Smithies, D. (2011). Attention Is Rational-Access Consciousness. In C. Mole, D. Smithies, and W. Wu (eds.), *Attention: Philosophical and Psychological Essays*. New York: Oxford University Press.

Styles, E. A. (1997). *The Psychology of Attention*. Hove, UK: Psychology Press/Erlbaum.

Treisman, A. M. (1960). Contextual cues in selective listening. *The Quarterly Journal of Experimental Psychology* 12(4): 242–48.

———. (1993). The perception of features and objects. In A. D. Baddeley, and L. Weiskrantz (eds.), *Attention: Selection, Awareness, and Control: A Tribute to Donald Broadbent*. New York: Oxford University Press.

Treisman, A. M., and Gelade, G. (1980). A feature-integration theory of attention. *Cognitive Psychology* 12(1): 97–136.

Wittgenstein, L. (1953). *Philosophical Investigations*. New York: Macmillan.

Wu, W. (2008). Visual attention, conceptual content, and doing it right. *Mind* 117(468): 1003–33.

CHAPTER 10

COMPUTATIONALISM

GUALTIERO PICCININI

COMPUTATIONALISM is the view that cognitive capacities have a computational explanation, or, somewhat more strongly, that cognition is (a kind of) computation. For simplicity, we will use these two formulations interchangeably. Most cognitive scientists endorse some version of computationalism and pursue computational explanations as their research program. Thus, when cognitive scientists propose an explanation of a cognitive capacity, the explanation typically involves computations that result in the cognitive capacity.

Computationalism is controversial but resilient to criticism. To understand why mainstream cognitive scientists endorse computationalism, we need to understand what computationalism says. That, in turn, requires an account of computation.

1. A SUBSTANTIVE EMPIRICAL HYPOTHESIS

Computationalism is usually introduced as an empirical hypothesis that can be disconfirmed. Whether computationalism has empirical bite depends on how we construe the notion of computation: the more inclusive a notion of computation, the weaker the version of computationalism formulated in its terms.

At one end of the continuum, some notions of computation are so loose that they encompass virtually everything. For instance, if computation is construed as the production of outputs from inputs, and if any state of a system qualifies as an input (or output), then every process is a computation. Sometimes, computation is construed as information processing, which is somewhat more stringent, yet the resulting version of computationalism is quite weak. There is little doubt that organisms gather and process information about their environment (more on this below).

Processing information is surely an important aspect of cognition. Thus, if computation is information processing, then cognition involves computation. But this does not tell us much about how cognition works. In addition, the notions of information and computation in their most important uses are conceptually distinct, have different histories, are associated with different mathematical theories, and have different roles to play in a theory of cognition. It is best to keep them separate (Piccinini and Scarantino 2011).

Computationalism becomes most interesting when it has explanatory power. The most relevant and explanatory notion of computation is that associated with digital computers. Computers perform impressive feats: they solve mathematical problems, play difficult games, prove logical theorems, etc. Perhaps cognitive systems work like computers. Historically, this analogy between computers and cognitive systems is the main motivation behind computationalism. The resulting form of computationalism is a strong hypothesis—one that should be open to empirical testing. To understand it further, it will help to briefly outline three research traditions associated with computationalism and how they originated.

2. THREE RESEARCH TRADITIONS: CLASSICISM, CONNECTIONISM, AND COMPUTATIONAL NEUROSCIENCE

The view that thinking has something to do with computation may be found in the works of some modern materialists, such as Thomas Hobbes (Boden 2006, 79). But computationalism properly so-called could not begin in earnest until a number of logicians (most notably Alonzo Church, Kurt Gödel, Stephen Kleene, Emil Post, and especially Alan Turing) laid the foundations for the mathematical theory of computation.

Turing (1936–37) analyzed computation in terms of what are now called Turing machines—a kind of simple processor operating on an unbounded tape. The tape is divided into squares, which the processor can read and write on. The processor moves along the tape reading and writing on one square at a time depending on what is already on the square as well as on the rules that govern the processor's behavior. The rules state what to write on the tape and where to move next depending on what is on the tape as well as which of finitely many states the processor is in.

Turing argued convincingly that any function that can be computed by following an algorithm (i.e., an unambiguous list of instructions operating on discrete symbols) can be computed by a Turing machine. Church (1936) offered a similar proposal in terms of general recursive functions, and it turns out that a function is general recursive if and only if it can be computed by a Turing machine. Given this extensional equivalence between Turing machines and general recursive functions, the thesis that any algorithmically computable function is computable by some Turing machine (or equivalently, is general recursive) is now known as the *Church-Turing thesis* (Kleene, 1952, §§ 62, 67).

Turing made two other relevant contributions. First, he showed how to construct *universal* Turing machines. These are Turing machines that can mimic any other Turing machine by encoding the rules that govern the other machine as instructions, storing the instructions on a portion of their tape, and then using the encoded instructions to determine their behavior on the input data. Notice that ordinary digital computers, although they have more complex components than universal Turing machines, are universal in the same sense (up to their memory limitations). That is, digital computers can compute any function computable by a Turing machine until they run out of memory.

Second, Turing showed that the vast majority of functions whose domain is denumerable (e.g., functions of strings of symbols or of natural numbers) are actually *not* computable by Turing machines. These ideas can be put together as follows: assuming the Church-Turing thesis, a universal digital computer can compute any function computable by algorithm, although the sum total of these Turing-computable functions is a tiny subset of all the functions whose domain is denumerable.

Modern computationalism began when Warren McCulloch and Walter Pitts (1943) connected three things: Turing's work on computation, the explanation of cognitive capacities, and the mathematical study of neural networks. Neural networks are sets of connected signal-processing elements ("neurons"). Typically, they have elements that receive inputs from the environment (input elements), elements that yield outputs to the environment (output elements), and elements that communicate only with other elements in the system (hidden elements). Each element receives input signals and delivers output signals as a function of its input and current state. As a result of their elements' activities and organization, neural networks turn the input received by their input elements into the output produced by their output elements. A neural network may be either a concrete physical system or an abstract mathematical system. An abstract neural network may be used to model another system (such as a network of actual neurons) to some degree of approximation.

The mathematical study of neural networks using biophysical techniques began around the 1930s (Rashevsky 1938, 1940; Householder and Landahl 1945). But before McCulloch and Pitts, no one had suggested that neural networks have something to do with computation. McCulloch and Pitts defined networks that operate on sequences of discrete inputs in discrete time, argued that they are a useful idealization of what is found in the nervous system, and concluded that the activity of their networks explains cognitive phenomena. McCulloch and Pitts also pointed out that their networks can perform computations like those of Turing machines. More precisely, McCulloch-Pitts networks are computationally equivalent to Turing machines without tape or finite state automata (Kleene 1956). Notice that modern digital computers are a kind of McCulloch-Pitts neural network (von Neumann 1945). Digital computers are sets of logic gates—digital signal-processing elements equivalent to McCulloch-Pitts neurons—connected to form a specific architecture.

McCulloch and Pitts's account of cognition contains three important aspects: an analogy between neural processes and digital computations, the use of mathematically defined neural networks as models, and an appeal to neurophysiological evidence to support their neural network models. After McCulloch and Pitts, many others linked computation and cognition, though they often abandoned one or more aspects of McCulloch and Pitts's theory. Computationalism evolved into three main traditions, each emphasizing a different aspect of McCulloch and Pitts's account.

One tradition, sometimes called *classicism*, emphasizes the analogy between cognitive systems and digital computers while downplaying the relevance of neuroscience to the theory of cognition (Miller, Galanter, and Pribram 1960; Fodor 1975; Newell and Simon 1976; Pylyshyn 1984; Newell 1990; Pinker 1997; Gallistel and King 2009). When researchers in this tradition offer computational models of a cognitive capacity, the models take the form of computer programs for producing the capacity in question. One strength of the classicist tradition lies in programming computers to exhibit higher cognitive capacities such as problem solving, language processing, and language-based inference.

A second tradition, most closely associated with the term *connectionism* (although this label can be misleading, see Section 6 below), downplays the analogy between cognitive systems and digital computers in favor of computational explanations of cognition that are "neurally inspired" (Rosenblatt 1958; Feldman and Ballard 1982; Rumelhart and McClelland 1986; Bechtel and Abrahamsen 2002). When researchers in this tradition offer computational models of a cognitive capacity, the models take the form of neural networks for producing the capacity in question. Such models are primarily constrained by *psychological* data, as opposed to *neurophysiological* and *neuroanatomical* data. One strength of the connectionist tradition lies in designing artificial neural networks that exhibit cognitive capacities such as perception, motor control, learning, and implicit memory.

A third tradition is most closely associated with the term *computational neuroscience*, which in turn is one aspect of theoretical neuroscience. Computational neuroscience downplays the analogy between cognitive systems and digital computers even more than the connectionist tradition. Neurocomputational models aim to describe actual neural systems such as (parts of) the hippocampus, cerebellum, or cortex, and are constrained by *neurophysiological* and *neuroanatomical* data in addition to psychological data (Schwartz 1990; Churchland and Sejnowski 1992; O'Reilly and Munakata 2000; Dayan and Abbott 2001; Eliasmith and Anderson 2003; Ermentrout and Terman 2010). It turns out that McCulloch-Pitts networks and many of their "connectionist" descendents are relatively unfaithful to the details of neural activity, whereas other types of neural networks are more biologically realistic. These include the following, in order of decreasing biological detail and increasing computational tractability: conductance-based models, which go back to Hodgkin and Huxley's (1952) seminal analysis of the action potential based on conductance changes; networks of integrate-and-fire neurons, which fire simply when the input current reaches a certain threshold (Lapicque 1907/2007; Caianiello 1961; Stein 1965; Knight 1972); and firing rate models, in which there are no individual action potentials—instead, the continuous output of each network unit represents the firing rate of a neuron or neuronal population (Wilson and Cowan 1972). Computational neuroscience offers models of how real neural systems may exhibit cognitive capacities, especially perception, motor control, learning, and implicit memory.

Although the three traditions just outlined are in competition with one another to some extent (more on this in Section 6), there is also some fuzziness at their borders. Some cognitive scientists propose hybrid theories, which combine explanatory resources drawn from both the classicist and the connectionist traditions (e.g., Anderson 2007). In addition, biological realism comes in degrees, so there is no sharp divide between connectionist and neurocomputational models.

3. THREE ACCOUNTS OF COMPUTATION: CAUSAL, SEMANTIC, AND MECHANISTIC

To fully understand computationalism, we need to understand what concrete computation is. Philosophers have offered three main families of accounts. (For a more complete survey, see Section 2 of Piccinini 2010a.)

3.1. The Causal Account

According to the causal account, a physical system S performs computation C just in case (i) there is a mapping from the states ascribed to S by a physical description to the states defined by computational description C, such that (ii) the state transitions between the physical states mirror the state transitions between the computational states. Clause (ii) requires that for any computational state transition of the form $s1 \rightarrow s2$ (specified by the computational description C), if the system is in the physical state that maps onto $s1$, its physical state *causes* it to go into the physical state that maps onto $s2$ (Chrisley 1995; Chalmers 1994, 1996; Scheutz 1999, 2001; see also Klein 2008 for a similar account built on the notion of disposition rather than cause).

To this causal constraint on acceptable mappings, David Chalmers (1994, 1996) adds a further restriction: a genuine physical implementation of a computational system must divide into separate physical components, each of which maps onto the components specified by the computational formalism. As Godfrey-Smith (2009, 293) notes, this combination of a causal and a localizational constraint goes in the direction of mechanistic explanation (Machamer, Darden, and Craver 2000). An account of computation that is explicitly based on mechanistic explanation will be discussed below. For now, the causal account simpliciter requires only that the mappings between computational and physical descriptions be such that the causal relations between the physical states are isomorphic to the relations between state transitions specified by the computational description. Thus, according to the causal account, computation is the causal structure of a physical process.

It is important to note that under the causal account, there are mappings between any physical system and at least some computational descriptions. Thus, according to the causal account, everything performs at least some computations (Chalmers 1996, 331; Scheutz 1999, 191). This (limited) *pancomputationalism* strikes some as overly inclusive.[1] By entailing pancomputationalism, the causal account trivializes the claim that a system is computational, for according to pancomputationalism, digital computers perform computations in the same sense in which rocks, hurricanes, and planetary systems do. This does an injustice to computer science—in computer science, only relatively few systems count as performing computations, and it takes a lot of difficult technical work to design and build systems that perform computations reliably. Or consider computationalism, which was introduced to shed new and explanatory light on cognition. If every physical process is a computation, computationalism seems to lose much of its explanatory force (Piccinini 2007a).

Another objection to pancomputationalism begins with the observation that any moderately complex system satisfies indefinitely many objective computational descriptions (Piccinini 2010b). This may be seen by considering computational modeling. A computational model of a system may be pitched at different levels of granularity. For example, consider cellular automata models of the dynamics of a galaxy or a brain. The dynamics of a galaxy or a brain may be described using an indefinite number of cellular automata—using different state transition rules, different time steps, or cells that represent spatial regions of different sizes. Furthermore, an indefinite number of formalisms different from cellular automata, such as Turing machines, can be used to compute the same functions computed by cellular automata. It appears that pancomputationalists are committed to the galaxy or the brain performing all these computations at once. But that does not appear to be the sense in which computers (or brains) perform computations.

---

[1] The *limited* pancomputationalism entailed by the causal account should not be confused with the *unlimited* pancomputationalism according to which everything implements *every* computation (Putnam 1988; Searle 1992). Unlimited pancomputationalism is a reductio ad absurdum of a naïve account of computation according to which any mapping from a computational description to a physical description of a system is enough to ascribe a computation to the system. The accounts of computation that are discussed in the text are intended to avoid unlimited pancomputationalism.

In the face of these objections, supporters of the causal account of computation are likely to maintain that the explanatory force of computational explanations does not come merely from the claim that a system is computational. Rather, explanatory force comes from the specific computations that a system is said to perform. Thus, a rock and a digital computer perform computations in the same sense. But they perform radically different computations, and it is the difference between their computations that explains the difference between the rock and the digital computer. As to the objection that there are still too many computations performed by each system, pancomputationalists have two main options: either to bite the bullet and accept that every system implements indefinitely many computations, or to find a way to single out, among the many computational descriptions satisfied by each system, the one that is ontologically privileged—the one that captures the computation performed by the system.

Those who are unsatisfied by these replies and wish to avoid pancomputationalism need accounts of computation that are less inclusive than the causal account. Such accounts may be found by adding restrictions to the causal account.

3.2. The Semantic Account

In our everyday life, we usually employ computations to process meaningful symbols, in order to extract information from them. The semantic account of computation turns this practice into a metaphysical doctrine: computation is the processing of representations—or at least, the processing of appropriate representations in appropriate ways (Fodor 1975; Cummins 1983; Pylyshyn 1984; Churchland and Sejnowski 1992; Shagrir 2006; Sprevak 2010). Opinions as to which representational manipulations constitute computations vary. What all versions of the semantic account have in common is that according to them, there is "no computation without representation" (Fodor 1981, 180).

The semantic account may be formulated (and is usually understood) as a restricted causal account. In addition to the causal account's requirement that a computational description mirror the causal structure of a physical system, the semantic account adds a semantic requirement. Only physical states that qualify as representations may be mapped onto computational descriptions, thereby qualifying as computational states. If a state is not representational, it is not computational either.

The semantic account of computation is closely related to the common view that computation is information processing. This idea is less clear than it may seem, because there are several notions of information. The connection between information processing and computation is different depending on which notion of information is at stake. What follows based on four important notions of information is a brief disambiguation of the view that computation is information processing.

1. Information in the sense of thermodynamics is closely related to thermodynamic entropy. Entropy is a property of every physical system.

Thermodynamic entropy is, roughly, a measure of an observer's uncertainty about the microscopic state of a system after she considers the observable macroscopic properties of the system. The study of the thermodynamics of computation is a lively field with many implications in the foundations of physics (Leff and Rex 2003). In this thermodynamic sense of "information," any difference between two distinguishable states of a system may be said to carry information. Computation may well be said to be information processing in this sense, but this has little to do with semantics properly so-called.

2. Information in the sense of communication theory is a measure of the average likelihood that a given message is transmitted between a source and a receiver (Shannon and Weaver 1949). This also has little to do with semantics.

3. Information in one semantic sense is approximately the same as "natural meaning" (Grice 1957). A signal carries information in this sense just in case it reliably correlates with a source (Dretske 1981). The view that computation is information processing in this sense is prima facie implausible, because many computations—such as arithmetical calculations carried out on digital computers—do not seem to carry any natural meaning. Nevertheless, this notion of semantic information is relevant here because it has been used by some theorists to ground an account of representation (Dretske 1981; Fodor 2008).

4. Information in another semantic sense is just ordinary semantic content or "non-natural meaning" (Grice 1957). The view that computation is information processing in this sense is similar to a generic semantic account of computation.

The semantic account is popular in the philosophy of mind and in cognitive science because it appears to fit its specific needs better than a purely causal account. Since minds and computers are generally assumed to manipulate (the right kind of) representations, they turn out to compute. Since most other systems are generally assumed not to manipulate (the relevant kind of) representations, they do not compute. Thus, the semantic account appears to avoid pancomputationalism and to accommodate some common intuitions about what does and does not count as a computing system. It keeps minds and computers *in* while leaving most everything else *out*, thereby vindicating computationalism as a strong and nontrivial theory.

The semantic account faces its share of problems too. For starters, representation does not seem to be presupposed by the notion of computation employed in at least some areas of cognitive science as well as computability theory and computer science—the very sciences that gave rise to the notion of computation at the origin of the computational theory of cognition (Piccinini 2008a). If this is correct, the semantic account may not even be adequate to the needs of philosophers of mind— at least those philosophers of mind who wish to make sense of the analogy between minds and the systems designed and studied by computer scientists and computability theorists. Another criticism of the semantic account is that specifying the kind of representation and representational manipulation that is relevant to computation seems to require a non-semantic way of individuating computations (Piccinini 2004). These concerns motivate efforts to account for concrete computation in non-semantic terms.

3.3. The Mechanistic Account

An implicit appeal to some aspects of mechanisms may be found in many accounts of computation, usually in combination with an appeal to causal or semantic properties (Chalmers 1996; Cummins 1983; Egan 1995; Fodor 1980; Glymour 1991; Horst 1996; Newell 1980; Pylyshyn 1984; Shagrir 2001; Stich 1983). Nevertheless, the received view is that computational explanations and mechanistic explanations are distinct and belong at different "levels" (Marr 1982, Rusanen and Lappi 2007). By contrast, this section introduces an explicitly mechanistic account of computation that does not rely on semantics (Piccinini 2007b; Piccinini and Scarantino 2011, § 3). According to this account, computational explanation is a species of mechanistic explanation; concrete computing systems are functionally organized mechanisms of a special kind—mechanisms that perform concrete computations.

Like the semantic account, the mechanistic account may be understood as a restricted causal account. In addition to the causal account's requirement that a computational description mirror the causal structure of a physical system, the mechanistic account adds a requirement about the functional organization of the system. Only physical states that have a specific functional significance within a specific type of mechanism may be mapped onto computational descriptions, thereby qualifying as computational states. If a state lacks the appropriate functional significance, it is not a computational state.

A functional mechanism is a system of organized components, each of which has functions to perform (cf. Craver 2007; Wimsatt 2002). When appropriate components and their functions are appropriately organized and functioning properly, their combined activities constitute the capacities of the mechanism. Conversely, when we look for an explanation of the capacities of a mechanism, we decompose the mechanism into its components and look for their functions and organization. The result is a mechanistic explanation of the mechanism's capacities.

This notion of mechanism is familiar to biologists and engineers. For example, biologists explain physiological capacities (digestion, respiration, etc.) in terms of the functions performed by systems of organized components (the digestive system, the respiratory system, etc.).

According to the mechanistic account, a computation in the generic sense is the processing of vehicles according to rules that are sensitive to certain vehicle properties, and specifically, to differences between different portions of the vehicles. The processing is performed by a functional mechanism, that is, a mechanism whose components are functionally organized to perform the computation. Thus, if the mechanism malfunctions, a miscomputation occurs.

When we define concrete computations and the vehicles that they manipulate, we need not consider all of their specific physical properties. We may instead consider only the properties that are relevant to the computation, according to the rules that define the computation. A physical system can be described at different levels of abstraction. Since concrete computations and their vehicles are described sufficiently abstractly as to be defined independently of the physical media that implement them, they may be called *medium-independent*.

In other words, a vehicle is medium-independent just in case the rules (i.e., the input-output maps) that define a computation are sensitive only to differences between portions of the vehicles along specific dimensions of variation—they are insensitive to any more concrete physical properties of the vehicles. Put yet another way, the rules are functions of state variables associated with a set of functionally relevant degrees of freedom, which can be implemented differently in different physical media. Thus, a given computation can be implemented in multiple physical media (e.g., mechanical, electromechanical, electronic, magnetic, etc.), provided that the media possess a sufficient number of dimensions of variation (or degrees of freedom) that can be appropriately accessed and manipulated and that the components of the mechanism are functionally organized in the appropriate way.

Notice that the mechanistic account avoids pancomputationalism. First, physical systems that are not functional mechanisms are ruled out. Functional mechanisms are complex systems of components that are organized to perform functions. Any system whose components are not organized to perform functions is not a computing system because it is not a functional mechanism. Second, mechanisms that lack the function of manipulating medium-independent vehicles are ruled out. Finally, medium-independent vehicle manipulators whose manipulations fail to accord with appropriate rules are ruled out. The second and third constraints appeal to special functional properties—manipulation of medium-independent vehicles in accordance with rules defined over the vehicles—that are possessed only by relatively few concrete mechanisms. According to the mechanistic account, those few mechanisms are the genuine computing systems.

Another feature of the mechanistic account is that it makes sense of miscomputation—a notion difficult to make sense of under the causal and semantic accounts. Consider an ordinary computer programmed to compute function f on input i. Suppose that the computer malfunctions and produces an output different from $f(i)$. According to the causal (semantic) account, the computer just underwent a causal process (a manipulation of representations), which may be given a computational description and hence counts as computing some function $g(i)$, where $g \neq f$. By contrast, according to the mechanistic account, the computer simply failed to compute, or at least failed to complete its computation. Given the importance of avoiding miscomputations in the design and use of computers, the ability of the mechanistic account to make sense of miscomputation gives it a further advantage over rival accounts.

The most important advantage of the mechanistic account over other accounts is that it distinguishes and characterizes precisely many different kinds of computing systems based on their specific mechanistic properties. Given its advantages, from now on we will presuppose the mechanistic account. The next section will describe some important kinds of computation.

4. KINDS OF COMPUTATION: DIGITAL, ANALOG, AND MORE

4.1. Digital Computation

The best-known kind of computation is digital computation. To a first approximation, digital computation, as we are using this term, is the kind of computation that was analyzed by the classical mathematical theory of computation that began with Alan Turing and other logicians and is now a well-established branch of mathematics (Davis et al. 1994).

Digital computation may be defined both abstractly and concretely. Roughly speaking, abstract digital computation is the manipulation of strings of discrete elements, that is, strings of letters from a finite alphabet. Here we are interested primarily in concrete computation, or physical computation. Letters from a finite alphabet may be physically implemented by what we call *digits*. To a first approximation, concrete digital computation is the processing of sequences of digits according to general rules defined over the digits (Piccinini 2007b). Let us briefly consider the main ingredients of digital computation.

The atomic vehicles of concrete digital computation are digits, where a digit is a macroscopic state (of a component of the system) whose type can be reliably and unambiguously distinguished by the system from other macroscopic types. To each (macroscopic) digit type, there corresponds a large number of possible microscopic states. Artificial digital systems are engineered so as to treat all those microscopic states in the same way—the one way that corresponds to their (macroscopic) digit type. For instance, a system may treat 4 volts plus or minus some noise in the same way (as a "0"), whereas it may treat 8 volts plus or minus some noise in a different way (as a "1"). To ensure reliable manipulation of digits based on their type, a physical system must manipulate at most a finite number of digit types. For instance, ordinary computers contain only two types of digits, usually referred to as "0" and "1." Digits need not mean or represent anything, but they can. For instance, numerals represent numbers, while other digits (e.g., "|," "\") need not represent anything in particular.

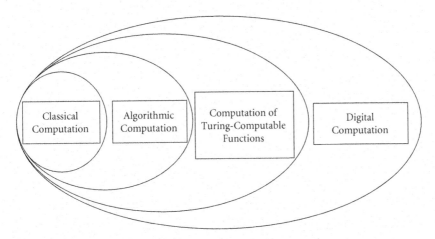

Figure 1. Types of digital computation and their relations of class inclusion.

Digits can be concatenated (i.e., ordered) to form sequences or *strings*. Strings of digits are the vehicles of digital computations. A digital computation consists in the processing of strings of digits according to rules. A rule for a digital computation is simply a map from input strings of digits, plus possibly internal states, to output strings of digits. Examples of rules that may figure in a digital computation include addition, multiplication, identity, and sorting.

Digits are unambiguously distinguishable by the processing mechanism under normal operating conditions. Strings of digits are ordered sets of digits (i.e., digits such that the system can distinguish different members of the set depending on where they lie along the string). The rules defining digital computations are, in turn, defined in terms of strings of digits and internal states of the system, which are simply states that the system can distinguish from one another. No further physical properties of a physical medium are relevant to whether its states implement digital computations. Thus, digital computations can be implemented by any physical medium with the right degrees of freedom.

To summarize, a physical system is a digital computing system just in case it is a system that manipulates input strings of digits, depending on the digits' type and their location on the string, in accordance with a rule defined over the strings (and possibly certain internal states of the system).

The notion of digital computation here defined is quite general. It should not be confused with three other commonly invoked but more restrictive notions of computation: classical computation in the sense of Fodor and Pylyshyn (1988), algorithmic computation, and computation of Turing-computable functions (see Figure 1).

Let us begin with the most restrictive notion of the three: classical computation. A classical computation is a digital computation that has two additional features. First, it manipulates a special kind of digital vehicle: sentence-like structures. Second, it is algorithmic: it proceeds in accordance with effective, step-by-step procedures that manipulate strings of digits and produce a result within finitely many steps. Thus, a classical computation is a digital, algorithmic computation whose algorithms are sensitive to the combinatorial syntax of the strings (Fodor and Pylyshyn 1988).

A classical computation is algorithmic, but the notion of algorithmic computation—digital computation that follows an algorithm—is more inclusive, because it does not require that the vehicles being manipulated be sentence-like.

Any algorithmic computation, in turn, can be performed by some Turing machine (i.e., the computed function is Turing-computable). This is a version of the Church-Turing thesis. But the computation of Turing-computable functions need not be carried out by following an algorithm. For instance, many neural networks compute Turing-computable functions (their inputs and outputs are strings of digits, and the input-output map is Turing-computable), but such networks need not have a level of functional organization at which they follow the steps of an algorithm for computing their functions; there may be no functional level at which their internal states and state transitions are discrete.

Here is another way to draw the distinction between algorithmic digital computation and digital computation more generally. Algorithmic digital computation is *fully* digital—digital every step of the way. Fully digital systems, such as McCulloch-Pitts networks (including, of course, digital computers), produce digital outputs from digital inputs by means of discrete intermediate steps. Thus, the computations of fully digital computing systems can be characterized as the (step-by-step) algorithmic manipulations of strings of digits. By contrast, digital computations more generally are only *input/output* (I/O) *digital*. I/O digital systems, including many neural networks, produce digital outputs from digital inputs by means of irreducibly continuous intermediate processes. Thus, the computations of merely I/O digital computing systems cannot be characterized as step-by-step algorithmic manipulations of strings of digits.[2]

Finally, the computation of a Turing-computable function is a digital computation, because Turing-computable functions are by definition functions of a denumerable domain—a domain whose elements may be counted—and the arguments and values of such functions are, or may be represented by, strings of digits. But it is equally possible to define functions of strings of digits that are not Turing-computable, and to mathematically define processes that compute such functions. Some authors have speculated that some functions over strings of digits that are not Turing-computable may be computable by some physical systems (Penrose 1994; Copeland 2000). According to the present usage, any such computations still count as digital computations. It may well be that only the Turing-computable functions are computable by physical systems—but whether this is the case is an empirical question that does not affect our discussion. Furthermore, the computation of Turing-

---

[2] Nevertheless, notice that in the connectionist literature the internal processes of I/O digital systems are often called *algorithmic*, using an extended sense of the term *algorithm*.

uncomputable functions is unlikely to be relevant to the study of cognition. Be that as it may—we will continue to talk about digital computation in general.

Many other distinctions may be drawn within digital computation, such as hardwired versus programmable, special purpose versus general purpose, and serial versus parallel (cf. Piccinini 2008b). These distinctions play an important role in debates about the computational architecture of cognitive systems, but we lack the space to discuss them here.

In summary, digital computation includes many types of neural network computations, including processes that follow ordinary algorithms, such as the computations performed by standard digital computers. Since digital computation is the notion that inspired the computational theory of cognition, it is the most relevant notion for present purposes.

4.2. Analog Computation

Analog computation is often contrasted with digital computation, but analog computation is a vaguer and more slippery concept. The clearest notion of analog computation is that of Pour-El (1974). Roughly, abstract analog computers are systems that manipulate continuous variables—variables that can vary continuously over time and take any real values within certain intervals—to solve certain systems of differential equations.

Analog computers can be physically implemented, and physically implemented continuous variables are different kinds of vehicles than strings of digits. While a digital computing system can always unambiguously distinguish digits and their types from one another, a concrete analog computing system cannot do the same with the exact values of (physically implemented) continuous variables. This is because continuous variables can take any real values, but there is a lower bound to the sensitivity of any physical system. Thus, it is always possible that the difference between two portions of a continuous variable is small enough to go undetected by the system. From this it follows that the vehicles of analog computation are not strings of digits and analog computations (in the present strict sense) are a different kind of process than digital computations.

Nevertheless, analog computations are only sensitive to the differences between portions of the variables being manipulated, to the degree that they can be distinguished by the system. Any further physical properties of the media implementing the variables are irrelevant to the computation. Like digital computers, therefore, analog computers operate on medium-independent vehicles.

There are other kinds of computation besides digital and analog computation. One increasingly popular kind, *quantum computation*, is an extension of digital computation in which digits are replaced by quantum states called *qudits* (most commonly, binary qudits, which are known as *qubits*). There is no room here for a complete survey of the kinds of computation. Suffice it to conclude that computation in the generic sense includes digital computation, analog computation, quantum computation, and more (Figure 2).

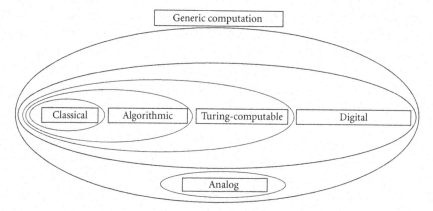

Figure 2. Types of computation and their relations of class inclusion.

5. COMPUTATION AND COGNITION: DIGITAL, ANALOG, OR A THIRD KIND?

Computationalism takes different forms depending on which type of computation it invokes to explain cognitive capacities. The weakest form of computationalism is *generic computationalism*, which simply says that cognition is a form of computation in the generic sense. Stronger versions of computationalism are *analog computationalism* and *digital computationalism*, according to which cognition is, respectively, a kind of analog or digital computation. Digital computationalism comes in more specialized versions that invoke different kinds of digital computation. A historically influential version of digital computationalism is *classical computationalism*, according to which cognition is a kind of classical computation. An important aspect of debates over the nature of cognition focuses on which type of computation cognition is. Let us begin with two reasons in favor of generic computationalism. Later we will mention some arguments for more specific versions of computationalism.

The vehicles of neural processes are medium-independent, so their manipulation constitutes a computation in the generic sense. As Edgar Adrian (1928; cf. Garson 2003) first showed, the vehicles transmitted and manipulated by neural processes are neuronal spikes (action potentials), and the functionally relevant aspects of neural processes depend on medium-independent aspects of the spikes—primarily, spike *rates*.[3] The functionally relevant aspects of spikes may be imple-

---

[3] There are other aspects of spikes that are thought to be functionally relevant at least in some cases, but such other aspects are still medium-independent.

mented either by neural tissue or by some other physical medium, such as a silicon-based circuit. Thus, spike trains—sequences of spikes produced by neurons in real time—appear to be medium-independent vehicles, in which case they qualify as proper vehicles for computations in the generic sense. Assuming that brains process spike trains and that spikes are medium-independent vehicles, it follows that brains perform computations in the generic sense.

A second reason for generic computationalism begins with the observation that cognition involves the processing of information in at least three important senses. First, cognition requires processing stochastic signals; this is information in Shannon's non-semantic sense (Shannon and Weaver 1949). Second, cognition requires processing signals that causally correlate with their sources; this may be called *natural semantic information* (cf. Dretske 1981). Third, at least some forms of cognition require processing full-blown representations (i.e., internal states that can represent either correctly or incorrectly); this may be called *non-natural semantic information* (Piccinini and Scarantino 2011).

For those who identify computation and information processing, the fact that cognition involves information processing is enough to conclude that cognition involves computation. As was pointed out, however, computation and information processing are best kept conceptually distinct, if for nothing else because identifying them obliterates important distinctions between different notions of computation and information.

Nevertheless, the fact that cognition involves information processing does entail that cognition involves computation. This is because information (in all senses of the term) is a medium-independent notion. It does not take any particular physical medium with any particular physical properties to possess certain statistical properties (Shannon information), causally correlate with a source (natural semantic information), or represent (nonnatural semantic information). Thus, processing information requires processing medium-independent vehicles. Recall that computation in the generic sense is the processing of medium-independent vehicles. Thus, information processing entails computation in the generic sense. (The converse does not hold because medium-independent vehicles need not carry any information; this is another reason to deny that computation is information processing.)

While it is safe to say that cognition involves computation in the generic sense, and that nervous systems perform computations in the generic sense, it is much harder to establish that cognition involves a more specific kind of computation.

We have already encountered McCulloch and Pitts's argument that cognition involves digital computation because of the all-or-none nature of the action potential. This suggestion encountered mighty resistance and is now abandoned by most neuroscientists. One reason is that unlike strings of digits properly so-called, action potentials do not appear to be organized into discrete time intervals of well-defined functional significance.

Shortly after McCulloch and Pitts argued that brains perform digital computations, others countered that neural processes may be more similar to analog computations (Gerard 1951; see also Rubel 1985). The evidence for the analog computational theory included neurotransmitters and hormones, which are released by neurons in degrees rather than in all-or-none fashion.

The claim that the brain is an analog computer is ambiguous between two interpretations. In the literal interpretation, "analog computer" is given Pour-El's (1974) precise meaning. The theory that the brain is an analog computer in this literal sense was never very popular. The primary reason is that although neural signals are a continuous function of time, they are also sequences of all-or-none signals.

In a looser interpretation, "analog computer" refers to a broader class of computing systems. For instance, Churchland and Sejnowski use the term *analog* so that it is sufficient for a system to count as analog if it contains continuous variables: "The input to a neuron is analog (continuous values between 0 and 1)" (1992, 51). Under such a usage, even a slide rule counts as an analog computer. Sometimes, the notion of analog computation is simply left undefined, with the result that by "analog computer" one refers to some kind of presumably non-digital but otherwise unspecified computing system.

The looser interpretation of "analog computer"—of which Churchland and Sejnowski's usage is one example—is not uncommon, but it may be misleading because it is prone to being confused with the literal interpretation of analog computation in the sense of Pour-El (1974).

Not all arguments for specific versions of computationalism are based on neuroscientific evidence. In fact, typical arguments for classical computationalism make no reference to the brain. One argument for classical computationalism begins with the observation that certain cognitive capacities exhibit productivity and systematicity—roughly speaking, they can generate an indefinite number of systematically related structured behaviors such as natural language sentences. The argument assumes that productive and systematic cognition requires the processing of language-like representations on the basis of their syntactic structure—that is, classical computation—and concludes that cognition is a kind of classical computation (Fodor and Pylyshyn 1988, Aizawa 2003). Like other arguments for specific versions of computationalism, this argument has encountered resistance, either from people who deny that cognition involves the processing of language-like representations or from those who deny that the relevant processing of language-like representations requires a classical computational system.

Another argument for classical computationalism is the one from cognitive flexibility. Human beings can solve an indefinite number of problems and learn an indefinite range of behaviors. How do they do it? Consider computers—the most flexible of artifacts. Computers can do mathematical calculations, derive logical theorems, play board games, recognize objects, control robots, and even engage in conversations to some degree. They can do this because they can execute different sets of instructions designed for different tasks. In Turing's terms, they approximate universal machines. Perhaps human beings are cognitively flexible because, like computers, they possess a general purpose processing mechanism that executes different instructions for different tasks (Fodor 1968; Newell 1990; Samuels 2010). The argument from cognitive flexibility is one of the most powerful, because there is no well-worked-out alternative explanation of cognitive flexibility, at least for high level cognitive skills such as problem solving (and according to some, even for lower level cognitive skills, cf. Gallistel and King 2009). Of course, the argument from cognitive flexibility is controversial too.

In conclusion, it is very plausible that the neural processes that explain cognitive capacities are computational in the generic sense, but it is difficult to determine which specific kinds of computation—classical, digital but non-classical, analog, etc.—are involved in which cognitive capacities. Whether any particular neural computation is best regarded as a form of digital computation, analog computation, or something else is a question that we cannot settle here.

6. CLASSICISM, CONNECTIONISM, AND COMPUTATIONAL NEUROSCIENCE REVISITED

The distinctions introduced in the preceding sections allow us to shed new light on the long-standing debates among classicism, connectionism, and computational neuroscience.

By the 1970s, McCulloch and Pitts were mostly forgotten. The dominant computational paradigm in cognitive science was classical or "symbolic" AI, aimed at writing computer programs that simulate intelligent behavior without much concern for how brains work (e.g., Newell and Simon 1976). It was commonly assumed that digital computationalism is committed to classicism, that is, the idea that cognition is the manipulation of language-like representations. Language-like representations have a constituent structure whose computational manipulation, according to classicism, explains productive and systematic psychological processes. In this view, cognition consists of performing computations on sentences with a logico-syntactic structure akin to that of natural languages, but written in the language of thought (Harman 1973; Fodor 1975).

It was also assumed that, given digital computationalism, explaining cognition is independent of explaining neural activity, in the sense that figuring out how the brain works tells us little or nothing about how cognition works: neural (or implementational, or mechanistic) explanations and computational explanations are at two different "levels" (Fodor 1975; Marr 1982).

During the 1980s, connectionism reemerged as an influential approach to psychology. Most connectionists deny that explaining cognition requires postulating a language of thought, and affirm that a theory of cognition should be at least "inspired" by the way the brain works (Rumelhart and McClelland 1986).

The resulting debate between classicists and connectionists (e.g., Fodor and Pylyshyn 1988; Smolensky 1988) has been somewhat confusing. Different authors employ different notions of computation, which vary in both their degree of precision and their inclusiveness. Specifically, some authors use the term *computation* only for classical computation—that is, at a minimum, algorithmic digital computation over language-like structures—and conclude that (non-classicist) connectionism falls outside computationalism. By contrast, other authors use a broader notion of computation along the lines of what we called computation in the generic sense, thus including connectionism within computationalism (see Piccinini 2008c for more details). But even after we factor out differences in notions of computation, more work is required to clarify the debate.

Classical computationalism and connectionism are often described as being at odds with one another because classical computationalism is committed to classical computation (the idea that the vehicles of digital computation are language-like structures) and—it is assumed—to autonomy from neuroscience, two theses flatly denied by many prominent connectionists. But many connectionists also model and explain cognition using neural networks that perform computations defined over strings of digits, so perhaps they should be counted among the digital computationalists (Bechtel and Abrahamsen 2002; Feldman and Ballard 1982; Hopfield 1982; Rumelhart and McClelland 1986; Smolensky and Legendre 2006).

Furthermore, both classicists and connectionists tend to ignore computational neuroscientists, who in turn tend to ignore both classicism and connectionism. Computational neuroscientists often operate with their own mathematical tools without committing themselves to a particular notion of computation (O'Reilly and Munakata 2000; Dayan and Abbott 2001; Eliasmith and Anderson 2003; Ermentrout and Terman 2010). To make matters worse, some connectionists and computational neuroscientists reject digital computationalism—they maintain that their neural networks, while explaining behavior, do not perform digital computations (Edelman 1992; Freeman 2001; Globus 1992; Horgan and Tienson 1996; Perkel 1990; Spivey 2007).

In addition, the very origin of digital computationalism calls into question the commitment to autonomy from neuroscience. McCulloch and Pitts initially introduced digital computationalism as a theory of the brain, and some form of computationalism or other is now a working assumption of many neuroscientists.

To clarify this debate, we need two separate distinctions. One is the distinction between digital computationalism ("cognition is digital computation") and its denial ("cognition is something other than digital computation"). The other is the distinction between classicism ("cognition is algorithmic digital computation over language-like structures") and neural-network approaches ("cognition is computation—digital or not—by means of neural networks").

We then have two versions of digital computationalism—the classical one ("cognition is algorithmic digital computation over language-like structures") and the neural network one ("cognition is digital computation by neural networks")—standing opposite to the denial of digital computationalism ("cognition is a kind of neural network computation different from digital computation"). This may be enough to accommodate most views in the current debate. But it still does not do justice to the relationship between classicism and neural networks.

A further wrinkle in this debate derives from the ambiguity of the term *connectionism*. In its original sense, connectionism says that behavior is explained by the changing "connections" between stimuli and responses, which are biologically mediated by changing connections between neurons (Thorndike 1932; Hebb 1949). This original connectionism is related to behaviorist associationism, according to which behavior is explained by the association between stimuli and responses. Associationist connectionism adds a biological mechanism to explain the associations: the mechanism of changing connections between neurons.

But contemporary connectionism is a more general thesis than associationist connectionism. In its most general form, contemporary connectionism, like computational neuroscience, simply says that cognition is explained (at some level) by neural network activity. But this is a truism—or at least it should be. The brain is the organ of cognition, the cells that perform cognitive functions are (mostly) neurons, and neurons perform their cognitive labor by organizing themselves in networks. In Section 5 we saw that some neural activity is computation at least in the generic sense. And in Section 2 we saw that even digital computers are just one special kind of neural network. So even classicists, whose theory is most closely inspired by digital computers, are committed to connectionism in its general sense.

The relationship between connectionist and neurocomputational approaches on one hand and associationism on the other is more complex than many suppose. We should distinguish between strong and weak associationism. Strong associationism maintains that association is the only legitimate explanatory construct in a theory of cognition (cf. Fodor 1983, 27). Weak associationism maintains that association is a legitimate explanatory construct along with others such as the innate structure of neural systems.

To be sure, some connectionists profess strong associationism (e.g., Rosenblatt 1958, 387). But that is beside the point, because connectionism per se is consistent with weak associationism or even the complete rejection of associationism. Some connectionist models do not rely on association at all—a prominent example being the work of McCulloch and Pitts (1943). And weak associationism is consistent with many theories of cognition, including classicism. A vivid illustration is Alan Turing's early proposal to train associative neural networks to acquire the architectural structure of a universal computing machine (Turing 1948, Copeland and Proudfoot 1996). In Turing's proposal, association may explain how a network acquires the capacity for universal computation (or an approximation thereof), while the capacity for universal computation may explain any number of other cognitive phenomena.

Although many of today's connectionists and computational neuroscientists emphasize the explanatory role of association, many of them also combine association with other explanatory constructs, as per weak associationism (cf. Smolensky and Legendre 2006, 479; Trehub 1991, 243–45; Marcus 2001, xii, 30). What remains to be determined is which neural networks, organized in what way, actually explain cognition and which role association and other explanatory constructs should play in a theory of cognition.

Yet another source of confusion is that classicism, connectionism, and computational neuroscience tend to offer explanations at different mechanistic levels. Specifically, classicists tend to offer explanations in terms of rules and representations without detailing the neural mechanisms by which the representations are implemented and processed; connectionists tend to offer explanations in terms of highly abstract neural networks, which do not necessarily represent networks of actual neurons (in fact, a processing element in a connectionist network may represent an entire brain area rather than an actual neuron); finally, computational neuroscientists tend to offer explanations in terms of mathematical models that represent concrete networks of neurons based on neurophysiological evidence. Explanations at different mechanistic levels are not necessarily in conflict with each other, but they do need to be integrated to describe a multilevel mechanism (Craver 2007). Integrating explanations at different levels into a unified multilevel mechanistic picture may require revisions in the original explanations themselves.

Different parties in the dispute among classicism, connectionism, and computational neuroscience may offer different accounts of how the different levels relate to one another. As we pointed out, one traditional view is that computational explanations are not even pitched at a mechanistic level. Instead, computational and mechanistic explanations belong at two independent levels (Fodor 1975; Marr 1982). This suggests a division of labor: computations are the domain of psychologists, while the implementing neural mechanisms are the business of neuroscientists. According to this picture, the role of connectionists and computational neuroscientists is to discover how neural mechanisms implement the computations postulated by (classicist) psychologists.

This traditional view has been criticized as unfaithful to scientific practices. It has been pointed out that (1) both psychologists and neuroscientists offer computational explanations (e.g., Piccinini 2006), (2i) far from being independent, different levels of explanation constrain one another (e.g., Feest 2003), and (3) both computational explanations (Churchland and Sejnowski 1992) and mechanistic explanations (Craver 2007) can be given at different levels.

One alternative to the traditional view is that connectionist or neurocomputational explanations simply replace classicist ones. Perhaps some connectionist computations approximate classical ones (Smolensky 1988), or perhaps not. In any case, some authors maintain that classicist constructs, such as program execution, play no causal role in cognition and will be eliminated from cognitive science (e.g., Churchland 2007).

A more neutral account of the relation among explanations at different levels is provided by the mechanistic account of computation. According to the mechanistic account, computational explanation is just one type of mechanistic explanation. Mechanistic explanations provide components with such properties and organization that they produce the explanandum. Computational explanation, then, is explanation in terms of computing mechanisms and components—mechanisms and components that perform computations. (Computation, in turn, is the manipulation of medium-independent vehicles.) Mechanistic explanations come with many levels of mechanisms, where each level is constituted by its components and the way they are organized. If a mechanistic level produces its behavior by the action of computing components, it counts as a computational level. Thus, a mechanism may contain zero, one, or many computational levels depending on what components it has and what they do. Which types of computation are performed at each level is an open empirical question to be answered by studying cognition and the nervous system at all levels of organization.

7. CONCLUSION

Computationalism is here to stay. We have every reason to suppose that cognitive capacities have computational explanations, at least in the generic sense that they can be explained in terms of the functional processing of medium-independent vehicles. Moreover, everyone is (or should be) a connectionist or computational neuroscientist, at least in the general sense of embracing neural computation. Nonetheless, much work remains to be done. More needs to be said about how different levels of explanation relate to one another, and how, in particular, computational explanations in psychology and neuroscience relate to each other and to other kinds of explanation. Ultimately, the computational study of cognition will require that we integrate different mechanistic levels into a unified, multilevel explanation of cognition.

Similarly, much work remains to be done to characterize the specific computations on which cognition depends: whether we ought to be committed to strong or to weak associationism; whether—and to what extent—the satisfactory explanation of cognition requires classical computational mechanisms as opposed to non-classical digital computation; and whether we need to invoke processes that involve non-digital computation, as some have maintained (Figure 3). It may turn out that one computational theory is right about all of cognition or it may be that different cognitive capacities are explained by different kinds of computation. To address these questions in the long run, the only effective way is to study nervous systems at all its levels of organization and find out how they exhibit cognitive capacities.

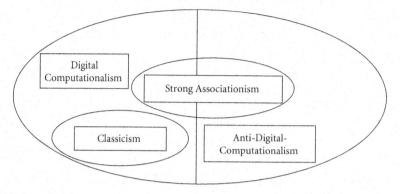

Figure 3. Some prominent forms of computationalism and their relations.

ACKNOWLEDGMENTS

Parts of this essay are revised and adapted descendants of parts of Piccinini 2010a and Piccinini and Scarantino 2011. Thanks to Susan Schneider, Eric Thomson, and the editors for helpful comments and to James Virtel for editorial assistance. This material is based in part upon work supported by the National Science Foundation under Grant No. SES-0924527.

REFERENCES

Adrian, E. D. (1928). *The Basis of Sensation: The Action of the Sense Organs*. New York: Norton.

Aizawa, K. (2003). *The Systematicity Arguments*. Boston: Kluwer.

Anderson, J. R. (2007). *How Can the Human Mind Occur in the Physical Universe?* Oxford: Oxford University Press.

Baars, B. J., Banks, W. P., and Newman J. B. (eds.) (2003). *Essential Sources in the Scientific Study of Consciousness*. Cambridge, MA: MIT Press.

Bechtel, W., and Abrahamsen, A. (2002). *Connectionism and the Mind: Parallel Processing, Dynamics, and Evolution in Networks*. Malden, MA: Blackwell.

Block, N. (1978). Troubles with Functionalism. In C. W. Savage (ed.), *Perception and Cognition: Issues in the Foundations of Psychology*. Minneapolis: University of Minnesota Press.

Boden, M. A. (2006). *Mind as Machine: A History of Cognitive Science*. Oxford: Oxford University Press.

Bringsjord, S., and Arkoudas, K. (2007). On the Provability, Veracity, and AI-Relevance of the Church-Turing Thesis. In A. Olszewski, J. Wolenski, and R. Janusz (eds.), *Church's Thesis after 70 Years*. Frankfurt: Ontos Verlag.

Caianiello, E. R. (1961). Outline of a theory of thought processes and thinking machines. *Journal of Theoretical Biology* 1: 204–35.

Chalmers, D. J. (1994). On implementing a computation. *Minds and Machines* 4.

———. (1996). Does a rock implement every finite-state automaton?. *Synthese* 108: 310–33.

Chrisley, R. L. (1995). Why everything doesn't realize every computation. *Minds and Machines* 4: 403–30.

Church, A. (1936). An unsolvable problem in elementary number theory. *The American Journal of Mathematics* 58: 345–63.

Churchland, P. M. (2007). *Neurophilosophy at Work*. Cambridge: Cambridge University Press.

Churchland, P. S., and Sejnowski, T. J. (1992). *The Computational Brain*. Cambridge, MA, MIT Press.

Copeland, B. J. (2000). Narrow versus wide mechanism: Including a re-examination of Turing's views on the mind-machine issue. *The Journal of Philosophy* 96(1): 5–32.

Copeland, B. J., and Proudfoot, D. (1996). On Alan Turing's anticipation of connectionism. *Synthese* 113: 361–77.

Craver, C. F. (2007). *Explaining the Brain*. Oxford: Oxford University Press.

Cummins, R. (1983). *The Nature of Psychological Explanation*. Cambridge, MA: MIT Press.

Davis, M. D., Sigal, R., and Weyuker, E. J. (1994). *Computability, Complexity, and Languages*. Boston: Academic.

Dayan, P., and Abbott, L. F. (2001). *Theoretical Neuroscience: Computational and Mathematical Modeling of Neural Systems*. Cambridge, MA: MIT Press.

Dennett, D. C. (1991). *Consciousness Explained*. Boston: Little, Brown.

Dretske, F. I. (1981). *Knowledge and the Flow of Information*. Cambridge, MA: MIT Press.

Edelman, G. M. (1992). *Bright Air, Brilliant Fire: On the Matter of the Mind*. New York: Basic Books.

Egan, F. (1995). Computation and content. *Philosophical Review* 104: 181–03.

Eliasmith, C., and Anderson, C. H. (2003). *Neural Engineering: Computation, Representation, and Dynamics in Neurobiological Systems*. Cambridge, MA: MIT Press.

Ermentrout, G. B., and Terman, D. H. (2010). *Mathematical Foundations of Neuroscience*. New York: Springer.

Feest, U. (2003). Functional analysis and the autonomy of psychology. *Philosophy of Science* 70: 937–48.

Feldman, J. A., and Ballard, D. H. (1982). Connectionist models and their properties. *Cognitive Science* 6: 205–54.

Fodor, J. A. (1968). The appeal to tacit knowledge in psychological explanation. *Journal of Philosophy* 65: 627–40.

———. (1975). *The Language of Thought*. Cambridge, MA: Harvard University Press.

———. (1980). Methodological solipsism considered as a research strategy in cognitive psychology. *Behavioral and Brain Sciences* 3(1): 63–109.

———. (1981). The mind-body problem. *Scientific American* 244: 124–32.

———. (1983). *The Modularity of Mind*. Cambridge, MA: MIT Press.

———. (2008). *LOT 2: The Language of Thought Revisited*. Oxford: Oxford University Press.

Fodor, J. A., and Pylyshyn, Z. W. (1988). Connectionism and cognitive architecture. *Cognition* 28: 3–71.

Freeman, W. J. (2001). *How Brains Make Up Their Minds*. New York: Columbia University Press.

Gallistel, C. R., and King, A. P. (2009). *Memory and the Computational Brain: Why Cognitive Science Will Transform Neuroscience*. Malden, MA: Wiley-Blackwell.

Garson, J. (2003). The introduction of information into neurobiology. *Philosophy of Science* 70: 926–36.

Gerard, R. W. (1951). Some of the Problems Concerning Digital Notions in the Central Nervous System. In H. v. Foerster, M. Mead, and H. L. Teuber (eds.), *Cybernetics: Circular Causal and Feedback Mechanisms in Biological and Social Systems. Transactions of the Seventh Conference.* New York: Macy Foundation.

Globus, G. G. (1992). Towards a noncomputational cognitive neuroscience. *Journal of Cognitive Neuroscience* 4(4): 299–310.

Glymour, C. (1991). Freud's Androids. In J. Neu (ed.), *The Cambridge Companion to Freud.* Cambridge: Cambridge University Press.

Godfrey-Smith, P. (2009). Triviality arguments against functionalism. *Philosophical Studies* 145(2): 273–95.

Grice, H. P. (1957). "Meaning." *The Philosophical Review* 66(3): 377–88.

Harman, G. (1973). *Thought.* Princeton, NJ: Princeton University Press.

Hebb, D. O. (1949). *The Organization of Behavior: A Neuropsychological Theory.* New York: Wiley.

Hodgkin, A. L., & Huxley, A. F. (1952). A quantitative description of membrane current and its application to conduction and excitation in nerve. *The Journal of Physiology* 117: 500–44.

Hopfield, J. J. (1982). Neural networks and physical systems with emergent collective computational abilities. *Proceedings of the National Academy of Sciences* 79: 2554–58.

Horgan, T., and Tienson, J. (1996). *Connectionism and the Philosophy of Psychology.* Cambridge, MA: MIT Press.

Horst, S. W. (1996). *Symbols, Computation, and Intentionality: A Critique of the Computational Theory of Mind.* Berkeley: University of California Press.

Householder, A. S., and Landahl, H. D. (1945). *Mathematical Biophysics of the Central Nervous System.* Bloomington, IN: Principia.

Kleene, S. C. (1952). *Introduction to Metamathematics.* Princeton, NJ: Van Nostrand.

———. (1956). Representation of Events in Nerve Nets and Finite Automata. *Automata Studies.* In C. E. Shannon and J. McCarthy (eds.), Princeton, NJ: Princeton University Press.

Klein, C. (2008). Dispositional implementation solves the superfluous structure problem. *Synthese* 165: 141–53.

Lapicque, L. (1907/2007). Quantitative investigation of electrical nerve excitation treated as polarization. *Biological Cybernetics* 97: 341–49. Translation from the French by N. Brunel and M. van Rossum.

Leff, H. S., and Rex, A. F. (eds). (2003). *Maxwell's Demon 2: Entropy, Classical and Quantum Information, Computing.* Bristol: Institute of Physics.

Knight, B. W. (1972). Dynamics of encoding in a population of neurons. *Journal of General Physiology* 59: 734–66.

Lycan, W. (1987). *Consciousness.* Cambridge, MA: MIT Press.

Machamer, P. K., Darden, L., and C. Craver. (2000). Thinking about mechanisms. *Philosophy of Science* 67: 1–25.

Marcus, G (2001). *The Algebraic Mind: Integrating Connectionism and Cognitive Science.* Cambridge, MA: MIT Press.

Marr, D. (1982). *Vision.* New York: Freeman.

McCulloch, W. S., and Pitts, W. H. (1943). A logical calculus of the ideas immanent in nervous activity. *Bulletin of Mathematical Biophysics* 7: 115–33.

Miller, G. A., Galanter, E. H., and Pribram, K. H. (1960). *Plans and the Structure of Behavior.* New York: Holt.

Newell, A. (1980). Physical symbol systems. *Cognitive Science* 4: 135–83.

———. (1990). *Unified Theories of Cognition*. Cambridge, MA: Harvard University Press.

Newell, A., and Simon, H. A. (1976). Computer science as an empirical enquiry: Symbols and search. *Communications of the ACM* 19: 113–26.

O'Reilly, R. C., and Munakata, Y. (2000). *Computational Explorations in Cognitive Neuroscience: Understanding the Mind by Simulating the Brain*. Cambridge, MA: MIT Press.

Penrose, R. (1994). *Shadows of the Mind*. Oxford: Oxford University Press.

Perkel, D. H. (1990). Computational Neuroscience: Scope and Structure. In E. L. Schwartz (ed.), *Computational Neuroscience*. Cambridge, MA: MIT Press.

Piccinini, G. (2003). Alan Turing and the mathematical objection. *Minds and Machines* 13(1): 23–48.

———. (2004). Functionalism, computationalism, and mental contents. *Canadian Journal of Philosophy* 34(3): 375–410.

———. (2006). Computational explanation in neuroscience." *Synthese* 153(3): 343–53.

———. (2007a). Computational modeling vs. computational explanation: Is everything a Turing machine, and does it matter to the philosophy of mind?. *Australasian Journal of Philosophy* 85(1): 93–115.

———. (2007b). Computing mechanisms. *Philosophy of Science* 74(4): 501–26.

———. (2008a). Computation without representation. *Philosophical Studies* 137(2): 205–41.

———. (2008b). Computers. *Pacific Philosophical Quarterly* 89(1): 32–73.

———. (2008c). Some neural networks compute, others don't. *Neural Networks* 21(2–3): 311–21.

———. (2009). Computationalism in the Philosophy of Mind." *Philosophy Compass* 4(3): 515–32.

———. (2010a). Computation in Physical Systems. In E. N. Zalta (ed.), *Stanford Encyclopedia of Philosophy* (Fall 2010 ed.). http://plato.stanford.edu/archives/fall2010/entries/computation-physicalsystems/.

———. (2010b). The Mind as neural software? Understanding functionalism, computationalism, and computational functionalism. *Philosophy and Phenomenological Research* 81(2): 269–11.

Piccinini, G., and A. Scarantino (2011). Information processing, computation, and cognition. *Journal of Biological Physics* 37(1): 1–38.

Pinker, S. (1997). *How the Mind Works*. New York: Norton.

Port, R., and van Gelder, T. (eds.) (1995). *Mind as Motion: Explorations in the Dynamics of Cognition*. Cambridge, MA: MIT Press.

Pour-El, M. B. (1974). Abstract computability and its relation to the general purpose analog computer (some connections between logic, differential equations and analog computers). *Transactions of the American Mathematical Society* 199: 1–28.

Putnam, H. (1988). *Representation and Reality*. Cambridge, MA: MIT Press.

Pylyshyn, Z. W. (1984). *Computation and Cognition*. Cambridge, MA: MIT Press.

Rashevsky, N. (1938). *Mathematical Biophysics: Physicomathematical Foundations of Biology*. Chicago: University of Chicago Press.

Rashevsky, N. (1940). *Advances and Applications of Mathematical Biology*. Chicago: University of Chicago Press.

Rey, G. (2005). Mind, intentionality and inexistence: An overview of my work. *Croatian Journal of Philosophy* 5(15): 389–415.

Rosenblatt, F. (1958). The perceptron: A probabilistic model for information storage and organization in the brain. *Psychological Review* 65: 386–408.

Rubel, L. A. (1985). The brain as an analog computer. *Journal of Theoretical Neurobiology* 4: 73–81.

Rumelhart, D. E., and McClelland, J. M. (eds.) (1986). *Parallel Distributed Processing: Explorations in the Microstructure of Cognition*. Cambridge, MA: MIT Press.

Rusanen, A.-M., and Lappi, O. (2007). The Limits of Mechanistic Explanation in Neurocognitive Sciences. In S. Vosniadou, D. Kayser, and A. Protopapas (eds.), *Proceedings of the European Cognitive Science Conference 2007*. Hove, UK: Lawrence Erlbaum Associates.

Samuels, R. (2010). Classical computationalism and the many problems of cognitive relevance. *Studies in History and Philosophy of Science* 41(3): 280–93.

Scheutz, M. (1999). When physical systems realize functions.…*Minds and Machines* 9(2): 161–96.

———. (2001). Causal versus computational complexity. *Minds and Machines* 11(4): 534–66.

Schwartz, E. L. (1990). *Computational Neuroscience*. Cambridge, MA: MIT Press.

Searle, J. R. (1980). Minds, brains, and programs. *The Behavioral and Brain Sciences* 3: 417–57.

———. (1992). *The Rediscovery of the Mind*. Cambridge, MA: MIT Press.

Shagrir, O. (2001). Content, computation and externalism. *Mind* 110(438): 369–400.

———. (2006). Why we view the brain as a computer. *Synthese* 153(3): 393–416.

Shannon, C. E., and Weaver, W. 1949. *The Mathematical Theory of Communication*. Urbana: University of Illinois Press.

Siegelmann, H. T. (2003). Neural and super-Turing computing. *Minds and Machines* 13(1): 103–14.

Smolensky, P. (1988). On the proper treatment of connectionism. *Behavioral and Brain Sciences* 11: 1–23.

Smolensky, P., and Legendre, G. (2006). *The Harmonic Mind: From Neural Computation to Optimality-Theoretic Grammar. Vol. 1: Cognitive Architecture; Vol. 2: Linguistic and Philosophical Implications*. Cambridge, MA: MIT Press.

Spivey, M. (2007). *The Continuity of Mind*. Oxford: Oxford University Press.

Sprevak, M. (2010). Computation, individuation, and the received view on representation. *Studies in History and Philosophy of Science* 41(3): 260–70.

Stein, R. (1965). A theoretical analysis of neuronal variability. *Biophysical Journal* 5: 173–94.

Stich, S. (1983). *From Folk Psychology to Cognitive Science*. Cambridge, MA: MIT Press.

Thompson, E. (2007). *Mind in Life: Biology, Phenomenology, and the Sciences of Mind*. Cambridge, MA: Harvard University Press.

Thorndike, E. L. (1932). *The Fundamentals of Learning*. New York: Teachers College, Columbia University.

Trehub, A. (1991). *The Cognitive Brain*. Cambridge, MA: MIT Press.

Turing, A. M. (1936–7 [1965]). On Computable Numbers, with an Application to the Entscheidungsproblem. In M. Davis (ed.), *The Undecidable*. M. Davis. Hewlett, NY: Raven Press.

Turing, A. M. (1948). Intelligent Machinery. Reprinted in D. Ince (ed.), *Mechanical Intelligence*. Amsterdam: North-Holland.

van Gelder, T. (1995). What might cognition be, if not computation? *Journal of Philosophy* 92(7): 345–81.

———. (1998). The dynamical hypothesis in cognitive science. *Behavioral and Brain Sciences* 21: 615–65.

von Neumann, J. (1945). First Draft of a Report on the EDVAC. Philadelphia, PA, Moore School of Electrical Engineering, University of Pennsylvania.

Wilson, H. R., and Cowan, J. D. (1972). Excitatory and inhibitory interactions in localized populations of model neurons. *Biophysical Journal* 12: 1–24.

Wilson, R. A. (1994). Wide computationalism. *Mind* 103: 351–72.

Wimsatt, W. C. (2002). Functional Organization, Analogy, and Inference. In A. Ariew, R. Cummins, and M. Perlman (eds.), *Functions: New Essays in the Philosophy of Psychology and Biology*. Oxford: Oxford University Press.

CHAPTER 11

....

REPRESENTATIONALISM

....

FRANCES EGAN

REPRESENTATIONALISM, in its most widely accepted form, is the view that the human mind is an *information-using* system, and that human cognitive capacities are to be understood as representational capacities. This chapter distinguishes several distinct theses that go by the name "representationalism," focusing on the view that is most prevalent in cognitive science. It also discusses some objections to the view and attempts to clarify the role that representational content plays in cognitive models that make use of the notion of "representation."

1. SOME REPRESENTATIONALIST THESES
....

The most persuasive argument for representationalism appeals to the fact that evolved creatures such as ourselves behave in ways that are well-suited to achieving their ends given their circumstances. The best explanation of this fact is that they are able to represent both their ends and their circumstances, and that these representational capacities are causally implicated in producing their behavior. Moreover, to the extent that an agent's behavioral capacities are flexible, in particular, to the extent that they are sensitive to its changing circumstances, it is plausible to think that its behavior is guided by representational states, at least some of which are themselves a causal product of those circumstances. Sterelny (1990) argues that behavioral flexibility *requires* representation:

> There can be no informational sensitivity without representation. There can be no flexible and adaptive response to the world without representation. To learn about the world, and to use what we learn to act in new ways, we must be able to

represent the world, our goals and options. Furthermore we must make appropriate inferences from those representations. (21)

Very roughly, the representation of a creature's ends constitutes its *desires*, and the representation of its circumstances its *beliefs*. Representationalism is sometimes understood as a thesis specifically about beliefs and desires. Let us call the view that propositional attitudes—beliefs, desires, fears, intentions, and their kin—are representational states of organisms *Representationalism$_{PA}$*. Representationalism$_{PA}$ is a widely held view, though behaviorists and eliminativists about the attitudes would, of course, deny it.

Strong Representationalism is the view that representational mental states have a specific form, in particular, that they are functionally characterizable relations to internal representations. Proponents of Strong Representationalism typically endorse the view that the system of internal representations constitutes a language with a combinatorial syntax and semantics. While strong representationalists typically construe mental representations as language-like, the essential point is that they are *structured* entities over which mental processes are defined. Braddon-Mitchell and Jackson (1996) argue that mental representations may be more analogous to maps than to sentences. Waskan (2006) argues that mental representations are akin to *scale models*. Proponents of strong representationalism include Fodor (1975, 1981, 1987, 2008), Gallistel and King (2009), Pinker (1997, 2005), and Pylyshyn (1984), among many others.

To complete our taxomony of representationalist theses, *Strong Representationalism$_{PA}$* is the view that propositional attitudes are to be understood as representational states of a specific sort, namely, as functionally characterizable relations to internal representations. According to Strong Representationalism$_{PA}$, to believe that Miles Davis was a genius is to be related in the way characteristic of belief to an internal representation that means *Miles Davis was a genius*.[1] Moreover, proponents of Strong Representationalism$_{PA}$ typically take these internal representations to have a language-like structure.

Several now-classic arguments have been offered in support of Strong Representationalism$_{PA}$. Harman (1972) claimed that logical relations hold among propositional attitudes, and that these relations are essential to their role in predictions and explanations of behavior. If the belief *that snow is white and grass is green* is true, then the belief *that snow is white* is true. In general, if the belief that *p & q* is true, then the belief that *p* is true. Generalizations of this sort presuppose that beliefs have sentential structure. Some beliefs are conjunctions, others conditionals, and so on. Beliefs (as well as desires, fears, and the other propositional attitudes), it is claimed, are part of a language-like system.

Harman's argument trades on the fact that belief *ascriptions* have sentential structure, but it fails to establish that propositional attitudes themselves have logical

---

[1] The most explicit statement of Strong Representationalism$_{PA}$ is Fodor's *Representational Theory of Mind* (RTM). See Fodor (1975, 1981, 1987, 2008) and Field (1978) for articulation and defense of RTM.

or sentential structure. We ascribe beliefs to subjects using sentences that are conjunctive or conditional, and we make use of the logical relations that hold among the *complements* of belief ascriptions as surrogates to reason about relations that hold among mental states themselves.[2] But it does not follow that the mental states so ascribed are conjunctions or conditionals, or that the relations that hold among these states are of the sort that hold among sentences (or propositions), that is, that they are logical relations. To assume that they are is just to assume what is at issue—that propositional attitudes have a language-like structure. In general, one must be careful not to attribute to thoughts themselves properties of the representational scheme that we use to talk and reason about them.

Fodor (1987) and Fodor and Pylyshyn (1988) argued that certain pervasive features of thought can only be explained by the hypothesis that thought takes place in a linguistic medium. Thought is *productive*: we can think arbitrarily many thoughts. It is also *systematic*: cognitive capacities are systematically related. If a subject can think the thought *John loves Mary*, then she can think the thought *Mary loves John*. The explanation for the productivity and systematicity of thought is that thoughts have a language-like structure. We can think arbitrarily many thoughts for the same reason that we can produce and understand arbitrarily many sentences. Thoughts, like sentences, are composed of a finite base of elements put together in regular ways, according to the rules of a grammar. Systematicity is explained in the same way: systematically related thoughts contain the same basic elements, just arranged differently.

Whether the argument succeeds in establishing that thought is language-like depends on two issues: (1) whether productivity and systematicity are indeed pervasive features of thought; and (2) if they are, whether they can be accounted for without positing an internal linguistic medium.

Thoughts are assumed to be productive in part because they are represented, described, and attributed by public language sentences, a system that is itself productive. But, as noted above, one must be careful not to attribute to thoughts themselves properties of the representational scheme that we use to think and talk about them. It would be a mistake to think that there are substances with infinitely high temperatures just because the scheme we use to measure temperature, the real numbers, is infinite. If thoughts are understood as internal states of subjects that are, typically, effects of external conditions and causes of behavior, then it is not obvious that there are arbitrarily many of them. The size of the set of possible belief-states of human thinkers, like the possible temperatures of substances, is a matter to be settled by empirical investigation.

When we turn to systematicity, the argument falls short of establishing that thought must be language-like. In the first place, it is not clear how pervasive systematicity really is. It is not generally true that if a thinker can entertain a proposition of the form aRb, then he can entertain bRa. One can think the thought *the boy parsed the sentence* but not *the sentence parsed the boy*. Moreover, it is a matter of

---

[2] See Swoyer (1991) and Matthews (2007) for a detailed argument of this point.

some dispute within the cognitive science community whether connectionist cognitive models, which do not posit a language of thought, might be capable of explaining the systematic relations that do hold among thoughts.[3]

Dennett's 1981 example of a chess-playing computer raises a potentially serious worry for any adherent of Strong Representationalism$_{PA}$. It is plausible to say of the machine that it believes that it should get its queen out early. Ascribing this propositional attitude to the machine allows us to predict and explain its play in a wide range of circumstances. But nowhere in the device's architecture is anything roughly synonymous with "Get the queen out early" explicitly represented, as required by Strong Representationalism$_{PA}$. The consequence of denying that the machine has the appropriate attitude on this ground alone[4] is that doing so would undermine our confidence, more generally, in attitude ascriptions based on the usual behavioral evidence. We have been ascribing propositional attitudes to agents on the basis of their behavior for millennia, without any knowledge of their internal functional or neural architecture. Of course, some of our attitude ascriptions may be false, but their falsity would be revealed, in the typical case, by additional information about the subject's behavioral dispositions—it does not believe it should get its queen out early after all, because when offered the opportunity to do so in a fairly wide range of circumstances, it does something else; she does not believe there is beer in the refrigerator after all, because she is making a special trip to the beer store—not by computational or neural considerations that remain well beyond the ken of ordinary attitude-ascribing folk.[5] Proponents of Strong Representationalism$_{PA}$ often promote it as promising a scientific vindication of our attitude-ascribing folk practices (see, e.g., Fodor 1987). On the contrary, in requiring that every genuine propositional attitude ascribable to a subject corresponds to, or is realized by, an explicit mental representation with the content of that very attitude, Strong Representationalism$_{PA}$ holds those practices hostage to the neural architectural details turning out a particular way, and thus invites an unreasonable eliminativism.

We may conclude that while (regular-strength) Representationalism$_{PA}$ is almost certainly true—propositional attitudes, as they figure in our commonsense predictive and explanatory practices, are conceived of as representational states of organisms—the case for Strong Representationalism$_{PA}$, the view that propositional attitudes are functionally characterizable relations to internal representations that

---

[3] See MacDonald and MacDonald (1996) for the classic papers on this issue, and Cummins (1996), Matthews (1997), and Aizawa (2006) for further discussion.

[4] One might, of course, deny that the chess-playing machine has *any* propositional attitudes, on the grounds that it is not conscious, too cognitively limited, etc. That is another matter.

[5] Certain far-fetched scenarios would also reveal the falsity of ordinary, behavior-based attitude ascriptions. If a subject's behavior was revealed to be caused by remote control by Martian scientists (see Peacocke 1983) then we would withdraw all attitude ascriptions as false. The conditions on correct attitude ascriptions are not *entirely* behavioral. See Egan (1995a) for defense of this view of the attitudes.

have linguistic structure, is inconclusive at best.[6] In any event, let us set Strong Representationalism$_{PA}$ aside for the remainder of this chapter and focus instead on the role that mental representation plays in the cognitive sciences. This strategy is appropriate for several reasons: In the first place, whether or not Strong Representationalism$_{PA}$ is true, cognitive scientists are not committed to its truth, and are not engaged in seeking the "vindication" of our folk practices that its truth would entail. Second, as Von Eckart (1995) notes,

> [W[hen cognitive scientists use the term "mental representation" they are not
> using it as extensionally equivalent with "propositional attitude". Rather they are
> using it to refer to computational entities…with representational properties,
> hypothesized in the context of a scientific research program. (164)

The issue that will concern us in the remainder of this chapter, then, is the role of mental representation in the cognitive sciences.

2. Representationalism in Cognitive Science—The Standard View

Representationalism, we said, construes the mind as a representational, or *information-using*, device. The notion of a "representational device" is given a precise meaning in cognitive science by Alan Newell's 1980 characterization of a *physical symbol system*.

A *physical symbol system* (hereafter, *PSS*) is a device that manipulates symbols in accordance with the instructions in its program. *Symbols* are objects with a dual character: they are both physically realized and have meaning or semantic content. A realization function f_R maps them to physical state-types of the system. A second mapping, the interpretation function f_I, specifies their meaning by pairing them with objects or properties in a particular domain. A given PSS is type-individuated by the two mappings f_R and f_I. By this we mean that if either f_R or f_I had been different, the device would be a different (type of) PSS.

The concept of a PSS gives precise meaning to two notions central to mainstream cognitive science: *computation* and *representation*. A *computation* is a sequence of physical state transitions that, under the mappings f_R and f_I, executes some specified task. A *representation* is an object whose formal and semantic properties are specified by f_R and f_I respectively.[7]

A PSS, Newell emphasizes, is a universal machine. Given sufficient, but finite, time and memory it is capable of computing any computable function. These

---

[6] See Stich (1983) and Matthews (2007) for detailed criticism of the RTM.

[7] Following Fodor's 1980 usage, "formal" here means *non-semantic*.

systems have what Fodor and Pylyshyn (1988) have called a "classical" architecture—
an architecture that preserves a principled distinction between the system's representations or data structures and the processes defined over them.

The *physical symbol systems hypothesis* is the idea that the mind is a specific sort
of computer, namely a device that manipulates (writes, retrieves, stores, etc.) strings
of symbols. The PSS hypothesis is a version of Strong Representationalism, the idea
that representational mental states[8]—mental states with semantic content—are
functionally characterizable relations to internal representations.

It is not hard to understand the attraction of the PSS hypothesis for philosophers of mind and psychology. Self-styled "computationalists" committed to the
view that mental processes are computational processes, notably, proponents of
Strong Representationalism$_{PA}$ such as Fodor (1975, 1981, 1987) and Pylyshyn (1984),
have hoped that computational models of cognitive processes will eventually mesh
with and provide a scientific basis for our commonsense explanatory practices.
These practices, as noted above, appeal to content-specific beliefs and desires. For
example, it is your belief that there is beer in the refrigerator, together with a content-appropriate desire (to drink a beer, or perhaps just to drink something cold),
that explains your going to the kitchen and getting a beer. Appealing to your belief
that there is beer at the local bar or your desire to win the lottery fails to provide any
explanation of your beer-fetching behavior. Moreover, this behavior is rational just
to the extent that it is caused by content-appropriate beliefs and desires. Similarly,
according to PSS-inspired computationalism, computational explanations of behavior will appeal to the contents of the symbol strings, or internal representations,
the manipulations of which are the causes of our intelligent behavior. But these
operations themselves respect what Fodor (1980) has dubbed the *formality condition*—they are sensitive only to *formal* (i.e., *non-semantic*) properties of the representations over which they are defined, not to their content.

The formality condition is often glossed by computationalists as the idea that
mental representations have their causal roles in virtue of their syntax. As Pylyshyn
(1984) puts the point,

> For every apparent, functionally relevant distinction there is a corresponding syn-
> tactic distinction. Thus, any semantic feature that can conceivably affect behavior
> must be syntactically encoded at the level of a formal symbol structure. By this
> means we arrange for a system's behavior to be describable as responding to the
> content of its representations —to what is being represented—in a manner com-
> patible with materialism. (74)

The idea of syntax and semantics marching in lockstep, to produce mechanical reasoning, is of course the fundamental idea underlying theorem proving in logic. But
Pylyshyn (1980) elaborates the view as follows:

[8] As opposed to perceptual experiences, pains, and other mental states characterized at least
in part by their "qualitative feel." For representational construals of qualitative mental states see
Byrne (2001), Dretske (1995, 2003), Lycan (1987, 1996), and Tye (1995, 2000, 2003).

> The formalist view requires that we take the syntactic properties of representa-
> tions quite literally. It is literally true of a computer that it contains, in some func-
> tionally discernible form...what could be referred to as a code or inscription of a
> symbolic expression, whose formal features mirror (in the sense of bearing a one-
> to-one correspondence with) semantic characteristics of some represented
> domain, and which causes the machine to behave in a certain way. (1980, 113)

Pylyshyn's tendency to overlook the fact that a representation has *both* its semantic
and syntactic properties only under interpretation—given by the mappings f_I and
f_R respectively—leads him to adopt a realist stance toward its syntactic properties
that he does not extend to its semantic properties. He says above that the syntactic
description of the device is *literally true*, seemingly suggesting that the semantic
description of the device is not. What Pylyshyn *should* say is that for any feature of
the system—semantic or syntactic—to affect the system's behavior, it must be realized in the device's physical states.[9] The idea that semantics and syntax march in
lockstep is, in effect, the idea that both semantics and syntax must be realized in the
physical states of the device, semantics in virtue of the dual mappings f_I and f_R, and
syntax in virtue of f_R alone. Neither semantics nor syntax is an intrinsic feature of
the device, although f_R can be seen as specifying the basic causal operations of the
device, since it specifies the physical organization relevant for understanding it as a
computing device.

Let us focus on the role of so-called "representational content" in computational models of cognitive capacities. Representationalists (of all stripes) tend to
endorse the following claims:

(1) The internal states and structures posited in computational theories of
cognition are *distally interpreted* in such theories; in other words, the
domain of the interpretation function f_I is objects and properties of the
external world.

(2) The distal objects and properties that determine the representational
content of the posited internal states and structures serve to *type-
individuate* a computationally characterized mechanism. In other words,
if the states and structures of the device had been assigned *different* distal
contents, then it would be a *different* computational mechanism.

(3) The relation between the posited internal states and structures and the
distal objects and properties to which they are mapped (by f_I)—what we
might call the *Representation Relation*—is a substantive, naturalistically
specifiable relation. So-called "naturalistic" theories of mental content
attempt to specify, in non-intentional and non-semantic terms, a sufficient condition for a mental representation's having a particular meaning.
The most popular proposals construe the relation as either information-
theoretic (see, e.g., Dretske 1981 and Fodor 1990) or teleological (see

---

[9] See Smith (unpublished) and Marras 1985 for arguments that the formality condition
imposes only a realizability constraint on computational theories.

Milliken 1984, Dretske 1986, 1995, and Papineau 1987, 1984). While the various proposals on offer face difficulties accounting for the fine-grainedness of mental content, and for misrepresentation,[10] it is nonetheless widely held among philosophers committed to representationalism that the relation between internal structures and their referents satisfies some such naturalistic constraint.

We shall call this package of commitments the *Essential Distal Content View*. Section 4 will present the argument that the Essential Distal Content View misconstrues both the nature of the interpretation function *fI* and the role of so-called "representational content" in computational accounts of cognition. First, though, let us turn to some well-known objections to the Standard View.

3. CHALLENGES TO THE STANDARD VIEW

3.1. The Gibsonian Challenge

Strong Representationalism, as exemplified by the PSS hypothesis, construes mental processes as operations on internal representations. The psychologist J. J. Gibson (see his 1966, 1979 work) held that visual perception is not mediated by representations, memories, concepts, inferences, or any other process characterizable in psychological terms. The difference between so-called "direct theorists" of perception, such as Gibson, and representationalists is often characterized as a disagreement over the richness of the retinal image, with direct theorists arguing that the stimulus contains more information than representationalists are willing to allow. Gibson held that the input to the visual system is not a series of static "time slices" of the retinal image, but rather the smooth transformations of the optic array as the subject moves about the environment, what Gibson (1979) called "retinal flow." There are important constancies in the stimulus that make unnecessary the positing of intervening inferences, calculations, or other processes defined over representations to account for either the subject's ability to perceive size and shape constancies or the richness of visual experience. In addition to intensity and wavelength information, properties directly picked up in the stimulus include, according to Gibson, higher-order properties that remain invariant through movement and changes in orientation. These higher-order invariants specify not only structural properties such as *being a cube* but what Gibson called "affordances," which are functionally significant aspects of the distal scene, such as the fact that an object is edible or could be used for cutting.

[10] See the papers in Stich and Warfield (1994) for discussion of some of these worries.

Two fundamental assumptions underlie Gibson's "ecological" approach to visual perception: (1) that functionally significant aspects of the environment structure the ambient light in characteristic ways, and (2) that the organism's visual system has evolved to detect these characteristic structures in the light. Both assumptions are controversial. With respect to (2), representationalists have complained that Gibson provides no account of the mechanism that allegedly detects salient higher-order invariants in the optical array. His claim that the visual system "resonates," like a tuning fork, to these properties is little more than a metaphor.[11] But it should be noted that in claiming that perception of higher-order invariants is direct, Gibson is proposing that the visual mechanism be treated as a black box from the point of view of psychology, because no inferences, calculations, representations, memories, beliefs, or other characteristically *psychological* entities mediate the processing. The *physiological* account of the mechanism's operation will no doubt be very complex. The claim might be plausible if assumption (1) is true—if there is a physically specifiable property (or set of properties) of the light corresponding to every perceptible affordance. But for all but the simplest organisms it seems unlikely that the light is structured in accordance with the organism's goals and purposes. More likely, the things that appear to afford eating or cutting or fleeing behavior structure the light in all kinds of different ways. This likelihood has led many perceptual theorists to claim that something like categorization—specifically, the bringing of an object identified initially by its shape, color, or texture under a further concept—is at work when an organism sees an object *as* food, *as* a cutting implement, or *as* a predator. Categorization is then construed as a process defined over representations.

3.2. The Challenge from "Embodied" and "Embedded" Cognition

More recently, work in robotics and "artificial life" has inspired a broad-based challenge to the standard view. The MIT researcher Rodney Brooks is the intellectual father of this movement. (See the papers collected in his 1999 work *Cambrian Intelligence.*) Brooks's robot Herbert, a self-propelled device equipped with various sensors and a moveable arm on top, was designed to motor around the MIT AI lab, collecting empty soda cans and returning them to a central bin. Herbert operates according to what Brooks calls "the subsumption architecture." Its systems decompose not into peripheral systems such as vision and motor systems on the one hand, and central systems such as memory on the other, with the latter systems subserving many different tasks, but rather into specialized activity-producing subsystems or skills. Herbert has one subsystem for detecting and avoiding obstacles in its path, another for finding and homing in on distant soda cans, another for putting its hand around nearby soda cans, and so on. What Herbert does not have is a general representation of its environment that subserves all these tasks. Intelligent behavior

---

[11] See Fodor and Pylyshyn (1981) for detailed discussion of this criticism.

emerges from the interaction of Herbert's various subsystems, but without the construction and manipulation of representations of its world. Brooks drew the following morals from his work:

> We have reached an unexpected conclusion (C) and have a rather radical hypothesis (H). (C) When we examine very simple level intelligence we find that explicit representations and models of the world simply get in the way. It turns out to be better to use the world as its own model.
>
> (H) Representation is the wrong unit of abstraction in building the bulkiest parts of intelligent systems. (1999, 80–81)

Other theorists, following Brooks, have developed similar models of intelligent behavior and drawn similar anti-representationalist conclusions.[12] The general idea is that since the model of some cognitive capacity does not employ representations, there is no reason to think that cognition requires representations.

It is a matter of some dispute whether these models are genuinely non-representational.[13] They count as "representational" in our weak sense—they are certainly *information-using* systems—but such representations as they do have typically differ from those posited in the standard view (Strong Representationalism) in at least two respects: (1) There is no principled demarcation between device and environment; any representations posited are not conceived of as contained entirely within the device, hence the idea that cognition, so modeled, is *embedded* or *situated*; and (2) There is no strict demarcation between the representation of the agent's circumstances and its goals, but more significantly, its representations are not passive structures over which computational operations are defined. Rather, they are "action-oriented," as Clark (1998) puts it—they *both* describe a situation and suggest an appropriate behavioral response to it—hence the idea that cognition is *embodied*.[14] Proponents of embedded or embodied cognition include Varela, Thompson, and Rosch (1991), Clancy (1997), Clark (1998), Gallagher (2005), and Noe (2004, 2009).[15]

So far, embodied/embedded devices are capable only of fairly simple behaviors. An interesting question is how fruitful Brooks's strategy of "using the world as its own model" will prove to be in modeling and understanding more complex cognitive tasks such as reasoning and problem solving, which seem to require the representation of alternative courses of action and counterfactual circumstances. Kirsh (1991), responding to early work, suggests that the embedded/situated approach is *inherently* limited:

---

[12] See, for example, Beer (1995), Harvey, Husbands, and Cliff (1994), van Gelder and Port (1995), and Wheeler (1996). See Chemero (2000) for a critical discussion of these anti-representationalist claims.

[13] Clark (1998), for example, thinks that they are; Gallagher (2005, 2008) thinks they are not.

[14] In these two respects, they are suggestive of Gibson's *affordances*, which are located in the environment and apt for appropriate behavior.

[15] See Anderson (2003) for an excellent review of the embodied cognition literature, and the papers in Robbins and Aydede (2008) for views for and against situated/embedded cognition.

Situationally determined activity has a real chance of success only if there are enough egocentrically perceptible cues available. There must be sufficient local constraint in the environment to determine actions that have no irreversibly bad downstream effects. Only then will it be unnecessary for the creature to represent alternative courses of actions to determine which ones lead to dead ends, traps, loops, or idle wandering. From this it follows that if a task requires knowledge about the world that must be obtained by reasoning or by recall, rather than by perception, it cannot be classified as situation determined. (171)

Clark (1998) also expresses skepticism that embedded/embodied accounts will be able to adequately model and explain higher cognitive processes, and suggests that a hybrid approach, incorporating both embedded/embodied processes and structured representations of the sort advocated by the standard view, is most promising. The hard question then is how the two schemes might be coordinated, given their disparate commitments about the fundamental nature of cognition.

3.3. The Chomskian Challenge

Noam Chomsky has argued in his recent work that the so-called "representational" states invoked in accounts of our cognitive capacities are not genuinely representational and that they are not correctly construed as about some represented objects or entities. Discussing computational vision theory he says,

There is no meaningful question about the "content" of the internal representations of a person seeing a cube under the conditions of the experiments...or about the content of a frog's "representation of" a fly or of a moving dot in the standard experimental studies of frog vision. No notion like "content", or "representation of", figures within the theory, so there are no answers to be given as to their nature. The same is true when Marr writes that he is studying vision as "a mapping from one representation to another..." (Marr, 1982, p. 31)— where "representation" is not to be understood relationally, as "representation of". (1995, 52–53)

The idea that "representation" should, in certain contexts, not be understood relationally as in "representation of x," but rather as specifying a monadic property, as in "x-type representation," can be traced to Goodman (1968).[16] So understood, the

---

[16] According to Goodman,

Saying that a picture represents a so-and-so is thus highly ambiguous as between saying what the picture denotes and saying what kind of picture it is. Some confusion can be avoided if in the latter case we speak rather of...a "Pickwick-picture" or "unicorn-picture" or "man-picture". Obviously a picture cannot, barring equivocation, both represent Pickwick and represent nothing. But a picture may be of a certain kind—be a Pickwick-picture or a man-picture—without representing anything." (1968, 22)

Goodman claims that the locution "representation of Pickwick" is syntactically ambiguous. On one reading it has the logical form of a one-place "fused" predicate—"Pickwick-representation"—where "Pickwick" is, in Quine's 1960 terminology, "syncategorematic." Chomsky is not committed to this syntactic thesis.

individuating condition of a given internal structure is not its relation to an "intentional object," there being no such thing according to Chomsky, but rather its role in cognitive processing. Reference to what looks to be an intentional object is simply a convenient way of type-identifying structures with the same role in computational processing.

The point applies as well to the study of the processes underlying linguistic capacities:

> [H]ere too we need not ponder what is represented, seeking some objective construction from sounds or things. The representations are postulated mental entities, to be understood in the manner of a mental image of a rotating cube, whether the consequence of tachistoscopic presentations or of a real rotating cube or of stimulation of the retina in some other way, or imagined, for that matter. Accessed by performance systems, the internal representations of language enter into interpretation, thought, and action, but there is no reason to seek any other relation to the world. (Chomsky 1995, 53)

Chomsky rejects the idea that intentional attribution—the positing of a domain of objects or properties to which internal structures stand in a *meaning* or *reference* relation—plays any explanatory role in cognitive science. Intentional construals of David Marr's 1982 theory of vision, such as Burge (1986), Chomsky claims, are simply a misreading, based on conflating the theory proper with its informal presentation. As Chomsky puts it, "The theory itself has no place for the [intentional] concepts that enter into the informal presentation, intended for general motivation" (1995, 55).

Chomsky himself has not spelled the argument out explicitly, though the motivation for his recent anti-representationalism is not hard to find.[17] As theories of our perceptual and linguistic capacities have become increasingly removed from commonsense, it becomes quite forced to say that the subject *knows* or *believes*, say, the *rigidity assumption* (Ullman 1979) or the *minimal link condition* (Chomsky 1995). Chomsky (1975) was willing to say that subjects "cognize" the principles posited in cognitive theories, but these contents—*that objects are rigid in translation* or *that derivations with shorter links are preferred over derivations with longer links*—do not look much like the sorts of things that subjects could plausibly be said to know, believe, etc. They are not inferentially promiscuous, not accessible to consciousness, and so on.

It is particularly unclear what independent objects the structures posited in accounts of our linguistic capacities represent. Among the candidates are elements of the public language,[18] elements of the speaker's idiolect, or, as Georges Rey (2003a, 2003b, 2005) has recently suggested, linguistic entities such as nouns, verb phrases,

---

[17] Though see Collins (2007) for the view that Chomsky has always been an anti-representationalist.

[18] Chomsky himself is skeptical of the notion of a "shared public language." See the papers in Chomsky (2000).

phonemes, and so on—what Rey calls "standard linguistic entities" (SLEs). SLEs, Rey argues, are to be understood as "intentional inexistents," objects of thought, akin to such fictional entities as Zeus or Hamlet, that do not exist. Discussion of the merits and demerits of these various proposals is beyond the scope of the present chapter. Chomsky, for his part, rejects them all, insisting that talk of represented objects is intended simply for informal exposition and plays no genuine role in the theory.

4. Rethinking the Standard View

Chomsky is, in effect, an *eliminativist* about representational content. He denies that the internal structures posited in computational theories are distally interpreted as representations of external objects and properties (Claim 1 of the Essential Distal Content View), and hence that computational mechanisms are type-individuated by a domain of external objects and properties (Claim 2).[19] Any reference to such a domain in computational accounts, he claims, is merely "informal presentation, intended for general motivation."

This section shall spell out a view of representation in computational cognitive theories according to which Chomsky is correct in denying the Essential Distal Content View, but nonetheless wrong in denying to representational content a genuine explanatory role. Chomsky's view fails to make clear the role played by the interpretation function f_I in computational accounts, and leaves mysterious how representational content could aid in the "informal motivation" of a computational theory. These points will be illustrated by reference to two computational models from different cognitive domains.[20]

David Marr's well-known explanatory hierarchy distinguishes three distinct levels at which a computational account of a cognitive capacity is articulated. Disputes about whether computational theories type-individuate the mechanisms they characterize by their representational content turn on how the level of description that Marr called the *theory of the computation* should be interpreted. The theory of the computation provides a *canonical description* of the function(s) computed by

---

[19] It is consistent with Chomsky's stated views that there *is* a substantive, naturalistically specifiable relation between posited structures and distal objects and properties (Claim 3 of the Essential Distal Content View), though Chomsky himself would regard the idea that theories of mind and language must respect such a "naturalistic constraint" as a manifestation of "methodological dualism," the idea that the study of language and mind, unlike scientific inquiry in other domains (which is allowed to be self-policing), should be held to independent, "philosophical" standards. See Chomsky (1994).

[20] See Egan (1995b, 1999, 2003) for defense of this account of the role of content in David Marr's theory, in particular.

the mechanism. It specifies what the device does. By a "canonical description," we mean the characterization that is decisive for settling questions of type-individuation or taxonomy. The canonical description is given by the interpretation function f_I. The canonical description is therefore a semantic characterization. But this is the important point: the canonical description of the function computed by a computationally characterized mechanism is a *mathematical* description. A couple of examples illustrate the point.

Marr (1982) describes a component of early visual processing responsible for the initial filtering of the retinal image. Although there are many ways to informally describe what the filter does, Marr is careful to point out that the theoretically important characterization, from a computational point of view, is a mathematical description: the device computes the Laplacean convolved with the Gaussian (1982, 337). As it happens, it takes as input light intensity values at points in the retinal image and calculates the rate of change of intensity over the image. But this *distal* characterization of the task is, as Chomsky might put it, an "informal" description, intended for general motivation. *Qua* computational device, it does not matter that input values represent *light intensities* and output values the rate of change of *light intensity*. The computational theory characterizes the visual filter as a member of a well-understood class of mathematical devices that have nothing essentially to do with the transduction of light.

The second and third levels of Marr's explanatory hierarchy describe a *representation and algorithm* for computing the specified functions, and the circuitry or neural hardware that implement the computations. Marr's account of early visual processing posits primitive symbol types—edges, bars, blobs, terminations, and discontinuities—and selection and grouping processes defined over them. It is at this second level that the theory posits symbol structures or representations and processes defined over them. These symbol structures (edges, bars, blobs, etc.) and the processes that operate on them are type-individuated by the mapping f_R, which (ideally, when fully specified) characterizes them at the level of physical states and processes, independent of the cognitive capacities that they subserve.

The second example, from an entirely different cognitive domain, is Shadmehr and Wise's 2005 computational theory of motor control. Consider a simple task involving object manipulation (see Figure 1). A subject is seated at a table with eyes fixated ahead. The hand or *end effector* (ee) is located at Xee, and the target object (t) at Xt. The problem is simply how to move the hand to grasp the object. There are an infinite number of trajectories from the hand's starting location Xee to the target at Xt. But for most reaching and pointing movements, the hand moves along just one of these trajectories; it typically moves along a straight path with a smooth velocity. Shadmehr and Wise (2005) describe one way in which the task might be accomplished.

The overall problem can be broken down into a number of subproblems. The first problem is: how does the brain compute the location of the hand? *Forward kinematics* involves computing the location of the hand (Xee) in visual coordinates from proprioceptive information received from the arm, neck, and eye muscles, and

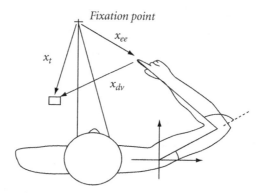

Figure 1. A motor control task.

information about the angles of the shoulder and elbow joints. Informally, this process coordinates the way the hand looks to the subject with the way it feels. The brain also has to compute the location of the target (Xt), using retinal information and information about eye and head orientation.

The second problem, computing a plan of movement, involves computing the *difference vector*, that is, the displacement of the hand from its current location to the target's location. But this "high level" plan specifies a displacement of the hand in visual coordinates. This visually oriented plan has to be transformed into a specification of the joint rotations and muscle forces required to effect the displacement. So, the third problem, involving the computation of *inverse kinematics and dynamics*, is how the high level motor plan, corresponding to a difference vector, is transformed into joint angle changes and force commands. Reaching and pointing movements involve continuous monitoring of target and hand location, with the goal of reducing the difference vector to zero. There are a number of complicating factors. For example, incidental eye and head movements require continuous updating of the situation. Deceleration of the hand should be smooth, to avoid knocking over the target.

Summarizing, the account decomposes the overall task into three computations, and specifies the function computed in each in precise mathematical terms:

(1) $f(\theta) = X_{ee}$, *forward kinematics*, the computation of hand location, in eye-centered coordinates, from propriocentric information and information about joint angles;

(2) $X_t - X_{ee} = X_{dv}$, the *difference vector*, the difference between the target location and initial hand position in eye-centered coordinates; and

(3) $f(X_{dv}) = \Delta\theta$, *inverse kinematics*, the computation from the high-level movement plan, in eye-centered coordinates to a required change of joint angles.

The motor control mechanism characterized by Shadmehr and Wise is not a physical symbol system; its operations are not interpreted in the account as manipulations of symbols. Nor does the account of the mechanism's implementation decompose neatly into representation and algorithm (Marr's level 2) and neural realization (Marr's level 3). Rather, the three computations that constitute the motor control mechanism are characterized as analog processes and realized in neural networks in the posterior parietal cortex, the premotor cortex, and the primary motor cortex respectively. The details need not concern us here.

The important point is that in both examples, the canonical description of the task executed by the device, the function(s) computed, is a mathematical description. As noted above, this description characterizes the mechanism as a member of a well-understood class of mathematical devices. A crucial feature of this characterization is that it is "environment neutral": the task is characterized in terms that prescind from the environment in which the mechanism is normally deployed. The mechanism described by Marr computes the Laplacean of the Gaussian whether it is part of a visual system or an auditory system, in other words, independently of the environment—even the *internal* environment—in which it is normally embedded. In fact, it is not implausible to suppose that each sensory modality has one of these same computational mechanisms, since it just computes a curve-smoothing function. The same point holds for the motor control mechanism characterized by Shadmehr and Wise. A mariner who knew the distance and bearing from his home port to his present location and the distance and bearing from his home port to a buried treasure could perform the same computation to compute the course from his present location to the treasure. In both cases, it is the abstract mathematical description that type-individuates the mechanism or process, not what Chomsky would call the "informal" description that characterizes the mechanism as computing *changes of light intensities* or the *displacement between target and hand location*.

To summarize: The characterization of a computational process or mechanism made available by the interpretation function f_I – the mapping that provides a canonical description of the function computed by the mechanism, and hence (along with the realization function f_R) serves to type-individuate it—is an abstract mathematical description. This semantic interpretation does not provide a *distal* interpretation of the posited internal states and structures; the specified domain is not external objects and properties in the subject's environment, but rather mathematical objects. The upshot is that the Essential Distal Content View mischaracterizes the semantic interpretation of a device given by the interpretation function f_I. Its domain is *not*, as Claim 1 holds, external objects and properties. And the representation relation determined by the mapping f_I does not, as Claim 3 holds, satisfy a constraint of the sort that proponents of a "naturalistic" semantics for mental representation have been hoping for. The interpretation maps the posited internal states and structures to a domain of *abstracta*, rather than external objects or properties required for an information-theoretic or teleological relation.

If this account is correct, then what should we make of the idea that visual states represent such visible distal properties as *depth* and *surface orientation*, and motor control states represent *hand location* and *shoulder angle*? Are such distal contents explanatorily idle, as Chomsky claims? And if they aid in "general motivation," how precisely do they do that?

Ordinary, distal representational contents do not serve to type-individuate a computational mechanism, as Claim 2 of the Essential Distant Content View holds, but they do serve several important explanatory functions. The questions that antecedently define a psychological theory's domain are usually couched in intentional terms. For example, we want a theory of vision to tell us, among other things, how the visual system can detect three-dimensional distal structure from information contained in two-dimensional images. A characterization of the postulated computational processes in terms of distal objects and properties enables the theory to answer these questions. This characterization tells us that states of the system co-vary, in the normal environment, with changes in depth and surface orientation. It is only under an interpretation of some of the states of the system as representations of depth and surface orientation that the processes given an environment-neutral, mathematical characterization by a computational theory are revealed as *vision*. Thus, content ascription plays a crucial *explanatory* role: it is necessary to explain how the operation of a mathematically characterized process constitutes the exercise of a cognitive capacity in the environment in which the process is normally deployed. The device would compute the same mathematical function in any environment, but only in some environments would its doing so enable the organism to see.

This is the most important function of representational content. Because the ascription of distal contents is necessary to explain how a computational process constitutes the exercise of a cognitive capacity in a particular context, we shall call the interpretation that enables the assignment of such distal contents the *cognitive interpretation*. The cognitive interpretation is to be sharply distinguished from the mathematical interpretation specified by f_I. Only the latter plays an individuative role.

To recap: When the computational characterization is accompanied by an appropriate cognitive interpretation, in terms of distal objects and properties, we can see how a mechanism that computes a certain mathematical function can, in a particular context, subserve a cognitive function such as vision or reaching and pointing. So when the input states of the Marrian filter are described as representing *light intensities* and the output states *changes of light intensity* over the image, we can see how this mechanism enables the subject to detect significant boundaries in the scene. When the input states of the mechanism that computes inverse kinematics are described as representing *displacement in visual space* and the output states *changes in joint angles*, we can see the role that the mechanism plays in the subject's successfully grasping the target object.

The account presented here draws a sharp distinction between the computational theory proper—the mathematical description made available by the mapping f_I, which (together with f_R) type-individuates the mechanism—and the distal characterization, made available by the cognitive interpretation that accompanies it and explains the contribution of the abstractly characterized mechanism to the larger cognitive life of the organism. We can also understand how representational content, specified by the cognitive interpretation, while not type-individuating a computational mechanism, can, as Chomsky puts it, provide "general motivation" for the theory.

The cognitive characterization is essentially a *gloss* on the more precise account of the mechanism provided by the computational theory. It forms a bridge between the abstract, mathematical characterization that constitutes the explanatory core of the theory and the intentionally characterized pretheoretic explananda that define the theory's cognitive domain. When the processes given a precise mathematical specification in the theory are construed, under interpretation, as representations of such distal properties as *edges*, or *joint angles*, the account can address the questions that motivated the search for a computational theory in the first place, such as *how are we able to see the three-dimensional structure of the scene from two dimensional images?*, or *how are we able to move our hand to grasp an object in sight?*

To call the cognitive characterization a "gloss" is not to suggest that the ascription of representational content is unprincipled. The posited states and structures are not interpretable as representations of distal visible properties (as, say, *object boundaries*, or *depth* or *surface orientation*) unless they co-vary with tokenings of these properties in the subject's immediate environment. It would be a mistake, though, to conclude that the structures posited in computational vision theories must (even in the gloss) represent their normal distal cause, and to find in these theories support for a *causal* or *information-theoretic* theory of content.[21] Some structures—zero-crossings in Marr's account, for example—are interpreted as representations of proximal features, in particular, as *discontinuities in the image*. The ascription of content is sometimes driven by purely expository considerations, such as allowing us to keep track of what the process is doing at points in the processing where the theory posits structures that do not correlate neatly with a salient distal property tokening. Even within a single cognitive interpretation, no single, privileged relation is assumed to hold between posited structures and the elements to which they are mapped. The choice of a cognitive gloss is governed by explanatory considerations, which we can, following Chomsky, characterize as "informal motivation."

An implication of the foregoing account of the role of representational content in computational models is that cognitive science has no need for a naturalistic semantics—the specification of non-intentional and non-semantic sufficient conditions for a mental state's having the meaning it does. Whatever the *philosophical* interest of such an account, it would hold little interest for the computational cognitive theorist. Content ascription in computational models is motivated primarily by explanatory considerations that a naturalistic semantics does not address.

---

[21] See Dretske (1981) and Fodor (1990) for examples of information-theoretic accounts of content.

5. Concluding Remarks on Strong Representationalism

Representationalism, recall, is the view that the human mind is an information-using system. So understood, representationalism is hard to deny. As Sterelny (1990) noted, it is hard to see how our various cognitive capacities could be explained except by positing states that are both sensitive to the world around us and causally involved in producing behavior.

Strong Representationalism goes further, construing human cognitive states as relations to internal representations, thereby positing a sharp distinction—inherent in the notion of a physical symbol system—between data structures or representations on the one hand, and the processes defined over them on the other. This view certainly has its attractions, not the least of which is that it purports to explain *how* thinkers can be information-using systems. They use information by manipulating representations of that information. This idea requires a distinction between the part of the system that uses representations and the representations themselves, which is exactly what the data structure/process distinction enforces.

A real attraction of cognitive models that conform to Strong Representationalism is their *explanatory transparency*. Symbols are structures ready-made for semantic interpretation—they just *are* objects with both formal and semantic properties, characterized by f_R and f_I respectively. Symbol structures—representations—are, in effect, "hooks" on which a semantic interpretation can be hung. Moreover, the information in physical symbol systems is *accessible for use*, encoded in exactly the features of the structures to which computational processes are sensitive. Thus, physical symbol systems are said to *explicitly represent* the information that they encode. And we can track the flow of information in the system by keeping track of the operations on the encoding structures.[22,23] This is an important explanatory role played by what we call the "cognitive interpretation."

The structure/process distinction inherent in Strong Representationalism is not inevitable. There are ways to eschew the distinction while preserving the central idea of (regular strength) representationalism—that the mind is an *information-using device*. The Shadmehr and Wise motor control mechanism described above is an example. It is not a physical symbol system; its computations are characterized as analog processes and realized in neural networks. Parallel distributed processing (PDP) systems, analog relaxation systems, massive cellular automata, and other kinds of computational mechanisms for which the structure/process distinction is not preserved are *not* ready-made for interpretation. Often, in PDP systems, no distinct state or part of the network serves to represent any particular object,

---

[22] This is the point of positing what Egan and Matthews (2006) call "intentional internals."

[23] Though see Kirsch (1990) for an argument that even PSS models may not be as transparent as commonly thought.

property, or proposition. Rather, the encoding of information is distributed over many units, connection strengths, and biases, with the result that the representation of any given object, property, or proposition is widely scattered throughout the network. It becomes quite forced to talk of "representations" in such systems,[24] given that our paradigm of a representation is the printed word, a discrete object that is spatially compact, movable, and, most important, *meaningful* (but only under interpretation). These systems are far from explanatorily transparent. One often cannot tell by looking at the computational and engineering details of the system which of its spatiotemporal parts are candidates for interpretation. It can be quite difficult to track the flow of information in these systems. But the point of interpretation is the same as for PSS systems—to make the computational processes perspicuous as *cognitive* processes.

It is even possible that mental representation is a *global* affair. A whole system might be sensitive to environmental changes and hence be involved in representing the world without any localizable part of the system doing any particular representational job. If human cognitive capacities are the result of processes of this sort, then understanding and modeling these capacities will be very challenging.

The explanatory transparency of systems that respect the structure/process distinction explains the attraction of Strong Representationalism, the view that human cognitive processes are to be understood as functionally characterizable relations to internal representations. While this may be a reason to hope that Strong Representationalism is true, it is not a reason to believe it.

REFERENCES

Aizawa, K. (2006). *The Systematicity Arguments*. Dordrecht, The Netherlands: Kluwer.

Anderson, M. (2003). Embodied cognition: A field guide. *Artificial Intelligence* 149: 91–130.

Beer, R. (1995). Computational and Dynamical Languages for Autonomous Agents. In Port and Van Gelder (eds.), *Mind as Motion*. Cambridge, MA: MIT Press.

Braddon-Mitchell, D., and Jackson, F. (1996). *Philosophy of Mind and Cognition*. Oxford: Blackwell.

Brooks, R. (1999). *Cambrian Intelligence*. Cambridge, MA: MIT Press.

Burge, T. (1986). Individualism and psychology. *The Philosophical Review* 95: 3–45.

Byrne, A. (2001). Intentionalism defended. *The Philosophical Review* 110: 199–240.

Chemero, A. (2000). Anti-representationalism and the dynamical stance. *Philosophy of Science* 67: 625–47.

Chomsky, N. (1975). *Reflections on Language*. New York: Pantheon Books.

———. (1994). Naturalism and dualism in the study of language and mind. *International Journal of Philosophical Studies* 2: 181–209. (Reprinted in Chomsky 2000.)

---

[24] Ramsey (2007) argues that it is not appropriate to characterize such systems as *representational*.

————. (1995). Language and nature. *Mind* 104: 1–61.

————. (2000). *New Horizons in the Study of Language and Mind*. Cambridge: Cambridge University Press.

Clancy, W. (1997). *Situated Cognition: On Human Knowledge and Computer Representation*. Cambridge: Cambridge University Press.

Clark, A. (1998). *Being There: Putting Brain, Body, and World Together Again*. Cambridge, MA: MIT Press.

Collins, J. (2007). Meta-scientific eliminativism: A reconsideration of Chomsky's review of Skinner's *Verbal Behavior*. *British Journal for the Philosophy of Science* 58: 625–58.

Cummins, R. (1996). Systematicity. *Journal of Philosophy* 93: 591–614.

Dennett, D. (1981). A Cure for the Common Code?, In *Brainstorms*. Cambridge, MA: MIT Press.

Dretske, F. (1981). *Knowledge and the Flow of Information*. Cambridge, MA: MIT Press.

————. (1986). Misrepresentation. In R. Bogdan (ed.), *Belief: Form, Content and Function*. New York: Oxford University Press.

————. (1995). *Naturalizing the Mind*. Cambridge, MA: MIT Press.

————. (2003). Experience as representation. In E. Villanueva (ed.), *Philosophical Issues*, 13: 67–82.

Egan, F. (1995a). Folk psychology and cognitive architecture. *Philosophy of Science* 62: 179–96.

————. (1995b). Computation and content. *The Philosophical Review* 104: 181–203.

————. (1999). In defense of narrow mindedness. *Mind and Language* 14: 177–94.

————. (2003). Naturalistic Inquiry: Where Does Mental Representation Fit In?. In L. Antony and N. Hornstein (eds.), *Chomsky and his Critics*, Oxford: Blackwell.

Egan, F., and Matthews, R. (2006). Doing cognitive neuroscience: A third way. *Synthese* 153: 377–91.

Field, H. (1978). Mental representation. In *Erkenntnis*, 13: 9–61.

Fodor, J. (1975). *The Language of Thought*. New York: Thomas Y. Crowell.

————. (1980). Methodological solipsism considered as a research strategy in cognitive science. *Behavioral and Brain Sciences* 3: 63–73.

————. (1981). *RePresentations: Philosophical Essays on the Foundations of Cognitive Science*. Cambridge, MA: MIT Press.

————. (1987). *Psychosemantics: The Problem of Meaning in the Philosophy of Mind*. Cambridge, MA: MIT Press.

————. (1990). *A Theory of Content and Other Essays*. Cambridge, MA: MIT Press.

————. (2008). *LOT 2: The Language of Thought Revisited*. Oxford: Oxford University Press.

Fodor, J., and Pylyshyn, Z. (1981). How direct is visual perception? Some reflections on Gibson's "ecological approach." *Cognition* 9: 139–96.

————. (1988). Connectionism and cognitive architecture: A critical analysis. *Cognition* 28: 3–71.

Gallagher, S. (2005). *How the Body Shapes the Mind*. New York: Oxford University Press.

————. (2008). Are minimal representations still representations? *International Journal of Philosophical Studies* 16: 351–69.

Gallistel, C. R., and King, A. P. (2009). *Memory and the Computational Brain*. Oxford: Wiley-Blackwell.

Gibson, J. (1966). *The Senses Considered as Perceptual Systems*. Oxford: Houghton Mifflin.

————. (1979). *The Ecological Approach to Visual Perception*. Oxford: Houghton Mifflin.

Goodman, N. (1968). *Languages of Art*. Indianapolis: Bobbs-Merrill.

Harman, G. (1972). *Thought*. Princeton, NJ: Princeton University Press.

Harvey, I., Husbands, P., and Cliff, D. (1994). Seeing the Light: Artificial Evolution, Real Vision. In D. Cliff (ed.), *From Animals to Animats 3*. Cambridge, MA: MIT Press.

Kirsh, D. (1990). When Is Information Explicitly Represented? In P. Hanson (ed.), *Information, Content, and Meaning*, Vancouver: University of British Columbia Press.

———. (1991). Today the earwig, tomorrow man. *Artificial Intelligence* 47: 161–84.

Lycan, W. (1987). *Consciousness*. Cambridge, MA: MIT Press.

———. (1996). *Consciousness and Experience*. Cambridge, MA: MIT Press.

Marr, D. (1982). *Vision*. New York: Freeman.

Marras, A. (1985). The Churchlands on methodological solipsism and computational psychology. *Philosophy of Science* 52: 295–09.

Matthews, R. (1997). Can connectionism explain systematicity? *Mind and Language* 12: 154–77.

———. (2007). *The Measure of Mind: Propositional Attitudes and Their Attribution*. Oxford: Oxford University Press.

Macdonald, C., and Macdonald, G. (1996), *Connectionism: Debates on Psychological Explanation*. Oxford: Blackwell.

Millikan, R. (1984). *Language, Thought, and Other Biological Categories*. Cambridge, MA: MIT Press.

Newell, A. (1980). Physical Symbol Systems. *Cognitive Science* 4: 135–83.

Noe, A. (2004). *Action in Perception*. Cambridge, MA: MIT Press.

———. (2009). *Out of Our Heads*. New York: Hill and Wang.

Papineau, D. (1987). *Reality and Representation*. Oxford: Blackwell.

———. (1993). *Philosophical Naturalism*. Oxford: Blackwell.

Peacocke, C. (1983). *Sense and Content*. Oxford: Oxford University Press.

Pinker, S. (1997). *How the Mind Works*. New York: Norton.

———. (2005). So how does the mind work? *Mind and Language* 20: 1–24.

Pylyshyn, Z. (1980). Computation and cognition: Issues in the foundation of cognitive science. *The Behavioral and Brain Sciences* 3: 111–32.

———. (1984). *Computation and Cognition: Toward a Foundation for Cognitive Science*. Cambridge, MA: MIT Press.

Quine, W. (1960). *Word and Object*. Cambridge, MA: MIT Press.

Ramsey, W. (2007). *Representation Reconsidered*. Cambridge: Cambridge University Press.

Rey, G. (2003a). Chomsky, Intentionality, and a CRTT. In L. Antony and N. Hornstein (eds.), *Chomsky and His Critics*. Oxford: Blackwell.

———. (2003b). Intentional Content and a Chomskian Linguistics. In A. Barber (ed.), *Epistemology of Language*. Oxford: Oxford University Press.

———. (2005). Mind, intentionality, and inexistence: An overview of my work. In *The Croatian Journal of Philosophy* 5: 389–415.

Robbins, P., and Aydede, M. (2008). *The Cambridge Handbook of Situated Cognition*. Cambridge: Cambridge University Press.

Shadmehr, R., and Wise, S. (2005). *The Computational Neurobiology of Reaching and Pointing: A Foundation for Motor Learning*. Cambridge, MA: MIT Press.

Smith, B. (unpublished). Semantic attribution and the formality condition.

Sterelny, K. (1990). *The Representational Theory of Mind*. Oxford: Blackwell.

Stich, S. (1983). *From Folk Psychology to Cognitive Science: The Case against Belief*. Cambridge, MA: MIT Press.

Stich, S., and Warfield, T. A. (1994). *Mental Representation: A Reader*. Oxford: Blackwell.

Swoyer, C. (1991). Structural representation and surrogative reasoning. *Synthese* 87: 449–508.

Tye, M. (1995). *Ten Problems of Consciousness*. Cambridge, MA: MIT Press.

———. (2000). *Consciousness, Color, and Content*. Cambridge, MA: MIT Press.

———. (2003). Blurry Images, Double Vision, and Other oddities: New Problems for Representationalism. In Q. Smith and A. Jokic (eds.), *Consciousness: New Philosophical Perspectives*, Oxford: Clarendon Press.

Ullman, S. (1979). *The Interpretation of Visual Motion*. Cambridge, MA: MIT Press.

van Gelder, T., and Port, R. (1995). It's about time. In R. Port and T. van Gelder (eds.), *Mind as Motion: Explorations in the Dynamics of Cognition*. Cambridge, MA: MIT Press, 1–44.

Varela, F., Thompson, E., and Rosch, E. (1991). *The Embodied Mind*. Cambridge, MA: MIT Press.

Von Eckart, B. (1995). *What is Cognitive Science?* Cambridge, MA: MIT Press.

Waskan, J. (2006). *Models and Cognition*. Cambridge, MA: MIT Press.

Wheeler, M. (1996). From Robots to Rothko. In M. Boden (ed.), *The Philosophy of Artificial Life*. Oxford: Oxford University Press.

CHAPTER 12

....

COGNITION AND
THE BRAIN

....

RICK GRUSH AND LISA DAMM

WHAT is the relationship between the brain and cognition? The answer to this question is likely to depend upon what exactly cognition is, and what sorts of tools, technological and conceptual, we have for studying the functioning of the brain. While many issues are relevant to these questions, we will focus this chapter by restricting attention to a few central topics. First, in Section 1, we will discuss the issue of what cognition is. This is not an easy question, since the traditional view according to which cognition is a matter of reasoning and decision making, understood as a kind of inferential process, has many attractions, but has also come under much pressure from various quarters.

Second, there is a question concerning what the study of the brain amounts to, and we will turn to this in Section 2. What kinds of insights or information have we been able to get, or can we expect to be able to get, from a study of the brain's operation? This question has both a technological aspect and a conceptual aspect, and we will focus largely on the latter.

Finally, we will turn to a question that is more philosophical. While Sections 1 and 2 discussed the relation between cognition and the brain, in Section 3 we turn to the where, if anywhere, the *mind* fits in to the picture. The reason is that there is, in many quarters, an implicit assumption to the effect that the mind *is* cognition, or *is* anything that cognizes in the sense of *solves a problem*. And if this is right, then in understanding the relation between cognition and the brain, we are thereby also understanding the relation between the mind and brain. We want to loosen this connection.

1. WHAT IS COGNITION?

1.1. A Traditional Taxonomy

There is a traditional taxonomy that attributes to the *mind*—psychology's putative object of investigation—a number of distinct faculties. While there are many ways to cut the pie, they typically revolve around the categories of perception/emotion, will/action, memory, and cognition. Cognition, in this view, is taken to be something on the model of formal inference, maybe with some statistics or Bayesianism or some other formalizable procedure thrown in for good measure. This sort of faculty psychology would tell the following rough story about the relation between the faculties: perception is a matter of gaining information about the world, memory stores information about the world, cognition employs this information to make decisions about what to do, then the result is effected by the will and action. While emotions may have some benefits, their contribution to this process is often thought to be one of disruption, especially at the cognition stage. Others might use the expression "cognition" to mean something that is anything other than perception, though perhaps even this is too restrictive. Cognitive science departments routinely have faculty who specialize in perception.

For purposes of this paper, we will focus on cognition in the sense of reasoning and problem solving. The assumptions made by the traditional taxonomy are assumptions to the effect that either (i) we know what reasoning is, in the sense that we know what formal system reasoning implements, or (ii) we at least have an idea of what the relevant alternatives are, and we can ascertain, through behavioral criteria, which one humans implement. Now if we already know what reasoning is as per (i), then the best way to study it will be neither neurophysiological nor behavioral. The studies of statistics, logic, and so forth, are formal studies with their own disciplines. That human beings implement a certain logical system might be interesting, but would be no more relevant for the study of that logical system itself than the fact that humans are subject to gravity is relevant to relativity physics.

If however the situation is as in (ii), then while we still stand to learn nothing about any of the alternatives by studying humans, there is at least a problem to be studied, which is which of the options is implemented by humans. By analogy, this position would be like stating that masses accelerate toward other masses either via gravitational attraction or electromagnetic attraction (we know the possibilities), and we are interested in which of these forces is responsible for humans falling toward Earth—and perhaps we strap some humans with magnets, or have others fall after we have increased Earth's mass but not its charge, and we get dissociative data that indicates that it is gravity, not electromagetism, that explains why humans fall. Obviously such a study tells us nothing about gravity or electromagnetism per se.

In the case of determining which cognitive system humans implement, the basic idea would be to set up a behavioral experiment such that the results will implicate one formal system over the competitors, and run the experiment. To take a toy example, suppose we have a student learning sentential logic, and we want to know whether their preferred method of conducting proofs is to use *modus ponens* and indirect proof a lot, or if instead they use a lot of *modus tollens* and conditional proof. We make an exam full of proofs, some of which are easy to do if one uses *modus ponens* and indirect proof, but are very difficult if one uses *modus tollens* and conditional proof, and others have the opposite pattern. We give the exam to the student, measure how long it takes her to do the proofs and look at the patterns of errors, and based on these results we have evidence that implicates the student's preferred method of doing proofs.

Now to move to a non-toy example. Philip Johnson-Laird's *Mental Models* (Johnson-Laird 1983), one of the most important books in cognitive psychology in the twentieth century, provides a huge wealth of evidence for attributing to humans a certain kind of reasoning architecture, and it does so without bringing in any considerations about the brain at all. For example, subjects will assess whether (Z) *Some As are Cs* follows from (X) *All As are Bs* and (Y) *Some Bs are Cs* by trying to construct a model such that (X) and (Y) hold of the model but (Z) does not. This contrasts with a view of the nature of human reasoning that sees such inferences as a matter of some sort of formal inference based on sentence-like representations of (X) and (Y) and attempts to use inference rules to derive (Z). The different theories make different predictions about how subjects will perform in various cases. One then presents subjects with carefully constructed types of inferences, and on the basis of what sorts of errors subjects make and different amounts of time taken, and so forth, one can get a grip on the mechanisms of reasoning. And all this without so much as checking to see whether there is a brain in the subject's head at all.

Of course anyone familiar with the literature in cognitive psychology will immediately point out that the cartoon picture we have painted is very much a cartoon. A huge amount of effort has been invested in questions of the implementation of formal inference, and the results are nowhere near decisive (for some recent research, see Goodwin and Johnson-Laird 2005; Van Heuveln 2006; Gyselinck et al. 2007)

To take one more of potentially dozens of theoretical possibilities for purposes of illustration, some researchers propose one or another sort of *dual process model*. The primary idea behind these models is that thoughts and behavior are driven by two separate and often conflicting mechanisms: the heuristic/associative and the systematic/rule-based (for review, see Evans 2003). Associative processing can be characterized as pattern completion and is described as the more intuitive, impulsive, and emotional form of processing. In contrast, rule-based processing uses symbolically represented knowledge to direct processing and is more effortful and more transparent to consciousness than associative processing. It is described as logical and rational. Some dual processes posit that processing is sequential, associative followed by rule, but other models set up the processing as parallel.

So far, we have been pointing out that if either (i) or (ii) is correct, then studying the brain will not teach us anything about reasoning per se. This does not mean, though, that if (i) or (ii) is true that we cannot advance our knowledge of *human* cognition (meaning, how humans implement their reasoning system) by doing brain science. First, even if behavioral studies are sufficient, it may be the case that doing them in concert with neurophysiological studies will be a quicker and more efficient way to tell which of the options humans implement. Second, even if one knows which system human brains implement, one might still be curious as to how the brain manages to implement it. Again to take a toy example, if there is a Tinkertoy machine that plays chess, one need not get into Tinkertoy parts in order to understand the game of chess. And moreover, one may not need to get into Tickertoy parts to decide what sort of strategy the mechanism is employing. One could in principle use a variety of behavioral data to determine whether it is using brute force search or some sort of piece-value system or a Lasker-Bauer combination. Nevertheless, one might still be very interested in *how* a collection of Tinkertoys can be a Lasker-Bauer-implementing mechanism, and so there would be a point in looking into the Tinkertoy mechanism itself. But one would not expect to learn anything new about Lasker-Bauer (or heuristic search or whatever) from such an endeavor. So in summary, if (i) or (ii) is right, then the study of the brain is not necessary for understanding cognition, though it might help, and it would be relevant to understanding how the brain implements whatever formal system it implements.

But the morass of current research indicates, if anything, that neither (i) nor (ii) is true. A third possibility is (iii) we have a rough idea of what the relevant options are, but behavioral studies alone cannot ascertain which apply. In this spirit, there are a host of neuroscientific studies aimed at using one or another kind of neuroscientific consideration to make a choice between various options (see, e.g., Sanfey et al. 2006; Goel 2005, 2007). But as we shall see in the remainder of this section, there is reason to think that all of (i)–(iii) are overly optimistic. While it is manifestly obvious that humans engage in reasoning and make decisions, it is quite far from obvious that we have any clear idea of what the relevant options are for how this is done. This will be explored in the remainder of Section 1.

1.2. Emotions

On the classical taxonomy emotion and reason are different taxa, and if anything emotion is viewed on this taxonomy as an impediment to reason. Nevertheless, there is growing evidence that this view of things is somewhere between impossibly simplistic and completely false. At the psychological level of explanation there is a good deal of research that seems to indicate that emotions provide information, anticipate future responses, influence reasoning strategy, index value, and direct attention toward particular objects. But few psychologists have attempted (successfully) to incorporate these results into an integrative general theory of cognition and emotion. We will discuss a number of proposals about the relationship between emotion and reason as a sort of case study about the way in which continuing research puts pressure on the traditional taxonomy.

Antonio Damasio's research program began with his observation that various patients with damage to their ventromedial cortex, including the famous Phineas Gage (see Damasio 1994), had normal intellect but exhibited impairment in their decision-making abilities coupled with abnormalities in their ability to express emotion and experience feelings. This led Damasio to formulate the *somatic marker* theory (Damasio 1998), which is a general theory of the relation between emotion and decision making. One point we shall shelve for now is that Damasio's original inspiration was neurophysiological, and there are neurophysiological aspects to this theory—we shall turn to the neurophysiological considerations in Section 2.

At the psychological level, Damasio claims that emotions are primarily representations of somatic states, including visceral and musculoskeletal. These representations can be perceptual in nature, when caused by the actual corresponding somatic states, or can be more imagistic in nature, when these same representations are induced by mechanisms other than the body, in a way analogous to how visual imagery produces a visual representation without the operation of the eyes (more about this below). These emotional states can be positive or negative in valence, and while they are often consciously experienced, can sometimes be entirely subconscious.

The theory of how emotions so understood interface with reason and decision making is this: People have experience with certain event types, and often these event types have emotional aspects; for example, touching a hot stove and being burned would lead to the subsequent negative emotional state (here a perception of an actual somatic state) brought about by the intense pain. The relationship between the event type and the associated emotional reaction is learned so that when the same type of event is encountered, or the same type of action considered, it can induce the corresponding emotion—perhaps an actual somatic state change, or perhaps imagery of that change—and the valance of that emotion can influence how the agent behaves in that situation. So for instance, when one considers touching the stove, the imagined scenario of touching the stove leads to the re-creation of the learned negative emotional response. This response can either be actually expressed in somatic changes, or only in an imagistic counterpart of such changes; furthermore, the emotional response may or may not be something of which the agent is explicitly aware. This emotional response, the change in somatic state or representation of somatic state, is a valenced marker, a *somatic marker*, attached to the image of the event or consideration of the action.

Damasio argues that somatic markers help facilitate reasoning by providing a rapid processing of potential decision outcomes based on immediate endorsement or rejection, which then helps constrain the decision-making space to a manageable size for which it becomes reasonable to employ more traditional means of evaluation such as cost-benefit analysis on the remaining options.

Damasio's theory of the nature of reason is motivated in large part by neurological considerations. Patients with ventromedial damage appear to show patterns of slow and error-prone decision making as well as deficits in emotional affect (but recent critics have conducted slight variations of the task and offered alternative explanations—see Maia and McClelland 2004; and Tomb et al. 2002).

But Damasio's theory is not universally accepted even among those who feel that emotions play a critical role in reason. Smith and Neumann (2005) have developed an account of emotion elicitation that is integrated with a dual process model of reasoning and incorporates emotions into rule-based processing as well as associative processing. The idea behind associative emotion elicitation is that once an emotion state consisting of the activation of multiple components has been triggered many times, any single component will then be capable of triggering the entire pattern of activation of all the components. So associatively elicited emotions are fast and automatic in contrast to rule-based elicitation, which can be more thoughtful and reflective and allow for more flexible activation of emotional response in ways that are more sensitive to social context. Smith and Neumann speculate that their theory can be well integrated with some of what we know about the physiology of emotion, and that it has interesting implications for explaining phenomena such as dissociations due to input stimuli and from cognitive beliefs.

In summary, there is a good deal of research aimed at questioning the traditional taxonomy according to which reason and emotion are distinct. But while in some sense this might be right, the position raises as many questions as it answers (not the least is: *what exactly is an emotion?*).

1.3. Sensorimotor Behavior

Another proposal about the nature of reason is one that might initially seem to be too silly to be taken seriously. The proposal is to assimilate cognition to sensorimotor behavior. Nobody doubts that cognition and motor behavior are closely related—one of the main things agents think about is what they should do, and moreover there is motor involvement in using language and playing chess. But the proposal on the table is radical in that it suggests that these phenomena—language, chess, reasoning—do not just *involve* sensorimotor control, but are *nothing but* sensorimotor control. In this view, the only difference between responding to a question and reacting to a tipping water glass is the degree of complexity. This contrasts with the traditional view that sees a difference *of kind* between scratching an itch and making a considered chess move, the difference being marked by the involvement in the latter case of inferences, or something like that. Rodney Brooks's 1991 seminal paper "Intelligence without representation" can be read an example of such a position, as can Timothy van Gelder's "What Might Cognition Be If Not Computation?" (1995). In both cases, a capacity to execute some behavior that seems to be sensorimotor in nature (avoiding obstacles during locomotion, adjusting the speed of a turbine) is taken as a paradigm of "intelligence" or "cognition," and accordingly mechanisms by which something executes this function is suggested as a mechanism underlying intelligence and cognition. And in both cases, the analysis is to lead us to accept that the mechanisms involved are purely sensorimotor.

How can such a proposal be assessed? Obviously if anyone were claiming that the only way to avoid obstacles or to control the angular velocity of a turbine was to use a classical computational architecture, then Brooks's and van Gelder's points would be decisive against that view. But it is not obvious that anyone has ever held such a view.

There is, however, a better way to read the challenge. Surely the main driving force behind the evolution of the nervous system over the vast bulk of its phylogenetic history has been sensorimotor coordination. Behaviors that prototypically involve reason or language have made only a recent appearance. While this fact by itself does not settle anything, it at the very least provides a motivation to explore the extent to which these recent developments (language, chess) can be understood as extensions of the earlier developments (building shelters, avoiding predators). And one way to follow this motivation is to explore the extent to which behaviors that have historically been conceived of as a matter of reasoning can be conceived of in other terms.

With this last point it becomes possible to see that two issues have become entangled in this debate. One concerns the involvement, or lack of involvement, of *representations* in some complex behavior. The other has to do with the involvement, or lack thereof, of mechanisms that are computational in the traditional sense, that is, as involving the manipulation of sentence-like representations according to syntactic rules. A great many of the arguments against representation are pitched against traditional computationalism and involve some sort of claim that the notion of representation only makes sense in the context of such systems (see Grush 2003). It remains to be seen whether anything that looks like higher cognition and reason can be representational without being computational in this narrow sense. But if it can, then to some extent higher cognition and sensorimotor control will be seen to be similar, but because sensorimotor control does involve manipulation of representations, not because higher functions do not. We will return to this topic in Section 2.3.

But for now we will close this section by mentioning one researcher whose work draws particularly strong connections between this approach and the previous topic, emotion. Berthoz (2006) rejects the traditional view that decision making is a rational process that occurred late in evolution. He believes that action and the "sentient body" (rather than logic and language) are at the heart of decision making, and he develops this view on two primary assumptions: (1) decision making is probably *the* fundamental property of the nervous system, and (2) its origin is action.

Berthoz argues that the brain is a simulator of action and a generator of hypotheses such that anticipating and predicting the consequences of actions based on the remembered past is one of the basic properties of the brain. Within this action-based framework, Berthoz claims that decision making involves simulated action, and emotion is the ultimate judge of decisions. Emotions function to target objects in the world, enable past experiences to guide future experiences via predicting and anticipating the future, and prepare the body for action. Berthoz attempts to develop what he calls a physiology of preference, which is a fairly detailed model of how different structures in the brain, through expressing preferences, contribute to the process of decision making. In this view emotion or mood influences perception and prepares the body for action before the body follows through on a decision by engaging in action. We will return to this in Section 2.3.

1.4. Discussion

What lessons can be learned form the considerations discussed above? The main one is that things are not as easy as we might have thought. If our goal is understanding the relation between cognition and the brain, we face the challenge that at least one of the relata, *cognition*, is not terribly well understood. There is nothing like agreement even at the most foundational level—as the debate about whether behaviors that qualify as "reason" require the involvement of emotion, or involve representation, or are perhaps best understood as a fancy sort of motor control, shows. And perhaps worse, even aside from these sorts of esoteric debates about conceptual foundations, there appears to be good reason to seriously question whether "reason" denotes anything like a natural mental kind amenable to study as such. There is a possibility that nervous systems have evolved a number of tricks and abilities that do not map neatly onto folk categories of "reason," "perception," "memory" and so forth, and the best we get is that some orchestrations of these capacities, in some contexts, approximate to some degree the prototype of "reason" or "long-term memory." But if this is the case, then progress might be best served by abandoning the folk taxonomy and its category of "reason."

That, anyway, is one possibility. But perhaps we can seek help on the question of the relation between cognition and the brain from the other *relatum*, the brain. It is to this that we turn next.

2. WHAT DOES THE BRAIN DO?

As we have seen, it is simply not clear what human cognition amounts to. The unclarity is present both at a semantic level—different people mean different things by "cognition"— and at the level of the phenomenon itself. To take one example, the answer to a question that would seem to be as basic as whether emotions are a hindrance to reason, or are an essential part of its correct functioning, is up for grabs. Human reasoning and decision making is messy when we are interested in why the driver changed lanes when he did. But even when we restrict attention to artificially tidy cases such as deciding whether Socrates is mortal when we are told that Socrates is a man and all men are mortal, it is simply not clear what is going on. So for some clarification maybe it is time to look at the thing responsible for implementing the decision making.

2.1. Neurons and Spikes

For the last century or so it has been fairly clear that one of the interesting things going on in the brain is neuron spiking. A typical neuron has many dendrites, or root-like appendages that change their electrical properties upon receipt of a chemical signal in the form of one or another kind of neurotransmitter. Under certain conditions, these changes to the electrical properties of the dendrites' membranes affects the electrical properties of the membrane of the cell's main body in such a way as to cause a *spike* in that neuron's axon—a fast-moving electrochemical wave that terminates at the end of the axon and causes the release of a neurotransmitter. This neurotransmitter is then picked up by the dendrites of other neurons (and in some cases, that same neuron), and can have an effect in turn on *its* spiking behavior. And the dominant view of brain function is that this process, or minor variants on it, is the core of the brain's information processing capacity.

While this simple picture is undoubtedly accurate as a portrayal of a lot of what is happening in the brain, it is nevertheless a simplification in many respects. We will mention only a few. First, there are many cells in the brain other than neurons. By many estimates, 90 percent of the cells in the brain are glial cells, typically thought to be cells that serve some sort of support function. However, some have thought that glial cells themselves may contribute directly, perhaps significantly, the brain's information processing (see, e.g., Allen and Barres 2005; Diamond 2006).

Second, some have claimed that individual neurons are themselves powerful information processors whose processing goes far beyond simply adding up dendritic inputs and producing a spike. If this is right, then the basic unit of computational power may not be the neuron, but subneuronal components (see, e.g., Priel et al. 2005; for a contrary view, see Litt et al. 2006).

2.2. Rates and Times

But there are still difficult questions to be answered even if we stick with the original picture according to which neuron firings are the fulcrum of the brain's information-processing prowess. One set of issues concerns what it is about the spiking that is relevant. The two major options here are *rate codes* and *time codes*. In rate code schemes, what is important about a spiking neuron is the rate at which it is firing, for example 10 spikes per second versus 80 spikes per second. The exact instant at which the spikes occur is, on this view, not terribly relevant. There are a number of motivations for this view. First, the physiology of dendritic membranes suggests that they act as low-pass filters mostly sensitive to the amount of neurotransmitter received over some small interval, and relatively insensitive to the exact timing of the receipt of each small bolus of neurotransmitter. Single presynaptic spikes, regardless of timing, rarely cause a post-synaptic action potential, while a lot of presynaptic activity, regardless of timing, can reliably cause post-synaptic action potentials. On the other hand, there is significant neurophysiological evidence that neurons can be extremely sensitive to the timing of events, both presynaptic events and their own post-synaptic activity. This leads to the view that what is important is not so much the rate at which a neuron is spiking, but the exact timing of individual spikes (for more on this issue, see VanRullen et al. 2005; Stein et al. 2005; Romo and Salinas 2003).

2.3. Questions of Architecture

The next set of questions has to do with what sort of information-processing archi-tecture the brain is implementing (whether or not neuron spiking is the main causal mechanism of its implementation, and whether or not rate codes or time codes or whatever are involved). There are several classes to consider.

First, the brain might implement some sort of classical computational architec-ture. While there are many flavors here, the basic proposal is that the level of descrip-tion at which it becomes clear how the brain is processing information is one at which the system is operating on symbols that are structured in some roughly sen-tence-like manner. Obviously this is the architecture of choice for those theorists who take cognition to be a matter of formal inference or general purpose problem solving. But the converse does not necessarily hold. For example, one could think that the information-processing architecture of the brain is classically computa-tional, but what the computations are doing is computing connectionist activations values, or computing the torque to apply to a joint in the vocal apparatus as per the proponents of cognition as sensorimotor control.

There is little *neurophysiological* reason to think that a classical computational architecture is implemented by the brain. Most of the considerations that speak in its favor are behavioral or otherwise top-down. Perhaps most famously, Fodor and Pylyshyn (1988) argued that only a classical architecture is capable of doing some of the things that we know human thinkers can do, things intimately related to proto-typically cognitive phenomena such as reasoning and language. For example, they claim that thought is systematic, in that any person capable of entertaining the con-tent *Mary loves John* will be able to entertain the content *John loves Mary*. This empiri-cal fact, they argue, shows that human cognition is a matter of constructing complex contents out of components that can be composed in various language-like ways.

The second class of options falls under the heading of connectionism (Rumelhart and McClelland 1986). The idea behind connectionist models is that information processing is a matter of a number of simple interconnected processing units such that information given to some of these units in the form of an activity level is passed along to other units via connections to intervening units, and finally results in activity in output units. Depending on details, these can all be the same units, and these connections might be recurrent. But what is common is the idea that infor-mation is coded in terms of scalar activation values of sets of these units, and it is processed by passing these activations values between these units. Learning is accomplished typically by changing the connection weights, which changes the details of how the activations spread through the web of units.

One point in favor of connectionist models is their prima facie biological plau-sibility. The formal structure of a standard connectionist unit maps closely onto the causal structure of a standard neuron. And furthermore one of the standard con-nectionist learning paradigms, Hebbian learning, seems to be active in neural sys-tems (Wörgötter and Porr 2005). Nevertheless, there isomorphism is less than perfect. First, while Hebbian learning is one connectionist learning paradigm, it is not the most well-known or powerful. Many connectionist models use some form of back-propagation, and it is not clear whether real neural systems can implement anything analogous to this.

The third class is dynamical systems theory and classical control theory. The *loci classici* for this approach is to be found in the cyberneticist movement of the early twentieth century. While the cognitive revolution of the mid-twentieth century pushed this approach from center stage, it has always been onstage even if off to the side, and had a brief return to the limelight in the 1990s. Dynamical systems theory is a mathematical framework for describing the evolution of certain kinds of complex systems (for an introduction, see Ashby 1952; Port and van Gelder 1995). The basic idea, applied to understanding cognition and behavior, is that complex behaviors, including cognitive behaviors, are the result of a system of causally interacting components.

Described this way, the framework is very general, and applies to classical and connectionist systems. Strictly speaking, almost everything is a dynamical system. But while this is strictly true, the proponents of this approach typically have something much more constrained in mind. In particular, while the proponents of this approach often describe it as based in dynamical systems theory, it is a fact that a good many of their examples, and their theoretical commitments, are more specific in that the dynamical systems they envisage are often kinds of feedback control schemes (see Grush 2003).

A familiar example of a feedback control scheme is a thermostat. This is a device that controls the heating and cooling apparatus in a room or building. It is a *feedback* control system because the causal mechanism is driven by feedback from the system being controlled. In the case of old-fashioned thermostats, this feedback was in the form of the room's temperature affecting a physical structure in the thermostat, which would then either open or close a circuit. The causal mechanism was designed such that if the temperature dropped below a certain point, the circuit turning on the heater would be engaged, and if the temperature rose above a certain point, circuit would be disengaged. The result was a causal mechanism, driven by feedback, which solved a sensorimotor(-ish) problem. The Watt Governor (an example used by both Ashby and van Gelder) is another example of a feedback control system.

The cybernetics movement was one whose inspiration was control theory, and in particular the emergence of tools for making systems that seemed to be able to control themselves (the term *cybernetics* is from the Greek word for *pilot*, as one of the chief applications of the control apparatus was in automatic piloting, including guided missiles).

The fourth is modern control theory and signal processing. Modern control theory like classical control theory is concerned *in principle* with describing how one system can interact fruitfully with, and in some cases control, some other system. However in modern control theory, the conceptual background is more sophisticated, taking into account various elements—such as noise, unpredictable disturbances, and deadtime (the time that it takes for a change to the control signal to become manifest in the feedback)—that were largely ignored in classical control theory. Along with this more sophisticated conceptual background come more sophisticated tools. Where classical control revolved around the creation of systems that would approach some sort of stable behavior (such as keeping the temperature of a room near a certain temperature), modern control theory is concerned, in addition, with filtering noise and jitter (temporal noise), estimating the state of the controlled system, and techniques for overcoming deadtime. Among the common tools for approaching these applications are inverse models and forward models.

On the topic of neural information processing architecture, it is the notion of the *forward model* that has seen the widest application. A forward model is a component of the controlling system whose function is to act as an "internal" model of the controlled system. A very rough analogy would be how the bridge of a ship maintains a model (implemented on a map) of the state of the controlled entity (the ship and its location) in order to assist it in controlling the ship. This model can be used for a variety of purposes, including overcoming deadtime and evaluating counterfactuals. That is the basic idea, but there are variations going under different names (e.g., system identifications, emulators, sensorimotor contingencies), and these variations are seldom clearly distinguished. (See Grush 2007, section 5.1, for a discussion of some of the different notions floating around the cognitive science literature and the differences between them; for a general introduction to uses of forward models in cognitive science, see Grush 2004).

Not surprisingly, given modern control theory's place in control and signal processing, the earliest applications were largely confined to motor control physiology (Ito 1970, 1984; see also Desmurget and Grafton 2000) and perceptual processing (Rao 1997). Perhaps the earliest attempt to explicitly use the notion of a forward model as a synthesizing representational framework for understanding many aspects of behavior perception and cognition was Grush (1995; see also 1997, 2004), and by the late 1990s it was a significant theoretical player in various areas of interdisciplinary cognitive science. See, for example, Hurley (forthcoming); Wolpert et al. (2003). One researcher whose work deserves special mention in this context is Chris Eliasmith. With his deep knowledge of the mathematics of modern control theory and neurophysiology, he has constructed dozens of detailed simulations of various aspects of brain function, including memory, navigation, the vestibulo-occular reflex, and many others (see Eliasmith et al. 2002; Eliasmith and Anderson 2003).

Theories exploiting modern control theory, especially those that attribute the use of forward models to one or more aspects of brain function, share some features with both sensorimotor accounts and with traditional representational/computational accounts. Like the sensorimotor accounts, theories based on modern control theory fit naturally with the idea that the primary selective pressure on the nervous system for the bulk of its evolution was better sensorimotor control. The functions the theorists attribute to the nervous system are closely tied to sensorimotor processing. But unlike sensorimotor theories and other approaches based on classical control theory, theories based on modern control theory are not forced to deny (or violently redescribe) the obvious fact that nervous systems represent. This is what they share with representational/computational accounts. But unlike these latter theories, those based on modern control theoretic tools do not take representations to be sentence-like or to be manipulated via formal rules, but rather take them to be model-like and "manipulated" by dynamic evolution of the model.

For an overview of the notion of the forward model and how it can be applied to various domains of brain function, see Grush 2004 (and for some discussion and replies to common misunderstandings, be sure to see the commentaries and replies). In this section we will confine ourselves to a few points that make contact with previous topics.

The first point concerns the relation between internal models and reasoning. Note that there is a natural affinity between a mental models approach to reasoning and the notion of the emulator or internal model (for discussion, see Grush 2004, section 6.2). Proponents of internal models or emulators claim that the nervous system implements a control scheme with the ability to construct and use internal models. Mental model theory claims that reasoning proceeds by attempting to construct models that are consistent with some descriptions and inconsistent with others. This obviously requires the capacity to construct models. Of course things are not quite that simple, since mental model theory requires more than just the ability to construct models: it requires mechanisms that decide what models to construct, and to decide what constitutes an inconsistency between models, and so forth. But even so, there is at least the possibility that tools from modern control theory might provide an implementation of at least one of the central constructs on one of the central theories of reasoning.

Second, note that two of the emotion theories we looked at in Section 1 involve forward models. Damasio's somatic marker theory posits what he calls an "as-if loop," which is a brain circuit involving the amygdala and hypothalamus that implements an internal model of the relevant bodily states. In other words, there are two ways in which an emotion, qua pattern of representations of somatic states, can be induced by recognition of an event type (say, the visual recognition of the presence of a dangerous animal): the body loop or the as-if loop. The primary difference is that in the body loop the somatic state actually changes and the changes in the body are relayed to the somatosensory cortices, whereas in the as-if loop the activation of the somatic-state representations bypasses the body and the signals are sent straight to the somatosensory cortices. Though Damasio does not use this terminology, the as-if loop is a forward model of the emotionally relevant somatic states (for more discussion, see Grush 2004, section 6.3). Berthoz's framework as well is built upon the idea that the brain is a simulator and a predictor, and that it executes this function by implementing models of the body and environment.

2.4. Discussion

We embarked on a discussion of these topics in hopes that looking at the brain might shed light on the relationship between the brain and cognition, since studies aimed at exploring cognition behaviorally seemed to raise more questions than they answered. And it seems that these hopes were dashed. In this domain too, many more questions were raised than answered, and if anything, the lesson is, as it was in the case of behavioral studies, that at this point we still lack a firm grasp of even the basic aspects of the phenomenon: what are the components that do the processing, what causal means are employed by these components to process information, and what information-processing architecture is being implemented.

We will close this section by pointing out something that seems to be less well appreciated than it ought to be. These debates are often pitched as ones concerning *the* information-processing architecture of the brain, or *the* model on which cognition is best conceived. The theoretical side of science is heavily influenced by scientists' attraction to simplicity and parsimony. But in the case of understanding the neural implementation of cognition, this drive may be creating more impediments than it clears. One possibility is that human information processing—including cognition, perception, decision making, motor control—is a messy amalgam of different special purpose and general purpose mechanisms, and maybe the same task, for example deciding whether Socrates is mortal, or chewing gum, might be executed by different mechanisms at different times. There may be cognitive phenomena for which connectionist systems provide the best explanation, and others for which a scheme based on modern control theory and emulation provides the best insight.

3. Cognition and the Mind

Human behavior, and perhaps the behavior of some other higher animals, is interestingly unlike the behavior of most of nature, and this difference can be gestured at by attributing *minds* to humans and some animals. The minds so attributed need not be conceived of as nonphysical soul-things, but could be physical things *qua* instantiators of certain kinds of abilities and properties. Though few explicitly recognize and grapple with this issue, proponents of this latter view take on the responsibility of specifying what these abilities and properties are. One historically popular position has been to identify *reason* and *cognition* as the key abilities. In subtly different ways, Aristotle, Descartes, Turing, and Newell and Simon all embraced versions of this—though sometimes the focus falls more on language or problem solving. So pervasive is this way of thinking that prominent philosophers have argued for conclusions about features of the *mind* by doing little more than trying to establish premises about properties of *problem solving*. A prime example is Clark and Chalmers's famous "Extended Mind" paper (Clark and Chalmers 1998), the aim of which is to tell us about the physical instantiation of the *mind*, though the bulk of the considerations involve describing the physical stuff involved in *problem solving*.

One conceptual tool that can be used to shed some initial light on what is a mind is to ask whether someone who had, or lacked, some ability would thereby have, or lack, a mind. Clearly blind people have minds, and so vision is not needed for mindedness. What about long-term memory? While the patient HM (see Corkin 1984) is famous for lacking any new long-term memory, it does not seem that he lacks a mind. It seems that he has a mind, but that his mind is curiously trapped in a recurring moment. What about short-term memory? Clearly the details are irrelevant. If there was a drug that gave subjects the capacity to keep two hundred things in short-term memory, or which eliminated primacy effects, this person would not thereby become unminded. In fact, it is interesting to run down the list of things that psychologists and cognitive neuroscientists study, and see how many (few) of them seem strongly related to having a mind.

Some are drawn to psychology or cognitive neuroscience because they feel a drive to understand the mind. But by the time they have finished their graduate studies, that question, that topic, has been flooded over with slightly different questions, and the change is not always noticed. One investigates vision, or recency effects, or which areas of the brain "light up" when solving a math problem, and the fact that one is no longer investigating the *mind*, but rather investigating some of the (arguably inessential) things that minds do, goes unnoticed.

Our point in bringing this up is not to answer any questions, but rather to put forth one that is too seldom raised. A paper on the brain and cognition might easily be read as addressing the physical implementation of mindedness. But these are not the same. And a good deal of conceptual confusion can be avoided by recognizing that they are not the same, or at least that they cannot just be unreflectively assumed to be the same without some sort of argument.

4. Conclusion

Whether one is interested in cognition strictly speaking (reasoning and decision making), or cognition in the sense of mindedness in general, or cognition in the broad sense of any spiffy thing humans or other higher animals do (including language, perception, motor control, emotion), the relationship between cognition and the brain is not nearly as straightforward as one might have thought. It is not at all clear what cognition is. Nor is it clear how the brain functions even for the simplest phenomena, such as purposefully twitching a finger. So it should not be surprising that when we turn to the question of the relation between cognition and the brain, there is little to be said with any confidence.

But while the current state of affairs lends itself to this sort of pessimistic summary, we can cast an optimistic light upon it. Our understanding of any unusual or complex phenomenon typically begins with metaphors, and depending on our ability to investigate the phenomenon, the metaphors may persist for centuries. For too long our understanding of both brain and cognition has relied on metaphors. And this has been so because metaphors are easy to understand, they are typically simple (and hence find encouragement in scientists' drive for simplicity and parsimony), and because until fairly recently we have lacked the methods, technology, and conceptual apparatus to investigate the phenomena of interest—cognitive behavior and brain function—in ways that are capable of getting beyond the metaphors to the phenomena themselves, or at least to getting to superior metaphors. We are on the verge of making tractable the last remaining intractable phenomenon in the natural world as we know it, even if to date most of that progress has been a matter of revealing the inadequacies of various simple pictures. Exciting times lie ahead.

REFERENCES

Allen, N. J., and Barres, B. A. (2005). Signaling between glia and neurons: Focus on synaptic plasticity. *Current Opinion in Neurobiology* 15(5):542–48.

Ashby, W. R. (1952). *Design for a Brain*. London: Chapman and Hall.

Berthoz, A. (2006). *Emotion and Reason: The Cognitive Neuroscience of Decision Making*, translated by G. Weiss. New York: Oxford University Press.

Brooks, R. (1991). Intelligence without representation. *Artificial Intelligence* 47(1–3): 139–59.

Clark, A., and Chalmers, D. (1998). The extended mind. *Analysis* 58(1): 7–19.

Corkin, S. (1984). Lasting consequences of bilateral medial temporal lobectomy: Clinical course and experimental findings in H. M. *Seminars in Neurology* 4(4): 249–59.

Damasio, A. (1994). *Descartes' Error: Emotion, Reason, and the Human Brain*. New York: Avon Books.

———. (1998). The Somatic Marker Hypothesis and the Possible Functions of the Prefrontal Cortex. In A. Roberts and T.W. Robbins et al. (eds.)., *The Prefrontal Cortex: Executive and Cognitive Functions*. London: Oxford University Press.

Desmurget, M., and Grafton, S. (2000). Forward modeling allows feedback control for fast reaching movements. *Trends in Cognitive Sciences* 4(11): 423–31.

Diamond, J. (2006). Astrocytes put down the broom and pick up the baton. *Cell* 125(4): 639–41.

Eliasmith, C., Westover, M. B., and Anderson, C. H. (2002). A general framework for neurobiological modeling: An application to the vestibular system. *Neurocomputing* 46: 1071–76.

Eliasmith, C., and Anderson, C. H. (2003). *Neural Engineering: Computation, Representation and Dynamics in Neurobiological Systems*. Cambridge MA: MIT Press.

Eliasmith, C. (2005). A new perspective on representational problems. *Journal of Cognitive Science* 6: 97–123.

———. (2003). Moving beyond metaphors: Understanding the mind for what it is. *Journal of Philosophy* 100(10): 493–520.

Evans, J. St. B. T. (2003). In two minds: Dual-process accounts of reasoning. *Trends in Cognitive Sciences* 7(10): 454–59.

———. (2008). Dual-processing accounts of reasoning, judgment and social cognition. *Annual Review of Psychology*. 59: 255–78.

Faber, J., Portugal, R., and Rosa, L. P. (2006). Information processing in brain microtubules. *BioSystems* 83(1): 1–9.

Fields, R. D., and Stevens-Graham, B. (2002). New insights into neuron-glia communication. *Science* 298(5593): 556–62.

Fodor, J., and Pylyshyn, Z. (1988). Connectionism and cognitive architecture: A critical analysis. *Cognition* 28(1–2): 3–71.

Goel, V. (2005). Cognitive Neuroscience of Deductive Reasoning. In K. Holyoak and R. Morrison (eds.), *Cambridge Handbook of Thinking and Reasoning*. Cambridge: Cambridge University Press.

Goel, V (2007). Anatomy of deductive reasoning. *Trends in Cognitive Sciences* 11(10): 435–41.

Goodwin, G., and Johnson-Laird, P. N. (2005). Reasoning about relations. *Psychological Review* 112(2): 468–93.

Grush, R. (1995). "Emulation and Cognition," PhD diss. University of California, San Diego UMI.

Grush, R. (1997). The architecture of representation. *Philosophical Psychology* 10(1): 5–25.

———. (2003). In defense of some "Cartesian" assumptions concerning the brain and its operation. *Biology and Philosophy* 18(1): 53–93.

———. (2004). The emulation theory of representation: Motor control, imagery, and perception. *Behavioral and Brain Sciences* 27(3): 377–96.

———. (2007). Skill theory v2.0: Dispositions, emulation, and spatial perception. *Synthese* 159(3): 398–16.

Gyselinck, V., De Beni, R., Pazzaglia, F., Meneghetti, C., and Mondoloni, A. (2007). Working memory components and imagery instructions in the elaboration of a spatial mental model. *Psychological Research* 71(3): 373–82.

Ito, M. (1970). Neurophysiological aspects of the cerebellar motor control system. *International Journal of Neurology* 7(2): 162–76.

———. (1984). *The Cerebellum and Neural Control*. New York: Raven Press.

Jessen, K. R. (2004). Glial cells. *The International Journal of Biochemistry & Cell Biology* 36(10): 1861–67.

Johnson-Laird, P. N. (1983). *Mental Models: Towards a Cognitive Science of Language, Inference, and Consciousness*. Cambridge, MA: Harvard University Press.

LeDoux, J. E. (2000). Emotion circuits in the brain. *Annual Review of Neuroscience* 23: 155–84.

Litt, A., Eliasmith, C. Kroona, F. W., Weinstein, S., and Thagard, P. (2006). Is the brain a quantum computer? *Cognitive Science* 30: 593–603.

Maia, T. V., and McClelland, J. L. (2004). A reexamination of the evidence for the somatic marker hypothesis: What participants really know in the Iowa gambling task. *Proceedings of the National Academy of Science* 101(45): 16075–80.

McGeer, V. (2007). Why neuroscience matters to cognitive neuropsychology. *Synthese* 159(3): 347–71.

Newman, E. A. (2003). Glial cell inhibition of neurons by release of ATP. *The Journal of Neuroscience* 23(5): 1659–66.

Noveck, I. A., Goel, V., and Smith, K. W. (2004). The neural basis of conditional reasoning with arbitrary content. *Cortex* 40: 1–10.

Priel, A., Tuszynski, J. A., and Cantiello, H. (2005). Electrodynamic signaling by the dendritic cytoskeleton: Toward an intracellular information processing model. *Electromagnetic Biology and Medicine* 24(3): 221–31.

Ranulfo, R., and Salinas, E. (2003). Flutter discrimination: Neural codes, perception, memory and decision making. *Nature Reviews Neuroscience* 4: 203–18.

Rao, R. (1997). Kalman filter model of the visual cortex. *Neural Computation* 9(4): 721–76.

Rolls, E. T. (2000). Precis of *The Brain and Emotion*. *Behavioral and Brain Sciences* 23(2): 177–91.

Rumelhart, D. E., McClelland, J. L., and the PDP Research Group. (1986). *Parallel Distributed Processing*, 2 vols. Cambridge, MA: MIT Press.

Sanfey, A. G., Loewenstein, G., McClure, S. M., and Cohen, J. D. (2006). Neuroeconomics: Cross-currents in research on decision-making. *Trends in Cognitive Sciences* 10(3): 108–16.

Smith, E., and Neumann R. (2005). Emotion Processes Considered from the Perspective of Dual-Process Models. In L. Feldman-Barrett, P. Niedenthal, and P. Winkielman (eds.), *Emotion and Consciousness*. New York: Guilford Press.

Stein, R. B., E. Gossen, R., and Jones, K. E. (2005). Neuronal variability: Noise or part of the signal? *Nature Reviews Neuroscience* 6: 398–97.

Tomb, I., Hauser, M., Deldin, P., and Caramazza, A. (2002). Do somatic markers mediate decisions on the gambling task? *Nature Neuroscience* 5(11): 1103–4.

van Gelder, T. (1995). What might cognition be if not computation? *Journal of Philosophy* 92(7): 345–81.

Van Heuveln, B. (2006). Using existential graphs to integrate mental logic theory and mental model theory. *Journal of Experimental & Theoretical Artificial Intelligence* 18(2): 149–55.

Van Rullen, R., Guyonneau, R., and Thorpe, S. J. (2005). Spike times make sense. *Trends in Neurosciences* 28(1): 1–4.

Wolpert, D. M., Doya, K., and Kawato, M. (2003). A unifying computational framework for motor control and social interaction. *Philosophical Transactions of the Royal Society of London B: Biological Sciences* 358(1431): 593–602.

Wörgötter, F., and Porr, B. (2005). Temporal sequence learning, prediction, and control: A review of different models and their relation to biological mechanisms. *Neural Computation* 17(2): 245–319.

CHAPTER 13

....

THE SCOPE OF THE
CONCEPTUAL

....

STEPHEN LAURENCE
AND ERIC MARGOLIS[1]

CONCEPTS are among the most fundamental constructs in cognitive science. Nonetheless, the question *"What is a concept?"* is a notoriously thorny one. In both philosophy and cognitive science, theorists disagree about what concepts are, what types of phenomena they explain, and even about whether concepts exist. The complexity of this issue is exacerbated by the fact that it encompasses a number of ongoing disputes. One is about the metaphysics of concepts, or the question of what sort of entity a concept is. Some theorists take concepts to be meaningful mental representations that combine to form whole thoughts (Fodor 1998; Margolis and Laurence 2007), while others take concepts to be the meanings themselves, understood as abstract entities that compose to form the propositional contents that thoughts have or express (Peacocke 1992; Zalta 2001). A different dispute concerns the structure of lexical concepts (i.e., concepts that correspond to words). Lexical concepts are variously taken to have definitional structure, prototype structure, exemplar structure, theory structure, no structure at all, or some more complex combination of these options (for reviews, see Laurence and Margolis 1999; Murphy 2002; Machery 2009). A third dispute—the one we will focus on in this chapter—concerns what we will refer to as *the scope of the conceptual*. Assuming that not all representations or meanings are on a par and that not all deserve to be designated as concepts, the question arises as to how the conceptual/nonconceptual distinction should be drawn. As we will see, many different answers have been given to this question. This

[1] This chapter was fully collaborative; the order of the authors' names is arbitrary. EM would like to thank Canada's Social Sciences and Humanities Research Council for supporting this research.

chapter will provide a critical overview of the main arguments that have guided recent philosophical thinking on these matters.

1. Preliminaries

Before we turn to the arguments that will be the focus of our discussion, we should emphasize that there really is no consensus on how to draw the conceptual/nonconceptual distinction or even on the factors that go into deciding how to draw it. And while debates about the conceptual/nonconceptual distinction have generated many interesting and productive ideas, major disagreements in the literature have led different theorists to use the terms *conceptual* and *nonconceptual* in different ways, sometimes talking at cross-purposes and often leaving the meanings of these crucial terms implicit in their discussions. All of these factors have meant that the large philosophical literature on this topic can be confusing for the uninitiated. For this reason, we think it is especially important to begin with a brief discussion of how we propose to frame the dispute.

One ground rule is that we wish to be initially neutral about whether there is even an important distinction to be made here. Different classes of mental states, for example, differ in many ways. But we shouldn't assume that they should be divided into conceptual versus nonconceptual states—that is, we shouldn't just assume that there is any distinction between these various states that is fundamental enough to warrant singling out some as conceptual and others as nonconceptual. In much the same spirit, we should also not assume that there is only one fundamentally important distinction to be made. We should be open to the possibility that there may be more than one type of fundamental distinction that needs to be drawn and that the field should adopt a richer nomenclature than a simple conceptual/nonconceptual split to keep track of these various distinctions.

Another matter that shouldn't be prejudged is whether the conceptual/nonconceptual distinction is about different types of *content* (or meaning) or about different types of representational *states* (Stalnaker 1998). Debates about the conceptual/nonconceptual distinction are often described as debates about the status of so-called *nonconceptual content* (the implicit assumption being that if there is an important distinction to be made, it has to do with there being two different types of content that mental states can have). But a content-based division is not mandatory. Mental states may well divide into two fundamentally different categories without doing so in virtue of possessing two fundamentally different types of content. Different propositional attitudes (e.g., intentions versus beliefs) are distinguished by differing functional roles, for example, and not by differing types of content.

Since not all differences among mental states are differences of content, but content differences between mental states can be taken as a special kind of state difference, we will adopt the neutral and inclusive terminology that contrasts *conceptual states* with *nonconceptual states*.[2] On the natural assumption that mental states exist and are the bearers of conceptual and (possibly) nonconceptual content, this way of talking should be harmless enough. Having adopted this terminological convention, we will allow ourselves to move freely from arguments and positions that have been characterized in the literature in terms of nonconceptual *content* to arguments and positions that are characterized in terms of nonconceptual *states*.

Now as we have noted, there are not many things that all philosophers in these debates agree upon. But one point of consensus is that the constituents of the representations or contents that are involved in paradigmatic belief states should count as concepts. Paradigmatic belief states include the consciously held beliefs of typical adults (typical in the sense that these adults have not suffered brain damage or abnormalities resulting in cognitive or linguistic impairments). When a typical adult consciously thinks to herself that the left front tire on her car needs air, the components involved in the thought (LEFT, FRONT, TIRE, etc.) are among the things we should take to be concepts.[3] Outside of paradigmatic cases like this are the border disputes regarding the scope of the conceptual. Our discussion will focus on two such border disputes. One is about the scope of the conceptual within the human mind:

> *Are all representational mental states of adult humans composed of concepts, or are some types of representational mental states (especially perceptual states) nonconceptual?*

The second is about the scope of the conceptual across different kinds of minds:

> *Are concepts unique to human adults or do animals and prelinguistic children have concepts as well?*

The most direct way to approach these questions would be to start with a firm criterion for what makes a state conceptual or nonconceptual, and then consider arguments that purport to establish which types of states fall under which designation and which organisms are the bearers of these states. However, as we have noted, the criteria for what makes a state conceptual or nonconceptual are often only implicit in discussions, and there is no single criterion that is widely agreed upon. Given this, and given the fact that it is the *arguments* for a conceptual/nonconceptual distinction that really drive these disputes, it may be more productive to work backwards, relying on the arguments and using these to illuminate various proposals regarding the nature of the distinction. In any event, this is how we will proceed. We won't be able to cover all of the important arguments that have been put forward or to go into any one argument in much detail. Our primary objective is to give

---

[2] This is in lieu of the rather more cumbersome *conceptual states or contents* versus *nonconceptual states or contents*.

[3] We adopt the convention of referring to conceptual and nonconceptual states using expressions in small caps.

readers an overall sense of the debate and to illustrate that, in many respects, it remains inconclusive.

With these preliminaries out of the way, we will now turn to the arguments that bear on the conceptual/nonconceptual distinction, beginning with arguments that are directed to potentially important differences between perceptual states and belief states.

2. WHICH KINDS OF MENTAL STATES ARE CONCEPTUAL?

In this section, we look at the scope of the conceptual as it pertains to adult human beings. Advocates of nonconceptual states argue that perceptual states can't always be assimilated to paradigmatic concept-involving states, such as beliefs, while critics of nonconceptual states contend that they can. We will review five of the most influential arguments that purport to show that some perceptual states are nonconceptual.[4] In addition to considering the question of how good these arguments are, we will also examine the (often implicit) conceptual/nonconceptual distinction that the arguments turn on.

2.1. Argument 1: Cross-Species Continuity

The *argument from cross-species continuity* begins with the supposition that animals can share our perceptual experiences even though they lack the concepts that figure in our beliefs about these experiences and in related beliefs. If this is so, then the concepts are not themselves required for having the experiences (Dretske 1995; Peacocke 2001). For example, a dog seeing a mobile phone may be supposed to have a visual experience similar to the one that an ordinary person has when seeing the phone, but it is doubtful that dogs have the concept MOBILE PHONE. Similarly, a bird hearing a guitar may have a similar auditory experience to the one that a human observer would have in the same situation, but birds don't have the concept GUITAR. Supposing this is right, then the concepts MOBILE PHONE and GUITAR are not necessary for having these experiences. The concepts are required for our having *beliefs* about mobile phones and guitars as such (e.g., the belief *that someone is talking on a mobile phone* or *that someone is playing a guitar*), but not for the perceptual states that underlie our experiences of these things.

---

[4] While the debate has largely focused on the question of whether perceptual states are nonconceptual and what this might mean, similar issues arise for other types of mental states, particularly those that have their home in modular processes that are inaccessible to conscious thought (Bermudez 2008).

This argument builds on two fairly intuitive claims. One is that animals and humans have relevantly similar perceptual experiences; the other is that animals lack the concepts in question. Of course, it is well-known that different species have distinct species-specific perceptual capacities. Dogs can hear frequencies above the normal range of human hearing, bees can see light in the ultraviolet range, sea turtles can sense Earth's magnetic field, etc. (Hughes 1999). Nonetheless, all that is needed for the argument is the claim that, at least in some cases, animals that lack the required concepts have relevantly similar perceptual experiences to human beings. This seems plausible enough.[5] Similarly, given sufficiently sophisticated concepts, such as MOBILE PHONE, the claim that animals lack these concepts shouldn't be especially controversial.

Does the argument from cross-species continuity establish that perceptual states are nonconceptual? It may seem somewhat surprising, but the answer is *no*. The most that the argument shows is that concepts such as MOBILE PHONE are not part of these perceptual states (the ones we share with animals), not that these perceptual states are anything but conceptual. It might be that the perceptual states are composed of simpler concepts, for example, concepts more like SILVER, SHINY, RECTANGULAR, and so on. For all this argument says, there is no reason to suppose that animals that share our perceptual states lack *these* concepts, or that the shared perceptual states are not composed of such concepts.[6] This objection illustrates a difficulty that is common to other arguments for nonconceptual states, and that amounts to a tempting yet mistaken form of reasoning. We call it the *conceptualization fallacy*. The fallacy is to suppose that when conceptualization occurs given a prior representational state that the prior state isn't itself conceptual. The reason that this is a fallacy is that the prior state might also be conceptual, so that the conceptualization involved needn't be based on an unconceptualized state, but might instead be a matter of reconceptualization (i.e., a move from one type of conceptualized representation to a different conceptualization). In the argument from cross-species continuity, it is assumed that when an adult goes from her perceptual state to the belief that she is seeing a mobile phone, her perceptual state isn't itself conceptual. But the fact that the belief state involves certain concepts not involved in the perception does not show that the perceptual state doesn't involve other concepts. Because of the conceptualization fallacy, the argument from cross-species continuity fails to establish that there are nonconceptual states.

There remains the question of what conception of the conceptual/nonconceptual distinction is at work in this argument. It is worth noting that nothing in the

---

[5] From a more critical perspective, one might wonder whether animals have any experiences at all or whether they have different experiences even where they have very similar perceptual capacities as human beings. But such skeptical worries are entirely general in that they apply to other human beings too (e.g., you can wonder whether anyone has the same experiences as you do). We will not be concerned with such general forms of skepticism here.

[6] We discuss the broader worry that animals may not have any concepts at all in Section 3 below.

argument suggests that there must be a special type of content (nonconceptual content) by virtue of which we can distinguish perceptual from conceptual states. There is, however, a somewhat enigmatic alternative characterization of the non-conceptual that has been suggested by a number of philosophers and which may be at work in the argument from cross-species continuity. According to this alternative, what makes perceptual states nonconceptual is that they involve content "that can be ascribed to a thinker even though that thinker does not possess the concepts required to specify that content" (Bermudez 1998, 49).[7] Unfortunately, this characterization isn't especially helpful as it stands, since it specifies the nonconceptual relative to an unexplained notion of the conceptual. So while this understanding of the nonconceptual might inform the argument from cross-species continuity, we need a prior understanding of the conceptual/nonconceptual distinction to make sense of it.

2.2. Argument 2: Fineness of Grain

The *fineness of grain argument* turns on the claim that perceptual states support discriminative capacities that are considerably more fine-grained than our inventory of concepts (see, e.g., Evans 1982; Peacocke 1992; Tye 1995; Heck 2000). For example, we are able to visually discriminate millions of different shades of color, but the number of color concepts is claimed to be far smaller. Sometimes natural language is used to give an approximate estimate of the number of color concepts. If we use English as our guide, the number of basic color terms ("red," "green," etc.) is about eleven. Even if we add in more esoteric terms, including those that rely on compound expressions (e.g., "lime green"), the number of color terms is orders of magnitude lower than the number of discriminable colors. When we believe that apples are red or that the sky is blue, arguably we are using representations that impose a conceptualization on the multitude of fine-grained representations employed in visual perception.

Whether this observation tells us that perceptual states are nonconceptual, though, is another matter. It would seem that what we have here is another instance of the conceptualization fallacy. Just because the belief state allows us to conceptualize a given perceptual experience doesn't mean that the states that are involved in the experiences are not conceptual too. Perhaps the perceptual states just draw upon different concepts. So at the very least it isn't clear that the fineness of grain argument shows that perceptual states should be deemed nonconceptual.

---

[7] This conception of the nonconceptual broadly corresponds to what Alex Byrne (2005) calls *state conceptualism* and Jeff Speaks (2005) calls *relative nonconceptual content*. Byrne and Speaks both distinguish something like this conception from one that is supposed to introduce a genuinely new and distinctive type of content (*content conceptualism* or *absolute nonconceptual content*). Interestingly, both argue that the various different arguments for nonconceptual content can't be seen as all arguing for the same type of nonconceptual content (state versus content, or relative versus absolute).

To get past this objection, the argument needs to be filled out with a substantial conception of the conceptual/nonconceptual distinction. One important suggestion along these lines comes from John McDowell (who is a *critic* of the fineness of grain argument, not an advocate). For McDowell, the main issue has to do with the way that a state is integrated with paradigmatic concept-involving states. "[I]t is essential to conceptual capacities…that they can be exploited in active thinking, thinking that is open to reflection about its own rational credentials" (1994, 47). Explaining precisely what McDowell's view amounts to is complicated enough to demand a chapter all by itself. But suppose that the representations involved in paradigmatic concept-involving states are indeed open to the sort of reflection he describes. The question then becomes whether the same thing can be said for perceptual states. In order for this to be the case, McDowell requires that the capacities that endow perceptual experiences with their contents "must also be able to be exercised in judgments, and that requires them to be rationally linked into a whole system of concepts and conceptions within which their possessor engages in a continuing activity of adjusting her thinking to experience" (47). He agrees that we do not have ready-made concepts such as GREEN and PURPLE for each of the many shades of colors we can experience, but argues that the contents of these experiences can be expressed conceptually all the same, and that this indicates that they are suitably integrated with the conceptual realm (56–57):

> In the throes of an experience of the kind that putatively transcends one's conceptual powers—an experience that *ex hypothesi* affords a suitable sample—one can give linguistic expression to a concept that is exactly as fine-grained as the experience, by uttering a phrase like "that shade", in which the demonstrative exploits the presence of the sample.

There are many questions one might raise about this picture of experiences, but we will confine ourselves to the question of whether McDowell is right that his demonstrative concepts (THAT SHADE) are as fine-grained as perceptual experiences. Sean Kelly (2001) argues that they are not, based on considerations that pertain to the context-sensitivity of perceptual experience. Kelly notes that an object with uniform color can be experienced differently under different conditions. For example, a uniformly white wall will look different in places where the wall is in shadow than in places where it is illuminated by direct sunlight. The problem for McDowell's claim is that the concepts associated with the different experiences (THIS COLOR and THAT COLOR) would pick out the very same property, and hence should make exactly the same contribution to experience. But then they could not explain the experiential difference associated with seeing the same color under different lighting conditions. As Kelly puts it, "the phrase 'that color' is unable to distinguish between that color as presented in the sun and that same color as presented in the shade. Because the relevant [experiential] difference is not a difference in color, no color term could make such a distinction" (2001, 607).

McDowell's suggestion regarding the conceptual/nonconceptual distinction isn't the only one that proponents of the fineness of grain argument may wish to draw upon. Starting with the same considerations we began with—the huge number of discriminable colors—Michael Tye remarks that "human memory is not up to the task of storing a different schema for each of these different shades" (1995, 66). And even if it were, the point remains that for most observers, the specific shades that they are capable of teasing apart don't correspond to stored color schema. Part of what Tye has in mind seems to be that concepts are stored, reusable mental representations. This is not enough, however, as there is a sense in which even the most fine-grained perceptual representations are stored and reusable as well. After all, our perceptual experiences have a representational basis, and that basis can be reactivated given the same overall external conditions. What seems to distinguish concepts for Tye is that we can reliably use these stored representations for purposes of re-identification. "I cannot see something as red_{29} [a maximally determinate shade of red] or recognize that specific shade as such; if I go into a paint store and look at a chart of reds, I cannot pick out red_{29}" (104).

One question we should ask about Tye's criterion is whether it is significant enough to warrant a distinction between the conceptual and the nonconceptual. Is fine-grainedness versus coarse-grainedness really a *fundamental* division among representations? Perhaps we can simply acknowledge that some concepts are more fine-grained than others, and leave it at that. Just as important is the question of how close the link is between coarse-grained representation and re-identifiablity and between fine-grained representation and the lack of it. Certainly it seems as though coarse-grained representations can also fail with respect to re-identifiability. Indeed, it is arguable that concepts such as RED fail the re-identification criterion depending on the environmental circumstances. Employing a variant on Kelly's argument, we might note that the same color looks different depending on the surrounding colors—a phenomenon known as *color contrast*. For example, given the choice of the labels "red" and "purple," the same colored circle may be deemed "purple" against a red background and "red" against a purple background. So RED and PURPLE raise their own re-identification problem, even though they are coarse-grained and are supposed to be clear candidates for falling on the concept side of the conceptual/nonconceptual divide.

Stepping back from the difficulties associated with McDowell's and Tye's views, we can see that there is no single way of drawing the conceptual/nonconceptual distinction that is at stake in the fineness of grain argument. McDowell draws the distinction in terms of a certain type of integration with paradigmatically conceptual states. Tye, on the other hand, draws the distinction in terms of stored representations that can reliably be used for purposes of re-identification. And interestingly, both of these ways of drawing the conceptual/nonconceptual distinction appear to differ from the ways of drawing the distinction that we encountered earlier (i.e., taking conceptual and nonconceptual states to involve fundamentally different types of content, or to be a matter of states that are attributable in the absence of conceptual representations).

2.3. Argument 3. The Richness of Experience

The *richness of experience argument* is based on a phenomenon that is closely related to the one at issue in the fineness of grain argument, and the two arguments are often used in conjunction with one another. The richness of experience argument focuses on the fact that perceptual states manage to take in an enormous amount of detail (see, e.g., Dretske 1981; DeBellis 1995; Carruthers 2000; Peacocke 2001). Imagine walking down a city street. You might see dozens of people bustling about, the buildings in the background, the cars parked on the side of the road, the blinking lights, the billboards, etc. The point is that the visual experience seems to simultaneously incorporate all of the determinate colors, shapes, textures, positions, etc. of these things, in stark contrast with the *belief* that you are walking down a bustling street, which abstracts from these details. Proponents of the argument suggest that perceptual states are so detailed that their content must exceed the resources of the conceptual system. The radical disparity between perceptual representations and paradigmatic concept-involving states in this respect is thought to argue for their being of fundamentally different kinds—perceptual states are *nonconceptual.*

One way to develop and refine the argument is to consider in more detail how the perceptual states differ from paradigmatic conceptual states. Christopher Peacocke, for example, has argued that the content that such perceptual states have is different in kind than the type of content associated with belief states. Peacocke introduces the idea of *scenario content* in characterizing the content associated with perceptual states. According to Peacocke (1992), such perceptual states have *scenario content*, which specifies how the space around a perceiver is filled in from a particular perspective. Scenario contents are given in terms of a representation of a filled space from a perceiver's perspective, where the space is represented as oriented around the perceiver by three spatial axes centered in a point of origin (e.g., in the perceiver's chest or head). For each point in the space, which is a certain distance and direction from the origin determined by the axes of orientation, the representation will specify "whether there is a surface there, and if so, what texture, hue, saturation, and brightness it has at that point, together with its degree of solidity" (63). This gives the flavor of Peacocke's scenario contents, which he goes on to specify in greater detail.[8] Notice that perceptual contents, on such an account, would differ from belief contents in terms of the semantic frames that underlie their composition. So perhaps we can say that while conceptual contents are associated with propositional semantic frames (e.g., a simple subject-predicate frame), nonconceptual contents are associated with non-propositional semantic frames such as those that scenario contents give us.[9]

---

[8] We are simplifying Peacocke's account in a number of ways here, among them that Peacocke goes on to suggest that there are other forms of nonconceptual content apart from scenario content.

[9] There is a great deal more that might be said about the possible relations between conceptual contents and scenario contents in terms of the types of semantic frames that they use, and this is related to further complications owing to the fact that not all theories of propositions take propositions to be structured. However, we lack to the space to go into these issues further.

The emphasis on propositional versus non-propositional semantic frames gives us one way to draw the conceptual/nonconceptual distinction, but there are reasons to question whether perceptual states are really so different from beliefs in this regard. For example, Jeff Speaks (2005) suggests that perceptual states can still be characterized in terms of propositional contents; we just have to allow that the proposition will be complicated in ways that reflect the details given in perception. According to Speaks, "If we think of the content of a given experience as a Russellian proposition, it will be a very complicated proposition indeed, which represents many objects as having a great many properties. But there is nothing implausible in the thought that the contents of perception are very complex" (356). In examining the case of visual sensation, Mohan Matthen (2005) has also noted that while there are differences between visual sensations and clear-cut vehicles of propositional contents (e.g., sentences), visual sensations nonetheless have a combinatorial structure that makes them suitable for expressing propositional contents. Moreover, he defends what he calls the *sensory classification thesis*, according to which sensations encode messages that particular individuals have various properties and are to be assigned to specific classes. None of this is to deny that perceptual experiences encompass more detail than beliefs. What remains in dispute, however, is whether this fact calls out for a new type of content—and with it a significant distinction between the conceptual and the nonconceptual— or whether the same type of content that works for beliefs is suitable for perceptual states as well.

Let's now turn briefly to the question of what conception of the conceptual/ nonconceptual distinction is at stake in the argument from richness of experience. Arguably, it is a version of one we mentioned earlier, namely, that nonconceptual states differ from conceptual states by virtue of the fact that nonconceptual states possess a fundamentally different type of content. Earlier we suggested that this conception is not the one at stake in the cross-species continuity argument. However, since scenario content is taken to be fundamentally unlike standard propositional content, and since it inherently specifies so much perceptual detail (filling in the space around a perceiver), this conception does seem to be a good fit for the richness of experience argument as we have interpreted it.

2.4. Argument 4: Contradictory Contents

The *argument from contradictory contents* has been raised in connection with a specific perceptual phenomenon that has been studied by psychologists. The phenomenon, known as *the waterfall illusion* or *the motion aftereffect illusion*, occurs when you stare at a scene that contains motion in one direction, and then shift your attention to a motionless object. The result is that the object appears to be moving in the opposite direction of the original motion and, at the same time, appears to remain still. Tim Crane (1988) has argued that this effect amounts to a visual experience that is inherently contradictory. Importantly, it's not that the object appears to move but that you know, contrary to appearances, that it's not moving. Rather, both the movement and the lack of movement are intrinsic to the experience; the very same object *looks* as if it is moving and not moving.[10]

Crane goes on to argue that the waterfall illusion establishes the existence of nonconceptual states. His strategy is to specify a principle that concepts are supposed to adhere to but that apparently does not hold up in cases where the illusion occurs. Here is the principle (Crane 1988, 144):

> *F* and *G* are different concepts if it is possible for a subject to rationally judge, of an object *a*, that *a* is *F* and that *a is not-G*.

Crane's point is that, against the background of this view of concepts, motion aftereffect can't be a matter of how one employs the concept MOTION (or some related concept). If the state that underlies the illusion were conceptual, then there would be a prohibition on predicating of an object that it is in motion and not in motion. But the illusion seems to do just this.

Crane is clearly motivated by the Fregean tradition in semantics. This tradition emphasizes that concepts should be individuated in a way that takes into account more than their referents in order to explain the varying cognitive significance of coreferential concepts. If all that mattered to conceptual identity were reference, then it would be a straightforward contradiction to think *that water quenches thirst* and at the same time to think *that H_2O does not quench thirst*. Instead, Fregeans distinguish these thoughts by claiming that their constituent concepts—WATER and H_2O—are themselves distinct. Though these concepts have the same referent, they present that referent in differing ways.[11]

What should we make of the argument from contradictory contents? We do not share Crane's sense about how the illusion that the argument relies on is best described. Our own sense is that in these experiences, there is no single thing that appears to both move and not move; rather objects appear to flow or expand or become distorted within their boundaries. In other words, the object as a whole remains stable, but certain of its features appear to be in motion. In this way of looking at the matter, it is simply not true that the object appears to be moving and not moving in the same respect, and so there is no contradiction; it may well be that there are different distinct representations involved in representing the object's movement (within itself) and its lack of movement (as a unit). And if that is so, then the argument from contradictory contents does not give us reason to suppose that

---

[10] Examples of the illusion are available online. See, e.g., www.michaelbach.de/ot/mot_adapt/index.html.

[11] Fregeans take these modes of presentations to be, or to be part of, distinct senses, which are themselves abstract objects. However, it is worth noting that the basic constraint—that concepts that differ in terms of mode of presentation are distinct—can also be endorsed by theorists who reject the Fregean ontology and maintain that concepts are mental representations. The difference is that in the mental representation view, when two concepts have different modes of presentation, the modes of presentation are taken to be realized as properties of mental representations and consequently to be in the head (Fodor 1998; Margolis and Laurence 2007).

the perceptual representations fail to satisfy Crane's principle or that they are nonconceptual.

Turning to the question of what marks the conceptual-nonconceptual divide, in this case it is clear since Crane is explicit about the principle he has in mind. Nonetheless, there are questions about how his way of drawing the distinction maps on to the various characterizations we have already encountered. His explicit account could be seen as aligned with several different versions of those that we have already encountered. For example, since the principle is closely tied to a Fregean conception of content, one might understand the division here in terms of fundamentally different types of content, with concepts having Fregean content and nonconceptual states some type of non-Fregean content. Alternatively, one might see the principle as showing that one can possess nonconceptual states in the absence of concepts (in accord with the second way of drawing the conceptual/nonconceptual distinction introduced in the discussion of the argument from cross-species continuity). Another alternative would be to see it in terms of an appropriate type of conceptual integration, along the lines suggested by McDowell above. However, it is also possible to see Crane's distinction as a new alternative, distinct from all the above suggestions.

2.5. Argument 5: Discursive versus Iconic

In arguing that perceptual representations are nonconceptual Jerry Fodor offers several different ways of characterizing the conceptual/nonconceptual distinction that he takes to be more or less equivalent (Fodor 2007, 2008). For example, he suggests that only conceptual representations involve *representing as* and thus only conceptual representations invariably distinguish between different ways of representing the same thing. He also suggests that only conceptual representations impose principles of individuation on what they represent.[12] But Fodor's primary characterization of the distinction is in terms of the contrast between what he calls *discursive* and *iconic* forms of representation. According to Fodor, the key difference between these two forms of representation concerns how a representation's various parts relate to the whole. Discursive representations are taken to have a canonical decomposition. This means that there is a correct way to subdivide a representation into its representational parts—not every way of dividing the representation into parts yields a division into parts that combine to produce the semantics of the whole. Natural language sentences are paradigmatic discursive representations. In "Sue put the book on the shelf," some subdivisions constitute canonical parts of the sentence (e.g., "Sue," "the book," "on the shelf"), however, other subdivisions do not (e.g.,

[12] Fodor explicates this notion by remarking that while it makes sense to ask which things, or how many things, a representation with explicit quantifiers picks out, it doesn't make sense to ask the same question given a perceptual representational system that lacks this apparatus (Fodor 2007, 110).

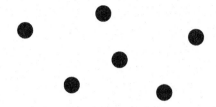

Figure 1. Representation-as without canonical decomposition.

"Sue put the," "book on"). In contrast, iconic representations for Fodor do not have canonical decompositions—any part of an iconic representation is on equal standing with any other part. For example, photographs, which are paradigmatic iconic representations for Fodor, can be cut into parts in any number of ways, and each part will depict a part of the scene that the photograph as a whole depicts. Here is Fodor's principle governing the decomposition of iconic representations (which he calls the *Picture Principle*): "if P is a picture of X, the parts of P are pictures of parts of X" (2007, 108). Of course, the issue for us is not about sentences and photographs. But Fodor's claim is that concept-involving states such as beliefs are discursive (like sentences), whereas perceptual states are iconic (like photographs) and hence nonconceptual.

We should note that it is by no means clear that Fodor's different ways of characterizing the conceptual/nonconceptual distinction are equivalent. Consider, for example, a simple conventional system of representation that uses dots to represent individual people. In that case, the representation in Figure 1 could be used to represent six people. Notice that this representation represents people as people (representation-as) and allows us to count the number of people represented (individuation). Nonetheless, the representation doesn't have a canonical decomposition. The whole can be decomposed by grouping the dots any way we like or by treating them individually. However we do it, each part will represent part of what the whole represents. It is also unclear how the various ways of marking the conceptual/nonconceptual distinction that Fodor suggests map onto the candidates that we have encountered above. Arguably, Fodor's suggestions are each distinct from one another and from the other suggestions discussed above.

In any case, we will focus on the core distinction that seems to matter most to Fodor, namely, the claim that conceptual representations contrast with perceptual representations in that only the former have canonical decompositions. Why think that perceptual representations are iconic in this way? Fodor's argument turns on his account of a familiar type of situation involving conceptualization. Imagine that you are engrossed in a project and a clock begins chiming in the background. After a few chimes, you might wonder what time it is, and only then start to attend to the number of chimes. If you are quick enough, however, you might be able to count the chimes that you weren't initially paying attention to. Fodor suggests that the likely psychological analysis of what is going on is that, in counting the chimes, you manage to conceptualize an iconic representation and that the reason you can count the chimes you were not originally attending to is because the iconic representation is briefly held in a special memory system. "Within the critical interval you can conceptualize...the chimes more or less at will. After that, the trace decays and you've lost your chance." (Fodor 2008, 188).

Unfortunately, Fodor's argument falls afoul of the conceptualization fallacy. Just because in counting the chimes one conceptualizes what one hears doesn't mean that auditory experience isn't itself fully conceptual. It could still represent things in terms of its own set of concepts. In that case, it would be as discursive as the belief that follows it.

A second argument that Fodor gives for thinking that perceptual representations do not have canonical decompositions appeals to experiments by George Sperling (1960). Sperling's subjects saw three rows of letters simultaneously appear on a screen for a brief period followed by an auditory cue indicating which one of the rows to report (e.g., a high tone to signal the top row, a medium tone the middle row, etc.). It turns out that under these conditions people can report all of the letters from any one of the rows even though they can't report all of the letters in the matrix and do not know in advance which row they will be queried about. Fodor suggests that "it is the cost of conceptualizing information in this memory, rather than the number of items that the memory is able to register, that bounds the subject's performance" (2007, 113). However, one might equally claim that it is the cost of *re*conceptualizing information that is already conceptual that bounds the subject's performance—another example of the conceptualization fallacy. Fodor also remarks that "Sperling's 'partial report' effect is *not* found when the items to be re-called are cued by category ('Report the numbers but ignore the letters'). This strongly suggests that representation...is indeed preconceptual" (2008, 189). This inference also commits the conceptualization fallacy. At best, one can infer that the initial representation is not conceptualized *in terms of letters and numbers*, not that it is not conceptualized at all. Further, none of the considerations Fodor mentions provides any reason to believe that perceptual representations fail to have canonical decompositions or that every part of a perceptual representation is a representation of a part of what is represented.

Fodor's primary way of drawing the conceptual/nonconceptual distinction is also squarely at odds with Peacocke's characterization of perceptual representation, which we encountered in the discussion of the richness of experience argument. Consider Peacocke's scenario contents. One part of such a representation is the representation of the horizontal axis of orientation. This part, however, is not a representation of a part of the space that is represented (pace Fodor's Picture Principle). Likewise, consider a portion of the content corresponding to a portion of filled space but independent of the axes of orientation and the rest of the filled space. This portion does not represent any particular part of the space since the axes of orientation are required to locate the portion of the space represented. Peacocke's scenarios aren't iconic representations in Fodor's sense of this term. Much the same could be said regarding Matthen's alternative conceptual account, which treats perceptual representations as discursive in Fodor's sense. The fundamental problem for Fodor, we suspect, is that he is working with an outdated conception of how visual perception works. There is no stage, not even an early stage, at which vision relies on representations that are akin to unanalyzed photographs (Matthen 2005).

3. WHO HAS CONCEPTS?

So far we have only looked at the issue of how the conceptual might be distinguished from the nonconceptual within the minds of ordinary adult human beings. We now turn to the contrast between the minds of these paradigmatic concept users and the minds of animals and infants. Much of the philosophical literature that defends such a contrast is not explicitly framed in terms of the claim that infants and animals have only nonconceptual states, but rather in terms of the claim that they lack genuine thought. Nonetheless, a major point of controversy for philosophers has been whether animals and infants are confined to representing the world using mental states that are significantly different from the conceptually articulated states that adult humans enjoy—that is, whether animals and infants have concepts. In this section, we will review five influential arguments that philosophers have given in support of the view that animals and infants possess only nonconceptual states, and thus fall outside the scope of the conceptual. Again, in addition to considering the question of how good these arguments are, we will examine the conceptions of the conceptual/nonconceptual distinction that the arguments seem to turn on.[13]

3.1. Argument 1: Limited to Current Perceptions

Michael Dummett (1993a, 1993b) argues that nonhuman animals are not capable of full-fledged conceptual thought but only a diminished form of thought, which he calls *proto-thought*. According to Dummett, animals are stuck in the here and now in that they are unable to detach themselves from their immediate perceived situation. The kind of thinking that this leaves them with is, at best, one in which they rely on "spatial images superimposed on [current] spatial perceptions" (1993b, 123). In contrast, because of their linguistic abilities, human beings can remove themselves from the moment and can rise above the confined world of current

---

[13] Some of these arguments are directed to the claim that natural language is required for possessing beliefs and other propositional attitudes. But since the arguments are generally understood to apply equally to the claim that natural language is required for concepts, we'll often let claims about the requirements for belief possession stand in for claims about the requirement for concept possession.

perceptions. Dummett gives the example of a man walking home only to find himself arriving at the solution to a mathematics problem that he had been working on earlier in the day. The man's immediate perceptions are of the road, the houses, etc., and have little to do with the solution that pops into his head.

One of the striking features of Dummett's discussion is that he offers absolutely no evidence to back up his views about what animals can and can't do. The impression he leaves is that his pronouncements are based entirely on casual personal observations. But it should go without saying that casual observations are not to be trusted, partly because they are unsystematic and partly because they are likely to simply reflect the theorist's biases. And, perhaps unsurprisingly, the scientific study of animal psychology doesn't support Dummett's claims at all. Among other things, most animals represent abstract, nonperceptual information such as information about time, remember information from the past, and bring this to bear on decision making about activities such as foraging (Gallistel 1990). Many animals, including birds and fish, can also represent other abstract properties, such as the approximate number of entities in a collection (Brannon 2005). Similarly, many species—even bees (Giurfa et al. 2001)—have abstract general representations of sameness and difference, which generalize both across and within sense modalities. Animals are also far more sophisticated than Dummett's remarks suggest in terms of the types of information processing of which they are capable. For example, recent work indicates that apes are capable of inferences by exclusion (Call 2006), elephants are capable of means-ends reasoning (Irie-Sugimoto et al. 2008), and rats are capable of reasoning about causal constraints (Beckers et al. 2006). Even insects appear capable of very sophisticated cognitive processing. In evaluating potential new nest sites, for example, ants (*Temnothorax albipennis*) compute complex algorithms weighing a wide range of factors (including floor size, headroom, entrance size, darkness level, hygiene of cavity, and the proximity of hostile ant groups) (Franks et al. 2005). A particularly vivid example of animals planning beyond the here and now is provided by recent studies of western scrub jays' caching behavior (Raby et al. 2007). Each morning for six days individual jays were alternately confined to one of two compartments, one of which had no food provided, while the other had food provided. Following this period, when the birds were given the opportunity during the evening to cache food in either of the two compartments, they cached significantly more food in the compartment where no food had been available, showing that they had learned that this compartment would have no food in the morning and that they were planning in advance for this contingency.[14]

[14] In a related experiment, the birds were provided with food in the morning in both compartments but only one type of food per compartment. Given the opportunity to cache either type of food, the birds preferentially cached the type of food that would not be available in the morning for the compartment that they would be in, again showing planning—this time planning to enable them to have multiple types of food in both compartments.

Dummett's way of marking the conceptual/nonconceptual distinction (or as he puts it, the distinction between genuine thought and proto-thought) does not obviously reduce to any of the previous conceptions we have considered. We have argued that his distinction can't do the work that Dummett wants it to do in providing a way of distinguishing humans from other animals; scientific studies of the sort that we have pointed to demonstrate that adult language-using humans are not unique in satisfying Dummett's criteria for genuine thought. Nonetheless, the distinction might still be salvageable if we see it as distinguishing between humans plus a few nonhuman animal species on the one hand and other nonhuman animal species on the other. This could be a viable position in the end, but on the face of it, Dummett's way of drawing the conceptual/nonconceptual distinction isn't particularly promising even if we abandon the idea that humans are alone in possessing genuine thought. The problem is that the distinction fails to establish a natural way of dividing up species given that humans, elephants, rats, scrub jays, and honeybees all end up on the same side of the divide.

3.2. Argument 2: The Opacity Argument

Donald Davidson is perhaps the most famous among contemporary philosophers for denying that animals are capable of conceptual thought. It is unclear if his claim is that animals can be interpreted in representational terms but don't really represent the world at all, or if it is that animals are capable of genuine forms of mental representation but that their minds are exclusively nonconceptual. For the purposes of this chapter, we will read him in the latter way.

One of Davidson's arguments turns on the intensionality of mental state attributions: in explaining people's behavior, we readily distinguish between thoughts and concepts that are coextensive. For example, we distinguish between the thought PAUL WANTS TO EAT THE APPLE THAT HE IS HOLDING from the thought PAUL WANTS TO EAT THE NEAREST APPLE WITH A WORM IN IT (even if the apple he is holding is the nearest one with a worm). Davidson suggests that we can only make sense of this distinction because ultimately we can interpret people's speech and not just their behavior. The problem for animals is that they don't have speech. Davidson illustrates the point by asking us to consider a dog that apparently knows that its master is home. But if the dog's master is also the president of the bank, can we say one way or another whether the dog knows that the president of the bank is home? According to Davidson, "We have no idea how to settle, or make sense of, these questions" (1975, 163).

Davidson's epistemological framing of the argument is unfortunate, since, even when dealing with fellow human beings, there is no way to guarantee that we are right about what they are thinking or even that they have thoughts. But putting aside Davidson's epistemological spin on this argument, we can read Davidson as appealing to one of the criteria for conceptual capacities that we have already come across: Fodor's proposal that only concepts involve representing-as. For Davidson, the claim is that only humans are capable of representing-as, and hence that only humans are capable of conceptual thought.

The main difficulty with this argument is Davidson's claim that language is necessary for representing-as. Granted, without language it is unlikely that anyone would be able think of an individual as a bank president, but BANK PRESIDENT is a particularly sophisticated concept and hence an unfair example. On the other hand, it is quite plausible that animals can represent the same individual in different ways—ways that matter to their own needs and interests. For example, sheep are known to be able to discriminate individual sheep by their faces and have been shown to retain knowledge of up to fifty photographed sheep faces for well over a year (Kendrick et al. 2001). Given that sheep, like us, are not able to recognize individuals in all possible circumstances, it is overwhelmingly likely that they will sometimes represent a given sheep as a specific individual, other times as another individual, and yet other times as simply another (unknown) sheep in the distance. Baboons are also known to represent individuals and are capable of representing conspecifics in terms of their place in both kinship and dominance hierarchies (Bergman et al. 2003). So a given baboon might represent the same individual as a specific individual, or as standing in a particular kinship relation to another baboon, or as standing in a particular dominance relation to another baboon. Arguably, much of the evidence discussed in the previous section is applicable here as well. For example, the compartment with no food in the morning is presumably represented *as* a compartment lacking food in the morning by the jays—that is, after all, why they stock it with food when given the opportunity. But they must have other ways of representing this same compartment, since they represented the compartment when they first encountered it, prior to knowing that it would lack food in the mornings. Much the same could be said for the bees representing a stimulus as the same as another stimulus versus representing it in terms of its perceptual features.

We have seen that the best way to make sense of Davidson's opacity argument is that it invites us to draw the conceptual/nonconceptual distinction in terms of the notion of representing-as, and in this way there is a connection between the opacity argument and one of Fodor's several suggestions regarding the nature of iconic states. We have also seen that Davidson fails to show that humans are unique in satisfying this standard of conceptuality. And, as with Dummett's criterion, Davidson's standard does not seem to provide a principled way of drawing the distinction when it comes to animal species, again clustering humans together with, among others, sheep, jays, and bees.

3.3. Argument 3: The Argument from Holism

The next argument, also due to Donald Davidson, turns on the claim that conceptual content requires a rich inferential network. Davidson asks us to consider a dog that has chased a cat up a tree, and that is sitting at the base of the tree and looking up. There is a natural inclination to say that the dog believes that the cat is in the tree, but Davidson suggests that there are grounds for questioning whether the dog can have the concepts that such a belief requires (for example, TREE or CAT). As Davidson puts it, the problem is that having the concept TREE requires having endlessly many general beliefs about trees, for example, that trees grow, that they need water, that they are combustible, and so on. "It seems to me that no matter where we start, we very soon come to beliefs such that we have no idea at all how to tell whether a dog has them, and yet such that, without them, our confident first attribution looks shaky" (Davidson 1982, 98).

We will put aside Davidson's epistemological way of framing this argument, as we did for the previous argument. The essence of the argument from holism is the claim that conceptual content is determined by a representation's role in inference and that representations have to be embedded in rich inferential networks to have any conceptual content at all. Since Davidson's dog is not able to draw appropriate inferences that connect TREE to GROWTH, WATER, etc., it can't have the concept TREE. And the suggestion is that this failure generalizes. The dog presumably doesn't have any networks of inferences rich enough for most ordinary concepts.

Davidson's holism argument is undermined by two serious difficulties. First, one can call into question Davidson's view that conceptual content is determined holistically and is a matter of conceptual role. Other theories of conceptual content have been proposed, and their proponents might even see it as an advantage of these alternatives that they don't imply that animals are incapable of having concepts (see, e.g., the different theories in Stich and Warfield 1994). But even if conceptual content were holistic in the way that Davidson claims, his conclusion wouldn't follow. Dogs may not have our concept TREE, but this doesn't mean that they don't have any concepts at all. The dog in Davidson's example might very well have a way of representing the tree, where the representation is embedded in a pattern of inferences that is appropriate to the dog's own mental life—maybe not the role that goes with the English word *tree* but a role that works just fine for the dog (Carruthers 1992; Graham 1998). Indeed, as Graham notes, most of us are hardly experts regarding most of our concepts. Western tree experts and people living in small-scale societies have far richer inferential networks regarding trees than most of us do (Atran and Medin 2008).

Davidson's way of drawing the conceptual/nonconceptual distinction in the argument from holism is reminiscent of the inferential integration criterion discussed in connection with McDowell's response to the fineness of grain argument. This way of drawing the conceptual/nonconceptual distinction seems unprincipled, though, in light of the objection to the argument from holism regarding the possibility of animal concepts being embedded in inferential networks of their own. To all appearances, the type of inferential integration that dogs' representations of trees have differs only in degree, not in kind, from the type of integration that our representations of trees have, much as the inferential integration of ordinary people's representations of trees differs in degree, not in kind, from that of tree experts' representations of trees.

3.4. Argument 4: Detection, Reasoning, and Kantian Spontaneity

Both Robert Brandom (1994, 2000) and John McDowell (1994) develop an argument that is related to the previous argument from Davidson but that raises the bar on what is required for having concepts even higher (see also Haugeland 1998; Davidson 2001). The heart of their argument is that conceptual thought requires more than a capacity for detection: it requires the ability to appreciate the reasons that would justify a given concept's application and use, and this, in turn, is inherently a social practice that is dependent on natural language. Animals do not have concepts, according to McDowell, because "a mere animal does not weigh reasons and decide what to do" (McDowell 1994, 115). Animals crucially lack "Kantian spontaneity, the freedom that consists in potentially reflective responsiveness to putative norms of reason" (McDowell 182). Brandom puts much the same point in blunter terms. As he sees it, animals (and infants) may have representational abilities, but these should be likened to the representational abilities of thermostats (2000, 162; italics in original):

> What is the difference between a parrot or a thermostat that represents a light as being red or a room as being cold by exercising its reliable differential responsive disposition to utter the noise "That's red" or to turn on the furnace, on the one hand, and a knower who does so by applying the concepts *red* and *cold*, on the other? What is the knower able to *do* that the parrot and the thermostat cannot. After all, they may respond differentially to *just* the same range of stimuli. The knower is able to use the differentially elicited response in *inference*. The knower has the practical know-how to situate that response in a network of inferential relations—to tell what follows from something being red or cold, what would be evidence for it, what would be incompatible with it, and so on. For the knower, taking something to be red or cold is making a move in the game of giving and asking for reasons—a move that can justify other moves, be justified still by other moves, and that closes off or precludes still further moves.

On any account that takes concepts seriously—any account that doesn't just treat concepts as a useful fiction or a manner of speaking—thermostats don't have concepts. If animals are cognitively no better off than thermostats, then it would be quite reasonable to suppose that they too shouldn't be placed in the category of beings who possess concepts. However, the analogy is deeply misleading. The main problem is that it suggests that animals are only passively responding to environmental features and are not able to recruit this information in subsequent processing that serves their purposes. But on the contrary, the evidence suggests that there is an enormous amount of internal processing that goes on in animals' minds and that, often enough, this involves a complex integration of information before settling on an appropriate course of action. We saw earlier that western scrub jays plan for the future. This sort of planning involves learning about and representing such environmental contingencies as when and where food will be available and using this information to adopt an appropriate caching strategy. Other experiments show these same birds to be capable of learning the rate of decay of foods of different types (when this rate was experimentally manipulated) and of combining this information with information about when and where foods of different types were cached (Clayton et al. 2003). In this way, western scrub jays can retrieve high-valued food items when these items have yet to decay, but not waste effort retrieving them when they have already decayed. If the birds were mere detectors, one would expect them to be limited to recovering a previously stored item upon recognizing a cue for a cache, or avoiding a decaying item by directly perceiving a telling odor. But the situation is actually far more complicated. The birds are capable of figuring out which stored items are best to recover given how much time has passed and given the decay rate of the items they have cached. They aren't simply reacting to an environmental stimulus.

Brandom and McDowell would probably object that the birds still are not reasoning in the relevant sense. They can't play "the game of giving and asking for reasons". Though a bird might be sensitive to whether something is a cricket (a tasty food item) and have cognitive processes capable of drawing inferences about when and where it is to be located, and even use all of this to plan for the future, the bird still is not in a position to appreciate the reasons that are needed to justify applying the concept CRICKET. The bird can't mull over the many implications that follow from something's being a cricket and consider the reasons that other birds might offer in an avian debate over cricket-centered norms.

At this point, though, it seems fair to ask why any of this really matters. As with the argument from holism, it's one thing to require that a concept have an inferential role, and another thing to require that for animals to have any concepts at all that these must be the same inferential roles as can be found in language-using adults. There is also room to question whether Brandom and McDowell's standard for possessing concepts is so high that it excludes a large number of adults. As Hilary Kornblith (2002) points out, not everyone is as disposed as Brandom and McDowell apparently are to reflect on their own and other people's reasons for how a word is to be used. In different cultures and historical periods, this would be considered unseemly behavior. And yet it would be bizarre to conclude that these people don't have concepts because they have effectively removed themselves from the game of reason giving. These objections point to a general worry about McDowell's and Brandom's criteria for singling out concept users as they do. It is always possible to pick a standard that elevates one group as the true concept users and that diminishes all other organisms. Given the anthropocentric viewpoint that human beings are inherently special and animals inherently inferior, it is a trivial matter to harp on something that we can do that they cannot.[15] But the distinction still has to be well-motivated, otherwise the conceptual/nonconceptual distinction amounts to little more than a vehicle for dignifying an arbitrary difference between humans and animals.

---

[15] For example, one could require that concept users be able to read French, or play chess, or appreciate a good philosophical argument. These too set (some of) us apart from animals, but they clearly involve arbitrary standards.

There is also another problem with the significance that McDowell and Brandom both attach to natural language when they insist that it is needed for giving and appreciating the reasons that are essential to concept possession. The problem, which McDowell himself notes, is that a radical disconnection between the minds of infants and adult humans introduces a significant challenge for explaining how infants ever become concept users (McDowell 1994, 125). After all, if infants are utterly incapable of true thought, how can they learn to see the world in terms of the required norms of reason? McDowell's answer is that language acquisition bridges this gap (125):

> This transformation risks looking mysterious. But we can take it in our stride if, in our conception of the *Bildung* that is a central element in the normal maturation of human beings, we give pride of place to the learning of language. In being initiated into a language, a human being is introduced into something that already embodies putatively rational linkages between concepts, putatively constitutive of the space of reasons, before she comes on the scene. This is a picture of initiation into the space of reasons as an already going concern; there is no problem about how something describable in those terms could emancipate a human individual from a merely animal mode of living into being a full-fledged subject, open to the world.

The problem with this answer is that the noises, marks, and gestures in which language is expressed do not in themselves exhibit rational linkages—they have to be *interpreted*. And interpretation of these noises, etc. isn't simply a matter of being surrounded by language. Houseplants are surrounded by language, but they don't learn English, as they lack the needed cognitive machinery. But what kind of machinery are we talking about? If McDowell is right that linguistic competence is itself dependent on an appreciation of the space of reasons, then children will require cognitive capacities for appreciating reasons in order to learn language itself. So McDowell's explanation does little or nothing to relieve the mystery of how a being with a "mere animal mode of living" can be transformed via exposure to natural language. If the infant is incapable of appreciating reasons, it remains mysterious how it grasps the rational linkages embodied in language, which it must grasp in order to learn the language.

3.5. Argument 5: The Metacognitive Argument

The final argument we will discuss is again owing to Donald Davidson and is perhaps the most famous argument against animals having conceptual thought. In Davidson's original formulation of the argument, it begins with the claim that having a belief requires having the concept of a belief. Davidson adds that having the concept of belief requires possession of a natural language. It follows, then, that to have a belief—any belief at all—requires facility with natural language (1975).[16]

---

[16] When we include Davidson's view that concepts are metaphysically dependent on propositional attitudes such as belief, the implication for concept possession is that we can't have any concept at all without having the concept of a belief.

While the overall structure of the argument is reasonably clear, the motivations behind the premises are considerably less so. Why think that having a belief requires having the concept of a belief? Davidson says little more than that the two are connected because having a belief requires the possibility of recognizing that the belief could be wrong. Presumably the idea is that beliefs, by their nature, are subject to correction, and that to correct a false belief requires representing that the belief is false. In other words, for Davidson, correcting a false belief requires having *a belief about a belief*, and this in turn implicates the concept of a belief, as well as concepts of truth and falsity. The link to language is also largely implicit, but Davidson seems to think that concepts of truth and falsity are dependent on language; they "can emerge," he says, "only in the context of interpretation" (1975, 170).[17] In a later related discussion, Davidson (1982) inserts one significant embellishment into this basic argument: he introduces the idea that the ability to be surprised is an indication of the ability to have beliefs about beliefs and hence to have any beliefs at all. He writes, "Surprise requires that I be aware of a contrast between what I did believe and what I come to believe. Such awareness, however, is a belief about a belief" (1982, 104).

Davidson's metacognitive argument raises some rather complex issues, but we will mention just a few potential lines of response. First, it is not clear that having a belief requires the concept of a belief. In general having an X doesn't require having a concept of X—you can have a pancreas without the concept of a pancreas, you can have a language-processing module without the concept of a language-processing module, etc. Davidson's principal argument that things are different when it comes to beliefs rests on a particular picture of belief revision, namely, that an agent must explicitly recognize that his belief is false in order to correct a false belief. But it is doubtful that this is the only way to correct a false belief. A far more natural model would simply appeal to the first-order causal organization of our belief-fixation mechanisms. For example, we see no reason why these mechanisms could not be structured in such a way that a conflict between a perceived event and an occurrent belief directly results in a disposition to update the belief. You think that your tennis racket is in the car, but when you look for it there, you don't see it. This, all by itself, causes you to no longer think that the racket is in the car. You don't have to think to yourself, as it were, MY PRIOR BELIEF THAT THE TENNIS RACQUET IS IN THE CAR IS FALSE. You just have to cease to believe that the racket is in the car and, as a result, entertain other places where it might be found. A similar response applies to Davidson's remarks about surprise. Sometimes surprise might occur as a result of a highly reflective process. But it might also occur as a result of an entirely first-order process.

---

[17] For a related argument that beliefs about beliefs require language, see Bermudez (2003). Bermudez's argument turns on the claim that a belief about beliefs requires a vehicle in which it occurs. According to Bermudez, the only vehicle that could do the job is a linguistic one—hence the need for natural language. We lack the space to discuss this argument in any detail, but we would suggest that Bermudez greatly underestimates the explanatory advantages of appealing to an internal system of representation for the vehicles of thought, that is, to something akin to Mentalese as opposed to English.

In that case, when one registers information that conflicts with a preexisting belief, there is a disposition to undergo a certain affective response that is linked with the processes of belief revision.[18] We should also point out that Davidson's claim that the concept of a belief requires language can also be challenged on empirical grounds. Recent research shows that infants can form beliefs about beliefs well before they have mastered a natural language (Onishi and Baillargeon 2005; Kovács, Téglás, and Endress 2010).

The characterization of the conceptual/nonconceptual distinction that is implicit in Davidson's metacognitive argument is a complex one involving a capacity for belief about beliefs, a concept of belief, and concepts of truth and falsity. This criterion is obviously distinct from all the other ways of marking the conceptual/nonconceptual distinction that we have discussed. It also stands out in that it has little motivation beyond the argument that Davidson cites on its behalf (the metacognitive argument), and, consequently, its plausibility turns entirely on the status of that argument. Given the difficulties that the metacognitive argument faces, Davidson's criterion does not seem to be well-motivated.

4. Conclusion

We have reviewed ten arguments for nonconceptual states—five that delve inside adult human minds and five that are meant to suggest a contrast between the cognitive lives of human adults, on the one hand, and animals and infants on the other. The philosophers associated with these arguments have identified numerous phenomena of interest. However, all of these arguments face serious objections and, as we have seen, there are almost as many different ways of drawing the conceptual/nonconceptual distinction as there are arguments for nonconceptual states. While much of value has come from this debate, the fact that there seem to be so many distinct and nonequivalent ways of dividing the class of representations into two subclasses highlights the question of why we should privilege any one of these distinctions as marking the conceptual/nonconceptual distinction. In addressing this question, we believe that philosophers should pay substantially more attention to the explanatory benefits of varying ways of demarcating the conceptual from the nonconceptual, and to the relevant bodies of science that bear on the distinction (including developmental psychology and animal psychology). Ultimately, how we draw the distinction between the conceptual and the nonconceptual should be a matter of the explanatory benefits of the classificatory scheme, and this ought to be

---

[18] It is somewhat ironic that Davidson places so much weight on the ability to be surprised, since psychologists now routinely use surprise as a tool for determining how prelinguistic children (and even animals, such as monkeys) represent the world (see, e.g., Baillargeon 2004; Hauser, MacNeilage, and Ware 1996).

informed by what our best science tells us about our own minds and about the minds of animals and infants.

REFERENCES

Atran, S., and Medin, D. (2008). *The Native Mind and the Cultural Construction of Nature*. Cambridge, MA: MIT Press.

Baillargeon, R. (2004). Infants' reasoning about hidden objects: Evidence for event-general and event-specific expectations. *Developmental Science* 7(4): 391–424.

Beckers, T., Miller R. R., De Houwer, J., and Urushihara, K. (2006). Reasoning rats: Forward blocking in Pavlovian animal conditioning is sensitive to constraints of causal inference. *Journal of Experimental Psychology: General* 135(1): 92–102.

Bergman, T. J., Beehner, J. C., Cheney, D. L, and Seyfarth, R. M. (2003). Hierarchical classification by rank and kinship in baboons. *Science* 302: 1234–36.

Bermudez, J. (1998). *The Paradox of Self-Consciousness*. Cambridge, MA: MIT Press.

———. (2008). Nonconceptual Mental Content. In *Stanford Encyclopedia of Philosophy*, http://plato.stanford.edu/entries/content-nonconceptual/ (accessed December 7, 2009).

Brandom, R. (1994). *Making It Explicit: Reasoning, Representing, and Discursive Commitment*. Cambridge, MA: Harvard University Press.

———. (2000). *Articulating Reasons: An Introduction to Inferentialism*. Cambridge, MA: Harvard University Press.

Brannon, E. M. (2005). What Animals Know About Number. In J. I. D. Campbell (ed.), *Handbook of Mathematical Cognition*. New York: Psychology Press, 85–108.

Byrne, A. (2005). Perception and Conceptual Content. In E. Sosa and M. Steup (eds.), *Contemporary Debates in Epistemology*. London: Basil Blackwell.

Call, J. (2006). Inferences by exclusion in the great apes: The effect of age and species. *Animal Cognition* 9: 393–403.

Carruthers, P. (1992). *The Animals Issue: Moral Theory in Practice*. Cambridge: Cambridge University Press.

———. (2000). *Phenomenal Consciousness: A Naturalistic Theory*. New York: Cambridge University Press.

Clayton, N., Bussey, T., and Dickinson, A. (2003). Can animals recall the past and plan for the future? *Nature Reviews: Neuroscience* 4: 685–91.

Crane, T. (1988). The waterfall illusion. *Analysis* 48(3): 142–47.

Davidson, D. (1975). Thought and Talk. In D. Davidson, *Inquiries into Truth and Interpretation*. Oxford: Oxford University Press.

———. (1982). Rational Animals. In D. Davidson, *Subjective, Intersubjective, Objective*. Oxford: Oxford University Press.

———. (2001) What Thought Requires. In J. Branquinho (ed.), *The Foundations of Cognitive Science*. Oxford: Oxford University Press.

DeBellis, M. A. (1995). *Music and Conceptualization*. Cambridge: Cambridge University Press.

Dretske, F. (1981). *Knowledge and the Flow of Information*. Cambridge, MA: MIT Press.

———. (1995). *Naturalizing the Mind*. Cambridge, MA: MIT Press.

Dummett, M. (1993a). *The Seas of Language*. Oxford: Oxford University Press.

———. (1993b). *The Origins of Analytical Philosophy*. Cambridge, MA: Harvard University Press.

Evans, G. (1982). *The Varieties of Reference*. Oxford: Oxford University Press.

Fodor, J. (1998). *Concepts: Where Cognitive Science Went Wrong*. Oxford: Oxford University Press.

———. (2007). Revenge of the Given. In B.P. McLaughlin and J. Cohen (eds.), *Contemporary Debates in Philosophy of Mind*. Oxford: Blackwell.

———. (2008). *LOT 2: The Language of Thought Revisited*. Oxford: Oxford University Press.

Franks, N. R., Hooper, J., Webb, C., and Dornhaus, A. (2005). Tomb evaders: House-hunting hygiene in ants. *Biology Letters* 1(2): 190–92.

Gallistel, C. R. (1990). *The Organization of Learning*. Cambridge, MA: MIT Press.

Giurfa, M., Zhang, S., Jenett, A., Menzel, R., and Srinivasan, M. (2001). The concepts of "sameness" and "difference" in an insect. *Nature* 410: 930–33.

Graham, G. (1998). *Philosophy of Mind: An Introduction,* 2nd ed. Oxford: Blackwell.

Haugeland, J. (1998). *Having Thought: Essays in the Metaphysics of Mind*. Cambridge, MA: Harvard University Press.

Hauser, M. D., MacNeilage, P., and Ware, M. (1996). Numerical representations in primates. *Proceedings of the National Academy of Sciences* 93: 1514–517.

Heck, R. G. (2000). Nonconceptual content and the space of reasons. Philosophical Review 109(4): 483–523.

Hughes, H. C. (1999). Sensory Exotica. Cambridge, MA: MIT Press.

Irie-Sugimoto, N., Kobayashi, T., Sato, T., and Hasegawa, T. (2008). Evidence of means–end behavior in Asian elephants (Elephas maximus), Animal Cognition 11(2): 359–65.

Kelly, S. (2001). The non-conceptual content of perceptual experience: Situation dependence and fineness of grain. Philosophy and Phenomenological Research 62(3): 601–8.

Kendrick, K., da Costa, A., Leigh, A., Hinton, M., and Peirce, J. (2001). Sheep don't forget a face. *Nature* 414: 165–66.

Kornblith, H. (2002). *Knowledge and Its Place in Nature*. Oxford: Oxford University Press.

Kovács, A., Téglás, E., and Endress, A. (2010). The social sense: Susceptibility to others' beliefs in human infants and adults. *Science* 330: 1830–34.

Laurence, S., and Margolis, E. (1999). Concepts and Cognitive Science. In E. Margolis and S. Laurence (eds.), *Concepts: Core Readings*. Cambridge, MA: MIT Press.

Machery, E. (2009). *Doing without Concepts*. New York: Oxford University Press.

Margolis, E., and Laurence, S. (2007). The ontology of concepts—abstract objects or mental representations? *Noûs* 41(4): 561–93.

Matthen, M. (2005). *Seeing, Doing, and Knowing: A Philosophical Theory of Sense Perception*. Oxford: Oxford University Press.

McDowell, J. (1994). *Mind and World*. Cambridge, MA: Harvard University Press.

Murphy, G. (2002). *The Big Book of Concepts*. Cambridge, MA: MIT Press.

Onishi, K. H., and Baillargeon, R. (2005). Do 15-month-old infants understand false beliefs? *Science* 308: 255–58.

Peacocke, C. (1992). *A Study of Concepts*. Cambridge, MA: MIT Press.

Peacocke, C. (2001). Does perception have a nonconceptual content? *Journal of Philosophy* 98(5): 239–64.

Raby, C. R., Alexis, D. M., Dickinson, A., and Clayton, N. S. (2007). Planning for the future by western scrub-jays. *Nature* 445: 919–21.

Speaks, J. (2005). Is there a problem about nonconceptual content? *The Philosophical Review* 114(3): 359–98.

Sperling, G. (1960). The information available in brief visual presentations. *Psychological Monographs*, 74(11): 1–29.

Stalnaker, R. (1998). What might nonconceptual content be? In E. Villanueva (ed.). *Philosophical Issues, Volume. 9: Concepts*. Atascadero, CA: Ridgeview.

Stich, S., and Warfield, T. (1994). *Mental Representation: A Reader*. Oxford: Blackwell.

Tye, M. (1995). *Ten Problems of Consciousness: A Representational Theory of the Phenomenal Mind*. Cambridge, MA: MIT Press.

Zalta, E. (2001). Fregean senses, modes of presentation, and concepts. *Philosophical Perspectives* 15: 335–59.

CHAPTER 14

··

INNATENESS

··

STEVEN GROSS AND
GEORGES REY

To what extent are the structures and contents of the mind innate, and to what extent are they learned or otherwise acquired from the environment? Versions of this question have shaped theorizing about the mind since the ancient Greeks and continue to divide researchers today. The debates concern a wide range of traits—for example, the capacity for color perception and discrimination, the ability to follow a gesture, and even a penchant for surprising people by sneezing in elevators (Holden 1987). Some touch upon matters of great general concern with possible implications for public policy, such as the nature of IQ, gender preference, and criminality. The relevant empirical methodologies are increasingly complex and varied: meta-statistical demographics, twin studies, examination of phenotypic correlations, study of early childhood development, etc. The debates concern not just what is innate, but also the prior questions of what innateness is, whether there is just one scientifically legitimate conception of innateness, and indeed whether this is any at all. Accordingly, we examine below several attempts to articulate a conception of innateness that could do some explanatory work (Section 2). We then focus on the philosophically salient case of whether *ideas*, or *concepts*, are innate (Section 3).

The topic of innateness has a rich history, an understanding of which illuminates the contemporary debates on which we will focus. We thus preface our discussion with some brief historical remarks (Section 1).

1. HISTORICAL BACKGROUND: RATIONALISM AND EMPIRICISM

The earliest and most famous argument for conceptual innateness occurs in Plato's *Meno*, where Socrates purports to demonstrate that a slave boy, who has received no explicit instruction, has sufficient ideas somehow available to him to understand a nonobvious proof in Euclidean geometry. Socrates concludes from this demonstration that the slave boy must be "recollecting" the relevant ideas from a previous incarnation. Plato's student, Aristotle, reacted against this suggestion with what is perhaps the first empiricist proposal. On his view, *all* ideas are derived from experience by a causal process in which "forms" (or properties of things) in the external world are transmitted into the mind (Aristotle 1968, 417–26). With the medieval resurgence of interest in Aristotle, the view was defended in highly influential writings of Aquinas (1266/1948, I, 87, a, 3; see Adams 1975, 73–74 for discussion), and is arguably a chief component of much of what passes today as common sense on the topic.

In the modern period, John Locke (1690/1975) also defended a strong form of empiricism about concepts, insisting that our simple ideas are derived from sensation, and all other ones are constructed from the simple ones by the mental operations of "compounding," "comparing," and "abstracting" (Book II, ch. 12). Interestingly, he also maintained that our ideas of at least secondary qualities, such as color and sound, are *not* caused by those very properties in the external world (Book II, ch. 8), inviting the suggestion, widely presupposed ever since, that at least these elementary perceptual ideas are innate in us (see Fodor 1981, 275–77 for discussion).

Full blown nativist proposals resurfaced in the modern period in the work of Herbert of Cherbury (1624/1937), Ralph Cudworth (1678/1999), and, most famously, René Descartes (1647/1911). Descartes is especially impressed by the fact that the geometric figures studied by mathematics are not physically possible objects of sensory perception (1641b/1970, 227). Even so basic an idea as that of an *enduring substance*, such as a portion of wax that seems to us to remain constant as it undergoes various physical transformations, seems to require an idea of *substance* that experience alone cannot provide (Descartes 1641a/1970, Meditation 2).

Descartes emphasized that the issue concerns not *occurrent*, but *dispositional* properties of a neonate—a point relevant to our discussion below. No one thinks that infants are born actually *entertaining* the full panoply of geometric ideas that might be innately available to them. Rather, innate ideas are like innate character traits and diseases, which people "are born with a certain disposition or propensity for contracting" (1647/199, 442), and which may require appropriate circumstances to be activated (442–43).

Following upon Darwin, nativism about "instincts" and other cognitive mechanisms received some support in psychology (James 1890/1983, ch. 24) and ethology (Lorenz 1957; and, for critical discussion, Lehrman 1953, 1970). But empiricism became the orthodoxy in at least Anglophone philosophy and psychology until the revival of rationalist approaches to the study of language in the work of Noam Chomsky (1965, 1966, 1968/2006) (see Chapter 15). Advancing what has come to be called a "poverty of stimulus" argument, he called attention to the inadequacy of the data to which children are standardly exposed for determining the elaborate grammars that they quickly, effortlessly, and universally acquire, an observation that cognitive scientists have since applied to many other domains, such as the understanding of objects, number, animals, artifacts, minds, and morals. (A range of contemporary nativist arguments and claims concerning cognition can be found in Mehler and Dupoux (1994) and Carruthers, Laurence, and Stich (2005, 2006, 2007).) The revival of *conceptual* nativism more specifically is most associated with the work of Jerry Fodor, which we will discuss at length in Section 3.[1]

2. WHAT IS INNATENESS?

But just what is at issue in innateness debates? Might it be, as some claim (e.g., Griffiths 1997), that we lack any scientifically legitimate conception of innateness at all? If so, are innateness debates in fact empty?

We can only focus on a few representative views here (see Mameli and Bateson 2006, for discussion of twenty-seven candidate conceptions of innateness). We begin (Section 2.1) with proposals that would subsume the innateness of psychological traits under a more general positive account of innateness in biological terms. We then turn (Section 2.2) to a proposal that focuses on *psychological* innateness, characterizing it negatively in terms of how innate psychological traits are *not* acquired—in particular, not by learning. Finally, we argue (Section 2.3) that, even if these proposals fail to sufficiently articulate scientifically legitimate conceptions of innateness, this need not undermine innateness debates—in particular, those concerning concepts, the topic of Section 3.[2]

---

[1] Most contemporary treatments diverge from many of the historical ones in not linking the issue of innate ideas to any claims of *a priori* knowledge, or knowledge justifiable independently of experience. It is now widely presumed there could be plenty of innate beliefs that are false (e.g., that space is Euclidean). Cf. Hart (1975).

[2] Our concern is not with the meaning of the vernacular term *innate*, or with folk conceptions of innateness. The question is what innateness may be, and what may be innate, from the perspective of our best current scientific theorizing. We nonetheless mark proposals' counterintuitive consequences. For, with sufficiently significant departures (for example, if what is learned were allowed to be innate), one may wonder whether we retain a conception of *innateness* at all—or at least whether it is advisable to retain the *word* "innate."

Griffiths (2002) suggests a distinct reason for avoiding the word. He maintains that the vernacular conception of innateness is an expression of a largely unreflective and automatic

2.1. Biological Conceptions

In the light of modern biology, it is natural to interpret nativism in terms of the contribution genes make to a trait's emergence. But a satisfactory formulation of this idea has proven elusive. Approaches that emphasize causal determination face the problem that prima facie practically all traits (among them innate traits) result from the *interaction* of genes and environment (that is, anything not part of the genome). Even the emergence of so seemingly paradigmatic an innate trait as eye color depends upon intricate genetic-environmental interactions. Moreover, it is unclear how to apportion causal responsibility. As Sober (1988) emphasizes, there is no "common currency" (312) with which to compare the relative contributions of genes and environment; unlike physical forces, biological determinants do not in general decompose into amounts of genetic versus nongenetic "force." Some other approach would be needed to identify when the genetic contribution is appropriately "critical" in a way that can underwrite a claim of innateness (cf. Waters 2007).

It might be thought that technical notions of *heritability* employed in population genetics might be of use here, where the genetic heritability of a trait within a population is *the proportion of phenotypic variation due to genetic variation*. But the heritability of a trait is not even defined for populations in which there is *no* variation—as with, for many populations, the intuitively innate trait of *having a head*. Moreover, intuitively innate traits—such as having five fingers—can exhibit low heritability: the majority of people lacking five fingers may be victims of accidents.[3] Another strategy would borrow ideas employed more generally in attempts to naturalize content (see Section 3.4.1 below) to cash out genetic determination in terms of a trait's being "coded" for in the genes. But scaling up such suggestions beyond the representation of amino acid triplets and proteins remains problematic (Godfrey-Smith 2007).

Alternatively, one might avoid the problem by rejecting the claim that *innate* traits result from the interaction of genes and environment—more specifically, by identifying innate traits with the properties of the genotype itself. If one does not require that such traits be characterized in molecular or non-relational terms, this identification can be less restrictive than it might appear at first, for it then can include dispositional properties, such as the disposition to have concept C activated

---

folk-biological essentialism that "acts as a sink…draw[ing] new, stipulative usages back towards the established use" which, confounding empirically dissociable properties, encourages illicit inferences. (Folk biology is itself significantly innate according to some, though not Griffiths. Cf. Medin and Atran 1999.) For some initial empirical investigation into the vernacular conception of innateness, see Griffiths et al.(2009). Note that, where they and others speak of concepts, we speak of *conceptions* so as to avoid prejudging whether elements of a conception are constituents of a concept—cf. Samuels (2007). For a pragmatic argument against eliminating either the term *innate* or the concept of innateness from scientific discussion, see Cowie (2009).

[3] The conflation of genetic heritability, genetic determination, and innateness has muddied public debate concerning IQ in particular. Cf. Lewontin (1974), Block (1995), and Sober (2001). For a discussion of *nongenetic* heritability, see Mameli (2004).

in appropriate circumstances (see Section 3). The restriction, however, precludes the possibility of innate traits that emerge in development: for example, having a head or secondary sexual characteristics (including their psychological aspects), as opposed to the disposition to develop such traits.

Some of these suggestions are worthy of further exploration. But for the remainder of this section we concentrate on two versions of another approach. These grant the indispensability of environmental contributions to the acquisition of innate traits, but emphasize the irrelevance of environmental variation.

2.1.1. *Invariance*

Sober formulates his "Invariance" conception of innateness as follows:

> (INV) a phenotypic trait is innate for a given genotype if and only if that phenotype will emerge in all of a range of developmental environments. (1988, 795)[4]

Such traits are said to have a "flat norm of reaction" for the genotype: plotting the trait as a function of the relevant environments yields a flat line. Invariance is easily confused with species universality. But where the former asks what *one* genotype would yield in varying environments, the latter asks what *various related* genotypes have in fact yielded in actual environments.

(INV) faces two major challenges: (i) to specify the relevant range of environments (a recurring problem, as we will see), and (ii) to accommodate invariance apparently owed to environmental stability. A third, more theory-laden challenge would be to accommodate innate traits that are not invariant because they are only present when triggered (see Section 2.2.2).

The basis of the first challenge is obvious enough: absent some restriction on the relevant range of environments—for example, to exclude conditions of extreme deprivation or radical intervention—practically no trait is innate according to (INV). But it is not an easy task to provide a specification with an independent scientific basis.

If we restrict the range to environments that are or have been statistically typical, we face the problem that many genotypes are tokened but once (note that a whole individual's genome tokens a specific genotype): their actual environment is thus their typical environment, rendering all their traits innate. If we advert to what is or has been typical for their species, there is the problem that environments intuitively extreme for one morph might be typical for another.[5] Appealing to the

---

⁴ Cf., for example, Tooby, Cosmides, and Barrett (2005, 323, fn. 7): "What we mean [by an innate trait] is that it reliably develops across the species' normal range of environments." Also, Goldin-Meadow (2003, 215): an innate trait is something "whose development is, if not inevitable, certainly one that each organism in the species is predisposed to develop in widely varying circumstances"—though note our remark below on species universality.

⁵ The formulation "what is or has been" includes proposals that advert to what evolutionary psychologists call the Environment of Evolutionary Adaptiveness. Cf. Tooby and Cosmides (1990).

typical environment of some smaller set of genotypes would not address the further problem that some intuitively innate traits in fact emerge in very few environments: most spiders fail to reach adulthood. Conversely, the occurrence of certain statistically rare environments seems to render some traits *non*-innate: bees differentiate into workers and queens depending on what they are fed in the larval stage, but very few bees are exposed to a queen's ration. We might instead try restricting the range to what is "normal" in some nonstatistical sense, but this only heightens the demand for an independent scientific basis for this restriction—an issue to which we return in Section 2.2.

Sober (1998) suggests that there might not be a single specification of relevant environments: one might need to fix the range "pragmatically" (795), presumably as it varies with (scientifically legitimate) explanatory interests—different interests picking out different ranges of environments, much as different interests might pick different ranges of circumstances to identify something as a solvent or a poison. If different interests could be in play on different occasions regarding one and the same trait, then innateness would be a relative property—or, alternatively, the term *innate* would express different properties in different contexts.

The second challenge is that, on any reasonable restriction of the environmental range, there seem to be intuitively non-innate invariant traits. Consider a person's belief that she has a nose. This belief is presumably learned and so not innate. But arguably this belief would have emerged in any of the environments relevant to (INV) (cf. Stich 1975b, 9; and Wendler 1996, 92–94). Sober's pragmatic reply to the first challenge seems only to heighten the difficulties here, since practically any trait will be innate at least relative to some range if there is some explanatory interest that has us hold it fixed (suppose we want to know why not everyone who learns grade school math can master the calculus).

A standard diagnosis of this difficulty is that (INV) fails to place proper constraints on the process by which a trait is acquired. It asks only whether it always *would* be acquired without distinguishing the roles of endogenous and exogenous contributions. But it is unclear how best to remedy this lack. Mallon and Weinberg (2006) add a requirement that the process be "closed" in the sense of normally leading to one outcome. But some learning-like processes are closed in this sense—for example, some forms of imprinting, as when some species of parasite normally imprints on a particular type of host (Mameli 2004 discusses some cases, albeit for a different purpose). An alternative attempt, to which we now turn, is Ariew's (1996, 1999, 2007) "Canalization" proposal.

2.1.2. *Canalization*

The term *canalization* was coined by the biologist C.H. Waddington (1957, 1975) to refer to a trait's relative insensitivity to genetic and environmental perturbations. The label comes from his comparison of an organism's development to a ball rolling down a grooved landscape. The grooves (or canals) represent the organism's genetically determined developmental possibilities. A trait is genetically well-canalized to the extent that genetic variation would not affect the canals that channel development in its direction. We can extend the metaphor to capture *environmental* canalization by allowing the landscape's topography to be determined in part by environmental factors as well. A trait then is environmentally well-canalized to the extent that changes to the *environment* would not affect the canals that channel development in its direction. Ariew proposes that we deploy Waddington's notion of environmental canalization to characterize innateness:

> (CAN) A trait is innate for a genotype to the degree to which its development is "insensitive to a range of environmental conditions." (Ariew 1999, 128)

The crucial difference between this proposal and (INV) is the appeal to "insensitivity." But how is this to be understood? It would not suffice to let talk of insensitivity simply mark the relevant difference between endogenous and exogenous invariance: the question in the first place was whether there is a scientifically legitimate way of articulating this distinction. Moreover, Ariew (1999, 123–26) follows Sober in emphasizing that there is no factoring out the comparative causal contributions of genes and the environment. At least in this sense, practically all traits are sensitive to both sorts of factors without one kind of factor intelligibly playing a larger or more significant role than the other.[6] If there is some other relevant sense of sensitivity, it must be supplied.

Some alternative attempts to cash out "insensitivity" come at a cost. Suppose, for example, that a trait's development is insensitive to the environment if it not only invariantly emerges, but moreover emerges in an invariant way—suggested perhaps by Collins's (2005, 167) discussion of "developmental implasticity." So construed, (CAN) clearly differs from (INV): if various developmental pathways can lead to one's believing one has a nose, then according to (CAN) it is not innate. But it is unclear why intuitively innate traits must have invariant developmental pathways (however such pathways are individuated). Many genotypes exhibit genetic redundancy: some other gene or genetic pathway can compensate for the inactivation of a gene otherwise central to the development of a trait. Waddington's epigenetic landscape in such cases would contain multiple branching canals that at some point flow back together toward their shared phenotypic goal. Why should the existence of "backup" developmental pathways preclude innateness? Similarly, if the backup involves learning: why should the *possibility* of acquiring a trait through learning undermine the innateness of a trait that as it happens is not learned? Consider the species of canary Ariew (2007, 572) mentions that can acquire its song either through learning or hormonal triggering.

These problems suggest another reading of (CAN), albeit one disavowed in Ariew (2007). Perhaps so far as innateness is concerned, what is crucial to how a trait develops is not that it develop in just one way, but that the developmental

---

[6] Mechanisms that arguably have evolved specifically to "buffer" a developmental pathway against environmental perturbations (see below) can work precisely because they *are* sensitive to the environment in this sense.

pathway that is in fact followed be *of the right kind.* In particular, it must involve a mechanism that evolved (or was co-opted) to have the function of "buffering" development against certain environmental contingencies. But here too we face several problems. First, we are forced to count genetic disorders as non-innate. Second, we are forced to count traits as non-innate if their presence is explained, not by the functions of evolved mechanisms, but by developmental or lower-level physical constraints, as in Cherniak (2005) on optimal neural wiring, and Chomsky (2005, 2007) on computational efficiency.[7] Finally, it is unclear that (CAN) so construed rules out learned traits: a trait could be canalized to be learned or otherwise acquired from experience. As Mameli and Bateson (2006, 172) point out, Sterelny (2003) argues that our folk biological ability to taxonomize animals is canalized in the current sense but is nevertheless in part culturally acquired.

This last worry recalls the objection to Invariance based on intuitively non-innate but invariantly acquired traits—for example, invariantly learned beliefs. Ariew's (2007) reply to such cases involves adducing environments in which the relevant environmental factors are absent. But these environments are arguably abnormal, and intuitively innate traits can be rendered non-innate in the same way: since both innate and invariant but non-innate traits have developmental pathways that would be disrupted in certain environments, a basis is needed for treating cases differently.

Ariew's response (2007—but see also 1999) is that what matters for innateness is whether a trait's emergence is sensitive to *certain specific kinds* of environmental factors, where the relevant factors can vary with the trait in question and indeed with one's explanatory interest. It is this that should fix the relevant range of environmental variation. Thus, *some* birdsong is innate because its emergence does not depend on song-like acoustic cues, and the Language Acquisition Device is innate insofar as its emergence does not depend on linguistic input—even if, as Ariew emphasizes, both depend on environmental contributions for emergence at all.[8] The belief that one has a nose, however, *is* dependent on particular sorts of experience relevant for assessing its innateness; so it is legitimate to consider an environmental range lacking such experiences—even if in some cases the environment is only "conceptually possible" (2007, 577). What gives content to Ariew's account then is not the particular contribution genes make—despite occasional remarks such as that a canalized trait is one whose development is under "strict genetic control" (1999, 134). Rather, it is a sense of what kinds of environmental contributions to disallow in particular cases—one presumably based on the explanatory success of particular developmental models and strategies. (See Griffiths 2009, section 5,

---

[7] It is not even clear whether *genetic* canalization is the result of selection pressures as opposed to, for example, genetic-developmental constraints. See Siegal and Bergman (2002) and Wilkin (2003).

[8] Ariew simplifies his examples in order to illustrate the idea. For example, the innate birdsong to which he refers remains highly schematic, although recognizably similar, in comparison to the normal song of conspecifics who have received the relevant cues; see Gould and Marler (1991).

however, for skepticism that research in developmental biology, including on bird-song, in fact vindicates the usefulness of such a conception of innateness.)

Arguably, Ariew's view mainly differs from Sober's in that it says more about the basis for pragmatically varying the relevant environmental range. Cashed out in terms of the exclusion of *specific* environmental factors, Canalization account also bears a close relation to the next conception we examine.

2.2. Psychological Primitivism

Fiona Cowie (1999) and Richard Samuels (1998, 2002, 2004) develop a conception of innateness that, like Ariew's, is designed to improve upon (INV) by factoring in the way an organism comes to possess a trait. There is a surface difference in that Ariew presents his characterization of innate traits positively in terms of how they *are* acquired (via a canalized developmental pathway), while Cowie and Samuels's Primitivism characterizes them negatively in terms of how they are *not* acquired. This difference vanishes, however, if Canalization must be cashed out in turn by reference to excluded environmental interactions. Primitivism then differs first and foremost in that it is restricted to a particular domain and that it locates its exclusion of particular environmental interactions at a disciplinary boundary. Thus Segal (2007) suggests that the two strategies simply reflect two sides of the same phenomenon, at least so far as psychological traits are concerned (more on this below).

Specifically, Primitivism builds on the thought that innate traits are not *learned*.[9] Of course, not being learned, even if necessary for innateness, intuitively does not suffice—consider sunburns. But the suggestion is that, suitably refined, it might suffice for the innateness of *psychological* traits. The idea is that, however such traits are acquired, so long as they are not acquired via a psychological process, psychologists can treat them as primitives so far as their own theorizing goes—and this is what is at issue at least in (many) innateness debates in the cognitive sciences.

Cowie's concern—so far as Primitivism as a conception of innateness goes—is to distinguish it from other conceptions and to identify a tradition based upon it that connects early modern debates with Fodor's work. Her critical discussion concentrates more on Fodor's claims concerning *what* is innate, so understood. Samuels's discussion, on the other hand, spends more time worrying whether this conception is itself scientifically viable; for that reason, we focus here on his development of the position (aspects of Cowie's critical discussion come up in Section 3 below).

---

[9] Cf., for example, Carey (2009, 453):

"Innate" simply means unlearned—not the output of an associative process, a hypothesis testing mechanism, or a bootstrapping process—that is, not the output of any process that treats information derived from the world as evidence.

2.2.1. *Primitivism and Overgeneration*

Samuels's initial statement of Primitivism is as follows:

> (PRIM)…a psychological structure is innate [for a genotype] just in case it is a
> psychological primitive.…[i.e.,] a structure posited by some correct scientific
> psychological theory [but such that] no correct scientific psychological
> theory…explains [its] acquisition…(2002, 246)

What counts as a correct scientific psychological explanation of acquisition—Samuels's
potentially less restrictive substitute for "learning"—is for science to say. But Samuels
mentions explanations that advert to perception, inference, or conditioning. Nonpsychological explanations would include neurobiological and molecular biological explanations that do not advert to such processes. It would include as well "brute-causal"
triggering by external factors, including triggering that follows upon experiential input,
a case particularly important for discussions of concept acquisition (see Section 3.1.2).
The distinction is no doubt unclear: for example, it is unclear what subpersonal computations count as inferential. But perhaps it is a virtue of the view that it reflects the
crux of many first-order debates about psychological innateness, and thus locates what
future research must clarify if this conception of innateness is to prove legitimate.

Samuels worries, however, that his account overgenerates, admitting intuitively
non-innate psychological traits whose acquisition is not explained psychologically.
Head trauma, stroke, or surgery, for example, can cause brain lesions with a variety
of psychological consequences (e.g., altered personality, memory, problem-solving
capacities, etc.). In one well-known case, the mind of Phineas Gage, a nineteenth
century railway worker, was dramatically transformed by an iron rod that passed
through his frontal lobes after he accidentally tamped it into fused gunpowder
(Damasio 1994). With good reason, Gage's friends reported that he was no longer
the man he was, not that his innate self had been triggered.

A possible reply is that closer examination would reveal a psychological component to such cases. The thought is not that a psychological process might intervene
between neural cell death and psychological end state (though that might indeed
cover some cases). Rather, it might be that Gage, for example, wound up as he did in
part because of earlier psychological acquisition processes: arguably, his psychological
end state resulted from that sort of cell death occurring in *that* sort of mind/brain.

This reply has the prima facie virtue of potentially helping as well with at least
some of a wide range of more mundane cases that Samuels does not discuss. Perceptual states seem to brute-causally yield non-innate psychological states in a variety of ways: consider the emotional effects of music or a warm bath, or the experience
of "love at first sight." Learned associations might provide a reply in some cases. A
broader list of psychological acquisition processes might cover others—so long, of
course, as the processes were reasonably distinguished from brute-causal triggering.
But, for some, one might need to invoke earlier psychological acquisition processes
that prepared the way for the mundane cases at issue.

On the other hand, the reply raises delicate questions of individuation: just
what should be included in a trait's acquisition process? Gage would not have

acquired those traits in the way he did had he not learned how to set a fuse in gunpowder. But presumably his having learned *that* does not suffice to render the acquisition process psychological. Note also that our more mundane cases all involve perception, explicitly deemed a psychological acquisition process by Samuels. So, one is tempted to count the emotion's acquisition process psychological as well by including the perceptual process as a part. But what would warrant doing so in such cases, but not in cases involving perceptual triggering? (See also the discussion in Section 3.4.2 below concerning the role learning a stereotype may play in concept acquisition.)

2.2.2. *Normal Development*

In any event, Samuels pursues a different reply to the overgeneration problem. He maintains that the cases that worry him—rods through heads, strokes, and presumably at least some forms of medical intervention—are clear instances of abnormal development, so he adds to (PRIM) a normalcy clause that would exclude them:

> (PRIM*) A psychological structure is innate for a genotype just in case it is a psychological primitive and would be "acquire[d]...in the normal course of events." (2002, 259)

Of course, the added clause does not address our mundane, clearly not abnormal cases—nor is it intended to—so they would have to be handled some other way, perhaps as above.

(PRIM*) might be read as requiring either (i) that the trait not be acquired abnormally (i.e., that it would emerge in *a* normal course of events) or (ii) that it would emerge in *all* normal courses of events (so that Samuels's view becomes a combination of unrefined Primitivism and Invariance restricted to normal environments). Questions concerning triggering arise either way.

The first reading has problems with some nonpsychological modes of acquisition capable of yielding different psychological traits in different normal environments. Clear cases are difficult to supply in the absence of empirical details, but the conceptual point is clear enough. Suppose that, along lines of Chomskyan linguistics, the specific grammatical rules a speaker respects are determined by the setting of certain "parameters," such as whether verbs precede or follow their objects (an "SVO" versus an "SOV" language), and these parameter settings are triggered by certain stimuli. Then one's speaking an SVO language as opposed to an SOV language would, implausibly, count as innate.[10]

The second reading would preclude innate psychological traits that happen to be triggered in only *some* normal courses of events, as one might hypothesize of,

[10] The supposition that grammatical parameter setting is a matter of triggering, however, is contrary to fact if it involves discerning statistical patterns, and if discerning statistical patterns, at least in the way the developing language faculty does, counts as psychological (cf. Yang 2004; Scholz and Pullum 2006).

say, some mathematical or musical ability. Of course, Samuels may diverge from intuition, but on this point, for better or for worse, his view would diverge as well from his understanding of Fodor's conceptual nativism—and Samuels holds that, all else being equal, an account of innateness should "preserve the standard categorization of central figures" (2002, 239). On Samuels's understanding of Fodor, concepts are innate because they are acquired by triggering. But such triggering need not invariantly occur: indeed, Samuels raises just this as a further objection to Sober's Invariance proposal. This divergence from Fodor is avoided, however, if the triggering is rather understood as only activating a concept already possessed—cf. Section 3.[11]

In addition, further refinement would be needed to accommodate traits that can be acquired in multiple ways, as discussed in Section 2.1.2. (PRIM*), for example, would not exclude abnormally acquired psychologically primitive traits that *would have been* acquired psychologically in some or all normal environment(s). This complication can be avoided by dropping Samuels's modal language: we might simply require that the primitive not *in fact* be acquired abnormally. But if there are triggered, non-innate psychological traits (the SOV example), then again we would need to consider as well what would emerge in other normal environments, contrary—as we just saw—to what can be allowed by conceptions that admit non-invariantly acquired innate traits.[12]

On either reading, if (PRIM*) is to articulate a scientifically legitimate conception of innateness, it is important that its notion of normalcy be neither evaluative—based on some conception of how we ought, or are supposed, to be—nor merely reflect a "folk" conception of our proper environments or course of development.[13] But there is room to question whether a scientifically legitimate alternative is available. We have already mentioned reasons for not adverting to what is statistically typical. Nor does a functional conception, invoking environments in which an

---

[11] Fodor (1998) later retreats from a commitment to concept nativism—see Section 3.4.2—allowing that concepts can fail to be learned but not thereby be innate, even in normal cases. Samuels's divergence from Fodor on *this* score might be mitigated by allowing for a relativization of innateness to specific developmental systems (discussed below): the concepts could be primitive for psychology, but not relative to some other developmental system of which they are the outcome.

[12] If there are innate genetic psychological disorders (perhaps autism), then their actual course of development must count as normal in the sense used here. This might sound counterintuitive, but perhaps only if one conflates a notion of genetic abnormality with what environments are normal for the genotype. That said, it may be unclear what should count as normal for an abnormal genotype.

[13] The evaluative sense is of course important and (we hope) guides our attempts to *improve* our environments—medically and otherwise. But this does not suffice for it to scientifically legitimate Samuels's conception of innateness. Consider Phenylketonuria (PKU), a genetic disorder that leads to mental retardation unless both the mother when pregnant and the child afterwards adhere to a severely restricted diet (cf. Kitcher 1996). If only "restricted diet" environments count as normal, then, according to (PRIM*), PKU-mental retardation is not innate and its absence is; if both

organism lives and thrives, seem to help. The natural nonevaluative explication would be in terms of fitness—so that normal environments were ones, roughly, that led to more offspring. But medical intervention (e.g., personality-altering brain surgery) can increase fitness.

Samuels does not expand on the notion of normalcy. In defense, he notes that *ceteris paribus* clauses that assume "some largely unarticulated set of normal conditions" (2004, 140) are a common feature of all sciences with the possible exception of physics, so that an appeal to normalcy in this case raises no special problems. Indeed, such an appeal may not be to any specific set of normal conditions so much as an exclusion of apparent exceptions to a law as due to independent interference (cf. Pietroski and Rey 1995; and, for general discussion of *ceteris paribus* laws, Earman, Glymour, and Mitchell 2003). But even if such a notion raises no special philosophical problems, it does not follow that it is empirically legitimate in this case. Like his psychological/non-psychological distinction, Samuels's normalcy clause places an empirical bet: that science will indeed find it fruitful to consider all of those developmental factors abnormal interferences. It is clear enough that we often have an interest in how the occurrence or absence of an accident, surgery, or stroke would affect outcomes. But then we are also interested in how the occurrence or absence of various dietary, genetic, educational factors, etc. would affect outcomes. What is not clear is that there is an independent *scientific* reason for deeming the former abnormal. It is perhaps noteworthy that, again, developmental biologists are arguably among the most skeptical that there exists a scientifically useful notion of innateness (cf. Griffiths and Gray 1994; Oyama 2000; Johnson 1997; Bateson 2000).

2.2.3. Generalizing Primitivism

Psychological Primitivism expressly limits its ambitions. Some might consider this a liability, others, a positive asset. In any event, it is worth noting, first, that nothing in Primitivism precludes the possibility of other scientifically legitimate conceptions of innateness. It is, for example, in principle consistent with any of a variety of biological conceptions (we return to Canalization in particular presently). But

---

restricted and unrestricted diets count, then neither presence nor absence is innate on reading one above, but both are on reading two; if only unrestricted diets count (what until recently would have been typical, perhaps universal), then PKU-mental retardation is innate but not its absence. It is clear what environments we want, but not clear what environments are abnormal in a (nonstatistical) scientifically useful sense. Similarly for other environmental alterations and innovations it is "normal" for us cultural creatures to introduce: clothing, improved diets, correction of vision, dental care, types of ornamentation, etc.—many of which have psychological consequences.

Regarding folk conceptions, note that Griffiths (2002) and Griffiths et al.(2009) maintain that the vernacular conception of innateness includes a nonscientific notion of *intended outcome*: "how the organism is *meant* to develop [so that] to lack the innate trait is to be malformed [and] environments that disrupt this trait are themselves abnormal." (2009, 609)

Primitivism does not require this: it could capture what is at stake in innateness debates concerning *psychological* traits even if biological innateness proved a chimera. It is also possible that biological innateness (assuming it is not a chimera) and psychological innateness as the primitivist conceives it could diverge—for instance, if an unlearned psychological trait were deemed biologically non-innate owing to the particular role of environmental factors such as diet (such considerations get raised in debates concerning, e.g., IQ and autism). Of course, the possibility of multiple legitimate innateness conceptions poses no problem apart from the risk of terminological confusion (cf. Cracraft 2000 on different conceptions of "species").

Second, the primitivist strategy might generalize, so one could view *Psychological* Primitivism as an instance of a more general template for generating distinct innateness conceptions. Psychological Primitivism isolates *psychological* traits and identifies those that are innate based on the absence in their normal acquisition of certain processes proprietary to the psychological: processes that involve proprietary interactions with, or input from, the environment. One might *mutatis mutandis* likewise identify innate traits in other domains. In famous work for which he received a Nobel Prize, Niels Jerne (1985) postulated that all the antibodies people ever develop in reaction to disease are already available prior to exposure, waiting to be activated by a pathogen. Whatever else might be true of their acquisition, such antibodies can be said to be innate in the sense that their acquisition does not require a process of exposure: they are primitives for the immune system. Other applications of the template—to language, concepts, the digestive system, or what have you—would be justified to the extent they cohered with successful scientific theorizing that posited primitives within a domain (including possibly, as the examples suggest, distinct applications to *sub*domains of the psychological).

We might alternatively characterize these primitives as what constitute the explanatorily relevant, normal *initial state* of some system of the organism. Such states need not be *temporally* initial: there could be elements that arise after a system's operations are engaged, but not *by* those operations, and then go on to serve as primitives for further operations. These elements could even be acquired as a result of a process that, while not part of the system, involves acquired elements of the system (again, we will return to the idea of acquired concepts triggered by learned stereotypes in Section 3.4.2 below). In the limiting case, the initial state would be the genotype itself (see above). But one need not suppose that the explanatorily interesting initial state for each such system—and thus what counts as innate for the system—is the same in each case or cannot include effects of the environment. In particular, on this view, the possibility arises again of one and the same trait being innate (primitive, initial) for one domain or system but not for another. It might be required, however, that no matter the domain or system, its initial state should not be itself acquired via learning, on pain of loosing contact with our pre-theoretic conceptions of innateness altogether. Such talk of domains, initial states, and proprietary processes (and their proprietary inputs) begs for clarification. But arguably they carry their weight to the extent that successful explanatory strategies find them useful.

Have we thus been led back to Canalization as cashed out negatively in terms of precluded environmental interaction? Recall the belief that one has a nose. (PRIM*) excludes this trait by dint of the process by which it is acquired in normal environments. (CAN) excludes it because it is not invariantly possessed in all *relevant* environments, where the nature of the trait and our explanatory project dictate that environments lacking experience necessary for acquiring this trait be deemed relevant (even if only conceptually possible). The basis for exclusion is thus differently characterized. But presumably, in the application of (CAN), the experience-deficient environment is deemed relevant precisely because such traits normally depend on acquiring the belief based on such experience—similarly for other cases. What underwrites innateness for both (CAN) and generalized (PRIM*) thus appears to be the same: strategies that succeed at explaining the development of systems from their initial states by identifying proprietary environment-involving processes. Arguably, however, the nontemporal notion of initial state derived from Primitivism differs from the more naturally temporal notion common to the developmental biological models that inspire Canalization.

2.3. Suppose We Do Not Know What Innateness Is

We have discussed several conceptions of innateness, suggesting indeed that they may not differ as much as it can seem at first glance. But suppose that upon even closer inspection none of them pan out, so that we do not currently possess an explanatorily useful conception of innateness. Would this render the concept, and debates involving it, empty or unintelligible? Not necessarily.

First, it is a common semantic externalist claim about natural kind concepts that possessing them is compatible with holding many false beliefs concerning the kind—and indeed with possessing practically no conception at all (cf. Section 3.4). If this is right, then it is entirely possible that our talk of innateness has sufficiently "locked on" to a real phenomenon *even if* we are currently unable to characterize that phenomenon satisfactorily, as in the case of ordinary talk of, for example, weight, germ, and jade. It is perhaps even possible that we have locked onto multiple distinct phenomena in different areas of inquiry. That said, a concept's explanatory utility is reduced to the extent we lack an articulated conception, since a greater understanding facilitates the integration of explanations employing the concept with other explanations and theories.[14]

Second, some "innateness" debates could retain their significance *even if* there is no *phenomenon* of innateness that they concern. For example, the idea behind

---

[14] Several authors (Mameli and Bateson 2006; Samuels 2007) have recently suggested a strategy worth exploring on this front. Instead of searching for informative necessary and sufficient conditions for innateness, one might attempt to uncover empirical correlations among properties that serve as evidence for innateness—including among properties that have been mistakenly identified, singly or jointly, with innateness. Cf. Boyd (1991) on natural kinds and homeostatic property clusters.

Psychological Primitivism—that innateness debates in psychology concern what is learned—retains its interest, even if there is no unified phenomenon of what is *not* learned. The point is not just that the complement of a natural kind need not itself be a natural kind in the way that *mammals*, may be, but *nonmammals* surely is not. Rather, the point is that the complement in this case—the psychological primitives—need not contain any scientifically interesting *sub*-kind of the innate. What really matters is whether a trait is acquired through learning, not whether—if it was not learned—the process was in some sense normal or abnormal.

We find support for this second point both in the main arguments made on behalf of psychological nativism and in the main replies. Consider Chomsky's "poverty of the stimulus" arguments for our linguistic competence possessing an innate component. These arguments stress how children come to possess a linguistic competence that far outstrips the evidence available in their experience, and that consequently that competence could not have been acquired on the basis of that experience alone. Replies typically consist in calling attention to further evidence available to the child after all, or in arguing that learning strategies can extract from experience more information than was realized. It does not matter, so far as such debates are concerned, whether the alternative to learning is usefully labeled "innate," or that the label picks out a natural kind (even if it matters to what extent the details of the alternative acquisition story can be supplied in a given case). It is enough that there is an initial state without which the further perceptual input to the system would be inadequate to explain the final stable state of language acquisition that a normal human being achieves.

This places great weight on the question: what counts as learning? But perhaps that is as it should be, at least for some innateness debates. Of course, not all innateness debates can be rescued in this manner, since (as we noted even for psychological traits) not all center on learning. If there is no innateness phenomenon but these debates are to retain their significance, it will have to be for other reasons. But Psychological Primitivism's focus on learning suits particularly well the subject of our next section—the innateness of concepts. As we will see, much depends on what counts as rationally learning a concept versus merely having it triggered consequent to a perceptual process.

3. INNATE IDEAS

As mentioned in Section 1, it was nativism regarding *concepts* that originally divided traditional rationalists such as Leibniz and Descartes, who claimed most of our concepts were innate, and empiricists, such as Locke, Berkeley, and Hume, who claimed they were "derived from experience." Although Chomsky's proposals revived Rationalist views generally, Jerry Fodor's (1975) seminal book, *The Language of Thought*, revived the specific Rationalist views about *conceptual* nativism. In chapter 2 of that book, he proposed the following, radical conceptual nativist hypothesis:[15]

> (RCN) To a first approximation, all concepts expressed by single morphemes in English are innate.

(Morphemes are the smallest linguistic units having meaning; thus "do" and "un-" are (mono-)-morphemes—MMs–in English; "undo" is *polymorphemic*.) Standard estimates of the MMs used by a standard English speaker are between fifty thousand and 250,000 words (see Bloom 2000), and, of course, more serious approximations would have to take account of concepts expressed by MMs in one language but not in another (e.g., "chic"), or not yet in any.

An understandable reaction to (RCN) is to reject it as obviously absurd. Over 250,000 innate concepts?! But it is not clear what entitles someone to this reaction. What does anyone know about precisely what concepts are and how they are acquired; or exactly what "learning" is, and how or whether it should be contrasted with being innate? One merit of Fodor's audacious view is that it calls attention to the need to think about these and related issues with a lot more care than has been traditionally bestowed. In a useful discussion, Laurence and Margolis (2002, 26–27) call it "Fodor's Puzzle of Concept Acquisition," and rightly compare it to puzzles about induction raised by Goodman and about translation raised by Quine. For these reasons, we will organize the discussion around Fodor's view, even though in the end we will express some sympathy with its critics. We will first set out a number of issues that are crucial to the debate (Section 3.1), turning then to the main arguments Fodor presents in his 1975 and 1981 discussions (Section 3.2). Those arguments will lead us to consider the main internalist (Section 3.3) and externalist (Section 3.4) views about the nature of concepts, and the role of prototypes as internal "schemata" relating concepts and percepts. We will conclude (Section 3.5) with a brief discussion of Fodor's (1998, 2008) latest views, and of the processes of learning versus triggering that seem to lie at the heart of the dispute.

3.1. Preliminary Issues

3.1.1. *What Concepts Are*

We shall presume that concepts are the constituents of the objects of the so-called "(propositional) attitudes," such as *think* or *expect*. Thus, people who think fish dream, are thinking the proposition [Fish dream], which they can do only if they have the concepts [fish] and [dream] (we designate propositions and concepts by enclosing in square brackets the words that express them). For simplicity, we shall

---

[15] That book was seminal for a number of important views, another being that there is a language of thought, which serves as the vehicle of computation in the brain. *Pace* Churchland (1986, 389), (RCN) is entirely independent of this latter hypothesis.

also assume that concepts also serve as the meanings of words, and we will acquiesce in Fodor's treatment of concepts as mental representations.[16]

A crucial feature that concepts arguably need to possess to be effective constituents of attitudes is what Fodor (1998) calls "publicity":[17] it must be possible for different people, and the same person at different times, to have the *same* concepts. Someone cannot share a thought with someone else, or even remember a thought of her own, unless some of the constituents remain the same across people and time. This is a nontrivial requirement, since it is not at all obvious which facts about cognitive life are in fact stable across people, whose attitudes are constantly changing as they experience and think about the changing world around them. On Fodor's view, concepts are *type* representations individuatable in part by their content, tokens of which occur in different brains and in the same brain at different times.

3.1.2. *Learning versus Triggering*

Virtually all parties to the debate agree that experience plays a role in arousing whatever innate dispositions people may have to form concepts. But central to the debate between Rationalists and Empiricists is the question of just what the character of that role may be. Learning as popularly understood can include both rational and more brute-causal effects of perceptual experience, and this distinction is crucial to understanding the difference between Rationalists and Empiricists, neither of whom want to deny that experience plays *some* role in the causation of conceptual activity. Empiricists typically want to claim that most of our concepts are in some *rational* fashion *learned* on the basis of experience; Rationalists claim that experience merely serves to "occasion" the activation of a concept that a person already possesses. Descartes (1641b/1970, 227), for example, was especially impressed by the fact that the geometric figures studied by mathematics are not physically possible objects of sensory perception, and so thought that the concept [triangle] couldn't possibly be learned from experience. What we see are various irregular figures that in a context activate, or "occasion," the concept, causing us to see the figures *as* triangles. In an extreme form, the proposal is sometimes that the activating stimuli need not be related to the activated representation in any way that is rational or relevant in terms of intentional content: all that experience does is to provide stimuli that "triggers" the innate conceptual disposition, in the way that, say, an innate immune reaction is triggered by exposure to a certain pathogen, a reaction that is presumably neither rational nor intentional (see Jerne 1985). As

---

[16] Not everyone acquiesces here, but space forbids discussing Peacocke's (1992), Zalta's (2001), and Rey's (2005) preferences for treating concepts as more like Frege's (1892/1966) "senses." See Margolis and Laurence (2007) for discussion. We also abstract from the related issue of "nonconceptual" content (see Crane 1992).

[17] An unfortunate term, suggesting that concepts need be *social*, as opposed to "private," along lines claimed by some interpreters of Wittgenstein (1953, §§ 258ff), a debate entirely orthogonal to the present one. A better term for what Fodor has in mind is *stability* (see Rey 1983).

Fodor (1981, 304) expresses the Rationalist creed: "Simple concepts which arise as the effects of such triggering are, no doubt, learned *in consequence* of experience; experiences are—directly or indirectly—among their causes. But they aren't learned *from* experience."[18]

3.1.3. *Evolutionary Worries*

Another common reaction to (RCN) is to worry about how 250,000-plus specific MM concepts could possibly have evolved. Steven Pinker (2007), for example, finds it "hard to see how an innate grasp of carburetors and trombones could have been useful hundreds of thousands of years before they were invented" (95). But here, as so often elsewhere, it is crucial to distinguish *evolutionary* theory from the specific mechanism of *natural selection*. There is not the slightest doubt that humans evolved from earlier primates, but just which traits can be explained by processes of selection is increasingly controversial (see Sterelny and Griffiths 1999). Humans display abundant capacities (e.g., in music, science, logic, and mathematics) that far exceed anything that selection itself plausibly required.

Everything depends here upon just how one counts "traits" (*stomach? stomach and intestines? stomach and intestines and heart?*), as well as the character of the genetic options available at the time of selection. Thus, bilateral symmetry and the structure of bodily organs may reflect more about the underlying physiochemical structures, or "laws of form" (Thompson 1917), than anything about the specific demands of the selecting environment.

Moreover, animals might be evolutionarily prepared in specific ways to deal with items and states that may not have been present at the time of the evolution. To take the example that we have already mentioned, and to which many have compared (RCN), Niels Jerne won the Nobel Prize for his work arguing that all the antibodies in the immune system are in some sense available innately, waiting to be triggered by antigens. Similarly, even concepts of things that were not remotely available on the savannah might be innately available: after all, perhaps people could only *invent* things that counted as carburetors and trombones because they were innately endowed with the concepts of these things in the first place, just as many birds sing only their species' songs, or spiders weave only certain kinds of webs.

Of course, many will still find an individual innately possessing 250,000-plus concepts/songs/dispositions outlandish, and demand that the burden be on the conceptual nativist to provide a positive account about how there could be *so many*, and *such specific* ones. However, until we have a much deeper understanding of the structure of our minds and how it might be related to underlying

---

[18] Fodor's wording here, "learned in consequence of experience," seems to imply that he thinks concepts are learned after all. We take this to be something of a verbal slip (or issue), the crucial distinction remaining whether there is a *rational* relation between a concept and the experiences that cause it.

genetic structures, it is difficult to know how to count these things, and generally what constraints selection places on psychological theory. In any case, conceptual nativists have presented serious arguments for their view, to which we now turn.

3.2. Initial Arguments for Conceptual Nativism

Poverty of stimulus arguments (see Section 2.3 above) provide one serious argument for (RCN): experience simply does not seem to supply enough information for children to learn the concepts of mathematics, geometry, or enduring substances, or the host of word meanings that they acquire with astonishing speed and uniformity in their first several years (see Bloom 2000). The ur-argument that pretty clearly establishes this stimulus poverty is really due to a famous problem raised by Chomsky's teacher, the philosopher Nelson Goodman, regarding the concept [grue]. This is an artificial concept he devised with the meaning "applies to all things examined before *t* just in case green but to other things just in case they are blue" (Goodman 1954/1983, 74). Now, all the emeralds people have so far examined have been green, and human beings seem to happily generalize this property to emeralds they have not examined, including those that will be examined only after AD 3000. But all the emeralds they have examined so far have in fact also been grue—since, after all, they have been examined before AD 3000. Why don't people generalize to *that* property instead of green, so that, after AD 3000 they would be surprised to find emeralds *green*? All the sensory data we receive at least before AD 3000 will not decide the matter. What is worrisome here is that there is an *infinitude* of such bizarre concepts (one for each increment of time beyond AD 3000). Clearly, childhood stimuli are in principle too impoverished to decide between them. So at least in that respect, at least a sensory concept such as [green] can seem to be innate.[19]

---

[19] There is a vast literature on this "new riddle of induction," as Goodman (1954/1983) called it (see, e.g., Stalker 1994 for a representative collection). It is worth noting that, although Goodman himself did not see it as inviting a nativist moral, it is this very problem that was partly responsible for the resurgence of interest in conceptual nativism, since Chomsky (1955/1975, 33–34; 1971, 6–8) saw his ideas about the innate biases of children to only a specific subset of possible languages to be simply another (rather more elaborate) instance of Goodman's riddle. Indeed, given that an infinite number of grue-like concepts can be constructed for *every* concept we have, it is worth wondering why their existence *alone* does not establish that *none* of the concepts we naturally employ are learned! Fodor does draw nativist conclusions from Goodman's riddle (1975, 39), but they concern the ordering of hypotheses; he does not directly connect such considerations to *conceptual* nativism—perhaps because he suspects that these unnatural concepts are ruled out by some general considerations that enter into concept construction. But the burden would be on the defender of conceptual learning to say what these might be.

It is surely no news to empiricists that at least dispositions to sensory percepts (e.g. example experiences of green) need to be innate, and so perhaps the concepts (e.g. [green]) derived directly from them. After all, the mind cannot really be an entirely "blank slate," lest it not be able to learn anything at all. So it is often conceded that there are, as Quine (1969) put it, innate "quality spaces," biasing a child to generalize to [green] rather than [grue], and these would then serve as the basis for constructing all the concepts the child acquires. But how does the child do this? How does she "construct" new concepts on the basis of experience? This is where Fodor's (1975, 1981) arguments become germane (we will return to their bearing on the "grue" problem in Section 3.5).

Fodor's initial (1975, 79ff; 1981) argument for the innateness of concepts was quite simple. He pointed out that standard accounts of learning treat it as a process of *hypothesis confirmation*: in a classic discrimination experiment, for example, an animal learns to respond differently to Rs versus not Rs presumably by confirming an hypothesis along the lines of

(L) x is R iff x is red and triangular

and it does this by keeping track of instances of R that are, and those that are not, red triangles. But although this may be a perfectly clear way for the animal to acquire the *belief* that (L), it does not seem to be a way to acquire the *concepts of* [red] and [triangular], or even the conjunction of them: for in order to confirm (L) the animal must already have the means to *entertain* it, which it surely cannot do *without already possessing its constituent concepts*. At best, a red and triangular stimulus could *trigger* or *occasion* the already available concept [R], not *introduce* it.

It is important here to distinguish three notions of concept possession that are not clearly distinguished in most discussions:

(i) Having a symbol expressing a certain content that is actually *activated* in some mental process;

(ii) Having a symbol expressing a certain content that has never actually been so activated, but is *lying in wait*;

and

(iii) Being able to *express* a concept by logical construction on symbols expressing certain contents, either already activated or lying in wait.

The above argument of Fodor's (1975) establishes that, if learning is hypothesis confirmation, then an organism cannot learn any concepts whose contents it cannot already *express* in the sense of (iii). Thus, he allows the obvious point that not all concepts are actually *activated*. However, if learning is hypothesis testing, then it requires, in the very hypothesis, the activation of a concept either already activated or lying in wait. In the above example, the organism learning that something is R iff it is red and triangular must in that sense already have the concept [R].

This conclusion, however, might only disturb an Aristotelian, who thought that expressive power was actually acquired by properties in the world being *transmitted to* (indeed, being instantiated in!) the mind (cf. Section 1 above). A more modern empiricist might argue that what the organism *can* learn are syntactically novel ways to construct representations of the same conceptual contents it might always have been *able* to express, which it then sometimes *abbreviates* as MM concepts for use in memory and thought, as when we "chunk" material for use in short-term memory (see Miller 1956).

In his (1975) and (1981), Fodor supposes that we can and do learn new *composite* concepts by logical construction, which seems to suggest that he thinks an organism can increase its *expressive power* in this way. He seems to suppose, for example, that one may well increase one's repertoire of concepts by thinking "that thing is both red and triangular." (We will see in Section 3.5 that in his (2008) he sees this supposition as confused, and rejects even the learning of composite concepts if that means increasing expressive power.) All that he is concerned to deny in his (1975) and (1981) is that such constructions offer an account of *MM concepts*, which he thinks are *not* acquired by abbreviatory chunking; even when not activated, they must be lying in wait.[20] And he thinks this because he finds defective all arguments and efforts to establish that MM concepts are such abbreviations. Indeed, the crucial part of Fodor's (1975, 1981) view is his rejection of the classical empiricist account of the "analysis" of MM concepts, whose history has in fact not been a happy one, and which we now briefly review.

3.3. Internal Constructions

3.3.1. *The Classical View*

The Classical View of concepts treated them roughly as representations that are in some important way decomposable into conditions that are individually necessary and jointly sufficient for satisfaction of the concept, and are known to any competent user. The standard example is the especially simple one of [bachelor], which seems to be analyzable as [eligible unmarried male]. More interesting ones that served as the inspiration of much of "analytic" philosophy were Weierstrauss's analysis of the concept of a mathematical limit and Frege's (1884/1950) analysis of the concept of number.

---

[20] It is worth bearing in mind that one of Fodor's (1975) most important intended targets was Piaget (1954), whom Fodor read as claiming that expressive power could be increased by learning, as a child progressed through various cognitive stages. Whether Piaget needs to be read this way or could be read as intending only this more modest actual activation view is an issue to which we will return in Section 4.

Some of Fodor's own formulations can suggest that triggering involves, or can involve, the *acquisition* of a concept, not just the *activation* of a concept already possessed. Cf. Section 2.2.2 above.

3.3.2. *Empiricism and Verificationism*

The Classical View, however, has always had to face the difficulty of *primitive* concepts in which a process of definition must ultimately end. As mentioned earlier, seventeenth-century empiricism had a simple solution: all the primitive concepts were *sensory*. In the work of Locke, Berkeley, and Hume, this was often thought to mean that concepts were somehow *composed* of introspectible mental items ("images," "impressions") by *associations* among basic sensory parts. Thus, Hume (1734/1978) analyzed the concept of [material object] as involving certain regularities in our sensory experience, and [cause] as involving spatiotemporal contiguity and constant conjunction (see Elman et al. 1996 for recent associationist proposals along more neurophysiological lines).

Mere association, however, is not really adequate to capture the stability of concepts or the roles they play in thought. One person's sensory associations with [justice] may differ vastly from another's without the two people failing to share the concept, and few if any such associations (e.g., a blindfolded woman) are candidates for the analysis of such an abstract concept. Moreover, after Frege's work in logic, it became clear that any account of how a complex concept might be constructed out of sensory ones had to include an account of *logical structure*. This is precisely what many Logical Positivists attempted to provide. They focused on logically structured propositions instead of images and associations, and transformed the empiricist claim into the famous Verifiability Theory of Meaning: the meaning of a proposition is the means by which it is confirmed or refuted, ultimately by sensory experience; the content of a concept is the means by which experiences confirm or refute whether something satisfies it. The theory gave rise to sustained reductionist programs, such as those of *phenomenalism* (reducing material objects claims to claims about sense experience) and *analytical behaviorism* (reducing claims about the mind to claims about physical behavior). Note that it is no accident that empiricism has tended to be the view *both* that concepts are verification conditions *and* that they are acquired from experience, since verification conditions, for empiricists, are ultimately sensory tests.

Verificationism, however, came under much attack in philosophy in the 1950s and 1960s. In the first place, few, if any, successful analyses of ordinary concepts (like [material object], [expect], [know]) in purely sensory concepts have ever been achieved (see, e.g., Ayer 1934 for some proposals, and Quine 1953 and Putnam 1962b/1975, for criticisms). And this is not surprising. The relation of a concept to the sensory evidence for its application is generally quite complex, and dependent on indefinite numbers of other things being (believed to be) in place: someone's *looking* ill confirms their being ill only if (it is believed that) the lighting is right, one's eyes and brain are in working order, and there has been no deceptive mischief. Change the background beliefs and one may well change the relevant sensory evidence, but without necessarily changing the concept (say, of being ill). As Quine (1953, 41) famously put it, "our beliefs confront the tribunal of experience only as a corporate body" (a view called "confirmation holism"), which circumstance he saw as undermining any effort to supply such analyses, or "analytic" truths, as distinct from ordinary, "synthetic" beliefs about the world.[21]

Moreover, there is the difficulty of drawing a principled limit to how deviant people can be about the inferences they draw with concepts they seem nonetheless entirely competent to use. *Conceptions* of a phenomenon may vary wildly while a *concept* remains the same. Philosophers are notorious for defending outrageous claims, for example, that material objects are ideas, that rocks are conscious, or that contradictions can be tolerated. Such cognitive states present a serious prima facie difficulty for a theory of concepts that claims their identity or possession involves specific connections to other concepts or experiences.

Perhaps the most important argument against the Classical View is one that is implicit in Quine (1953), but has been increasingly explicit in the work of Chomsky (1968/2006), Harman (1964), and Lipton (2004), drawing upon Peirce (1903/1998, 287): the role of "*abduction*," or "inference to the best explanation" in both ordinary and scientific reasoning. Abduction is a form of nondeductive inference that, unlike *induction*, may involve the introduction of terms *not in the observational (especially sensory) vocabulary*. Thus, when physicists infer the existence of elementary particles and subatomic forces from ordinary macroscopic data, or even when a jury finds a defendant guilty on the basis of the evidence presented in a trial, they are often not merely making some sort of statistical generalization about that data, but leaping to the activation of what seem to be concepts such as [quark] or [corporate conspiracy] that manifestly involve commitments that go far beyond the concepts being deployed in characterizing the evidence.

In view of these issues about confirmation, it is unlikely that concepts in general could be defined in terms of the evidence adduced for them, and insofar as ordinary concepts are deployed to explain regularities in experience, it has seemed similarly unlikely to many philosophers that those concepts could be defined in terms of that experience.

There have been a number of responses to these problems with the Classical View, internalist and externalist. The internalist ones, to be discussed in the next three subsections, simply alter the Classical methods of internal mental "construction" of concepts and so continue to support Empiricism. It is because Fodor is equally sceptical of these further methods that he follows a number of philosophers in advocating an Externalist theory, to which we will in turn in Section 3.4, and which we will see affords a basis for his extreme Rationalism.

---

[21] A verificationist could go on to insist, as Quine (1960, 1969) is standardly read, that meaning would be similarly holistic ("meaning holism"). But, as Fodor and Lepore (1992) point out, this threatens to undermine the *stability* of concepts, given that no two people (or different stages of one person over time) are likely to share anything like the *totality* of their beliefs (only "threatens," since meaning holism—although perhaps not verificationism—can be defended in a variety of ways against this worry; see Greenberg and Harman (2006) and Pagin (2006)).

3.3.3. *Prototype Theories*

"Prototype" theories are sometimes taken to be a kind of definition theory (e.g., something is an F to the degree that it resembles a prototype), but more usually as a rejection of the demand for strict definitions, replacing them with descriptions of typical instances, or traits shared by typical instances. Rosch (1973) and Smith and Medin (1981/1999) showed that people respond differently (in terms of response time and other measures) to questions about whether, for example, penguins as opposed to robins are birds, in a fashion that suggests that concept membership is a matter not of possessing a Classical analysis, but of distance from a prototype or typical exemplars, robins being thereby better birds than are penguins.

Prototypes are clearly music to an Empiricist's ears, since constructing a prototype is an activity very much rooted in one's experience, and the character of the examples of a concept that one happens to have encountered. In her response to (RCN), Cowie (1999, 146–47) proposes that concepts do have, at least partly, a prototypical structure, and that it is this aspect of a concept that is learned from experience (see also Prinz (2002) on "proxytypes").

A number of objections have been raised against prototype theories of concepts:

  (i) Not all concepts have prototypes (consider [carburetor] or [not a cat]), and, even in the case of those that do, it just does not seem plausible to insist that a competent user of a concept be acquainted with one, much less the *same* one for everyone: Spaniards and Australians may have different prototypes of [bird] but nonetheless share the concept;

 (ii) Many concepts cannot be identified with prototypes, since competent users of a concept know full well there can be perfectly good instances of a concept (e.g. [bird]) that are not prototypical (e.g., penguins), and, moreover, that something could satisfy the prototype (feathered, chirps) without being an instance (e.g., fancy toys);

(iii) In order for prototypes to figure effectively as an account of concepts, some "distance" metric among them would need to be specified (how unlike a typical bird can something be and still be a bird?), and it is not clear how this is provided without an independent characterization of the concept;

(iv) Prototypes are not *compositional*: (knowing) the prototype of [pet] (e.g., a dog) and the prototype of [fish] (e.g., a trout) does not determine (knowing) the prototype of [pet fish], and someone might know the prototype of [pet fish] without knowing the prototypes of [pet] or [fish] (see Fodor 1998).

Prototypes arguably play a role in how people quickly *tell* whether something satisfies a concept, that is, they provide good *evidence* for its application, and may figure centrally in a person's *conception* of its extension, but, as we saw in rejecting verificationism, evidence and conceptions are one thing, conceptual content quite another (see Rey 1983/1999, 1986; and Fodor 1998, 2001 for further discussion).

3.3.4. *Non-Sensory Primitives*

There is no intrinsic reason for either the Classical or maybe even Prototype views to be committed to either a sensory or other verificationist theory of concept construction, and there have been both philosophers and cognitive scientists who have thought that certain very general concepts, such as [agent], [object], [property], [number], [cause]—sometimes called "framework concepts"—might be primitive and innate, and provide a better basis than mere sensory experience for conceptual construction (see Miller and Johnson-Laird 1976; Moravcsik 1975; Pustejovsky 1995; Jackendoff 1983, 1992; Pinker 2007, ch. 3, for numerous proposals along these lines, and Fodor 1998, chs. 3–4 for criticisms).

A problem for this approach is to specify the relevant framework concepts and, more importantly, provide an account of what (constitutively) determines their content. Proponents will maintain that this is settled simply by whatever primitives their successful explanatory programs posit, but it is not clear how those programs as they are currently pursued will suffice. Fodor (1998, 49ff) complains that in order for such explanations to go through, the framework concepts must be understood univocally, and this univocality needs to be established. It is not enough merely to appeal to homophonic English words. Thus, Fodor argues, Jackendoff's (1992, 37–39) example of [keep] seems on the face of it either ambiguous or polysemous in "keeping a bird," "keeping time," "keeping a crowd happy," and "keeping an appointment." Perhaps it isn't, and [keep] can be analyzed univocally into some complex such as [cause a state that endures over time]—that is, if [keep] is not a framework concept after all. But now we have to ask the same question about *its* constituents: is [cause] or [endure] univocal? Does an appointment *endure* in the same sense as money does? Are they both *caused* in the same sense? What sense is that?

A related general problem with any approach that looks to analyses of any sort is to provide a basis for claiming that some proposed construction really is the *correct* account or "analysis" of a concept, as opposed to simply some banal or deeply entrenched belief about the world. Is it part of the concept [cat] that cats are animals? Or is this merely a belief that everyone takes for granted? Would thinking that cats are robots controlled from Mars be as incoherent as thinking that there are cats that are not cats? This is a form of the challenge already mentioned that Quine (1953, 1956) raised against the analytic/synthetic distinction, to which no generally accepted reply has yet been made (see Putnam, 1962a/1975; Katz 1990, 216ff; but also Horwich 1998, 2005; and Rey 2009, for recent proposals to meet this challenge that are related to Fodor's own proposals).

3.3.5. *Other Methods of Definition*

It is notable that many of Fodor's (1981) examples of constructions are *Boolean* (composed of simple truth functions such as [and] and [or]). But there is no reason to suppose that mental constructions might not be more sophisticated, employing complex quantifications and modal and probability operators. For example, concepts such as [energy], [mass], [force], and [space] are likely best defined *together* by setting out the complex laws of physics in which they all occur, and adding "and that's all there is to being any one of these things." A raft of concepts is thereby introduced in terms of the roles they play together in explaining some domain. The philosopher Frank Ramsey (1929) developed a technical proposal, now called "ramsification," that allows this to be done with a great deal of precision (roughly, a "Ramsey sentence" says that there does indeed exist a number of things, say, energy, mass, etc. that satisfy the terms that a theory introduces with the conjunction of all its claims; each of the individual terms can then be defined by its role with the others in that long conjunction; see David Lewis (1972/1980) for a lucid exposition). Such clauses in a theory might serve for what Carnap (1952) called "meaning postulates," or principles that are set out as defining of the terms being introduced by them (Murphy and Medin, 1985/1999; Block 1986). This would seem to be in part what Susan Carey (2009, chs. 8 and 11) has in mind in defending a "bootstrapping" proposal she finds in Quine (1960, 1969) and other philosophers of science. According to her proposal, "place holder" symbols are generated by assimilating a lot of information about a domain in the form of diagrams, lists, stray claims, and sometimes some serious theory: one could think of all this material as being expressed by one long Ramsey sentence that conjoins all of it, and allows one to refer to the single phenomenon (if any) in the world that satisfies it. The gradual assimilation of this material permits a learner to gradually master new concepts by slowly grasping partially now one conjunct and now another until they all fit together to characterize a stable concept.

One prima facie problem with this approach, however, is still the above Quinean one of deciding which clauses in a theory are to be included as meaning constitutive, and which as merely empirical claims about the world. Moreover, people are constantly creating and revising new theories involving old concepts, without ipso facto changing the content of those concepts. When Darwin proposed that humans are a kind of primate, he may have changed the prevailing *conception* (i.e., common beliefs) about human beings, but he did not thereby change the *content* of the concept. Indeed, it was precisely because he was employing a concept, [human], with the very same content as creationists that the latter were so upset!

More modestly, however, ramsifications might serve as "reference fixers," or descriptions that *in a particular context* serve to fix the reference of a term without being synonymous with it (see Jackson, 1998). This idea has its origins in Kripke's (1972/1980) discussion of proper names, where he points out that the descriptions people commonly associate with a proper name (e.g., "the discoverer of America" with "Columbus," "the author of *Moby Dick*" with "Melville") are certainly not synonymous with the name, and, indeed, could turn out to be false of its referent. He argued that proper names and many natural kind terms are "rigid designators" that name the same thing in all possible worlds, including those in which the common descriptions are not necessarily true (for example, it was perfectly possible for

Columbus not to discover America): the common descriptions serve merely to "fix the reference" of a term in a particular context, and not as definitions (an idea to which we will return below in considering Laurence and Margolis's (2002) "kind syndromes").

Kripke's proposal caused a minor revolution in philosophy, for it reinforced a suspicion that had been sown by Quine's attack on the analytic that *no* internal, *epistemic* condition would suffice for a theory of meaning. So, again, the Classical empiricist view of concept acquisition seemed implausible. Rather, meaning must in the first instance essentially involve some sort of relation to phenomena *external* to the mind and brain of a thinker.

3.4. Externalism

3.4.1. *Causal Theories*

Kripke's idea that reference is constituted at least in part by external facts was already suggested by some of the views of the later Wittgenstein (1953), and was independently proposed by Hilary Putnam (1975). Tyler Burge (1979) applied it not only to the meanings of both natural ("water") and artifactual ("sofa") kind terms, but also to their corresponding concepts. They pointed out that such terms and concepts do not standardly involve definitions known to their users, but rather (causal) relations with their actual referents and/or the social community in which they are used.

Now, of course, if possessing a concept *did* depend upon having a certain historical connection to an environment, certainly a certain facile form of empiricism would be vindicated: what concepts one had would depend directly upon what world one had experienced. However, while these causal intuitions provide an interesting challenge to the Classical View, they are quite inadequate as a theory of conceptual competence. Mere causal interaction in a certain community and environment cannot be enough, since surely not *all* sentient beings in New York have the concepts of a Columbia University physicist they have bumped into on the subway! But if mere causation is not enough, and definitions are not available, what does determine whether someone has a specific concept?

A number of writers have proposed varieties of *counterfactual* causal links: x has the concept [y] iff some state of x did/would causally co-vary with the worldly phenomenon y; for example, x did/would *discriminate* instances of y under certain (ideal, normal, evolutionarily significant) conditions, as a matter of nomological necessity. Thus, someone has the concept [horse] iff she could under certain conditions tell the horses from the non-horses. This is the idea behind "informational" (or "co-variational") theories of the sort proposed by Dretske (1980, 1987) and Stalnaker (1984). Fodor (1991, 1998) developed the most sophisticated of these accounts with his "asymmetric dependency" analysis, whereby a symbol x means that p iff its application to non-p cases depends on its application to p cases, but not vice versa (thus, one calls a misperceived cat "a dog" only because one calls a dog "a dog," but one does not call a dog "a dog" because one calls a misperceived cat one). In such cases, Fodor (1998) talks of symbols "locking" onto their referents. (Notice that if the causal connection between a concept and its meaning-constitutive referent were *counterfactual*, then the above facile argument for empiricism would no longer be available: the causal relation would not depend upon *actual* history, much less experience.)

There are numerous problems with informational theories. Once one departs from the relatively straightforward cases of perceptual concepts that even empiricists would be happy to regard as primitive, it is by no means obvious that there really are the kinds of genuine laws linking the brain symbols to worldly phenomena in a way that would provide those symbols with their conceptual content (what are the laws for [cause]? [object]? [space]?, or for logical concepts such as [not] or [if]?) (see Loewer 1996). Particularly troublesome cases are "empty" concepts, such as [ghost], [angel], [soul], and (at least for some) [triangle] and [square], for which there are arguably no *possible* worldly phenomena with which brain symbols could co-vary.[22] Fodor (1998) crucially claims "there can be no *primitive* concept without a corresponding property for it to lock to" (165), but then it would appear that he would have to allow such empty concepts to be "constructed" after all. But, if them, why not others?[23]

These and other considerations have led a number of philosophers and psychologists to embrace "two factor" theories that claim that conceptual content is determined by *both* causal relations to the world *and* internal computational roles (see Block 1986; Carey 2009, 514ff). For example, the content of [bird] may consist both of a role component that figures in accounts of internal psychological processing—for example, common conceptions and prototypes associated with birds—*and* an external causal component that determines what phenomena in the world is picked out. However, many of the problems we mentioned with respect to these components considered separately would appear to persist in their combination: it seems possible for prototypes and conceptions to vary across people without variation in their concepts, and it is not clear how internal factors can figure in a compositional semantics.

In his (1998), Fodor proposes an interesting intermediate view that, while not treating prototypes as *constitutive* of concepts, treats them as playing a crucial role in the *triggering* of them.

3.4.2. *Fodor's (1998) Prototype Triggering View*

Fodor (1998, 127ff) considers an important problem for his view that concepts are triggered and not learned, what he calls the "doorknob/DOORKNOB" problem: why is it that [doorknob] is regularly triggered by doorknobs, and not, say, by

---

[22] Such cases motivate an appeal to Fregean senses, cf. fn 16.

[23] One might reply here that in the case of empty terms, such as "elf," "phlogiston." or "Vulcan," there is, indeed, nowhere to retreat but to the claims by which they are introduced, but that in the cases where terms *do* succeed in referring, the introductory material gets trumped by the referent, so that, e.g., "water" or "polio" is no longer tied to it (see Rey 2005 for discussion).

giraffes or French flags (see also Sterelny, 1989)? The fact that it is suggests that a concept perhaps is not merely "triggered" as a brute causal process, but, in Descartes's phrase, is "occasioned" by virtue of some kind of rational, confirmatory relation between a concept and perceptual stimuli, and so might involve a kind of learning after all.

To solve this problem, Fodor adopts a quasi-Kantian metaphysics of the very items so picked out. He exploits the prototype views we mentioned earlier, *not* toward making some point about the content of *concepts*, but instead about the metaphysics of, for example, *doorknobs*! The reason doorknobs occasion activation of the [doorknob] concept is that *doorknobs just are the kind of thing to which humans generalize when presented with prototypical doorknobs*. That is, he treats such cases on the model of the "response-dependent" properties that many philosophers have proposed for secondary properties such as color. (Unlike Kant, however, Fodor does not claim that *all* concepts are of response-dependent properties; in the next section we will return to the "scientific" ones that are not.) Whether such a story can really be told generally in conjunction with an informational theory of content without circularity remains controversial (the proposal skirts perilously close to the claim that possessing the concept of doorknob requires a representation being locked onto—well, just those things to which the representation is locked! See Cowie 1999, 96–99). However, Fodor (1998, 138) argues that circularity is avoided by the fact that at least many prototypes are known to be specifiable independently of the concepts they occasion (as in Rosch 1973).

Whether or not his view avoids circularity, Fodor sees it as permitting him to retreat from his initial (1975, 1981) controversial view that all MM *concepts* are innate. His 1975 argument too quickly assumed that if a concept was not learned, it was innate. Recalling the distinction between rational and brute-causal effects of perceptual experience, on the present view concepts can be *acquired* from experience by a nonrational triggering mechanism without being learned, and so need not be either learned *or* innate. As he puts it in his still later 2008 book:

> You can't infer from a concept's not being learned to its being innate; not, at least, if "innate" means something like "not acquired in consequence of experience." There would appear to be plenty of ethological precedents—from "imprinting" to "parameter setting" [in Chomskyan linguistic theories] inclusive—where it's implausible that the acquisition of a concept is mediated by a rational process like inductive inference, but where concept acquisition is nevertheless highly sensitive to the character of the creature's experience. (2008, 144–45)

This leads him to refine his 1975 position:

> What's learned (not just acquired) are stereotypes (statistical representations of experience). What's innate is the disposition to grasp such and such a concept (i.e., to lock to such and such a property) in consequence of having learned such and such a stereotype. (2008, 162)

Indeed: "the kind of nativism about [doorknob] that an informational atomist has to put up with is perhaps not one of concepts but of *mechanisms*" (1998, 142), the processes of which he takes to be nonrational and even nonintentional. They are simply "brute causal," not psychological processes.[24]

Slightly revising the distinctions we drew earlier between activation, lying in wait, and expressive power, we might put Fodor's (1998) view as follows: concepts themselves are not innate and do not lie in wait, unactivated. What is innate are highly specific dispositional mechanisms to acquire a concept upon exposure to the concept's prototype. These *mechanisms* lie in wait in a creature's brain, and the actual acquisition of a concept consists in a mental representation being produced by the activation of the mechanism, and coming to stand in a certain counterfactually specified locking relation to the real world phenomenon that constitutes its content. The notion of a conceptual system's *expressive power* is expanded from merely concepts activated or lying in wait to the range of concepts these innate mechanisms can produce in this way and by logical combination. Fodor's main claim remains that expressive power in this sense is not increased by learning. The relation of a prototype to the concept it activates through these mechanisms is brute causal.

Incidentally, Fodor was not the first to despair of an intentional solution to relating percepts to concepts. In his *Critique of Pure Reason*, Kant (1787/1934, A138ff) postulated his "schemata" to mediate between the two,[25] and resignedly claimed that

> the schematism of our understanding, in its application to appearances and their mere form, is an art concealed in the depths of the human soul, whose real modes of activity nature is hardly likely ever to allow us to discover. (Kant, 1787/1934, A141)

For Fodor (1998, 2008), it is stereotypes that play the role of Kant's schemata, triggering innate dispositions to lock onto a property. But, as we saw above (Section 3.3.3), they do not exhaust a typical concept that has commitments far beyond them. So there is still the question why a specific stereotype triggers one concept rather than another, for example, [green] rather than [grue]. This is what Fodor (2008) thinks is determined by just brute causation.

Note that, *pace* Cowie (1999, 72), Fodor's despair here does not imply any "non-naturalism" on his part. Fodor's view is not that there is no *natural* explanation of acquisition; there is just no *intentional* one. Explanation of *most* of nature is not intentional. Why should it be surprising if concept acquisition turns out not to be so?

---

[24] This emphasis on whether the processes are brute causal or psychological recalls Samuels's conception of innateness (Section 2.2). But, of course, on that conception the concepts would therefore be innate as well, which Fodor here is allowing that they "perhaps" are not. (Cf. fn. 11.) It is unclear what of importance would be lost if Fodor embraced Samuels's view and continued to claim that concepts are innate as well.

[25] Unlike Fodor, Kant was here mainly concerned only with what he called "pure" concepts, such as [object] and [cause], and not with how to acquire a concept from experience, but how to apply one to it. And, of course, Kant's project is to determine transcendentally what is constitutive of experience, not to provide a naturalistic account of it. But the similarity in problems is striking.

Still, one might find such a view puzzling and unsatisfying: there remain those 250,000-plus *specific dispositions* to acquire concepts by specifically linked stereotypes, and this might seem even more profligate than merely 250,000-plus concepts by themselves. And so one might well want to continue to investigate further ways concepts could be constructed from experience without merely triggering specific innate dispositions. Fodor (2008) deals with this response.

3.5. Fodor's (2008) View

In his most recent discussion of the issues, Fodor (2008) provides a further argument against concept learning to those already noted above, one based simply on the familiar problem of distinct coextensional concepts, for example [morning star]/[evening star], [creature with a heart]/[creature with a kidney], [triangle]/[trilateral] (cf. Frege 1892/1966). How, Fodor wonders, can experience by itself provide a basis for learning the one concept and not the other? Only, he replies, by "representing the same experiences in different ways" (2008, 135), which again requires that one already has the very concept that experience has been recruited to teach.

Moreover, he scorns his earlier views as being "too modest" with regard to complex representations:

> What I should have said is that it's true and a priori that the whole notion of concept learning is per se confused.... *no* concept can be learned, primitive or complex. (2008, 130 and 138)

Certainly, continuing the line of his (1975), construction of complex representations is not a way of acquiring an *expressive capacity* that one's nervous system did not already possess (we will return shortly to weaker, mere activation notions of concept possession, with which he may have been confusing expressive capacity in his (1975)).

Fodor (2008) also replies to some plausible proposals advanced by Margolis (1998) and Laurence and Margolis (2002), who argue that many ordinary natural kind concepts are acquired by children rather in the way that Fodor (1998) reserves for "scientific" ones. As we mentioned in the last section, in his (1998, 150–62) Fodor, unlike Kant, does allow that not all concepts are of mind-dependent phenomena. In particular, he allows that natural kind concepts, for example, [water], [gold], apply to things in the world that, unlike doorknobs, enter into laws, independently of what humans generalize to. Fodor claims, however, that these are "a late and sophisticated achievement," attained only "in the context of the scientific enterprise" (1998, 159). Although children and other proto-scientific people do in fact have such concepts, Fodor does not think they play a scientific role in their thought—they are for such people concepts of natural kinds but are not yet treated by them *as* natural kind concepts.

It is hard to see, however, why the intellectual attitudes (if not all the skills) of scientists are not *sometimes* available to nonscientists. In the work mentioned earlier on proper names and natural kind terms, Kripke (1972/1980) and Putnam (1975) drew attention to what would seem to be a property of much of the *ordinary* use of natural kind terms, such as "water" and "gold," whereby people are prepared to take such terms to refer to things with a "hidden essence" precisely along the lines Fodor reserves for scientists. Lest one suppose that perhaps Kripke and Putnam are securing agreement about these terms only from scientifically minded philosophers, it is worth noting that a number of cognitive scientists have produced evidence of similar intuitions in children and non-Westerners (see Macnamara 1986; Keil 1989; Medin and Atran 1999; Gelman 2003).[26]

In relation to concepts, Laurence and Margolis (2002, 38) develop a suggestion of Putnam (1975) and propose a quite general function (they call it a "kind syndrome...a collection of properties that is highly indicative of a kind yet is accessible in perceptual encounters.... [e.g.,] typical shape, motions, markings, sounds, colors, etc.") that, in conjunction with the essentialist attitude, takes an eliciting description to the nearest, contextually natural or explanatory kind that includes the material in the syndrome. Thus, [water] might be constructed as the nearest natural kind that includes the odorless, tasteless stuff that we drink, is found in streams and ponds, and is the stuff of rain. One might well extend this move beyond merely natural kind concepts to whatever is the "best account" of the reference of a term: thus, a nonnatural kind concept, [viol], might be constructed as the nearest kind that plays an explanatory role in the history of the modern violin (see Rey 1983). Carey (2009, 519) tries to generalize this account with her above-mentioned bootstrapping proposal, which she argues is not restricted to or dependent upon an antecedent notion of natural kind. Although such a kind syndrome is not *constitutive* of the concepts of, say, [water] or [viol], plausibly the only way someone could lock on to the corresponding properties is by knowing such clauses. It is, as Margolis (1998) puts it, a "sustaining mechanism" that explains the locking and keeps it in place. So, why not suppose that learning those clauses provides an account of ordinary folk *learning* a concept, even if it is not defining of it?

Fodor (2008, 144) notes in reply that "'You can learn (not just acquire) A' and 'Learning A is sufficient for acquiring B' just doesn't imply 'You can learn B.'" Again, B may merely be *triggered* by the learning of A. Remember, as we noted in discussing

---

[26] Recently, Machery et al. (2004) have found evidence of intuitions purportedly contrary to those of Kripke and Putnam among some non-Western peoples. Such findings tend, however, to be beside the point, since what is at issue is not whether *all* uses of such terms *always* involve reference to a hidden essence, but only that a significant class are so for at least some people other than scientifically minded philosophers, which this research in fact confirms: many of the non-Westerners shared the Kripke/Putnam intuitions. For further criticisms, see Devitt (2011).

Fodor (1998, 154–55) argues that the evidence that children have natural kind concepts as such is all interpretable merely as evidence that they make an appearance/reality distinction: the evidence of which he is aware does not establish that they think of a hidden essence as being the *cause* of appearances. This latter presumption is questionable: it is not at all clear that hidden essences must be (believed to be) causal in this way; many appearances (e.g., the color of the sea) might be (believed to be) due entirely to factors other than water being H2O. But in any case, in reply to Fodor, Gelman (2003, ch. 5, esp. 135) provides just such evidence of causal thinking.

abduction, the concepts we deploy in experience often involve commitments far beyond the evidence they explain, and so the question is just what takes us from the evidence to *these* and *not other* further commitments.[27]

A way to bring out the gap that Fodor is positing between the kind syndrome and conceptual locking is to consider the problem of saying just what the "nearest" kind is: point to a cat and you point *inter alia* to, say, its ear, a *flea* on its ear, to a *domestic cat*, to a *feline*, to a *mammal*, and so on. It is here that Goodman's "grue" problem we discussed in Section 3.2 returns with a vengeance. There are an infinite number of possible concepts that are compatible with any finite set of data a creature will have encountered: not merely an infinitude of grue predicates, in which "observed before *t*" is incremented by one year, but an infinitude of other analogous combinations, for example, [gred], [grellow], [catiraffe] (a cat and observed before *t* or otherwise a giraffe), etc. (Note that we cannot rely on human beings always referring to genuine natural kinds: consider [race], [humors], [earth], [air], [angel], even [red] and [green] themselves, which depend on idiosyncracies of our visual system.) Devitt and Sterelny (1987/1999) call this general problem the "*qua* problem": *qua*, or as, *what* do you refer when you point to the cat? Spelling out "nearest natural kind"—perhaps in terms of contextually relevant contrasts—would be a difficult, but perhaps not entirely hopeless research program. Fodor, though, is not holding his breath. He thinks the answer lies not in some rational connection between the relevant concept and the stimuli (even in the context), but in a brute-causal fact by which the concept is simply perceptually triggered by these stimuli.

4. CONCLUSION

Proposals such as Laurence and Margolis's (2002) proposal seem to come as close as one can to capturing the pre-theoretic notion of "learning a concept." According to them, at least with natural kind concepts, new primitive concepts can be acquired by learning, because there can be cases where one's experience "initiates a process where information is collected, stored, and manipulated in a way that controls a representation so that it tracks" what it is in fact about (2002, 43). On their view, because such concepts are learned, they are not innate. Fodor, on the other hand, would insist that only the *collected information* is learned, denying that the concept associated with it is therefore learned. The disagreement here can be seen as one concerning whether brute perceptual triggering is required in addition to the

---

[27] An interesting possibility raised by Fodor's view here is that someone might create a Ramsey sentence that *does not* trigger an innate (disposition to acquire a) concept. Perhaps this is the situation we find ourselves in with respect to, for example, the phenomenon of *light* as described in modern physics, which behaves both like a wave *and* like a stream of particles, while it is difficult for (most of) us to conceive how something could do both.

rationally assimilated information, but also in part as one concerning what should count as "learning," or at least learning a concept: can *it* include mere brute triggering by collected rationally relevant information or not?

Put aside the pre-theoretic notion of learning, as well as Fodor's last proposal to retreat merely to dispositions to acquire concepts upon exposure to prototypes. Remembering the distinctions we originally drew in Section 3.2 between *expressive power, lying in wait,* and *actual activation,* a boringly (or maybe radically) ecumenical view suggests itself. Some concepts could be *both* innate and learned at least in the following sense: concepts would be *innate* insofar as they cannot be constructed from or defined in terms of experience, but *learned* insofar as they are activated in an effort to confirm their application ultimately on the basis of experience.[28]

What is given at birth, or as a consequence of brute non-intentional influences on the brain, is a system with a certain expressive power: a set of primitives that have their content by virtue of their causal relations to the world, and principles of logical combination that permit the construction of an indefinite variety of logically complex representations out of them. These primitives and their logical combinations determine what the creature can *possibly* think, and this, if Fodor is right, can never be increased by any rational process.[29]

However, the *activation* or *deployment* of a concept can often depend upon learning about the world, laboriously constructing stereotypes or Ramsey sentences that are designed to pick out the nearest explanatory kinds that offer, via abduction, the best explanation of relevant phenomena. The process could proceed much along the lines of a "hypothetical deductive" (HD) model of explanation, according to which explanation consists in deducing descriptions of the target phenomena from general hypotheses that are thereby confirmed (see Hempel 1965). Such a process is not the less confirmatory should the hypotheses, or their constituent concepts themselves, not have their *source* in experience, or not be reducible to it. (Of course, not all activation is by learning: some may be brute, and some may, as Hume (1734/1978, 10) emphasized, be merely imaginative—and here, indeed, one wonders what the source of imagined concepts could be other than the native repertoire, if they are neither triggered nor constructed from experience.)

There are, of course, constraints on how easily people actually can perform the logical constructions. To invoke a nice distinction from Chomsky (1965), there may be "performance" issues (e.g., short-term memory resources, motivation) constraining the activation of an underlying conceptual "competence." It is relatively easy to construct [viol] and [prime number]; much harder, [curved space-time]. The difficulties may even fall into patterns: construction of representations of

---

[28] The data to be explained in learning a concept in this way need not be evidence about whether a concept actually *applies to the world*. Often we come to grasp concepts by explaining merely the remarks of another speaker, as when we understand the concepts, say, [karma] or [phlogiston], of some theory we presume to be false.

[29] Note that this ecumenical view could be adapted also to the expanded notion of expressive power introduced in Section 3.4.2.

abstract concepts beyond perceptual appearances seems to require sometimes enormous effort (as in the case of the physics concepts of *heat* and *entropy*); and perhaps the difficulties are subject to developmental stages of the sort suggested by Piaget (1954, 1980). Indeed, even if concepts themselves (and not merely the dispositions to acquire them) were innate, Piaget could be right about childhood inabilities, say, to think about rational numbers. A neonate may well have a system capable of expressing every concept she will ever learn, but simply lack the further memory capacities or motivation to activate them.

The idea that one and the same trait could be both innate and learned might seem counterintuitive, if not downright paradoxical (cf. fn. 2). But it will perhaps seem less so when one considers the grounds for thinking of concepts in this way. The suggestion is that it is the full repertoire of concepts that is innate, with learning serving to select from the set of concepts those that are suitable for explaining the relevant data of experience. On this view, we thus must distinguish two levels of possession: concepts are innately had, but can also be had in a further, activated sense as a result of learning. Learning a concept thus turns out to be a kind of acquisition that involves an innate component—the concept itself—just as, if Jerne is right, acquiring certain immunities involves being exposed to certain pathogens that activate innate antibodies. Our ordinary, intuitive notion of innateness would simply turn out to pick out a phenomenon that is more intricate that it initially appears to be.

Just how much this conception of learning ought to satisfy conceptual empiricists depends upon both the degree to which concept activation *does* involve something such as an HD model of explanation of purely sensory material and the degree to which the HD model is an adequate model of learning. Neither of these questions can remotely be regarded as settled. If the locking of a representation onto an external content were caused merely by some purely "accidental" reference fixer—for example, a child who happens to lock onto quarks as a result of learning the description "what father wrote his book about" and using it to explain his father's scholarly behavior—it would seem pretty far from counting as *learning*.[30] But is locking as a result of mastering some facts in a textbook in principle any different? Even if they are not definitions, some stereotypes, reference fixers, and Ramsey sentences seem more "rationally connected" than others to the concepts they may trigger, but it is not at all clear how to specify just what that rational connection might be. Indeed, what *constrains* the further commitments a concept involves beyond the evidence that triggered it? Why is it—if it is—more rational for a child to jump to the concept [green] and not [grue] in encountering grass and pistachio nuts?

Conceptual nativism thus turns on a number of issues that may not have been evident at the outset:

---

[30] Consider a child who might be genetically incapable of understanding quantum physics: on some views, he might be said to think "his father works on quarks" simply by virtue of "reference borrowing" among members of a language group (cf. Kripke, 1972/80), but without his actually having *learned* the concept.

   (i) whether empiricists can succeed in providing successful *analyses* of some
       significant portion of MM concepts from some set of primitives, despite
       their persistent failures to do so (see Section 3);

  (ii) whether some sort of externalist, "informational" theory of content can
       be sustained on behalf of concepts lacking internalistic analyses (Section
       3.4.1);

 (iii) whether and the degree to which concept acquisition involves an HD or
       other "rational" process (and whether that process counts as "learning"),
       or, instead, a brute-causal, nonrational mere "triggering" process, lost in
       the mysteries of neurophysiology (Section 3.5);

and

  (iv) whether there is any independent justification for supposing some
       250,000-plus concepts, or specific dispositions to acquire them, are
       innate.

At this point, it would be foolhardy to suppose that any of these issues are anywhere
near being settled. The possibilities of information theories of content and various
means of concept construction have yet to be adequately explored, and, as we have
emphasized, the psychological versus non-psychological processes of occasioning
versus triggering a concept on which the question of conceptual nativism seems to
turn, have yet to be understood.[31]

REFERENCES

Adams, R. (1975). Where Do Our Ideas Come From? In S. Stich (ed.), *Innate Ideas.*
    Berkeley: University of California Press.
Aquinas, T. (1266/1948). *Summa Theologica*, translated by Fathers of the English Dominican
    Province. New York: Benziger Bros.
Ariew, A. (1996). Innateness and canalization. *Philosophy of Science* 63: S19–S27.
———. (1999). Innateness Is Canalization. In V. Hardcastle (ed.), *Where Biology Meets
    Psychology.* Cambridge, MA: MIT Press.
———. (2007). Innateness. In M. Matthen and C. Stephens (eds.), *Handbook of the
    Philosophy of Biology.* Amsterdam: Elsevier.
Aristotle (1968). *On the Soul (De Anima)*, Books II–III, translated by D.W. Hamlyn.
    Clarendon Aristotle Series, Oxford: Oxford University Press.
Ayer, A.J. (1934). *Language, Truth and Logic.* Mineola, NY: Dover.
Bateson, P. (2000). Taking the Stink Out of Instinct. In H. Rose and S. Rose (eds.), *Alas,
    Poor Darwin.* London: Jonathan Cape.

[31] For helpful discussion, we are grateful to Jonathan Adler, Greg Ball, Paul Bloom, John
Collins, Jerry Fodor, Eckart Forster, Barbara Landau, Joseph Levine, Edouard Machery, Yitzhak
Melamed, Eric Margolis, Richard Samuels, and participants in a Philosophy of Linguistics
Workshop at the University of Maryland.

Block, N. (1986). Advertisement for a semantics for psychology. In P. French, T. Uehling, and H. Wettstein (eds.), *Studies in the Philosophy of Mind*, vol. 10 of *Midwest Studies in Philosophy*. Minneapolis: University of Minnesota Press.

Block, N. (1995). How heritability misleads about race. *Cognition* 56: 99–120.

Bloom, P. (2000). *How Children Learn the Meanings of Words*. Cambridge, MA: MIT Press.

Boyd, R. (1991). Realism, anti-foundationalism and the enthusiasm for natural kinds. *Philosophical Studies* 61: 127–148.

Burge, T. (1979). Individualism and the mental. *Midwest Studies in Philosophy* 4: 73–121.

Carey, S. (2009). *The Origin of Concepts*. Oxford: Oxford University Press.

Carnap, R. (1952). Meaning postulates. *Philosophical Studies* 3: 65–73.

Carruthers, P., Laurence, S., and Stich, S. (eds.) (2005). *The Innate Mind: Volume 1: Structure and Contents*. Oxford: Oxford University Press.

———. (2006). *The Innate Mind, Volume 2: Culture and Cognition*. Oxford: Oxford University Press.

———. (2007). *The Innate Mind, Volume 3: Foundations and the Future*. Oxford: Oxford University Press.

Cherniak, C. (2005). Innateness and Brain-Wiring Optimization: Non-Genomic Nativism, In A. Zilhao (ed.), *Evolution, Rationality and Cognition*. London: Routledge.

Chomsky, N. (1955/1975). *The Logical Structure of Linguistic Theory*. Chicago: Chicago University Press.

———. (1965). *Aspects of the Theory of Syntax*. Cambridge, MA: MIT Press.

———. (1966). *Cartesian Linguistics*. New York: Harper and Row.

———. (1968/2006). *Language and Mind*. Berkeley: University of California Press.

———. (1971). *Problems of Knowledge and Freedom: The Russell Lectures*. New York: Pantheon.

———. (2005). Three factors in language design. *Linguistic Inquiry* 36: 1–22.

———. (2007). Biolinguistic explorations: Design, development, evolution. *International Journal of Philosophical Studies* 15: 1–21.

Churchland, P. (1986). *Neurophilosophy*. Cambridge, MA: MIT Press.

Collins, J. (2005). Nativism: In defense of a biological understanding. *Philosophical Psychology* 18: 157–77.

Cowie, F. (1999). *What's Within: Nativism Reconsidered*. Oxford: Oxford University Press.

———. (2009). Why Isn't Stich an ElimiNativist?. In D. Murphy and M. Bishop (eds.), *Stephen Stich and His Critics*. Oxford: Blackwell.

Cracraft, J. (2000). Species Concepts in Theoretical and Applied Biology: A Systematic Debate with Consequences. In Q. Wheeler and R. Meier (eds.), *Species Concepts and Phylogenetic Theory*. New York: Columbia University Press.

Crane, T. (1992). The Non-Conceptual Content of Experience. In T. Crane (ed.), *The Contents of Experience*. Cambridge: Cambridge University Press.

Cudworth, R. (1678/1964). *The True Intellectual System of the Universe*. London, 1678. Facsimile reprint: Stuttgart-Bad Canstatt: Friedrich Frommann Verlag.

Damasio, A. (1994). *Descartes' Myth: Emotion, Reason and the Human Brain*. New York: Putnam.

Descartes, R. (1641a/1911). Meditations. In E. Haldane and G. Ross (eds. and trans.), *Philosophical Works of Descartes*, vol. 1. Cambridge: Cambridge University Press.

———. (1641b/1911). Author's Reply to Fifth Set of Objections. In E. Haldane and G. Ross (eds. and trans.), *Philosophical Works of Descartes*, vol. 2. Cambridge: Cambridge University Press.

Descartes, R. (1647/1911). *Notes Against a Program*. In E. Haldane and G. Ross, G. (eds. and trans.), *Philosophical Works of Descartes*, vol 1. Cambridge: Cambridge University Press.

Devitt, M. (2011). Experimental semantics. *Philosophy and Phenomenological Research* 82: 418–35.

Devitt, M., and Sterelny, K. (1987/1999). *Language and Reality: An Introduction to the Philosophy of Language*, rev. 2nd ed. Cambridge, MA: MIT Press.

Dretske, F. (1980). *Knowledge and the Flow of Information*. Cambridge, MA: MIT Press.

———. (1987). *Explaining Behavior*. Cambridge, MA: MIT Press.

Earman, J., Glymour, C., and Mitchell, S. (eds.) (2003). *Ceteris Paribus Laws*. Berlin: Springer.

Elman, J., Bates, E., Johnson, M., Karmiloff-Smith, A., Parisi, D., and Plunket, K. (1996). *Rethinking Innateness: A Connectionist Perspective on Development*. Cambridge, MA: MIT Press.

Fodor, J. (1975). *The Language of Thought*. Cambridge, MA: Harvard University Press.

———. (1981). The present status of the innateness controversy. In J. Fodor, *Representations: Essays on the Foundations of Cognitive Science*. Cambridge, MA: MIT Press.

———. (1991). *A Theory of Content*. Cambridge, MA: MIT Press.

———. (1998). *Concepts: Where Cognitive Science Went Wrong*. Oxford: Oxford University Press.

———. (2001). Doing without what's within: Fiona Cowie's critique of nativism. *Mind* 100: 99–148.

———. (2008). *LOT2: The Language of Thought Revisited*. Oxford: Oxford University Press.

Fodor, J., and Lepore, E. (1992). *Holism: A Shopper's Guide*. Oxford: Blackwell.

Frege, G. (1884/1950). *The Foundations of Arithmetic*, translated by J. L. Austin. Oxford: Blackwell.

———. (1892/1966). On Sense and Reference. In P. Geach and M. Black (eds.), *Translations from the Works of Gottlob Frege*. Oxford: Blackwell.

Gelman, S. (2003). *The Essential Child: Origins of Essentialism in Everyday Thought*. New York: Oxford University Press.

Godfrey-Smith, P. (2007). Innateness and genetic information. In P. Carruthers, S. Laurence, and S. Stich (eds.), *The Innate Mind, Volume 3: Foundations and the Future*. Oxford: Oxford University Press.

Goldin-Meadow, S. (2003). *The Resilience of Language: What Gesture Creation in Deaf Children Can Tell Us about How All Children Learn Language*. New York: Psychology Press.

Goodman, N. (1949). On likeness of meaning. *Analysis* 10: 1–7.

———. (1954/1983). *Fact, Fiction and Forecast*, 3rd ed. Cambridge, MA: Harvard University Press.

Gould, J., and Marler, P. (1991). Learning by Instinct. In D. Mock (ed.), *Behavior and Evolution of Birds*. San Francisco: Freeman.

Greenberg, M., and Harman, G. (2006). Conceptual Role Semantics. In B. Smith and E. Lepore (eds.), *Oxford Handbook of Philosophy of Language*. Oxford: Oxford University Press.

Griffiths, P. (1997). *What Emotions Really Are*. Chicago: University of Chicago Press.

———. (2002). What is innateness? *The Monist* 85: 70–85.

———. (2009). The distinction between innate and acquired characteristics. *Stanford Encyclopedia of Philosophy*. <http://plato.stanford.edu/entries/innate-acquired/>

Griffiths, P., and Gray, R. (1994). Developmental systems and evolutionary explanations. *Journal of Philosophy* 91: 277–304.

Griffiths, P., Machery, E., and Lindquist, S. (2009). The vernacular concept of innateness. *Mind & Language* 24: 605–30.

Hart, W. D. (1975). Innate Ideas and A Priori Knowledge. In S. Stich (ed.), *Innate Ideas*. Berkeley: University of California Press.

Hempel, C. (1965). *Aspects of Scientific Explanation and Other Essays in the Philosophy of Science*. New York: Free Press.

Herbert of Cherbury (1624/1937). *De Veritate*, translated by M. H. Carré. Bristol, UK: University of Bristol Press.

Holden, C. (1987). The genetics of personality. *Science* 237: 598–601.

Horwich, P. (1998). *Meaning*. Oxford: Oxford University Press.

———. (2005). *Reflections on Meaning*. Oxford: Oxford University Press.

Hume, D. (1734/1978). *A Treatise of Human Nature*, 2nd ed., edited by L. A. Selby-Bigge and P. H. Nidditch. Oxford: Oxford University Press.

Jackendoff, R. (1983). *Semantics and Cognition*. Cambridge, MA: MIT Press.

———. (1992). *Languages of the Mind: Essays on Mental Representation*. Cambridge, MA: MIT Press.

Jackson, F. (1998). *From Metaphysics to Ethics: A Defence of Conceptual Analysis*. Oxford: Oxford University Press.

James, W. (1890/1983). *The Principles of Psychology*. Cambridge, MA: Harvard University Press.

Jerne, N. (1985). The generative grammar of the immune system. *Science* 229: 1057–59.

Johnson, M. (1997). *Developmental Cognitive Neuroscience*. Oxford: Blackwell.

Kant, I., (1787/1934). *Critique of Pure Reason*, 2nd ed., Translated by Norman Kemp Smith. New York: St. Martin's Press.

Katz, J. (1990). *The Metaphysics of Meaning*. Oxford: Oxford University Press.

Keil, F. (1989). *Concepts, Kinds, and Cognitive Development*. Cambridge, MA: MIT Press.

Kitcher, P. (1996). *The Lives to Come*. New York: Simon & Schuster.

Kripke, S. (1972/80). *Naming and Necessity*. Cambridge, MA: Harvard University Press.

Laurence, S., and Margolis, E. (2002). Radical concept nativism. *Cognition* 86: 25–55.

Lehrman, D. (1953). A critique of Konrad Lorenz's theory of instinctive behaviour. *The Quarterly Review of Biology* 28: 337–63.

———. (1970). Semantic and Conceptual Issues in the Nature-Nurture Problem. In L. Aronson, E. Tobach, D. Lehrman, and J. Rosenblatt (eds.), *Development and Evolution of Behavior*. San Francisco: Freeman.

Lewis, D. (1972/1980). Psychophysical and Theoretical identifications. In N. Block (ed.), *Readings in the Philosophy of Psychology*, vol. II. Cambridge, MA: Harvard University Press.

Lewontin, R. (1974). The analysis of variance and the analysis of causes. *American Journal of Human Genetics* 26: 400–411.

Locke, J. (1690/1975). *An Essay Concerning Human Understanding*. Oxford: Oxford University Press.

Loewer, B. (1996). A Guide to Naturalizing Semantics. In C. Wright and B. Hale (eds.), *A Companion to Philosophy of Language*. Oxford: Blackwell.

Lorenz, K. (1957). The Nature of Instincts. In C. H. Schiller (ed.), *Instinctive Behavior*. New York: International University Press.

Machery, E., Mallon, R., Nichols, S, and Stich, S. (2004). Semantics, cross-cultural style. *Cognition* 92: 1–12.

Macnamara, J. (1986). *Border Dispute: The Place of Logic in Psychology*. Cambridge, MA: MIT Press.

Mallon, R., and Weinberg, J. (2006). Innateness as closed process invariance. *Philosophy of Science* 73: 323–44.

Mameli, M. (2004). Nongenetic selection and nongenetic inheritance. *British Journal for the Philosophy of Science* 55: 35–71.

Mameli, M., and Bateson, P. (2006). Innateness and the sciences. *Biology and Philosophy* 21: 155–88.

Margolis, E. (1998/1999). How to acquire a concept. Reprinted in E. Margolis and S. Laurence (eds.), *Concepts: Core Readings.* Cambridge, MA: MIT Press.

Margolis, E. and Laurence, S. (eds.) (1999). *Concepts: Core Readings.* Cambridge, MA: MIT Press.

———. (2007). The ontology of concepts—abstract objects or mental representations? *Noûs* 41: 561–93.

Medin, D., and Atran, S. (eds.). (1999). *Folkbiology.* Cambridge, MA: MIT Press.

Mehler, J., and Dupoux, E. (1994). *What Infants Know: The New Cognitive Science of Early Development,* translated by P. Southgate. Oxford: Blackwell.

Miller, G. (1956). The magical number seven, plus or minus two: Some limits on our capacity for processing information. *Psychological Review* 63: 81–97.

Miller, G. and Johnson-Laird, P. (1976). *Language and Perception.* Cambridge, MA: Harvard University Press.

Millikan, R. (1984). *Language, Thought and Other Biological Categories.* Cambridge, MA: MIT Press.

Moravcsik, J. (1975). *Aitia* as generative factor in Aristotle's philosophy. *Dialogue* 14: 622–36.

Murphy, G., and Medin, D. (1985/1999). The role of theories in conceptual coherence. In E. Margolis and S. Laurence (eds.), *Concepts: Core Readings.* Cambridge, MA: MIT Press.

Oyama, S. (2000). *Evolution's Eye: A Systems View of the Biology-Culture Divide.* Durham, NC: Duke University Press.

Pagin, P. (2006). Meaning holism. In B. Smith and E. Lepore (eds.), *Oxford Handbook of Philosophy of Language.* Oxford: Oxford University Press.

Peacocke, C. (1992). *Concepts.* Cambridge, MA: MIT Press.

Piaget, J. (1954). *The Construction of Reality in the Child.* New York: Basic Books.

———. (1980). Language within Cognition: Schemes of Action and Language Learning. In M. Piatelli-Palmerini (ed.), *Language and Learning.* Cambridge: Harvard University Press, 163–83.

Peirce, C. (1903/1998). A syllabus of certain topics of logic. In *Essential Peirce,* vol. 2. Bloomington: Indiana University Press.

Pietroski, P., and Rey, G. (1995). When other things aren't equal: Saving *ceteris paribus* laws from vacuity. *British Journal for the Philosophy of Science* 46: 81–110.

Pinker, S. (2007). *The Stuff of Thought.* New York: Viking.

Plato (1975). *Meno.* Relevant passages reproduced in S. Stich, *Innate Ideas.* Berkeley: University of California Press .

Prinz, J. (2002). *Concepts and Their Perceptual Basis.* Cambridge, MA: MIT Press.

Pustejovsky, J. (1995). *The Generative Lexicon.* Cambridge, MA: MIT Press.

Putnam, H. (1962a/1975). It ain't necessarily so. *Journal of Philosophy* 54. Reprinted in his *Philosophical Papers,* vol. 2. Cambridge: Cambridge University Press.

———. (1962b/1975). Dreaming and depth grammar. Reprinted in his *Philosophical Papers,* vol. 2. Cambridge: Cambridge University Press.

Quine, W. (1953). Two Dogmas of Empiricism. In W. Quine, *From a Logical Point of View and Other Essays.* Cambridge, MA: Harvard University Press.

———. (1956/1976). Carnap and Logical Truth. Reprinted in W. Quine, *Ways of Paradox and Other Essays* (revised ed.).

———. (1960). *Word and Object.* Cambridge, MA: MIT Press.

———. (1969). Natural Kinds. In W. Quine, *Epistemology Naturalized.* New York: Columbia University Press.

Ramsey, F. (1929/1990). Theories. In F. Ramsey, *Philosophical Papers*, edited by D. Mellor. Cambridge: Cambridge University Press.

Rey, G. (1983/1999). Concepts and stereotypes. In E. Margolis and S. Laurence (eds.), *Concepts: Core Readings.* Cambridge, MA: MIT Press.

———. (1985). Concepts and conceptions. *Cognition* 19: 297–303.

———. (2005). Philosophical Analyses as Cognitive Psychology. In H. Cohen and C. Lefebvre (eds.), *Handbook of Categorization in Cognitive Science.* Amsterdam: Elsevier.

———. (2008). The analytic/synthetic distinction. *Stanford Encyclopedia of Philosophy.* <http://plato.stanford.edu/entries/analytic-synthetic/>.

———. (2009). Concepts, Defaults, and Internal Asymmetric Dependencies: Distillations of Fodor and Horwich. In N. Kompa, C. Nimtz, and C. Suhm (eds.), *The A Priori and Its Role in Philosophy.* Paderborn: Mentis, 185–204.

Rosch, E. (1973). On the Internal Structure of Perceptual and Semantic Categories. In T. E. Moore (ed.), *Cognitive Development and Acquisition of Language.* Amsterdam: Academic Press.

Samuels, R. (1998). What brains won't tell us about the mind: A critique of the neurobiological argument against representational nativism. *Mind & Language* 13: 548–70.

———. (2002). Nativism in cognitive science. *Mind & Language* 17: 233–65.

———. (2004). Innateness in cognitive science. *Trends in Cognitive Sciences* 8: 136–41.

———. (2007). Is Innateness a Confused Concept? In P. Carruthers, S. Laurence, and S. Stich (eds.), *The Innate Mind, Volume 3: Foundations and the Future.* Oxford: Oxford University Press.

Scholz, B., and Pullum, G. (2006). Irrational Nativist Exuberance. In R. Stainton (ed.), *Contemporary Debates in Cognitive Science.* Oxford: Blackwell.

Segal, G. (2007). Poverty of Stimulus Arguments Concerning Language and Folk Psychology. In P. Carruthers, S. Laurence, and S. Stich (eds.), *The Innate Mind, Volume 3: Foundations and the Future.* Oxford: Oxford University Press.

Siegal, M., and Bergman, A. (2002). Waddington's canalization revisited: Developmental stability and evolution. *PNAS* 9: 10528–32.

Smith, E., and Medin, D. (1981/1999). The Exemplar View (ch. 3 of their *Categories and Concepts*). In E. Margolis and S. Laurence (eds.), *Concepts: Core Readings.* Cambridge, MA: MIT Press.

Sober, E. (1988). Apportioning causal responsibility. *Journal of Philosophy* 85: 303–18.

———. (1998). Innate Knowledge. In W. Craig (ed.), *Routledge Encyclopedia of Philosophy, Volume 4.* London: Routledge.

———. (2001). Separating Nature and Nurture. In D. Wasserman (ed.), *Genetics and Criminal Behavior.* Cambridge: Cambridge University Press.

Stalker, D. (ed.) (1994). *Grue: The New Riddle of Induction,* Chicago: Open Court.

Stalnaker, R. (1984). *Inquiry.* Cambridge, MA: MIT Press.

Sterelny, K. (1989). Fodor's nativism. *Philosophical Studies*, 55: 119–141.

———. (2003). *Thought in a Hostile World: The Evolution of Human Cognition.* Oxford: Blackwell.

Sterelny, K., and Griffiths, P. (1999). *Sex and Death*. Chicago: University of Chicago Press.

Stich, S. (ed.) (1975a). *Innate Ideas*. Berkeley: University of California Press.

———. (1975b). The Idea of Innateness. In S. Stich (ed.), *Innate Ideas*. Berkeley: University of California Press.

Thompson, D. (1917). *On Growth and Form*. Cambridge: Cambridge University Press.

Tooby, J., and Cosmides, L. (1990). The past explains the present: Emotional adaptations and the structure of ancestral environments. *Ethology and Sociobiology* 11: 375–424.

Tooby, J., Cosmides, L., and Barrett, C. (2005). Resolving the Debate on Innate Ideas: Learnability Constraints and the Evolved Interpretation of Motivational and Conceptual Functions. In P. Carruthers, S. Laurence, and S. Stich (eds.), *The Innate Mind, Volume 1: Structure and Contents*. Oxford: Oxford University Press.

Waddington, C. H. (1957). *The Strategy of the Gene*. London: Routledge.

———. (1975). *The Evolution of an Evolutionist*. Ithaca, NY: Cornell University Press.

Waters, K. (2007). Causes that make a difference. *Journal of Philosophy*, 104: 551–79.

Wendler, D. (1996). Innateness as an explanatory concept. *Biology and Philosophy* 11: 89–116.

Wilkin, A. (2003). Canalization and Genetic Assimilation. In B. Hall and W. Olson (eds.), *Keywords and Concepts in Evolutionary Developmental Biology*. Cambridge, MA: Harvard University Press.

Wittgenstein, L. (1953). *Philosophical Investigations*. New York: Macmillan.

Yang, C. (2004). Universal grammar, statistics or both? *Trends in Cognitive Sciences* 8: 451–56.

Zalta, E. (2001). Fregean senses, modes of presentation, and concepts. *Philosophical Perspectives* 15: 333–59.

CHAPTER 15

··

THE LANGUAGE FACULTY

··

PAUL PIETROSKI

AND STEPHEN CRAIN

A review essay on the human visual system can safely presuppose that there is such a system, with homologs in other species, and get on with describing what scientists have learned about how retinal images lead to detailed and informative representations of distal scenes. Discussions of the language faculty are more controversial from the outset. It is not obvious what the alleged faculty does, or whether its operations are specific to language. Indeed, it is not clear what language is, or what languages are. So at least initially, it is hard to say what would make a cognitive system specifically linguistic. Nonetheless, while other animals can communicate to some degree, children go though what seems to be a special kind of linguistic metamorphosis. Therefore, as a way of gaining insight in this domain, in this chapter we focus on some remarkable facts about how children acquire languages. Our conclusion is that humans have a language faculty—a cognitive system that supports the acquisition and use of certain languages—with several core properties. This faculty is apparently governed by principles that are logically contingent, specific to human language, and innately determined. Moreover, at least some of these principles are grammatically pervasive. They are manifested in diverse constructions, and they unify linguistic phenomena that are superficially unrelated.[1]

---

[1] There are many other ways of organizing this kind of discussion. See, e.g., Chomsky (1981, 1986), Fodor (1983), Hornstein and Lightfoot (1981), Pinker (1994), Hauser, Chomsky, and Fitch (2002), Collins (2004), Jenkins (2000), Pinker and Jackendoff (2005), Reinhart (2006). Our primary aim is to illustrate several kinds of arguments, which in our view remain unrebutted, for positing a substantive language faculty; though cf. Cowie (1999), Elman et.al. (1996), Pullum and Scholz (2002). Given space constraints, we focus on facts concerning the relation of syntax to semantics. (For relevant discussions of phonology, see Dresher (2005) and Halle ([2002]). We have also tried to abstract away from various debates among those who assume a substantive language

Every biologically normal child acquires thousands of words and a capacity to understand endlessly many complex expressions, given an ordinary and relatively brief course of experience. Indeed, a child can easily acquire more than one language—for example, Japanese and Mohawk, or English and ASL. As we discuss below, these languages respect certain logically contingent constraints, and each such language is acquirable for the most part by the age of three. This suggests that human children come into the world prepared to acquire languages of a certain sort, much as caterpillars are born ready to grow and acquire the traits of butterflies. Of course, speakers of Japanese and Mohawk differ saliently, in ways that depend on experience. But across the languages that children naturally acquire, there are commonalities that reflect experience-independent aspects of human cognition that seem to be specific to language acquisition. Children may also employ more general capacities when formulating and testing generalizations; acquisition of a language such as English is presumably a complex phenomenon, arising from the interaction of various cognitive systems. But this phenomenon, not manifested in birds or bees or chimpanzees, requires something distinctive that lets human children acquire and use the languages they do.

1. PROCEDURAL MATTERS

We begin with some observations and terminology used by Chomsky (1957, 1965, 1986) and others to motivate a three-part idea. First, children acquire *procedures* for associating signals—such as the sounds of spoken English, or the signs of ASL—with concepts or other interpretations. Second, these procedures, often called I-languages, respect substantive *constraints* on how signals can be associated with interpretations. Third, in many respects acquiring an I-language is *not* a process of generalizing from experience, if only because a child's experience often provides no basis for the relevant generalizations. Instead of basing generalizations on experience, children employ an innately determined procedure that *projects* I-languages in response to their experience. One can think of the language faculty as the biological system that somehow implements this "metaprocedure" that projects procedures for associating human linguistic signals with interpretations. As we will stress, if one wants to know which aspects of human languages are determined by mental architecture that is specific to human language acquisition, one must consider the best candidates for principles that govern the I-languages acquired by children. And this requires, in turn, attention to both linguistic theory and psycholinguistic experiments.

We do not want to argue about what counts as a language. So let us say, over-generously, that a language is anything that associates signals of some kind with interpretations of some kind. And let us say that an expression of a language associ-

---

faculty concerning its nature and the best vocabulary for describing it. While such debates often reveal facts that bolster the case presented here, we want to avoid the impression that arguments in this domain always rely on a particular and tendentious theoretical perspective.

ates a signal (type) with an interpretation (type), allowing for indirect associations. An expression might pair a given sound with a dog, a property, or a concept. And a complex sound might be paired with a complex concept via some compositional instruction. This allows for various conceptions of what languages are: *sets* of expressions, *procedures* or algorithms for generating expressions, physical *implementations* of such procedures, classes of similar sets/procedures/implementations, and so forth. But in any case, one can focus on "finite-yet-unbounded" languages that have infinitely many complex expressions that can be characterized recursively. Such languages can associate endlessly many signals with interpretations, and endlessly many interpretations with signals.

Let a *naturally acquirable human language* (Naturahl) be a finite-yet-unbounded language with two further properties: its signals are overt sounds or signs, and it can be acquired by a biologically normal human child, given an ordinary course of human experience. For reasons that will become clear, we want to assume as little as possible about the relevant interpretations. (Though one might well take them to be mental representations of some kind.) In what follows, we ignore the relatively minor variations in Naturahls that can pervade a community of mutually intelligible speakers. Dialects of English can differ, as can the idiolects of neighbors. But let us idealize and say that common language names, such as "English" and "French," signify Naturahls that are acquired and shared by many speakers.

Any biologically normal human child can acquire any Naturahl, given an ordinary course of experience with users of that language. And different children, who undergo different courses of experience, can acquire the same language. This leaves room for many views about what the acquisition device is. But some aspect of human infant cognition supports the acquisition of implemented *procedures* for associating sounds/signs with interpretations. In acquiring a Naturahl, a child acquires a capacity to understand novel expressions and express novel thoughts. This capacity requires something like an algorithm that determines how complex sounds are associated with complex interpretations in endlessly many cases. The language faculty is posited as providing a procedure for acquiring such algorithms; see Chomsky (1957), Lasnik (1999).

It is useful to follow Chomsky (1986) in applying an extensional/intensional distinction—regarding mappings from inputs to outputs—to the study of Naturahls and the faculty that lets children acquire these languages. We can think of functions as *sets* of ordered pairs (extensions) or *procedures* (intensions) that determine outputs given inputs. Consider the set of ordered pairs $<x, y>$ such that x is a whole number, and y is the absolute value of $x - 1$. This infinite set, $\{\ldots(-2, 3), (-1, 2), (0, 1), (1, 0), (2, 1)\ldots\}$, can be characterized in many ways. One can use the notion of absolute value, and say that $F(x) = |x - 1|$. But one can also use the notion of a positive square root, and say that $F(x) = {}^+\sqrt{(x^2 - 2x + 1)}$. Intuitively, these two characterizations of the same set correspond to different procedures for computing a value given an argument; see Frege (1892), Church (1956). A mind might be able to execute one algorithm but not the other. Chomsky likewise contrasted E-languages with I-languages. An E-language is a set of signal-interpretation pairs, while an I-language is a procedure that pairs signals with interpretations. And if human minds are unable to execute (or acquire) certain procedures, then characterizing Naturahls as *sets* of expressions may fail to capture important generalizations.

The I-languages that children acquire are biologically *implementable*, since they are actually implemented in human biology. This obvious truth is a potential source of theoretical constraint: only implementable algorithms can be I-languages that children can acquire.[2] And as Chomsky and others have stressed, one can adopt this perspective without overworking the analogy with intensions. In particular, a function has a unique value for each argument, but Naturahls admit the possibility of ambiguity. A single sound (type), such as that of "bank," might be paired with more than one interpretation (type). In section two, we discuss more interesting examples of homophony. Moreover, an I-language need not be a procedure that determines a set of "well-formed" expressions that include only the expressions that a competent speaker can understand. For example, speakers of English know that (1) is meaningful but defective.

(1)   *The child seems sleeping

The asterisk indicates that (1) is judged to be anomalous: there is something wrong with using the string of words in (1) as a sentence. By contrast, (2) and (3) are fine English sentences.

(2)   The child seems to be sleeping
(3)   The child seems sleepy

Yet even though (1) is somehow ungrammatical, speakers hear it as having the interpretation of (2) *and not* the interpretation of (3); see Higginbotham (1985). The defect does not preclude understanding (1), which is not incomprehensible word salad such as (4).

(4)   *Be seems child to sleeping the

Such examples challenge the idea that Naturahls can be described in any theoretically interesting way as E-languages. Since (1) is defective, the corresponding signal-interpretation pair is not an element of any set that is plausibly the set of English expressions. Yet when a child acquires English, the algorithm/procedure/I-language acquired associates the sound of (1) with an interpretation, though one that can be signaled "better" with (2). More generally, the acquired I-language may fail to determine *any* set/E-language that can be identified with English. So if the goal is to describe Naturahls and how children acquire them, it seems that theorists must describe Naturahls in I-language terms. From this perspective, the human faculty for language is a biologically implemented device for acquiring I-languages.

---

[2] For Chomsky, "I-" has further connotations. He takes I-languages to be *individualistic*—as opposed to public/conventional languages that a speaker might grasp only imperfectly (cf. Dummett (1986))—and *internalistic*, as opposed to languages that are individuated (externalistically) with reference to things in the mind/language-independent world (cf. Burge (1989)). See George (1989) for further discussion and distinctions.

2. Constrained Homophony

In what follows, we present several reasons for thinking there is such a faculty, implemented somehow in ways determined by the human genome and how it unfolds in embryology. We focus mainly on the fact that children acquire procedures that associate sounds with interpretations in ways that permit lots of ambiguity, while also imposing exacting limits on how ambiguous strings of words can be. In short, ambiguity is ubiquitous but highly constrained in Naturahls. To the extent that constraints on linguistic ambiguity are not due to other cognitive systems, these constraints can reveal features of the posited faculty. And specifically linguistic constraints on ambiguity are good candidates for innate specification, since it is so unlikely that children *learn* that word interpretations *cannot* be combined in ways that would yield more ambiguity than adults permit. To do what children do, given their experience, they need a faculty that projects I-languages in distinctive ways that are largely independent of experience.

To illustrate, let us begin with the string of words in (5). This string can be understood in either of the two ways indicated with (5a) and (5b).

(5)   The man called the woman from Paris
(5a) The man called the woman, and the woman was from Paris
(5b) The man called the woman, and the call was from Paris

In this sense, (5) is a homophone like "bank" or "bear." But (5) is homophonous even holding fixed the interpretations associated with the word sounds. Here, the ambiguity is said to be structural, in that "from Paris" can be understood as modifying the noun "woman" or the verb phrase "called the woman." But the string of words in (5) *cannot* be understood in a third way.

(5c)   #The man called the woman, and the man was from Paris

Note that (5c) can be used to express a coherent thought, which could easily be constructed from concepts indicated with the words in (5), in a way that parallels an intuitive subject-predicate division: the man is an individual x such that x called the woman, and x is from Paris.[3] Yet the word string in (5) cannot have the interpretation of (5c). So (5) is *two-but-not-three* ways ambiguous. We can invent a language in which string (5) has all three readings, or just one. But acquiring English, as opposed to some such invented language, is a matter of acquiring a procedure that is permissive in treating (5) as ambiguous and yet restrictive in a specific way.

This point is pervasive. Given any string of words that has at least one interpretation (in some Naturahl), that string will be *n-but-not-n+1* ways ambiguous, where *n* may be 0. Even unambiguous strings can be interesting, because pairs of superficially similar strings can be unambiguous in different ways. Borrowing a famous

---

[3] That is, the man is both one who called the woman *and one who is* from Texas. Compare (5b), according to which the man called the woman who is from Paris. Similar points apply to "The cook saw the thief with binoculars."

example from Chomsky, (6) can only be understood as in (6a), even though the relation of (6) to (6b) seems no less direct than the relation of (6) to (6a). By contrast, (7) can only be understood as in (7b).

(6)   John is eager to please   (7)   John is easy to please
(6a) John is eager that he please us   (7a) #It is easy for John to please us
(6b) #John is eager that we please him   (7b) It is easy for us to please John

These observations invite the following suggestion: in both (6) and (7), "please" takes a covert subject and a covert object; though because "eager" and "easy" differ semantically, "eager" takes a (nonpleonastic) subject that is also understood as the subject of "please," while "easy" can take a pleonastic subject ("It") or a subject that is understood as the object of "please." Explaining these facts will require interacting assumptions about how "eager" and "easy" are related to the concepts they indicate, and how words can be combined in Naturahls. But whatever the details, it is hard to see how children could figure out what (6) and (7) *cannot* mean without help from a faculty that is largely responsible for these facts—which go unnoticed by most speakers, and which remain puzzling even for trained linguists with access to lots of data.

It is striking that speakers *agree* about many such unambiguities, once they are prompted to reflection. Repeatedly, and despite considerable variation in experience, children in English-speaking countries grow up to be adults for whom (6–7) are unambiguous in the particular ways that these strings are unambiguous for other competent speakers of English. One can speculate that this regularity of acquisition is, somehow, a by-product of a general learning mechanism applied to the sound-interpretation pairs that any normal child acquiring English will encounter. But absent at least a sketch of a proposal, such speculations are implausible, given that Naturahls allow for lots of ambiguity. For it is hard to learn from experience that a string *cannot* have an additional (coherent and composable) interpretation.

Indeed, a string of words can have an interpretation that seems crazy, compared with an equally composable interpretation that the string cannot have. Consider (9), which must be understood as the bizarre question (10), as opposed to the more reasonable question (11).

(9)   Was the boy who fed the waffles fed the horses?
(10) Yes-or-No: the boy who fed the waffles was fed the horses?
(11) #Yes-or-No: the boy who was fed the waffles fed the horses?

In (9), the auxiliary verb "Was" is construed as related to "fed" in the *main* clause, corresponding to the passive phrase "was fed the horses" in (10)—as opposed to "fed" in the *relative* clause, corresponding to the more expected passive "was fed the waffles" in (11); compare (12).

(12) Was the boy who fed the horses fed the waffles?

Even if one knows that the boy fed the horses, and that he was fed the waffles, (9) remains perversely unambiguous. And this is not because Naturahls abhor ambiguity.

Still, one wants to know if "negative" facts concerning the absence of readings are restricted to the I-languages of adults. For if so, one might suspect that some kind of experience-dependent learning procedure—as opposed to an innate faculty largely responsible for linguistic metamorphoses in human children—plays a major explanatory role in accounting for the emergence of such facts in human languages. So it is important to know the ages at which children recognize various kinds of nonambiguity. And at least in many cases, children seem to be fully competent (adult-like) speakers in these respects by the age of three. If this is correct, it considerably narrows the window of opportunity for learning, thereby bolstering the case for an innate and substantive acquisition faculty. We end this section with one cluster of examples, though the literature contains many others.[4]

Young children know that pronouns such as "he" and "him" can have deictic or anaphoric interpretations, and that pronouns like "himself" must be understood as anaphoric (i.e., as having an antecedent in the same sentence). But more interestingly, children know that certain anaphoric interpretations are impossible. In (13), the antecedent of "himself" must be "Grover," while the antecedent of "he" cannot be "Grover." These facts can be represented as in (14), ignoring any deictic readings of pronouns, with coindexing indicating anaphoric dependence.

(13)   Kermit said he thinks Grover should wash himself
(14)   Kermit$_1$ said he$_{1/*2}$ thinks Grover$_2$ should wash himself$_{*1/2}$
In (15) the antecedent of "him" cannot be "Grover",
(15)   Kermit$_1$ said he$_{1/*2}$ thinks Grover$_2$ should wash him$_{1/*2}$

though the antecedent can be "he"/Kermit". Attending to many such examples might suggest a generalization: an antecedent for "himself" must be associated with a "nearby" referential expression, perhaps within the smallest sentential clause containing the pronoun, while "he" and "him" cannot take a nearby antecedent. One can test this hypothesis by presenting adult speakers with unusual constructions such as (16) and (17). But as indicated, the facts suggest a subtler generalization.

(16)   Kermit$_1$ expected to feed Grover$_2$ and wash himself$_{1/*2}$
(17)   Kermit$_1$ expected to feed Grover$_2$ and wash him$_{1*/2}$

In (16–17), the possibilities for "him" and "himself" *reverse* the possibilities in (14–15). Linguists can use such data to revise their theories. But it seems unlikely that all children who acquire English *encounter and use* such data in determining how pronouns can/cannot be understood. Yet three-year-olds know which interpretations are available for examples such as (14–17). This suggests that children never consider superficially simpler algorithms for interpreting pronouns.

To be sure, young children cannot just report that a string of words fails to have a certain reading. But in suitably constructed experiments, their behavior reveals an

---

[4] See, e.g., Crain (1991); Crain and Pietroski (2001, 2002); Crain and Thornton (1998), Crain, Goro, and Thornton (2006); Guasti (2002); Goodluck (1991); Pietroski and Crain (2005).

I-language that is adult-like with regard to which interpretations are allowed and which are not allowed. By age three, children are adept at saying whether a puppet has correctly or incorrectly described a scenario that has just been played out. If the puppet uses an ambiguous string that is true on a salient reading, but false on another reading, children (like adults) say the puppet was right. Other things equal, children say the puppet was wrong only when they understand the sentence as being false on each relevant reading. By constructing scenarios appropriately, one can have the puppet say something that is *false* on every interpretation available for adults, but *true* on a logically possible interpretation that would be especially *salient* if it were available. If children consistently say the puppet is wrong in such cases, this is evidence that children do not assign the "extra" interpretation to the string. And if children consistently demonstrate adult-like competence in appreciating but not overgenerating ambiguities, this is evidence that the I-languages of children are adult-like in this respect. (See Crain and Thornton (1998).)

As an illustration, we consider one study involving examples such as (18–20).

(18)   The Ninja Turtle$_1$ danced while he$_{1/2}$ ate pizza.
(19)   While he$_{1/2}$ ate pizza, the Ninja Turtle$_1$ danced
(20)   He$_1$ danced while the Ninja Turtle$_{*1/2}$ ate pizza.

Note that in (20), "He" cannot take "the Ninja Turtle" as its antecedent. Crain and McKee (1985) began their investigation of children's knowledge of this constraint by first showing that children do not rule out anaphoric relations between pronouns and potential antecedents in examples such as (19). Such strings were presented in two contexts. For example, in one context the Ninja Turtle was dancing and eating pizza; in a second context, another salient male character was eating pizza while the Ninja Turtle was dancing. Children accepted (19) in both contexts about two-thirds of the time. This established a baseline for how often children permit anaphoric relations absent a linguistic constraint.

The same children were then tested on sentences such as (20). The crucial contexts corresponded to the illicit referential dependence—for example, a situation in which the Ninja Turtle was dancing and eating pizza, but another salient male character refused to dance. If children respect a constraint *prohibiting* anaphora in (20), one would expect them to *reject* (20) as a description of this situation. But if children permit an anaphoric link between "He" and "the Ninja Turtle," one would expect them to accept the sentence about two-thirds of the time, as they did in response to (19). In fact, children overall (average age 4;2) *rejected* examples such as (20) 88 percent of the time in these contexts. Even the youngest children (n=7, average age 3;1) rejected examples such as (20) 79 percent of the time. This strongly suggests that, by age three, children already respect the relevant constraint on anaphoric relations.[5]

---

[5] In experiments of this kind, children almost never reject sentences 100 percent of the time. A "noise" rate of about 10 percent is standard and hardly surprising. When children are confused by a question, they are more likely to say "yes" than "no." Thus, 90 percent rejection is very good evidence that children understand the puppet's claim as wrong on any salient reading.

As we indicated at the outset, another property of the human faculty for language is its conformity to "deep" principles that unify phenomena that, on the surface, appear to be unrelated. For example, Chomsky (1981) argues that the same constraint against coreference manifested in (20)—repeated below—is related to the striking contrast between (21) and (22).

(20)   He$_1$ danced while the Ninja Turtle.$_{1/2}$ ate pizza.
(21)   Who did he say Luisa had criticized
         for which x: he said Luisa had criticized x
(22)   Who said he had criticized Luisa
         for which x: x said x had criticized Luisa

In (21), the pronoun "he" must be used deictically, to refer to some male individual. But the pronoun can be interpreted as bound in (22), where it is anaphorically linked to "Who." As Chomsky (1981) also points out, the contrasts in (18–20) and in (21–22) are apparently related to the contrast illustrated in (23) and (24). Again, the pronoun "he" must be used deictically in (23). But in (24), "he" can be bound by the quantificational expression "everybody."

(23)   He said Luisa had criticized everyone
         for every x: he said Luisa had criticized x
(24)   Everybody said he had criticized Luisa
         for every x: x said x had criticized Luisa

These facts about the licensing of anaphoric relations are in turn unified by the structural notion of c-command: anaphors/variables are c-commanded by their antecedents/binders. Think of phrase markers as sets of points partially ordered by a dominance relation. Then at least to a first approximation, one node c-commands another iff every node that dominates the first also dominates the second. In the following tree, node 6 c-commands each of 1–5; 4 c-commands 1–3; 1 and 2 c-command each other.

```
                    7
                   / \
                  6   5
                     / \
                    4   3
                       / \
                      2   1
```

This structural notion figures in the description and explanation of *many* linguistic phenomena, including those illustrated with (6–24). And experimental studies reveal childrens' sensitivity to relations of c-command; see below. This makes sense *if* children have a language faculty that plays two roles: first, it leads children to hear strings such as (6–24) as structured expressions whose constituents exhibit such relations; second, it treats these relations as semantically important, in a way that severely constrains how children can associate the syntactic structures with interpretations. To the extent that domain general learning mechanisms fail to detect the abstract structural property of c-command, or fail to treat this property as interpretively relevant in the right ways, these mechanisms fail to provide a unified account of the various phenomena tied together by c-command and the corresponding facts about acquisition.

In which case, the conclusion invited is that these abstract structures and constraints reflect the operation of a human language faculty. Of course, this is only the start of an explanation. One wants to know how human biology implements such a faculty and what, exactly, is implemented. But ignorance on this score is no argument that the constraints are, despite evidence to the contrary, somehow acquired in response to experience.[6]

3. PATTERNS IN PATTERNS

We assume that the language faculty is not a mere hodgepodge of constraints that happen to be enforced by the biological mechanisms children use to associate signals with interpretations. If current theories sometimes make it look this way, that is because linguists do not yet know what the *basic* operations of the language faculty are, or how these operations interact to yield what may seem to be a disparate assortment of constraints. Put another way, if children have a distinctively human and specifically linguistic faculty of language, it presumably employs a small number of basic operations that manifest in various ways (e.g., as constraints stated in terms of c-command).[7] In this section we further illustrate the more general point, beginning with some humble observations about logical connectives and quantificational words.

---

[6] Much discussed "poverty of stimulus arguments"—for recent reviews, see Crain and Pietroski (2001), Laurence and Margolis (2001), and the replies to Pullum and Scholz (2002)—can reveal that the facts to be explained have their source in a substantive language faculty, whose nature remains unclear, as opposed to general learning mechanisms whose nature remains unclear.

[7] In terms of the evolutionary time scale and underlying biology, the difference between humans and our nearest primate cousins is small. So somehow, a small and recent change had dramatic effects; see Hauser, Chomsky, and Fitch (2002), Hornstein and Pietroski (2009).

Before children are three, they know that the sound of "or" signifies *inclusive* disjunction (Crain and Khlentzos 2008; see also Chierchia et al. 2004; Crain, Gualmini, and Meroni 2000; Crain, Goro, and Thornton 2006). To be explicit, let the ampersand and wedge have their usual meanings: "*P & Q*" is true iff both "*P*" and "*Q*" are true; "*P v Q*" is false iff both "*P*" and "*Q*" are false. Then "*v*" signifies inclusive disjunction. Let '°' signify exclusive disjunction: "*P ° Q*" is true iff "*P v Q*" is true and "*P & Q*" is false. Examples such as (29), as naturally used by speakers of English, might initially suggest that (the sound of) "or" has the exclusive interpretation of "°."

(29)   You may have cake or ice cream

But (30) is *not* understood as an instruction to refrain from an exclusive disjunction.

(30)   Don't kick the dog or pull its tail

Contexts of negation—which, as we will see, are special cases of a more general pattern—strongly suggest that for adults, the sound of "or" is associated with the inclusive interpretation of "*v*." Intuitions to the contrary, prompted by examples such as (29), can be diagnosed as reflections of pragmatic implicatures as in Grice (1975). And this familiar point about adult comprehension raises questions about acquisition.

One can imagine a language with a sentential connective that sounds like "or," yet has the interpretation of "°." Indeed, utterances of sentences such as (29) might well confirm the hypothesis that English is such a language for any children who considered this hypothesis without yet encountering enough disconfirming examples such as (30).[8] Nonetheless, as noted above, young children know that "or" has an inclusive meaning. In an illustrative and ongoing long-term study of four 2-year-olds, we have presented children with negated disjunctions and recorded their responses (behavioral and verbal). On a typical trial in one condition, children were shown three dogs: a white dog, a brown one, and a black one. Kermit the Frog, who was being manipulated by the experimenter, indicated that he wanted to play. The experimenter held up the three dogs, and Kermit said, "I don't want to play with the white dog or the brown dog." If children understand "or" as inclusive disjunction, then one expects them to consistently give Kermit the black dog—as opposed to sometimes giving Kermit the white and brown dogs. (Negating an exclusive disjunction would signify that Kermit did not want *just one* of the white and brown dogs.) Responses are adult-like, as expected.

---

[8] If one starts with the *concept* of inclusive disjunction, constructing a concept of inclusive disjunction to associate with "or" would not be hard, compared with children's other cognitive accomplishments. And even if exclusive interpretations are implicatures, they must be represented somehow, raising the question of why children do not directly/semantically associate such representations with "or." See Chierchia and McConnell-Ginet (2000) for some introductory discussion that bridges Grice (1975) to relevant literature in linguistics.

In another condition, negated disjunctions were used in Wh-questions such as "Who doesn't have A or B?" On a typical trial, various characters were introduced: some with yo-yos, some with sponge balls, and some with strawberries. The question posed to children might be "Who doesn't have a yo-yo or a sponge ball?" One of the youngest children, at age 2;3, consistently responded in conformity with the conjunctive entailment—identifying characters who failed to have a yo-yo *and* failed to have a sponge ball—beginning on the very first trial. Other children produced consistent adult-like responses later than this. But all four children demonstrated knowledge by age 2;10 that negated disjunctions yield conjunctive entailments.

The transcripts of parental input suggest that children get little evidence that disjunction is inclusive. The vast majority of the input is consistent with an exclusive interpretation. So if this interpretation were possible in human languages, many children should adopt it. Yet two-year-olds have already concluded that the disjunction word in English has the inclusive meaning of "*v*." This is already striking. Moreover, given this conclusion, children must treat "or"-statements in *unnegated* contexts such as (29) as having a more restrictive "secondary" interpretation: (P v Q) & not(P & Q). So they seem to be *presupposing* some kind of semantics/pragmatics distinction in their understanding of "or"-statements. And they seem to presuppose that the core meaning for a disjunction word is inclusive.

One can imagine a different linguistic scheme in which "or" has the exclusive meaning of "°," and *negated* statements have a more restricted secondary meaning: not(P ° Q) & not(P & Q). A speaker of such a language would know that (30) fails to semantically entail that (just) kicking the dog is disallowed, while also knowing that an utterance of this sentence pragmatically implicates that each of the two actions is disallowed. But this is not how English works.[9] Children know this: the secondary implications are associated with the unnegated grammatical context. And this fact, concerning children's knowledge, has wider implications.

For adult speakers of English, "Ted didn't order sushi or pasta" typically has a conjunctive entailment: Ted didn't order sushi, *and* Ted didn't order pasta. We doubt that children note such facts and use them to acquire English. But evidence of an inclusive meaning for disjunction words seems to be even poorer in other languages. For example, adult speakers of Japanese do not judge (31) to have conjunctive entailment. Rather, adult speakers of Japanese hear (31) as the claim that Ted did not order sushi *or* Ted did not order pasta (not-S v not-P).

---

[9] For example, pragmatic implications are cancelable. One can say "He sang or danced, and he may have done both" without contradiction. And if you bet that Chris will sing or dance, you win if Chris does both. (Likewise, if you promise to sing or dance, you keep your word if you do both.) But it is a contradiction to say "He didn't kick the dog or pull his tail, but he may have done both." And if the sign says "No parking or loitering," you cannot beat the ticket by saying that you parked and loitered.

(31)   Ted ga sushi ka pasuta o tanomanakatta.
        Ted NOM sushi or past ACC order-NEG-PAST
        'it's sushi or pasta that Ted did not order (but I don't know which)'

Yet in contrast to adults, Japanese-speaking children interpret disjunction in simple negative sentences as having conjunctive implications, like English-speakers (young and old). Using a Truth Value Judgment Task, Goro, and Akiba (2004) tested thirty Japanese-speaking children (mean age 5;3) as well as a control group of Japanese-speaking adults. On a typical trial, subjects were asked to judge whether (31) was an accurate description of a situation in which Ted had eaten sushi but not pasta. Japanese-speaking adults uniformly accepted the target sentences, whereas the overwhelming majority of Japanese-speaking children consistently rejected them. To repeat, Japanese-speaking children interpreted negated disjunctions as licensing conjunctive entailments, whereas local adults do not. This is further evidence that across languages, children understand disjunction words inclusively, despite experience that invites exclusive interpretation.

This raises the question of why Japanese adults hear (31) as they do. But once again, we think the answer lies with the nonobvious grammatical structure projected by the language faculty, as opposed to learning from experience. Adults understand (31) with the disjunction operator "ka" having scope over the negation, despite surface appearances. In this respect, "ka" in (31) is like "some" in (32), on the reading that *does not* imply that Ted did not eat *any* kangaroo.

(32)   Ted did**n't** eat **some** kangaroo.
        Possible Meaning: There is **some** *kangaroo that Ted did**n't** eat*

On the indicated reading, "some" is a positive polarity item (PPI): an expression interpreted as having scope *over* a negation in the same clause.[10] But negation in a higher clause still takes scope over a PPI. In (33), the negation clearly has widest scope.

(33)   You did**n't** convince me that Ted ate **some** kangaroo.
        *Mandatory Meaning: You did**n't** convince me that Ted ate* **any** *kangaroo.*

So if the Japanese disjunction-word "ka" is a positive polarity item (for adults), then "ka" should generate a conjunctive entailment in sentences with negation in a higher clause, as in (34).

---

[10] See Szabolcsi (2002) for a discussion of PPIs, building on a traditional discussion of *negative* polarity items like "any" discussed below (see Ladusaw (1996) and Ludlow (2002)). Note that "Ted ate any kangaroo" is odd, and "Ted did**n't** eat **any** kangaroo" is unambiguously the negation of "Ted ate some kangaroo," suggesting that "any" is best interpreted in the scope of negation. This makes the PPI reading of "Ted did**n't** eat **some** kangaroo" noteworthy.

(34)   Gen ga Ted ga sushi ka pasuta o tanomu no o minakatta
       Gen NOM Ted NOM sushi or pasta ACC order-Prt Nmlzr ACC see-NEG-PAST
       (Prt: Present, Nmlzr: Nominalizer)
       'Gen didn't see Ted order sushi or pasta'

And this is the case. In both Japanese and English (as with all other languages, as far as we know), disjunction words in the scope of negation generate conjunctive entailments. But as "ka" reveals, this generalization can be obscured in the simple constructions that are likely sources of data for children. The generalization may be "manifest" only in structurally complex examples such as (34).

It is *very* unlikely, however, that children acquiring Japanese infer that "ka" signifies inclusive disjunction based on exposure to sentences such as (34). Such sentences are too exotic to ensure that every language learner is exposed to sufficiently many to guarantee convergence on the local adult grammar. So it seems that Japanese children understand "ka" as inclusive disjunction, despite input from parents who treat "ka" as a PPI in simple negative sentences. Human language acquisition is, for whatever reasons, constrained in this way. And this has cascading effects for other aspects of interpretation. In the next section, we develop this point to illustrate how appeal to an innate language faculty can help explain why apparently disparate phenomena are not acquired separately.

4. RELATED GENERALIZATIONS

As noted by medieval logicians, quantificational expressions such as "every," "some," and "no"—which combine with a predicate like "dog" to form a phrase that can combine with a predicate like "barked" to form a sentence—can be characterized in terms of the inferences they license between pairs of predicates such that one is (known to be) more restrictive than the other. Consider "dog" and "brown dog." If some brown dog barked, then some dog barked. The direction of valid inference is from the more restrictive to the less restrictive predicate: replacing "brown dog" with "dog" preserves truth. Likewise, if some dog barked loudly, then some dog barked. The inference is from the more restrictive "barked loudly" to the less restrictive "barked."

Replacing "some" with "no" reverses this pattern: if no dog barked, then no brown dog barked. Put another way, "no" inverts the "default" direction of inference, so that replacing "dog" with "brown dog" preserves truth. Interestingly, "every" is like "no" with regard to the first predicate, and like "some" with regard to the second predicate: if every dog barked, then every brown dog barked; and if every dog barked loudly, then every dog barked. A grammatical context is said to be *downward entailing* if that context licenses inferences from *less to more* restrictive predicates. So "every" is said to be downward entailing in its first (nominal) argument position, while "no" is downward entailing in both argument positions. Children know such facts.[11] But even more interestingly, their acquisition of Naturahls respects correlations between other linguistic properties.

One phenomenon correlated with "downward entailingness" involves so-called negative polarity items such as "ever" and "any." Negative polarity items (NPIs) are expressions that must appear in downward entailing contexts, as illustrated in (35), with the first predicates in bold.

(35)   No/Every **cow that ever ate a vegetable** became ill
(36)   No **cow that became ill** ever ate a vegetable
(37)   *Some **cow that ever became ill** ate a vegetable
(38)   *Some/Every **cow that became ill** ever ate a vegetable

A second phenomenon involves the interpretation of disjunction words. Consider (39) and (40).

(39)   Every **cow that ate broccoli or asparagus** became ill
(40)   Every **cow that became ill** ate broccoli or asparagus

In (39), the disjunction is in the downward entailing argument of *every*, and there is a conjunctive implication: every cow that ate broccoli became ill, *and* every cow that ate asparagus became ill. In (40), the disjunction is not in a downward entailing argument, and there is no conjunctive implication.

This trio of facts—concerning direction of inference, NPI licensing, and the implicational effect of disjunction—presumably has a common source. But one can imagine language acquirers who figure this out late, if at all, after passing through stages at which only one or two of the generalizations are respected. To the extent that domain general learning mechanisms allow for this kind of "partial" knowledge of linguistic constraints, such mechanisms are unlike children. (Compare the earlier remarks about c-command; see also Marcus (1998).)

Using Truth Value Judgment tasks, several studies have shown that children have an adult-like understanding of disjunction in scope of "every." In one study by Gualmini, Meroni, and Crain (2003), four- and five-year-old children consistently accepted (41) as a description of a situation in which each woman bought eggs, and none of the women bought bananas; but the same children rejected (42) in a scenario in which women who bought eggs received a basket, but women who bought bananas did not.

(41)   Every **woman** bought eggs or bananas
(42)   Every **woman who bought eggs or bananas** got a basket.

This asymmetry in responses reveals children's knowledge of the corresponding asymmetry in the argument positions of "every": downward entailing *only* with

---

[11] See Crain, Gualmini, and Pietroski (2005) for a discussion of experiments; and see note 10.

respect to the first (nominal) argument position. Such experiments also confirm that children understand "or" inclusively. And this connects with the fact that "or" is often pragmatically associated with an exclusive-disjunction implicature—but *not* in downward entailing environments.

A domain general learning procedure might help children learn the environments in which NPIs *can* appear. But acquiring the constraint on where such expressions *cannot* appear is another matter. Unless such a procedure reflects whatever deep property unites the various phenomena related to downward entailment, there is no reason to expect such a mechanism to yield acquisition of an I-language that relates NPIs to contexts in which disjunction words have conjunctive implications, yet NPIs are restricted from appearing in contexts in which disjunction words lack conjunctive implications.

We conclude this section with one last point about disjunction and how its interpretation is related to c-command. The point is simple, theoretically, but striking from the perspective of acquisition, since it seems so unlikely that a child would ever "find" the generalization. As illustrated in (43–46), a disjunction word generates a conjunctive entailment only if c-commanded by a downward entailing expression.

(43)   The news that George won did**n't** surprise Karl **or** Jeb
(44)   The news that George did**n't** win surprised Karl **or** Jeb
(45)   The news that George won did**n't** surprise any of the judges
(46)   *The news that George did**n't** win surprised any of the judges

In (43) but not (44), negation c-commands disjunction, and in (43) but not (44), there is a conjunctive entailment. Indeed, in (44) there is an exclusive pragmatic implicature. Likewise, the NPI "any" is licensed in (45) but not (46).

In a study of five-year-olds, Gualmini and Crain (2005) had a puppet use sentences such as (47) or (48) as a description of a story about two girls who had each lost a tooth.

(47)   The girl who stayed up late did **not** get a dime **or** a jewel.
(48)   The girl who did **not** go to bed got a dime **or** a jewel.

Disjunction resides in the scope of "not" in (47). But in (48), "not" is embedded in a relative clause, and so fails to have scope over "or." This difference in structure results in a difference in interpretation. For both adults and children, (47) has a conjunctive entailment, while (48) does not. In the story, one girl went to sleep, but one girl stayed up to see the tooth fairy. The girl who was asleep received both a dime and a jewel from the tooth fairy, but the girl who had stayed awake was only given a jewel. At the end of the story, 87 percent of children presented with (48) accepted the sentence. By contrast, 92 percent of children presented with (47) *rejected* the sentence, on the grounds that the girl who stayed up late had received a jewel.

For these children, (47) has a conjunctive entailment. This requires sensitivity to the fact in (47), "not" c-commands "or," which signals inclusive disjunction. This is a lot to learn, all at once. But such facts are unsurprising if children have a language faculty that leaves them with no other interpretive options. But if children are born able to adopt other linguistic options, and must somehow learn that these options happen not to be exploited in English, then the rapid convergence on adult grammar is very surprising indeed.

5. LIMITED VARIATION

Given a human faculty for language, one expects many aspects of adult grammar to be determined independent of experience and manifested at an early age. From a theoretical perspective, innate linguistic principles define a space of possible human languages: a space the child explores, influenced by her environment, until she stabilizes on a grammar equivalent to that of adults in her linguistic community. Languages outsisde this space will go "untried." But correlatively, at any stage of acquisition, children are employing a possible state of the language faculty (a possible I-language), just not the one being used by local adults. So even if the known adult grammars constitute only some of the possible human grammars, one expects to find children trying out grammars with features found in adult languages elsewhere on the globe. If this expectation is confirmed, it provides dramatic support for a human language faculty that determines the linguistic options available to children. We conclude this chapter by briefly noting an example of children acquiring an I-language that diverges from those of local adults. For even if such cases are in some sense unusual, they provide vivid evidence of children projecting I-language in accordance with constraints imposed by their language faculty and in the absence of relevant experience.[12]

Using an elicited production task, Thornton (1990) evoked wh-questions from three- and four-year-old children. Thornton found that about one-third of these children (of English-speaking parents) she interviewed consistently inserted an "extra" Wh-word in their long-distance questions, as illustrated in (49).

(49) What do you think what's in the box?

For example, a child might use (49) repeatedly when prompted to inquire after a certain puppet's views concerning the contents of the box. (The experimenter always says "Ask him what he thinks is in the box," with only one occurrence of "what.") But interestingly, children who diverge from adult usage in this way do *not* produce utterances such as (50) or (51).

---

[12] In cases of creolization, children acquire I-languages that have the grammatical properties of an ordinary Naturahl, despite experiences with adults who use a pidgin language that lacks many of these properties. This shows that children can project an I-language that is not already implemented in their linguistic community, and such projection cannot be a species of *learning* in any traditional sense. But discussion of creoles would take us too far afield. Here, we offer an example closer to home of children projecting an I-language not implemented by any adults.

(50)   Which Smurf do you think *which Smurf* is wearing roller skates?
(51)   Who do you want *who* to win?

This suggests that the children in question have an I-language that lets them form questions with a "medial wh-word" subject to a constraint. And this suggestion is bolstered by the fact that many adults have such an I-language. For example, (52) is acceptable in Bavarian; see McDaniel (1986). But (53), with a medial wh-*phrase*, is defective.

(52)   Wer glaubst du wer nach Hause geht?
       Who do you think who goes home
(53)   *Wessen Buch glaubst du *wessen Buch* Hans liest?
       Whose book do you think *whose book* Hans is reading?

   Likewise, one cannot have a medial subject of an infinitive. In a question such as (54), the question word appears only once.

(54)   Wen versucht Hans anzurufen?
       Who is Hans trying to call?

The contrasts in (52–54) raise questions about how children who acquire medial-wh languages figure out when *not* to use a medial wh-expression. But even if children could learn the relevant constraints, given experience with those who use a medial-wh language, it seems impossible for children to learn a constraint on a construction that does not exist in the local adult language.

   Instead, it seems that the language faculty makes it possible to acquire an I-language that permits questions with a medial-wh, even if one does not encounter such questions. Yet whatever I-language one acquires, that way of associating signals with interpretations is subject to whatever constraints the language faculty imposes on these procedures. And for whatever reason, expressions such as (50–51) and (53) are verboten. From this perspective, it is not surprising that when American children use a medial-wh, they are governed by the same constraints as Bavarian children. But if one does not assume that all children project I-languages in accordance with constraints imposed by a distinctive faculty, then it is quite surprising that children who use (49) *do not* use (50) or (51), even when prompted to do so. For further examples, see Thornton (1996), Crain and Thornton (1998), and Crain, Gualmini, and Pietroski (2005).

6. Conclusion

We have tried to illustrate, with examples that may have some independent interest for philosophers, that humans have a faculty for acquiring procedures (I-languages) that associate linguistic signals with interpretations in constrained ways, where these constraints are not reflections of more general constraints on thought and/or the experience that leads children to acquire a particular procedure. The constraints, which lie at a far remove from ordinary perception, govern superficially disparate constructions. Yet they are respected by young children. As Chomsky and many others have long argued, this suggests that language acquisition is made possible by a distinctive human faculty for generating I-languages.

REFERENCES

Burge, T. (1989). Wherein Is Language Social. In A. George (ed.), *Reflections on Chomsky*. Blackwell: Oxford.

Chierchia, G., Guasti, M. T., Gualmini, A., Meroni, L., and Crain, S. (2004). Semantic and Pragmatic Competence in Children and Adult's Interpretation of "or." In I. Noveck and S. Wilson (eds.), *Experimental Pragmatics*. Palgraves: London.

Chierchia, G., and McConnell-Ginet, S. (2000). *Meaning and Grammar: An Introduction to Semantics*, 2nd ed. Cambrudge, MA: MIT Press

Chomsky, N. (1957). *Syntactic Structures*. The Hague: Mouton.

Chomsky, N. (1965). *Aspects of the Theory of Syntax*. Cambridge, MA: MIT Press.

Chomsky, N. (1981). *Lectures on Government and Binding*. Dordrecht: Foris.

Chomsky, N. (1986). *Knowledge of Language: Its Nature, Origin and Use*. New York: Praeger.

Church, A. (1956). *Introduction to Mathematical Logic*. Princeton, NJ: Princeton University Press.

Collins, J. (2004). Faculty disputes. *Mind and Language* 19: 503–33.

Cowie, F. (1999). *What's Within: Nativism Reconsidered*. New York: Oxford University Press.

Crain, S. (1991). Language acquisition in the absence of experience. *Behavioral and Brain Sciences* 4: 597–650.

Crain, S., Goro, T., and Minai, U. (2007). Hidden Units in Child Language. In A. Schalley and D. Khlentzos (eds.), *Mental States: Nature, Function and Evolution*. Amsterdam: John Benjamins.

Crain, S., Goro, T. and Thornton, R. (2006). Language acquisition is language change. *Journal of Psycholinguistic Research* 35: 31–49.

Crain, S., and Khlentzos, D. (2008). Is logic innate? *Biolinguistics* 2(1): 24–56.

Crain, S., Gualmini A., and Meroni, L. (2000). The acquisition of logical words. *Logos and Language* 1: 49–59.

Crain, S., Gualmini, A. and P. Pietroski (2005). Brass tacks in linguistic theory: Innate grammatical principles. *Proceedings of the First Annual AHRB Conference on Innateness and the Structure of the Mind*. New York: Oxford University Press.

Crain, S., and C. McKee (1985). The Acquisition of Structural Restrictions on Anaphora. In S. Berman, J. Choe, and J. McDonough (eds.), *Proceedings of 16th North Eastern Linguistics Society*. Amherst, MA: GLSA, 94–110.

Crain, S., and Pietroski, P. (2001). Nature, nurture and Universal Grammar. *Linguistics and Philosophy* 24: 139–86.

———. (2002). Why language acquisition is a snap. *The Linguistic Review* 19: 163–83.

Crain, S., and Thornton, R. (1998). *Investigations in Universal Grammar: A Guide to Experiments in the Acquisition of Syntax and Semantics*. Cambridge, MA: MIT Press.

Dresher, E. (2005). Chomsky and Halle's Revolution in Phonology, In J. McGilvray (ed.), *The Cambridge Companion to Chomsky*. Cambridge: Cambridge University Press.

Dummett, M. (1986). A Nice Derangement of Epitaphs: Some Comments on Davidson and Hacking. In E. LePore (ed.), *Truth and Interpretation: Perspectives on the Philosophy of Donald Davidson*. Oxford: Basil Blackwell.

Elman, J., Bates, E., Johnson, M., Karmiloff-Smith, A., Parisi, D., and Plunkett, K. (1996). *Rethinking Innateness: A Connectionist Perspective on Development*. MIT Press: Cambridge, MA.

Fodor, J. A. (1983). *The Modularity of Mind*. Cambridge, MA: MIT Press.

Frege, G. (1892). On Concept and Object. In P. Geach and M. Black (trans.), *Translations from the Philosophical Writings of Gottlob Frege* (Oxford: Oxford University Press, 1952).

George, A. (ed.) (1989). How Not to Become Confused About Linguistics. In *Reflections on Chomsky*. Blackwell: Oxford.

Goro, T., and Akiba, S. (2004). The acquisition of disjunction and positive polarity in Japanese. *Proceedings of the 23rd West Coast Conference on Formal Linguistics*. Somerville, MA: Cascadilla Press.

Goodluck, H. (1991). *Language Acquisition: A Linguistic Introduction*. Blackwell, Oxford.

Grice, H. P. (1975). Logic and Conversation. In P. Cole and J. Morgan (eds.), *Syntax and Semantics 3: Speech Acts*. New York: Academic Press.

Gualmini, A., Meroni, L., and Crain, S. (2003). Children's Asymmetrical Responses. In Y.Otsu (ed.), *Proceedings of the Fourth Tokyo Conference on Psycholinguistics*. Tokyo: Hituzi Syobo.

Gualmini, A., and Crain, S. (2005). The structure of children's linguistic knowledge. *Linguistic Inquiry* 36: 463–74.

Guasti, M.T. (2002). *Language Acquisition: The Growth of Grammar*. Cambridge, MA: MIT Press.

Halle, M. (2002). *From Memory to Speech and Back*. Berlin: Walter de Gruyter.

Hauser, M., Chomsky, N., and Fitch, W. (2002). The faculty of language: What is it, who has it, and how did it evolve? *Science* 298: 1569–79.

Higginbotham, J. (1985). On semantics. *Linguistic Inquiry* 16: 547–93.

Hornstein, N., and Lightfoot, D. (eds.) (1981). Introduction to *Explanations in Linguistics: The Logical Problem of Language Acquisition*. London: Longman, 9–31.

Hornstein, N., and Pietroski, P. (2009). Basic operations. *Catalan Journal of Linguistics* 8: 113–39.

Jenkins, L. (2000). *Biolinguistics*. Cambridge: Cambridge University Press.

Ladusaw, W. (1996). Negation and Polarity items. In S. Lappin (ed.), *Handbook of Contemporary Semantic Theory*. Oxford: Blackwell.

Lasnik, H. (1999). *Syntactic Structures Revisited*. Cambridge, MA: MIT Press.

Laurence, S., and Margolis, E. (2001). The poverty of the stimulus argument. *The British Journal for the Philosophy of Science* 52: 217–76.

Ludlow, P. (2002). LF and Natural Logic. In G. Preyer and G. Peter (eds.), *Logical Form and Language*. Oxford: Oxford University Press.

McDaniel, D. (1986). "Conditions on wh-chains," PhD diss. City University of New York.

Marcus, G. (1998). Rethinking eliminative connectionism. *Cognitive Psychology* 37: 243–82.

McGilvray, J. (ed.) (2005). *The Cambridge Companion to Chomsky*. Cambridge: Cambridge University Press.

Pietroski, P., and Crain, S. (2005). Innate Ideas. In J. McGilvray (ed.), *The Cambridge Companion to Chomsky*. Cambridge: Cambridge University Press.

Pinker, S. (1994). *The Language Instinct*. New York: William Morrow.

Pinker, S., and Jackendoff, R. (2005). The faculty of language: What's special about it? *Cognition* 95: 201–36.

Pullum, G., and Scholz, B. (2002). Empirical assessment of the stimulus poverty argument. *Linguistic Review* 19: 9–50.

Reinhart, T. (2006). *Interface Strategies: Optimal and Costly Computations*. Cambridge, MA: MIT Press.

Szabolcsi, A. (2002). Hungarian disjunctions and positive polarity. In I. Kenesei and P. Siptar (eds.), *Approaches to Hungarian* 8: 217–41. Budapest: Akademiai Kiado.

Thornton, R. (1990). "Adventures in Long-Distance Moving: The Acquisition of Complex Wh-questions," PhD diss. University of Connecticut, Storrs.

Thornton, R. (1996). Elicited Production. In D. McDaniel, C. McKee, and H. S. Cairns (eds.), *Methods for Assessing Children's Syntax*. Cambridge, MA: MIT Press.

CHAPTER 16

LANGUAGE IN COGNITION

PETER CARRUTHERS

THIS chapter reviews some of the ways in which natural language might be implicated in human cognition. After some initial ground clearing, it discusses the views of Whorf and Vygotsky, together with some of their contemporary adherents, before discussing some proposals that have been made for the language-dependence of certain classes of concept (for natural kinds, for mental states, and for numbers). The chapter then discusses the alleged role of language in integrating the outputs of different conceptual modules and in realizing so-called "System 2" cognitive processes.

1. INTRODUCTION

Our question in this chapter concerns the degree of involvement—or lack thereof—of natural language in human cognition. In what ways, if any, do human thought processes involve language? To what extent is human thinking dependent upon possession of one or another natural language? The answers that people have returned to these questions range along a spectrum of claims of varying strength, with the variations reflecting changes in the quantifiers and/or the modality with which the claim is made. At one extreme sits the assertion made by some philosophers that it is *conceptually necessary* that *all* thought is dependent upon language. At the other extreme is the claim that all thought is, not only conceptually, but also metaphysically and causally, *independent* of natural language. And in between these two poles lie a multitude of possible claims that *most*, *some*, or *specific types* of thought are dependent upon natural language, where the dependence in question can be conceptual, metaphysical (that is, constitutive), or causal.

It is unclear whether anyone has ever really endorsed the thesis of the independence of thought from language in its most extreme form. For even those who, such as Fodor (1975), picture natural language as but an input-output system for central cognitive processes of thinking and reasoning will allow that there are many thoughts (both tokens and types) that we would never have entertained in the absence of language. Everyone allows that the utterances of other people can have a significant impact on the thoughts that occur to us at any given moment. Hence there are some thought *tokens* that we would never have entertained in the absence of language. And everyone allows that the testimony of other people is the source of many of our beliefs as well. Hence there are some thought *types* that we would never have entertained if we had been incapable of comprehending what people say to us. These obvious points are taken for granted by all parties in the debate.

While the strongest thesis that thought (or all propositionally structured forms of thought)[1] is conceptually dependent upon language has been defended by some philosophers (Davidson 1973, 1975; Dummett 1981, 1989; McDowell 1994), this is not a view that we need to take seriously for the purposes of this chapter. At any rate, we do not propose to do so for such views are not given any credence among cognitive scientists. Not only are carefully considered attributions of thought to nonlinguistic creatures rife within cognitive science, but it is taken for granted that for any given type of thought, it will be an open empirical question whether such thoughts might be entertained by a creature that lacks a natural language (with the trivial exception of thoughts that are explicitly *about* natural language, of course).

The discussion that follows will focus on the space between the two extremes. We shall begin (in Sections 2 and 3) with a discussion of some historically influential claims for certain sorts of dependence of thought upon language, made by Whorf (1956) and Vygotsky (1961) respectively, together with some contemporary variants. We shall then in Section 4 discuss some ways in which specific types of concept might be claimed to be language-dependent. Section 5 discusses the role that language might play in unifying and combining the outputs of different central/conceptual "modules" (Hermer-Vazquez, Spelke, and Katsnelson 1999; Carruthers 2002). Finally, Section 6 considers the role that language might play within so-called "dual systems theories" of human reasoning processes (Evans and Over 1996; Frankish 2004; Carruthers 2009a).

2. OF WHORF AND WHORFIANISM

The zeitgeist of the second quarter of the twentieth century was behaviorism. It was widely assumed that all animal behavior could be explained in terms of conditioned responses to stimuli, and that the same forms of explanation could, ultimately, be

[1] Dummett (1981), for example, allows that animals might be capable of what he calls "proto-thoughts," which lack the conceptual-compositional structure of genuine human thoughts.

extended to explain most if not all of human behavior as well (Watson 1924; Skinner 1957). It was against this background that the anthropologist and linguist Benjamin Lee Whorf made his proposals about the ways in which natural language serves to structure and shape human cognition. (Many of Whorf's articles are collected together in Whorf 1956.) And that background no doubt played a significant role in winning such wide acceptance for Whorf's views.

Whorf, like most other anthropologists before and since, was impressed by the immense variety displayed by human cultures; he was likewise, as a linguist, impressed by the variety of grammatical forms and modes of conceptualization displayed by the world's natural languages. Some languages, for example, have no words for "left" and "right," and instead describe spatial relationships exclusively by means of geocentric coordinates such as "north" and "south," and/or object-centered coordinates such as "between the river and the sea." Some languages, such as English, have multiple color terms, whereas some, such as Dani, have just two terms meaning roughly "light" and "dark." And famously, the Eskimos were supposed to have many more words available to them than do other people for describing types of snow.[2] What Whorf proposed is that these differences have significant effects on the cognitive processes of the people in question, leading them to apprehend the world quite differently.

While Whorf's views continue to be popular in some areas of the social sciences and (especially) the humanities under the banner of "the social construction of reality," they fell into disrepute among cognitive scientists through much of the second half of the twentieth century. In part this resulted from the cognitive revolution in psychology and surrounding disciplines that took place in the early years of this period. And in part it resulted from a well-known experimental study demonstrating the *lack* of influence of color vocabulary on color vision, color memory, and color categorization. Let us comment briefly on each.

Within a behaviorist framework it perhaps did not seem so implausible that an important new set of stimuli (the linguistic utterances of other people) and behavioral responses (one's own speech) should have a reorganizing influence on previously existing input-output pairings. Hence it did not seem implausible that acquiring one sort of language rather than another might make a difference in how subjects apprehend the world more generally. But once we take seriously that the mind is real, and really organized into different faculties for perception, inference, and action, it immediately becomes problematic to understand how and why linguistic structures should have any significant reorganizing effects outside of the language faculty itself.

As for color, the story begins with an influential study by Berlin and Kay (1969), who investigated color vocabulary in a wide array of languages. What they found were systematic relationships suggestive of a set of underlying universals that reflect the fixed structure of the visual system. In particular, as languages introduce additional

---

[2] This last claim has since been discredited. See Martin (1986) and Pullum (1991).

color terms, they always do so in a specific order, suggesting an underlying universal structure of relative color salience. Heider and Olivier (1972) then followed up with an experimental study of color naming and color memory in speakers of English (which has eleven basic color terms) and Dani (which has just two). It turned out, as expected, that English speakers use a far greater variety of color terms when asked to name a set of color chips, but they found no differences between the two groups in their capacity to remember and re-identify a color chip over a thirty-second interval. This seemed to many people to be a decisive refutation of one strand in the Whorfian account of the relationship between language and thought.[3]

Since the early 1990s, however, Whorfianism has been undergoing something of a revival, albeit in a weakened form (Hunt and Agnoli 1991; Lucy 1992a, 1992b; Gumperz and Levinson 1996). What has been argued in this new wave of research is no longer that language has a *structuring* effect on cognition (meaning that the absence of language makes certain sorts of thoughts, or certain sorts of cognitive process, completely unavailable to people). The main claim, rather, is that one or another natural language can make certain sorts of thought and cognitive process more *likely*, and more *accessible* to people.

The basic point can be expressed in terms of Slobin's (1987) idea of "thinking for speaking." If your language requires you to describe spatial relationships in terms of compass directions, for example, then you will continually need to pay attention to, and compute, geocentric spatial relations; whereas if descriptions in terms of "left" and "right" are the norm, then geocentric relations will barely need to be noticed. This might be expected to have an impact on the efficiency with which one set of relations is processed relative to the other, and on the ease with which they are remembered (Levinson 1996). Likewise in respect of motion events, if you speak a language, such as English, that employs an extensive and often-used vocabulary for *manner* of motion ("walk," "stride," "saunter," "hop," "skip," "run," "jog," "sprint," etc.), then you will continually need to pay attention to, and encode, such properties. In languages such as Spanish and Greek, in contrast, manner of motion is conveyed in an auxiliary clause ("He went into the room *at a run*"), and it often goes unexpressed altogether. One might then predict that speakers of such languages would be both slower at recognizing, and poorer at remembering, manner of motion (Slobin 1996). This claim has been subjected to careful experimental scrutiny by Papafragou et al. (2002), however, who are unable to discover any such effects.

Levinson's claims for the effects of spatial language on spatial cognition have also been subject to a lively controversy (Levinson 1996, 2003; Li and Gleitman,

---

[3] Roberson et al. (2000) have more recently undertaken a replication and extension of Heider and Olivier's study, and claim to find a significant influence of language on memory after all. But as Munnich and Landau (2003) point out, the subjects in the Roberson et al. study engaged in overt speech rehearsal of color names during the thirty-second interval before their memory was tested. The task was therefore a verbally mediated one. And that language should have an impact upon verbally mediated tasks is not at all surprising, and lends no support to the Whorfian view that languages have an important effect on *non*linguistic forms of cognition.

2002; Levinson et al. 2002; Li et al. 2005; Papafragou 2007). Let us pull out just one strand from this debate for discussion. Levinson (1996) had tested Tenejapan Mayans—who employ no terms meaning *left* and *right*—on a spatial reversal task. They were confronted with an array of four items on a desk in front of them, and told to remember the spatial ordering of three of the items. They were then rotated through 180 degrees and walked to another table, where they were handed the three items and told to "make them the same." The Mayans turned out to employ geocentric rather than egocentric coordinates when complying with the instruction, just as the hypothesis of "thinking for speaking" would predict.

In the course of their critique, however, Li and Gleitman (2002) point out that the task is plainly ambiguous. The instruction, "make them the same," can mean "lay them out similarly in respect of egocentric space" or "lay them out similarly in respect of geocentric space." (And indeed, Westerners who are given these tasks will notice the ambiguity and ask for clarification.) Li et al. (2005) therefore reasoned that Levinson's results might reflect, not an effect of language upon thought, but rather an effect of language upon language. Since the instruction is ambiguous, subjects are presented with the problem of disambiguating it before they can respond appropriately. And since geocentric descriptions are overwhelmingly more likely in the society to which the Mayans belong, they might naturally assume that the instruction is intended geocentrically, and act accordingly. It does not follow that they would have had any particular difficulty in solving the task in an egocentric fashion if cued accordingly. And for all that the experiment shows, they might routinely deploy egocentric concepts in the course of their daily lives (if not in their daily speech).

To test this, Li et al. (2005) devised a series of unambiguous spatial tasks that admit of only a single correct solution. In one of these, for example, the subjects had to match a card containing two differently sized circles to one of four cards of the same sort, but variously oriented. Once they were familiar with the task, they were allowed to study the card at one table before being rotated 180 degrees and walked to a second table where four cards were laid out for them to match against. But they did this under one of two conditions. In one, the card was covered and carried to the other table while they watched without its orientation relative to Earth being changed. (This is the geocentric condition.) In the other, the card was placed in their hands and covered before they turned around through 180 degrees to face the other table. (This is the egocentric condition.) Contrary to Levinson's predictions, the subjects did just as well or better in the egocentric condition. And when the task demands were significantly increased (as when Li et al. had subjects recall and trace out one particular path through a maze under two conditions similar to those described above), the Mayan subjects actually did significantly *better* in the egocentric condition (80 percent correct versus 35 percent correct; see Papafragou 2007).

Therefore, the claim that different natural languages have differing effects on nonlinguistic cognition is still unproven. While the idea has a certain intuitive plausibility, and while evidence has been presented in its support, it has also been successfully criticized in a number of studies. Whether any sustainable version of weak

Whorfianism will emerge from the ongoing process of testing and debate is open to serious doubt.

3. Vygotsky and Linguistic Scaffolding

At around the same time that Whorf was writing, the Soviet psychologist Lev Vygotsky was developing his ideas on the interrelations between language and thought, both in the course of child development and in mature human cognition. These remained largely unknown in the West until his book *Thought and Language* was first published in translation in 1962 (with portions omitted). This attracted significant attention, and a number of further works were translated through the 1970s and 1980s (Vygotsky 1971, 1978; Wertsch 1981, 1985).

One of Vygostky's ideas concerned the ways in which language deployed by adults can *scaffold* children's development, yielding what he called a "zone of proximal development." He argued that what children can achieve alone and unaided is not a true reflection of their understanding. Rather, we also need to consider what they can do when scaffolded by the instructions and suggestions of a supportive adult. Moreover, such scaffolding not only enables children to achieve with others what they would be incapable of achieving alone, but plays a causal role in enabling children to acquire new skills and abilities. Relatedly, Vygotsky focused on the overt speech of children, arguing that it plays an important role in problem solving, partly by serving to focus their attention, and partly through repetition and rehearsal of adult guidance. And this role does not cease when children stop accompanying their activities with overt monologues, but just disappears inwards. Vygotsky argued that in older children and adults *inner* (subvocal) speech serves many of the same functions.

Many of these ideas have been picked up by later investigators. For example, Diaz and Berk (1992) studied the self-directed verbalizations of young children during problem-solving activities. They found that children tended to verbalize more when the tasks were more difficult, and that children who verbalized more often were more successful in their problem solving. Likewise Clark (1998) draws attention to the many ways in which language is used to support human cognition, ranging from shopping lists and Post-it notes, to the mental rehearsal of remembered instructions and mnemonics, to the performance of complex arithmetic calculations on pieces of paper. And by writing an idea down, for example, you can present yourself with an object of further leisured reflection, leading to criticism and further improvement.

The thesis that language plays such roles in human cognition is not—or should not be—controversial. But in Vygotsky's own work, it goes along with a conception of the mind as being to an important extent socially constructed, developing in plastic ways in interactions with elements of the surrounding culture, guided and supported by adult members of that culture. These stronger views—such as the similar constructionist views of Whorf—are apt to seem implausible when seen from the perspective of contemporary cognitive psychology. But as we shall see in Section 6, a restricted version of them can survive as an account of a certain *level* of thinking and reasoning within the human mind.

4. LANGUAGE-DEPENDENT CONCEPTS

As we noted in Section 1, everyone allows that some types of thought are dependent upon language, at least to the extent that language is needed for, or is de facto the cause of, their existence. There are two ways in which one might seek to strengthen this claim. One would be to maintain that not only the acquisition of some *thought-types*, but also some types of *concept* (and hence entire classes of thought involving those concepts), are dependent upon language. Another would be to maintain that language is not just *necessary* for the acquisition of certain types of thought and/or concept, but is actually *constitutive of* the thoughts/concepts so acquired. Many different proposals of these two kinds have been made in recent decades. Here we shall briefly survey three of the more interesting ones.

4.1. Language-Learning and Kind-Concepts

No one in cognitive science doubts that prelinguistic children possess a great many concepts for the kinds of thing and substance that they encounter. And there is a copious body of evidence that older (language-using) children *essentialize* natural kinds—believing that each natural kind has an underlying "essence" that determines its nature (Gelman 2003). Children think that an object that possesses the essence of a kind belongs to that kind, even if it is very different in appearance and behavior. And likewise they think that something can fail to belong to a kind, despite sharing the superficial features of members of that kind, by failing to have the right sort of inner constitution or essence.

     One suggestion—defended by Xu (2002; Xu, Cote, and Baker 2005)—is that it is the evidence provided by adult naming practices that tells children which of their concepts, from the wider array of kind-concepts that they possess, they should essentialize. This makes a good deal of sense from an evolutionary standpoint, since the naming practices of the adults in a given society can be thought of as representing the accumulated wisdom of the group, especially concerning which patterns of classification of items in the environment can undergird robust inductive inferences from one circumstance to another. Children might then be predisposed, when learning a new noun referring to a natural kind, to assume that its referent has an underlying essence. If they previously lacked any concept for the kind of thing in question, then they form one. But even if they had previously possessed such a concept, what they now do is essentialize it, thereafter assuming that there is an underlying essence to the kind that determines category membership.

4.2. That-Clauses and Theory of Mind

The hypothesis just canvassed concerns a causal role for language in the acquisition of kind concepts. Others have put forward a similar hypothesis concerning the mental-state concepts that lie at the heart of the "theory of mind" or "mindreading" abilities of human beings. More specifically, it has been proposed that it is by engaging in conversation with a partner that children first come to realize that there are different epistemic perspectives on the world besides their own (Harris 1996). And the evidence does suggest that language use boosts the capacity for mindreading, at least (Perner, Ruffman, and Leekham 1994). Moreover, deaf children who are significantly delayed in their acquisition of language show a corresponding delay for mindreading (Peterson and Siegal 1995). However, the existing data are consistent with a view of mental-state concepts as embedded in an innately structured mind-reading faculty or module, whose maturation is boosted by the challenges involved in the interpretation of speech, but whose development is not strictly dependent upon language (Siegal and Surian 2006).

Some have put forward the stronger claim that some mental-state concepts (specifically the concept of *false belief* and its cognates) aren't just causally dependent upon language, but are *constituted* by aspects of the latter (Segal 1998; de Villiers and de Villiers 2000, 2003). The idea is that we only come to be able to think about beliefs, as potentially false representational states of a thinker, by virtue of mastering the clausal structure of natural language *that*-clauses. It is by acquiring competence with such sentences as, "John said that it is cold" and, "Mary believes that it is warm" that children acquire mastery of the concept of false belief; and natural language *that*-clauses remain constitutive of such mastery thereafter.

There exists powerful evidence against this strong constitution-thesis, however. For there are cases of severe agrammatic aphasia in which subjects seem to remain normal in their mind-reading performance (Varley 1998; Varley et al. 2001).[4] These patients have undergone extensive left-hemisphere damage, and as a result have significant problems with language. One such patient has matching comprehension and production deficits, suggesting that there is an underlying deficit in linguistic competence. He has lost almost all capacity to comprehend and to use verbs (while retaining some nouns); and he has certainly lost any capacity to formulate or comprehend *that*-clauses. But he is adept at communicating via pantomime, and performed as normal on a battery of false-belief tasks of the sort often administered to children (explained to him via a combination of one-word instruction and pantomime).

---

[4] Likewise, there exist cases of temporary paroxysmal aphasia in which language comprehension and production is completely shut down, but in which meta-cognitive skills and mind reading seem to remain fully intact. See Lecours and Joanette (1980).

While these data count powerfully against the thesis that natural language *that*-clauses are constitutive of the mind-reading capacities of adults, they are consistent with the claim that *that*-clause comprehension is at least a necessary condition of the *development* of mind-reading in children, as de Villiers and de Villiers (2000, 2003) also claim. Experiments conducted with Cantonese-speaking and German-speaking children count against this developmental claim, however (Perner et al., 2003; Cheung et al., 2004). So, too, does the fact that there exist many sign languages and some spoken Australian Aboriginal languages that contain no *that*-clause construction at all (Mark Baker, personal communication), since no one has suggested that such people are incapable of mind-reading. (In such languages, instead of saying, "John said that it is cold," subjects would use a clausal adjunct, saying, "It is cold, John said [it].") Moreover, the recent finding that infants perform successfully in non-verbal false belief tasks long before they become capable of using *that*-clauses speaks strongly against the proposal (Onishi and Baillargeon 2005; Southgate, Senju, and Csibra 2007; Surian, Caldi, and Sperber 2007; Song et al. 2008; Scott and Baillargeon 2009; Buttelmann, Carpenter, and Tomasello 2009).

4.3. Number Words and Exact Number Concepts

There is now extensive evidence that some numerical concepts are independent of language (Gallistel 1990; Dehaene 1997; Xu and Spelke 2000; Lipton and Spelke 2003). Specifically, both nonhuman animals and pre-verbal infants as young as six months can make judgments of approximate numerosity. They can judge the rough size of a set, and they can recognize that one set is larger or smaller than another, provided that the sizes of the two sets are sufficiently far apart. (What counts as "sufficient" here is partly a product of learning and experience—older children can make finer numerosity discriminations than can young infants—but it also depends upon the size of the sets in question—numerical discrimination becomes harder as the sets become larger.) Moreover, animals can effect simple numerical computations over their approximate numerosity representations (addition, subtraction, multiplication, and division). For example, they can calculate the rate of return from a foraging source (Gallistel 1990), which requires dividing the total quantity obtained from the source by the time spent foraging there.

Other researchers have claimed, furthermore, that animals and pre-verbal infants possess a capacity for exact small-number judgment and comparison, for numbers up to three or four (Wynn 1992a, 1992b; Dehaene 1997). For example, infants who see two objects placed behind a screen will be surprised if there is only one object remaining when the screen is lifted, or if there are three objects there instead. These judgments seem, on the face of it, to be numerical ones, since they are independent of the identities and properties of the objects in question. For example, a young infant will *not* be surprised if it first sees a toy clown placed behind the screen, but a toy truck is there instead when the screen is lifted; but it *will* be surprised if when the screen is lifted there are *two* toy trucks there. There is an emerging consensus, however, that these capacities are underlain by an attentional mechanism that opens a fresh *object-file* for each new object introduced (which can be thought of as an arbitrary label, "x," "y," etc.; note that the file itself can be empty of information), rather than requiring any strictly numerical concepts (Simon 1997; Leslie et al. 1998; Leslie, Gallistel, and Gelman 2007). Thus an infant that has seen two clowns placed behind a screen will have two object-files ("x" and "y") open. If two trucks are then observed it can match the "x" to one truck and the "y" to the other (provided that the object-files are empty, or at least contain no information about object category); but if three trucks are observed, then one of them will be left without any associated object-file label, causing the infant to be surprised.

When it comes to exact numerical concepts for numbers larger than four (*five, six, seven; seventeen; sixty-four; five million;* and so forth), most researchers accept that their acquisition is dependent upon language, specifically on the mastery of count-word lists ("five," "six," "seven," "eight," and so on) together with the procedures for counting. So this is a case where a whole class of concepts appears to be *developmentally* dependent upon language, at least. It remains very much in dispute how the acquisition process is supposed to occur, however. Some have argued that the basic procedure involves mapping the counting list onto the pre-existing approximate numerosity system (Gallistel and Gelman 1992; Wynn 1992a, 1992b). Others have thought that children bootstrap their way to a conception of exact large number by aligning the first few items on their count list with the representations in the object-file system, and then making a sort of inductive leap (Carey 2004; Feingensen, Dehaene, and Spelke 2004). However, Leslie et al. (2007) review these and other proposals and conclude that children require, in addition, an innate concept of the number *one*, together with an innate concept of recursion. (See also Laurence and Margolis 2007.)

There is also some evidence that natural language number-words might be constitutive of our adult possession and deployment of exact number concepts, in addition to being developmentally necessary for their acquisition. For Spelke and Tsivkin (2001) found that bi-lingual subjects trained on new number facts in one language recalled those facts more swiftly and accurately when tested in the language of teaching than when tested in their other language. In contrast, no such effect was found for new approximate number information, nor for new geographical and historical facts. This suggests that the latter are represented and stored independently of natural language, whereas exact number information is stored along with its natural language encoding.

5. LANGUAGE AS CONTENT COMBINER

A somewhat different proposal concerning the role of natural language in cognition is that it enables us to combine together the outputs of conceptual "modules," some of which would not otherwise get combined (Hermer and Spelke 1994, 1996;

Carruthers 1998, 2002; Hermer-Vazquez et al. 1999; Shusterman and Spelke 2005). Language is thus said to underpin the flexibility and *conjoinability* of content that is distinctive of human thought processes. (It is manifest to ordinary introspection that any concept that we possess can be conjoined with any other. Indeed, some philosophers have wanted to make this a constraint on concept possession *tout court*, claiming that only creatures capable of doing it can count as genuine concept users. This is the so-called "Generality Constraint" on concept possession—Evans 1982; see Carruthers 2009b, for a critique.) Explaining this proposal will require some background, however, before the evidence adduced in its support can be outlined and evaluated.[5]

Some psychologists—especially those who might describe themselves as "evolutionary"—have argued that in addition to specialized input systems (vision, audition, etc.) the mind contains a large number of specialized conceptual systems for forming new beliefs, for creating new motivations, and for decision making (Barkow et al. 1992; Sperber 1996, 2002; Pinker 1997; Carruthers 2006). Thus, in addition to the approximate numerosity system described in Section 5.3 above, there might be systems for forming beliefs about other people's mental states ("mind-reading"), for making judgments of physical causation ("folk-physics"), for keeping track of who owes what to whom in social exchanges ("cheater-detection"), and so on and so forth. Different theorists differ over the properties that they attribute to these conceptual "modules," and in particular about whether or not they are *encapsulated* (that is, closed off from information held elsewhere in the mind). But the important point for our purposes is that there is general agreement that conceptual modules will have limited connectivity with each other. It will often be the case that two or more modules routinely pass their outputs to a third, "down-stream," module, which may then be capable of combining those outputs into a single thought. But for thoroughgoing modularists there are unlikely to be any systems that are capable of receiving output from *all* conceptual modules—with the notable exception of the language faculty, which has presumably evolved to be capable of receiving, conjoining, and reporting information deriving from *any* conceptual module.

The main difficulty for those wishing to test this proposal lies in our ignorance of the detailed patterning of conceptual-module connectivity. For without knowing which modules connect up with which others independently of language, it is impossible to make predictions about the sorts of combinations of concepts that we should never observe in the absence of language. Fortuitously, however, data unearthed by Cheng (1986) concerning the spatial cognition of rats seemed to provide a plausible way of testing the idea. What Cheng found is that when rats are shown the location of some food hidden in one of the corners of an oblong rectangular space, and are

---

[5] It is worth noting that the proposal receives some indirect theoretical support from linguistics, where it has been claimed that the language faculty forces us to re-represent all concepts, from whatever source—even singular concepts such as proper names—as predicates. The result is that all atomic sentences can then take the form of existential quantifications over event descriptions that have been built up using conjunctions of the resulting predicates alone. See Pietroski (2005).

then removed from that space and disoriented before being returned, they rely *only* on geometric information when searching for the food. This leads the rats to search with equal frequency in the two geometrically equivalent corners, hence failing to go to the correct corner on 50 percent of occasions. All cues other that geometry are ignored. One of the walls of the container can be strikingly patterned or heavily scented, for example, but the rats ignore this information (even though it would enable them to succeed, and even though they can use it in other contexts for other purposes), and rely only on geometry.[6]

Hermer and Spelke (1994, 1996) had the idea of testing this phenomenon in young children, with identical results. When children are disoriented in a confined[7] rectangular space they, too, rely only on the geometry of the space when searching for the location of a hidden toy that they had been shown previously, even though one of the walls of the room may be bright red while the others are white. Older children and adults, in contrast, can solve these reorientation problems, utilizing both color *and* spatial information to locate the hidden target.

What Hermer and Spelke then did was to examine whether or not there is something about the children's language abilities that predicts success. And indeed, they found that the *only* predictor of success (but a highly reliable one) is productive use of the terms "left" and "right." Since color vocabulary is acquired much earlier, their suggestion is that it is only with the acquisition of appropriate spatial vocabulary that children first become capable of combining object-property information with geometric information. And in a later study, Shusterman and Spelke (2005) established that the relationship is causal and not just correlational, since *training* children in the use of "left" and "right" has a marked impact on their success in the disorientation tasks thereafter.

Even more dramatically, Hermer-Vazquez et al. (1999) extended these results in experiments with adults. It turns out that if adults are required to "shadow" speech during the experiment (continuously repeating back words that they hear through headphones), then their performance collapses to that of the younger children and rats—they, too, search equally often in the two geometrically equivalent corners. (In contrast, adult performance is unaffected by the requirement to shadow a complex rhythm, which places them under equivalent cognitive load.)

Although these results are striking, and although they show that language is certainly doing *something* in cognition, they don't demonstrate that language is enabling the combination of the outputs from different conceptual modules

---

[6] One fact that might explain this puzzling phenomenon is that symmetrical spaces are very rarely found in nature. On the contrary, the geometries of spaces "in the wild" are almost always unique. Hence an evolved disposition to default to geometric information when disoriented might have proven both reliable and efficient.

[7] Why the size of the space should matter is puzzling, but Spelke (personal communication) has obtained evidence that in a larger space, the children treat the red wall as a sort of directional beacon, like a distant tree or a line of hills, and that this takes priority over their default-to-the-geometry heuristic.

(namely, one concerned with geometry and one concerned with object properties). This is because "left" and "right" aren't geometric terms. And although the sentence, "It is left of the red wall" combines *spatial* information with color information, it doesn't combine *geometric* information with color information. (This is the sort of sentence that adults say, on introspective grounds, that they repeat to themselves when solving the tasks successfully.) Moreover, even the terms "left" and "right" aren't strictly necessary to enable subjects to solve the tasks. Simply drawing young children's attention to the red wall will actually work just as well. ("Look, I'm placing the doll near the *red* wall.") For if the child first orients towards the red wall, then the geometry of the visible space (long wall on the left, short wall on the right, say) can be used to select the correct corner thereafter.

The best explanation of the role of language in these experiments is as follows. Adults and older children formulate a sentence like, "It is left of the red wall" and see that it encodes all of the information that they need to solve the task. They therefore rehearse that sentence to themselves (if they aren't shadowing speech) while undertaking the task. When they reach the search phase, they then treat the rehearsed sentence somewhat like an instruction for action ("Go to the corner that is left of the red wall"), the following of which enables them to by-pass or pre-empt what would otherwise have been their default inclination to look only at the geometry of the space. Younger children who lack the word "left," on the other hand, might try out for themselves the sentence, "It is *near* the red wall." But since this manifestly doesn't encode all of the information that they need—it doesn't tell them *which side* of the red wall to go to—they don't bother to rehearse it. But in fact (unknown to them), had they done so—or had they just rehearsed, "Go to red"—they would actually have succeeded, since once oriented towards red their geometric knowledge would have kicked in to take them to the correct corner.

On this account, it turns out that the role of language in cognition isn't to unify the outputs of some otherwise unconnected modules (or it isn't in this instance, at least—the wider question remains open). Rather, language is playing a quasi-executive function, serving to manipulate the subject's attention and on-line goals. This is the sort of role that we will explore more systematically in Section 6.

6. DUAL SYSTEMS THEORY AND LANGUAGE

Most researchers who study human reasoning, and the fallacies and biases to which it is often subject, have converged on some or other version of a *dual systems* account (Evans and Over 1996; Evans 2008; Sloman 1996, 2002; Stanovich 1999, 2009; Kahneman 2002). Most now agree that System 1 is really a *collection* of different systems that are fast and unconscious, operating in parallel with one another. (For a modularist, these can be identified with the set of conceptual modules.) The principles according to which these systems function are, to a significant extent, universal to the human species, and they are not easily altered (e.g., by verbal instruction). Moreover those principles are, for the most part, heuristic in nature ("quick and dirty"), rather than deductively or inductively valid. It is also generally thought that most of the mechanisms constituting System 1 are evolutionarily ancient and shared with other species of animal.

System 2, in contrast, is slow, serial, and conscious. The principles according to which it operates are variable (both across cultures and between individuals within a culture), and can involve the application of valid norms of reasoning (although System 2, too, can involve the use of heuristics). These System 2 principles are malleable and can be influenced by verbal instruction, and they often involve normative beliefs (that is, beliefs about how one *should* reason). Moreover, System 2 is generally thought to be uniquely human, and some researchers, at least, emphasize the role that representations of natural language sentences (in so-called "inner speech") play in the operations of System 2 (Evans and Over 1996; Evans 2008).

Many writers are not fully explicit about how System 1 and System 2 relate to one another. But the general picture seems to be that they exist *alongside* each other, competing for control of the person's behavior. This is puzzling, however, for one wants to know what evolutionary pressures could have produced such wholesale cognitive change—in effect, creating a whole new system for forming beliefs and goals, and for decision making, alongside of a set of systems that already existed for those very purposes. Why would evolution not have tweaked, or extended, or added to the already-existing System 1 architecture, rather than starting over again afresh with System 2? Frankish (2004, 2008), in contrast, develops an account of System 2 processes as *realized in* those of System 1. So what we have are two *levels* or *layers* of cognitive processes, with one dependent upon the operations of the other, rather than being wholly distinct. This has the advantage that evolution need not have added anything much to the System 1 architecture for System 2 to come into existence. Rather, those System 1 systems needed to be orchestrated and utilized in new ways.

Carruthers (2006, 2009a) deepens and extends this picture, arguing that System 2 begins with the custom of mental rehearsal of action schemata (which is probably already present among some of the other great apes). This utilizes back-projecting pathways from motor cortex to the various perceptual systems, which evolved in the first instance for the swift on-line fine-tuning of action (Wolpert and Ghahramani 2000; Wolpert and Flanagan 2001; Wolpert, Doya, and Kawato 2003). But with overt action suppressed, these pathways can be used to generate visual and other images of the action in question. These images are then "globally broadcast" (Baars 1988, 2002) to the full suite of System 1 modules for forming predictions and new motivations. Inferences are drawn, and the broadcast image can be elaborated accordingly. Moreover, the agent's emotional systems react to the results as they would to the real thing (albeit more weakly, perhaps). These emotional reactions are monitored by the subject and—depending on their strength and valence—the subject's motivation to perform the original rehearsed action is then adjusted up or down accordingly (Damasio 1994, 2003).

On this account, some other species of animal already possess the beginnings of System 2 (although it is perhaps rarely used). But in the course of human evolution, the addition of a number of other System 1 systems—for language production and comprehension, for mindreading and higher-order thinking, and for normative reasoning and motivation—together with a disposition to engage in *creative* activation and rehearsal of action schemata (Carruthers 2007) led to a transformation in the character of System 2. Action schemata for *speech* could now be rehearsed with overt action suppressed, issuing (normally) in auditory (sometimes articulatory or visual) imagery of the sentence in question. The resulting image gets globally broadcast, and is received as input by the language comprehension system inter alia, which sets to work and attaches a conceptual content to it. The result is also received as input by the mindreading faculty, which might then issue in thoughts *about* the rehearsed sentence or thought, setting off a train of higher-order thinking and reasoning. Normative beliefs about what one *should* think or how one *should* reason might also be evoked, leading the subject to have beliefs about what he or she ought to think/say next. Since such beliefs come with built-in motivation, the result might be the formulation and rehearsal of another sentence of the appropriate sort. And since speech actions, like other actions, can be generated and rehearsed creatively, sometimes the contents entertained within System 2 will be radically new, not resulting directly from any thoughts that are entertained at the System 1 level.

On this account, it is because System 2 consists in *cycles* of operation of System 1 that it is comparatively slow, it is because (roughly speaking) only one action can be rehearsed at a time that it is serial, and it is because the resulting images are globally broadcast that System 2 (or rather, this aspect of its operations) is conscious. Moreover, because System 2 is action-based, it can be influenced in any of the ways that actions in general can be. Hence someone can *tell* you how you should tackle some problem, and you can comply, responding just as you would to any other bit of action-advice. For example, your logic teacher might tell you, "In order to evaluate a conditional, you need to look for a case where the antecedent is true and the consequent is false." When confronted by a Wason selection task, you might then ask yourself, "How do I tell whether a conditional is true or false?," evoking a memory of this advice and leading you to turn over both the "P" and the "not-Q" cards.

Likewise you can *imitate* the public reasoning processes of others, such as might be displayed in a philosophy lecture or in a scientific lab meeting, extracting abstract schemata for the sorts of sequences of moves that one should go through when confronted by certain types of problem, and then replicating them. And of course any normative beliefs that you form about the ways in which one should reason (perhaps from a course in logic, or a course in scientific method) can influence System 2, just as they might influence any other type of behavior. Moreover, individuals can vary along any of the above dimensions (e.g., in their normative beliefs), as well as in their dispositions to engage in mental rehearsal in the first place (which we call "being reflective").

On this account, then, natural language plays an important constitutive role in distinctively human (System 2) thought processes. This is not to say that *only* language plays such a role, however. On the contrary, other sorts of action can be rehearsed, leading to global broadcasts and transformations of visual and other forms of imagery, which can in turn issue in decision making. Indeed, one of the dimensions along which individual people differ from one another concerns the extent to which they are verbal or visual thinkers. But everyone uses language in their thinking some of the time, and many people use it a very great deal (Hurlburt 1990). And without language, not only are there many token thoughts that we would never entertain, but a large swathe of our cognitive lives, and our subsequent public behavior, would be very different indeed.

7. CONCLUSION

Partly in reaction to the more extreme views of Whorf and Vygotsky, most cognitive scientists have been inclined to play down the importance of language in human cognition outside of certain limited domains (such as exact number concepts, perhaps). And they are surely correct that a great many cognitive processes are independent of language, many of them shared with other animals. But if the account sketched in Section 6 is even remotely along the right lines, then representations of natural language sentences have an important role to play in certain aspects of distinctively human thinking and reasoning.

REFERENCES

Baars, B. (1988). *A Cognitive Theory of Consciousness*. Cambridge: Cambridge University Press.
———. (2002). The conscious access hypothesis: Origins and recent evidence. *Trends in Cognitive Science* 6: 47–52.
Barkow, J., Cosmides, L., and Tooby, J. (eds.) (1992). *The Adapted Mind: Evolutionary Psychology and the Generation of Culture*. Oxford: Oxford University Press.
Berlin, B. and Kay, P. (1969). *Basic Color Terms: Their Universality and Evolution*. Berkeley: University of California Press.
Buttelmann, D., Carpenter, M., and Tomasello, M. (2009). Eighteen-month-old infants show false belief understanding in an active helping paradigm. *Cognition* 112: 337–42.
Carey, S. (2004). Bootstrapping and the origin of concepts. *Daedalus* 113 (Winter 2004): 59–68.
Carruthers, P. (1998). Thinking in Language?: Evolution and a Modularist Possibility. In P. Carruthers and J. Boucher (eds.), *Language and Thought*, Cambridge: Cambridge University Press.
———. (2002). The cognitive functions of language. & Author's response: Modularity, language, and the flexibility of thought. *Behavioral and Brain Sciences* 25: 657–719.
———. (2006). *The Architecture of the Mind: Massive Modularity and the Flexibility of Thought*. Oxford: Oxford University Press.
———. (2007). The Creative-Action Theory of Creativity. In P. Carruthers, S. Laurence, and S. Stich (eds.), *The Innate Mind: Volume 3: Foundations and the Future*. Oxford: Oxford University Press.

————. (2009a). An Architecture for Dual Reasoning. In J. Evans and K. Frankish (eds.), *In Two Minds: Dual Processes and Beyond.* Oxford: Oxford University Press.

————. (2009b). Invertebrate Concepts Confront the Generality Constraint (and Win). In R. Lurz (ed.), *The Philosophy of Animal Minds: New Essays on Animal Thought and Consciousness.* Cambridge: Cambridge University Press.

Cheung, H., Hsuan-Chih, C., Creed, N., Ng, L., Ping-Wang, S., and Mo, L. (2004). Relative roles of general and complementation language in theory-of-mind-development: evidence from Cantonese and English. *Child Development* 75: 1155–70.

Cheng, K. (1986). A purely geometric module in the rat's spatial representation. *Cognition* 23: 149–78.

Clark, A. (1998). Magic words: How language augments human computation. In P. Carruthers and J. Boucher (eds.), *Language and Thought.* Cambridge: Cambridge University Press.

Damasio, A. (1994). *Descartes' Error: Emotion, Reason and the Human Brain.* London: Papermac.

————. (2003). *Looking for Spinoza: Joy, Sorrow, and the Feeling Brain.* New York: Harcourt.

Davidson, D. (1973). Radical interpretation. *Dialectica* 27: 313–28.

————. (1975). Thought and Talk. In S. Guttenplan (ed.), *Mind and Language.* Oxford: Oxford University Press.

Dehaene, S. (1997). *The Number Sense.* Oxford: Oxford University Press.

De Villiers, J., and de Villiers, P. (2000). Linguistic Determinism and the Understanding of False Beliefs. In P. Mitchell and K. Riggs (eds.), *Children's Reasoning and the Mind.* New York: Psychology Press.

————. (2003). Language for Thought: Coming to Understand False Beliefs. In D. Gentner and S. Goldin-Meadow (eds.), *Language in Mind.* Cambridge, MA: MIT Press.

Diaz, R., and Berk, L. (eds.) (1992). *Private Speech: From Social Interaction to Self-Regulation.* Hillsdale, NJ: Erlbaum.

Dummett, M. (1981). *The Interpretation of Frege's Philosophy.* London: Duckworth.

————. (1989). Language and Communication. In A. George (ed.), *Reflections on Chomsky.* Oxford: Blackwell.

Evans, G. (1982). *The Varieties of Reference.* Oxford: Oxford University Press.

Evans, J., and Over, D. (1996). *Rationality and Reasoning.* New York: Psychology Press.

Evans, J. (2008). Dual-processing accounts of reasoning, judgment, and social cognition. *Annual Review of Psychology* 59: 255–78.

Feigenson, L., Dehaene, S., and Spelke, E. (2004). Core systems of number. *Trends in Cognitive Sciences* 8: 307–14.

Fodor, J. (1975). *The Language of Thought.* Brighton, UK: Harvester Press.

Frankish, K. (2004). *Mind and Supermind.* Cambridge: Cambridge University Press.

————. (2008). Systems or levels? Dual-process theories and the personal-subpersonal distinction. In J. Evans and K. Frankish (eds.), *In Two Minds: Dual Processes and Beyond.* Oxford: Oxford University Press.

Gallistel, R. (1990). *The Organization of Learning.* Cambridge, MA: MIT Press.

Gallistel, R., and Gelman, R. (1992). Preverbal and verbal counting and computation. *Cognition* 44: 43–74.

Gelman, S. (2003). *The Essential Child: Origins of Essentialism in Everyday Thought.* Oxford: Oxford University Press.

Gumperz, J., and Levinson, S. (eds.). (1996). *Rethinking Linguistic Relativity.* Cambridge: Cambridge University Press.

Harris, P. (1996). Desires, Beliefs, and Language. In P. Carruthers and P. Smith (eds.), *Theories of Theories of Mind.* Cambridge: Cambridge University Press.

Heider, E., and Olivier, D. (1972). The structure of the color space in naming and memory for two languages. *Cognitive Psychology* 3: 337–54.

Hermer, L., and Spelke, E. (1994). A geometric process for spatial reorientation in young children. *Nature* 370: 57–59.

———. (1996). Modularity and development: The case of spatial reorientation. *Cognition* 61: 195–232.

Hermer-Vazquez, L., Spelke, E., and Katsnelson, A. (1999). Sources of flexibility in human cognition: Dual-task studies of space and language. *Cognitive Psychology* 39: 3–36.

Hunt, E., and Agnoli, F. (1991). The Whorfian hypothesis: A cognitive psychology perspective. *Psychological Review* 98: 377–89.

Hurlburt, R. (1990). *Sampling Normal and Schizophrenic Inner Experience.* New York: Plenum Press.

Kahneman, D. (2002). Maps of bounded rationality: A perspective on intuitive judgment and choice. Nobel laureate acceptance speech. Available at: http://nobelprize.org/economics/laureates/2002/kahneman-lecture.html

Laurence, S., and Margolis, E. (2007). Linguistic determinism and the innate basis of number. In P. Carruthers, S. Laurence, and S. Stich (eds.), *The Innate Mind: Volume 3: Foundations and the Future.* Oxford: Oxford University Press.

Lecours, A., and Joanette, Y. (1980). Linguistic and other aspects of paroxysmal aphasia. *Brain and Language* 10: 1–23.

Levinson, S. (1996). Frames of Reference and Molyneux's Question: Cross-Linguistic Evidence. In P. Bloom, M. Peterson, L. Nadel, and M. Garrett (eds.), *Language and Space.* Cambridge, MA: MIT Press.

Levinson, S. (2003). *Space in Language and Cognition: Explorations in Cognitive Diversity.* Cambridge: Cambridge University Press.

Levinson, S., Kita, S., Haun, D., and Rasch, B. (2002). Returning the tables: Language effects and spatial reasoning. *Cognition* 84: 155–88.

Leslie, A., Gallistel, C., and Gelman, R. (2007). Where Do the Integers Come From? In P. Carruthers, S. Laurence, and S. Stich (eds.), *The Innate Mind: Volume 3: Foundations and the Future.* Oxford: Oxford University Press.

Leslie, A., Xu, F., Tremoulet, P., and Scholl, B. (1998). Indexing and the object concept: Feveloping "what" and "where" systems. *Trends in Cognitive Sciences* 2: 10–18.

Li, P., and Gleitman, L. (2002). Turning the tables: Language and spatial reasoning. *Cognition* 83: 265–94.

Li, P., Abarbanell, L., and Papafragou, A. (2005). Spatial reasoning skills in Tenejapan Mayans. In *Proceedings from the 27th Annual Meeting of the Cognitive Science Society.* Hillsdale, NJ: Erlbaum.

Lipton, J., and Spelke, E. (2003). Origins of number sense: Large number discrimination in human infants. *Psychological Science* 14: 396–401.

Lucy, J. (1992a). *Grammatical Categories and Cognition: A Case Study of the Linguistic Relativity Hypothesis.* Cambridge: Cambridge University Press.

———. (1992b). *Language Diversity and Thought: A Reformulation of the Linguistic Relativity Hypothesis.* Cambridge: Cambridge University Press.

Malt, B., Sloman, S., and Gennari, S. (2003). Speaking versus Thinking About Objects and Actions. In D. Gentner and S. Goldin-Meadow (eds.), *Language in Mind.* Cambridge, MA: MIT Press.

Martin, L. (1986). Eskimo words for snow: A case study in the genesis and decay of an anthropological example. *American Anthropologist* 88: 418–23.

McDowell, J. (1994). *Mind and World.* Cambridge, MA: MIT Press.

Munnich, E., and Landau, B. (2003). The effects of spatial language on spatial representation: Setting Some Boundaries. In D. Gentner and S. Goldin-Meadow (eds.), *Language in Mind*. Cambridge, MA: MIT Press.

Onishi, K. and Baillargeon, R. (2005). Do 15-month-olds understand false beliefs? *Science* 5719: 255–58.

Papafragou, A. (2007). Space and the Language-Cognition Interface. In P. Carruthers, S. Laurence, and S. Stich (eds.), *The Innate Mind: Volume 3: Foundations and the Future*. Oxford: Oxford University Press.

Papafragou, A., Massey, C., and Gleitman, L. (2002). Shake, rattle, "n" roll: The representation of motion in language and cognition. *Cognition* 84: 189–219.

Perner, J., Ruffman, T., and Leekham, S. (1994). Theory of mind is contagious: You catch it from your sibs. *Child Development* 65: 1228–38.

Perner, J., Sprung, M., Zauner, P., and Haider, H. (2003). *Want that* is understood well before *say that, think that,* and false belief: A test of de Villier's linguistic determinism on German-speaking children. *Child Development* 74: 179–88.

Peterson, C., and Siegal, M. (1995). Deafness, conversation, and theory of mind. *Journal of Child Psychology and Psychiatry* 36: 459–74.

Pietroski, P. (2005). *Events and Semantic Architecture*. Oxford: Oxford University Press.

Pinker, S. (1997). *How the Mind Works*. London: Penguin Press.

Pullum, G. (1991). *The Great Eskimo Vocabulary Hoax and Other Irreverent Essays on the Study of Language*. Chicago: University of Chicago Press.

Roberson, D., Davies,I., and Davidoff, J. (2000). Color categories are not universal: Replications and new evidence from a Stone-Age culture. *Journal of Experimental Psychology: General* 129: 369–98.

Scott, R. and Baillargeon, R. (2009). Which penguin is this? Attributing false beliefs about object identity at 18 months. *Child Development* 80: 1172–96.

Segal, G. (1998). Representing Representations. In P. Carruthers and J. Boucher (eds.), *Language and Thought*. Cambridge: Cambridge University Press.

Siegal, M., and Surian, L. (2006). Modularity in language and theory of mind: What is the evidence? In P. Carruthers, S. Laurence, and S. Stich (eds.), *The Innate Mind: Volume 2: Culture and Cognition*. Oxford: Oxford University Press.

Simon, T. (1997). Reconceptualizing the origins of number knowledge: A non-numerical account. *Cognitive Development* 12: 349–72.

Shusterman, A., and Spelke, E. (2005). Investigations in the Development of Spatial Reasoning: Core Knowledge and Adult Competence. In P. Carruthers, S. Laurence, and S. Stich (eds.), *The Innate Mind: Structure and Contents*. Oxford: Oxford University Press.

Skinner, B. (1957). *Verbal Behavior*. East Norwalk, CT: Appleton-Century-Crofts.

Slobin, D. (1987). Thinking for speaking. *Proceedings of the Berkeley Linguistics Society* 13: 435–45.

——. (1996). From "Thought and Language" to "Thinking for Speaking." In J. Gumperz and S. Levinson (eds.), *Rethinking Linguistic Relativity*. Cambridge: Cambridge University Press.

Sloman, S. (1996). The empirical case for two systems of reasoning. *Psychological Bulletin* 119: 3–22.

——. (2002). Two Systems of Reasoning. In T. Gilovich, D. Griffin, and D. Kahneman (eds.), *Heuristics and Biases: The Psychology of Intuitive Judgment*. Cambridge: Cambridge University Press.

Song, H., Onishi, K., Baillargeon, R., and Fisher, C. (2008). Can an actor's false belief be corrected by an appropriate communication? Psychological reasoning in 18.5-month-old infants. *Cognition* 109: 295–315.

Southgate, V., Senju, A., and Csibra, G. (2007). Action anticipation through attribution of false belief by 2-year-olds. *Psychological Science* 18: 587–92.

Spelke, E., and Tsivkin, S. (2001). Language and number: A bilingual training study. *Cognition* 78: 45–88.

Sperber, D. (1996). *Explaining Culture: A Naturalistic Approach*. Oxford: Blackwell.

———. (2002). In Defense of Massive Modularity. In I. Dupoux (ed.), *Language, Brain and Cognitive Development*. Cambridge, MA: MIT Press.

Stanovich, K. (1999). *Who is Rational? Studies of Individual Differences in Reasoning*. Hillsdale, NJ: Erlbaum.

Stanovich, K. (2009). *What Intelligence Tests Miss: The Psychology of Rational Thought*. New Haven, CT: Yale University Press.

Surian, L., Caldi, S., and Sperber, D. (2007). Attribution of beliefs by 13-month old infants. *Psychological Science* 18: 580–86.

Varley, R. (1998). Aphasic Language, Aphasic Thought. In P. Carruthers and J. Boucher (eds.), *Language and Thought*. Cambridge: Cambridge University Press.

Varley, R., Siegal, M., and Want, S. (2001). Severe impairment in grammar does not preclude theory of mind. *Neurocase* 7: 489–93.

Vygotsky, L. (1961). *Thought and Language*. Cambridge, MA: MIT Press. (First published in Russian in 1934; republished and translated, with portions that were omitted from the first English translation restored, in 1986 by MIT Press.)

———. (1971). *The Psychology of Art*. Cambridge, MA: MIT Press.

———. (1978). *Mind in Society: The Development of Higher Psychological Processes*. Cambridge, MA: Harvard University Press.

Watson, J. (1924). *Behaviorism*. New York: Norton.

Wertsch, J. (ed.) (1981). *The Concept of Activity in Soviet Psychology*. Armonk, NY: Sharpe.

———. (ed.) (1985). *Culture, Communication, and Cognition*. Cambridge: Cambridge University Press.

Whorf, B. (1956). *Language, Thought, and Reality*. London: Wiley.

Wolpert, D., and Flanagan, R. (2001). Motor prediction. *Current Biology* 11: 729–32.

Wolpert, D., and Ghahramani, Z. (2000). Computational principles of movement neuroscience. *Nature Neuroscience* 3: 1212–17.

Wolpert, D., Doya, K., and Kawato, M. (2003). A unifying computational framework for motor control and social interaction. *Philosophical Transactions of the Royal Society of London*, B 358.

Wynn, K. (1992a). Children's acquisition of the number words and the counting system. *Cognitive Psychology* 24: 220–51.

———.(1992b). Evidence against empiricist accounts of the origins of numerical knowledge. *Mind and Language* 7: 315–32.

Xu, F. (2002). The role of language in acquiring object kind concepts in infancy. *Cognition* 85: 223–50.

Xu, F., Cote, M., and Baker, A. (2005). Labeling guides object individuation in 12 month old infants. *Psychological Science* 16: 372–77.

Xu, F., and Spelke, E. (2000). Large number discrimination in 6-month-old infants. *Cognition* 74: B1–B11.

CHAPTER 17

......

THEORY OF MIND

......

ALVIN I. GOLDMAN

"Theory of Mind" (ToM) refers to the cognitive capacity to attribute mental states to self and others. Other names for the same capacity include "commonsense psychology," "naïve psychology," "folk psychology," "mindreading" and "mentalizing." Mental attributions are commonly made in both verbal and nonverbal forms. Virtually all language communities, it seems, have words or phrases to describe mental states, including perceptions, bodily feelings, emotional states, and propositional attitudes (beliefs, desires, hopes, and intentions). People engaged in social life have many thoughts and beliefs about others' (and their own) mental states, even when they do not verbalize them.

In cognitive science the core question in this terrain is: How do people execute this cognitive capacity? How do they, or their cognitive systems, go about the task of forming beliefs or judgments about others' mental states, states that are not directly observable? Less frequently discussed in psychology is the question of how people self-ascribe mental states. Is the same method used for both first-person and third-person ascription, or entirely different methods? Other questions in the terrain include: How is the capacity for ToM acquired? What is the evolutionary story behind this capacity? What cognitive or neurocognitive architecture underpins ToM? Does it rely on the same mechanisms for thinking about objects in general, or does it employ dedicated, domain-specific mechanisms? How does it relate to other processes of social cognition, such as imitation or empathy?

This chapter provides an overview of ToM research, guided by two classifications. The first articulates four competing approaches to (third-person) mentalizing, viz., the theory-theory, modularity theory, rationality theory, and simulation theory. The second classification is the first-person/third-person contrast. The bulk of the discussion is directed at third-person mindreading, but the final section addresses self-attribution. Finally, our discussion provides representative coverage of the principal fields of investigators of ToM: philosophy of mind, developmental psychology, and cognitive neuroscience. Each of these fields has its distinctive research style, central preoccupations, and striking discoveries or insights.

1. THE THEORY-THEORY

Philosophers began work on ToM, or folk psychology, well before empirical researchers were seriously involved, and their ideas influenced empirical research. In hindsight one might say that the philosopher Wilfrid Sellars (1956) jump-started the field with his seminal essay, "Empiricism and the Philosophy of Mind." He speculated that the commonsense concepts and language of mental states, especially the propositional attitudes, are products of a proto-scientific *theory* invented by one of our fictional ancestors. This was the forerunner of what was later called the "theory-theory." This idea has been warmly embraced by many developmental psychologists. However, not everyone agrees with theory-theory as an account of commonsense psychology, so it is preferable to avoid the biased label "theory of mind." In much of my discussion, therefore, I opt for more neutral phraseology, "mindreading" or "mentalizing," to refer to the activity or trait in question.

The popularity of the theory-theory in philosophy of mind is reflected in the diversity of philosophers who advocate it. Jerry Fodor (1987) claims that commonsense psychology is so good at helping us predict behavior that it is practically invisible. It works well because the intentional states it posits genuinely exist and possess the properties generally associated with them. In contrast to Fodor's intentional realism, Paul Churchland (1981) holds that commonsense psychology is a radically false theory, one that ultimately should be eliminated. Despite their sharp differences, these philosophers share the assumption that naïve psychology, at bottom, is driven by a science-like theory, where a theory is understood as a set of law-like generalizations. Naïve psychology would include generalizations that link (1) observable inputs to certain mental states, (2) certain mental states to other mental states, and (3) mental states to observable outputs (behavior). The first type of law might be illustrated by "Persons who have been physically active without drinking fluids tend to feel thirst." An example of the second might be "Persons in pain tend to want to relieve that pain." An example of the third might be "People who are angry tend to frown." The business of attributing mental states to others consists of drawing law-guided inferences from their observed behavior, stimulus conditions, and previously determined antecedent mental states. For example, if one knows that Melissa has been engaged in vigorous exercise without drinking, one may infer that she is thirsty.

Among the developmental psychologists who have championed the theory-theory are Josef Perner, Alison Gopnik, Henry Wellman, and Andrew Meltzoff. They seek to apply it to young children, who are viewed as little scientists who form and revise their thinking about various domains in the same way scientists do (Gopnik and Wellman 1992; Gopnik and Meltzoff 1997). They collect evidence, make observations, and change their theories in a highly science-like fashion. They generate theories not only about physical phenomena but about unobservable mental states such as belief and desire. As in formal science, children make transitions from simple theories of the phenomena to more complex ones.

The most famous empirical discovery in the developmental branch of theory of mind is the discovery by Wimmer and Perner (1983) of a striking cognitive change in children between roughly three and four years of age. This empirical discovery is that three-year-olds tend to fail a certain *false-belief task* whereas four-year-olds tend to succeed on the task. Children watch a scenario featuring puppets or dolls in which the protagonist, Sally, leaves a chocolate on the counter and then departs the scene. In her absence Anne is seen to move the object from the counter to a box. The children are asked to predict where Sally will look for the chocolate when she returns to the room, or alternatively where Sally "thinks" the chocolate is. Prior to age four children typically answer incorrectly, that is, that Sally thinks it is in the box (where the chocolate really is). Around age four, however, normal children answer as an adult would, by specifying the place where Sally left the chocolate, thereby ascribing to Sally (what they recognize to be) a false belief. What happens between three and four that accounts for this striking difference?

Theory-theorists answer by positing a change of theory in the minds of the children. At age three they typically have conceptions of desire and belief that depict these states as simple relations between the cognizer and the external world, relations that do not admit the possibility of error. This simple theory gradually gives way to a more sophisticated one in which beliefs are related to propositional representations that can be true or false of the world. At age three the child does not yet grasp the idea that a belief can be false. In lacking a representational theory of belief, the child has—as compared with adults—a "conceptual deficit" (Perner 1991). This deficit is what makes the three-year-old child incapable of passing the false-belief test. Once the child attains a representational theory of belief, roughly at age four, she passes the location-change false-belief test.

A similar discrepancy between three- and four-year olds was found in a second type of false-belief task, the deceptive container task. A child is shown a familiar container that usually holds candy and is asked, "What's in here?" He replies, "candy." The container is then opened, revealing only a pencil. Shortly thereafter the child is asked what he thought was in the container when he was first asked. Three-year-olds incorrectly answer "a pencil," whereas four-year-olds correctly answer "candy." Why the difference between the two age groups, despite the fact that memory tests indicate that three-year-olds have no trouble recalling their own psychological states? Theory-theorists again offered the same conceptual-deficit explanation. Since the three-year-olds' theory does not leave room for the possibility of false belief, they cannot ascribe to themselves their original (false) belief that the container held candy, so they respond with their current belief, namely, that it held a pencil.

This explanation was extremely popular circa 1990. But several subsequent findings seriously challenge the conceptual-deficit approach. The early challenges were demonstrations that various experimental manipulations enable three-year-olds to pass the tests. When given a memory aid, for example, they can recall and report their original false prediction (Mitchell and Lacohee 1991). They can also give the correct false-belief answer when the reality is made less salient, for instance, if they are told where the chocolate is but do not see it for themselves (Zaitchik 1991). Additional evidence suggests that the problem with three-year-olds lies in the area of inhibitory control (Carlson and Moses 2001). Inhibitory control is an executive ability that enables someone to override "prepotent" tendencies (i.e., dominant or habitual tendencies, such as the tendency to reference reality as one knows it to be). A false-belief task requires an attributor to override this natural tendency, which may be hard for three-year-olds. An extra year during which the executive powers mature may be the crucial difference for four-year-olds, not a change in their belief concept. A meta-analysis of false-belief task findings encourages Wellman, Cross, and Watson (2001) to retain the conceptual-deficit story, but this is strongly disputed by Scholl and Leslie (2001).

Even stronger evidence against the traditional theory-theory time line was uncovered in 2005, in a study of fifteen-month-old children using a nonverbal false-belief task. Onishi and Baillargeon (2005) employed a new paradigm with reduced task demands to probe the possible appreciation of false belief in fifteen-month-old children, and found signs of exactly such understanding. This supports a *much* earlier picture of belief understanding than the child-scientist form of theory-theory ever contemplated.

A final worry about this approach can now be added. A notable feature of professional science is the diversity of theories that are endorsed by different practitioners. Cutting-edge science is rife with disputes over which theory to accept—disputes that often persist for decades. This pattern of controversy contrasts sharply with what is ascribed to young children in the mentalizing domain. They are said to converge on one and the same theory, all within the same narrow time-course. This bears little resemblance to professional science.

Gopnik takes a somewhat different tack in recent research. She puts more flesh on the general approach by embedding it in the Bayes-net formalism. Bayes nets are directed-graph formalisms designed to depict probabilistic causal relationships between variables. Given certain assumptions (the causal Markov and faithfulness assumptions), a system can construct algorithms to arrive at a correct Bayes net causal structure if it is given enough information about the contingencies or correlations among the target events. Thus, these systems can learn about causal structure from observations and behavioral interventions. Gopnik and colleagues (Gopnik et al. 2004; Schulz and Gopnik 2004) report experimental results suggesting that two- to four-year-old children engage in causal learning in a manner consistent with the Bayes net formalism. They propose that this is the method used to learn causal relationships between mental variables, including relationships relevant to false-belief tasks.

Here are several worries about this approach. Can the Bayes net formalism achieve these results without special tweaking by the theorist, and if not, can other formalisms match these results without similar "special handling"? Second, if the Bayes-net formalism predicts that normal children make all the same types of causal inferences, does this fit the scientific inference paradigm? We again encounter the problem that scientific inference is characterized by substantial diversity across the community of inquirers, whereas the opposite is found in the acquisition of mentalizing skills.

2. THE MODULARITY-NATIVIST APPROACH TO THEORY OF MIND

In the mid-1980s, other investigators found evidence supporting a very different model of ToM acquisition. This is the *modularity* model, which has two principal components. First, whereas the child-scientist approach claims that mentalizing utilizes domain-general cognitive equipment, the modularity approach posits one or more domain-specific modules, which use proprietary representations and computations for the mental domain. Second, the modularity approach holds that these modules are innate cognitive structures, which mature or come on line at preprogrammed stages and are not acquired through learning (Leslie 1994; Scholl and Leslie 1999). This approach comports with nativism for other domains of knowledge, such as those subsumed under Spelke's (1994) idea of "core knowledge." The core-knowledge proposal holds that infants only a few months old have a substantial amount of "initial" knowledge in domains such as physics and arithmetic: knowledge that objects must trace spatiotemporally continuous paths through space, or that one plus one yields two. Innate principles are at work that are largely independent of and encapsulated from one another. Modularists about mentalizing endorse the same idea. Mentalizing is part of our genetic endowment that is triggered by appropriate environmental factors, just as puberty is triggered rather than learned (Scholl and Leslie 2001).

Early evidence in support of a psychology module was reported by Simon Baron-Cohen, Alan Leslie, and Uta Frith in two studies, both concerning autism. The first (Baron-Cohen et al. 1985) compared the performance of normal preschool children, Down syndrome children, and autistic children on a false-belief task. All children had a mental age of above four years, although the chronological age of the second two groups was higher. Eighty-five percent of the normal children, 86 percent of the Down syndrome children, but only 20 percent of the autistic children passed the test. In the second study (1986) subjects were given scrambled pictures from comic strips with the first picture already in place. They were supposed to put the strips in order to make a coherent story, and were also supposed to tell the story in their own words. The stories were of three types: mechanical, behavioral, and mentalistic. The autistic children all ordered the mechanical strips correctly and dealt adequately with the behavioral script. But the vast majority of autistic children could not understand the mentalistic stories. They put the pictures in jumbled order and told stories without attribution of mental states.

The investigators concluded that autism impairs a domain-specific capacity dedicated to mentalizing. Notice that the autistic children in the 1986 study were not deficient on either the mechanical or the behavioral script, only on the mentalistic one. Conversely, the Down syndrome children, despite their general retardation, were not deficient on the false-belief task. Thus autism seems to involve an impairment specific to mentalizing, whereas mentalizing need not be impaired by general retardation as long as the ToM-dedicated module remains intact.

These conclusions, however, are not entirely secure. Some children with autism pass ToM tasks, including false-belief tests. The number who pass varies from one study to the next, but even a small percentage calls for explanation. If autism involves a failure to develop a ToM, how could these participants with autism pass the tests? Others therefore argue that failure on tasks that tap mentalizing abilities may be more directly interpreted in terms of domain-general deficits in either executive functions or language (Tager-Flusberg 2000).

Nativist modularists adduce additional evidence, however, in support of their view, especially evidence for an appreciation of intentional agency in preverbal infants. A variety of cues are cited as evidence for the attribution of intentionality, or goal-directedness, in infancy, including joint attention behaviors (gaze-following, pointing, and other communicative gestures), imitation, language and emotional referencing, and looking-time studies.

In one study of gaze-following, Johnson, Slaughter, and Carey (1998) tested twelve-month-old infants on a novel object, a small, beach ball-sized object with natural-looking fuzzy brown fur. It was possible to control the object's behavior from a hidden vantage point so that when the baby babbled, the object babbled back. After a period of familiarization, an infant either experienced the object reacting contingently to the infant's own behavior or merely random beeping or flashing. Infants followed the "gaze" of the object by shifting their own attention in the same direction under three conditions: if the object had a face, or if the object beeped and flashed contingent on the infant's own behavior, or both. These results were interpreted as showing that infants use specific information to decide when an object does or does not have the ability to perceive or attend to its surroundings, which seems to support the operation of a dedicated input system (Johnson 2005). Woodward (1998) used a looking-time measure to show that even five-month-olds appear to interpret human hands as goal-directed relative to comparable inanimate objects. They looked longer if the goal-object of the hand changed, but not if the hand's approach path to the goal-object changed. This evidence also suggests an early, dedicated system to the detection of goal-oriented entities.

All of the above findings post-date Alan Leslie's (1994) postulation of a later-maturing cognitive module: the "theory-of-mind mechanism" (ToMM). Leslie highlighted four features of ToMM: (1) it is domain specific, (2) it employs a proprietary representational system that describes propositional attitudes, (3) it forms the innate basis for our capacity to acquire ToM, and (4) it is damaged in autism. ToMM uses specialized representations and computations; it is fast, mandatory, domain specific, and informationally encapsulated, thereby satisfying the principal characteristics of modularity as described by Fodor (1983).

An initial problem with the modularity theory is that ToMM, the most widely discussed module postulated by the theory, does not satisfy the principal criteria of modularity associated with Fodorian modularity. Consider domain specificity. Fodor says that a cognitive system is domain specific just in the case when "only a restricted class of stimulations can throw the switch that turns [the system] on" (1983, 49). It is doubtful that any suitable class of stimulations would satisfy this condition for ToMM (Goldman 2006, 102–4). A fundamental obstacle facing this proposal, moreover, is that Fodor's approach to modularity assumes that modules are either input systems or output systems, whereas mindreading has to be a central system. Next consider informational encapsulation, considered the heart of modularity. A system is informationally encapsulated if it has only limited access to information contained in other mental systems. But when Leslie gets around to illustrate the workings of ToMM, it turns out that information from other central systems is readily accessible to ToMM (Nichols and Stich, 2003, 117–21). Leslie and German (1995) discuss an example of ascribing a pretend state to another person, and clearly indicate that a system ascribing such a pretense uses real-world knowledge, for example, whether a cup containing water would disgorge its contents if it were upturned. This knowledge would have to be obtained from (another) central system. Perhaps such problems can be averted if a non-Fodorian conception of modularity is invoked, as proposed by Carruthers (2006). But the tenability of the proposed alternative conception is open to debate.

3. The Rationality-Teleology Theory

A somewhat different approach to folk psychology has been championed by another group of philosophers, chief among them Daniel Dennett (1987). Their leading idea is that one mind reads a target by "rationalizing" her, that is, by assigning to her a set of propositional attitudes that make her emerge—as far as possible—as a rational agent and thinker. Dennett writes:

> [I]t is the myth of our rational agenthood that structures and organizes our attributions of belief and desire to others and that regulates our own deliberations and investigations.... Folk psychology, then, is *idealized* in that it produces its predictions and explanations by calculating in a normative system; it predicts what we will believe, desire, and do, by determining what we ought to believe, desire, and do. (1987, 52)

Dennett contends that commonsense psychology is the product of a special stance we take when trying to predict others' behavior: the *intentional stance*. To adopt the intentional stance is to make the default assumption that the agent whose behavior is to be predicted is rational, that her desires and beliefs, for example, are ones she rationally ought to have given her environment and her other beliefs or desires.

Dennett does not support his intentional stance theory with empirical findings; he proceeds largely by thought experiment. So let us use the same procedure in evaluating his theory. One widely endorsed normative principle of reasoning is to believe whatever follows logically from other things we believe. But attributors surely do not predict their targets' belief states in accordance with such a strong principle; they do not impute "deductive closure" to them. They allow for the possibility that people forget or ignore many of their prior beliefs and fail to draw all of the logical consequences that might be warranted (Stich 1981). What about a normative rule of inconsistency avoidance? Do attributors assume that their targets conform to this requirement of rationality? That too seems unlikely. If an author modestly thinks that he must have made some error in his book packed with factual claims, he is caught in an inconsistency (this is the so-called "paradox of the preface"). But would attributors not be willing to ascribe belief in all these propositions to this author?

These are examples of implausible consequences of the rationality theory. A different problem is the theory's incompleteness: it covers only the mindreading of propositional attitudes. What about other types of mental states, such as sensations like thirst or pain and emotions like anger or happiness? It is dubious that rationality considerations bear on these kinds of states, yet they are surely among the states that attributers ascribe to others. There must be more to mindreading than imputed rationality.

Although first inspired by armchair reflection, rationality theory has also inspired some experimental work that—at least at first blush—seems to be supportive. Gergely et al. (1995) performed an intriguing experiment they interpreted as showing that toddlers take the intentional stance at twelve months of age. They habituated one-year-old infants to an event in which a small circle approaches a large circle by jumping over an obstacle. When the obstacle is later removed, the infants show longer looking-times when they see the circle take the familiar jumping path as compared with a straight path toward the target. Apparently, infants expect an agent to take the most rational or efficient means to its goal, so they are surprised when it takes the jumping path, although that is what they have seen it do in the past.

The title of Gergely et al.'s (1995) paper, "Taking the Intentional Stance at 12 Months of Age," conveyed the influence of Dennett's rationality theory. The authors' first interpretation of the results articulated this theme, viz., that infants attribute a causal intention to the agent that accords with a rationality principle. Toward the end of their paper, however, they concede that an infant can represent the agent's action as intentional without attributing a mental representation of the future goal state to the agent's mind. Thus, the findings might simply indicate that the infant represents actions by relating relevant aspects of reality (action, goal-state, and situational constraints) through a principle of efficient action, which assumes that actions function to realize goal-states by the most efficient means available. Indeed, in subsequent writings the authors switch their description of infants from the "intentional" stance to the "teleological" stance, an interpretational system for actions in terms of means-ends efficiency (Gergely and Csibra 2003). The teleological stance is a qualitatively different but developmentally related interpretational system that is supposed to be the precursor of the young child's intentional stance. The two stances differ in that teleological interpretation is nonmentalistic—it makes reference only to actual and future states of reality. Developmentally, however, teleological interpretation is transformed into causal mentalistic interpretation by "mentalizing" the explanatory constructs of the teleological stance (232).

This approach raises three problems. First, can the teleological stance really be transformed into the full range of mentalistic interpretation in terms of rationality principles? One species of mindreading involves imputing beliefs to a target based on inferential relations to prior belief states. How could this interpretational system be a transformation of an efficiency principle? Inference involves no action or causal efficiency. Second, the teleological stance might equally be explained by a rival approach to mentalizing, namely, the simulation theory. The simulation theory might say that young children project themselves into the shoes of the acting object (even a circle) and consider the most efficient means to its goal. They then expect the object to adopt this means. Third, as already noted above, there are kinds of mental states and mindreading contexts that have nothing to do with rationality or efficiency. People ascribe emotional states to others (fear or delight, disgust or anger) based on facial expressions. How could these ascriptions be driven by a principle of efficiency? We do not have the makings here of a general account of mindreading, but rather at most, a narrow segment of it. And even this narrow segment might be handled just as well by a rival theory (viz., the simulation theory).

4. The Simulation Theory

A fourth approach to commonsense psychology is the *simulation theory*, sometimes called the "empathy theory." Robert Gordon (1986) was the first to develop this theory in the present era, suggesting that we can predict others' behavior by answering the question, "What would *I* do in *that* person's situation?" Chess players playing against a human opponent report that they visualize the board from the other side, taking the opposing pieces for their own and vice versa. They pretend that their reasons for action have shifted accordingly. Thus transported in imagination, they make up their mind what to do and project this decision onto the opponent.

The basic idea of the simulation theory resurrects ideas from a number of earlier European writers, especially in the hermeneutic tradition. Dilthey wrote of understanding others through a process of "feeling with" others (*mitfuehlen*), "reexperiencing" (*nacherleben*) their mental states, or "putting oneself into" (*hineinversetzen*) their shoes. Similarly, Schleiermacher linked our ability to understand other minds with our capacity to imaginatively occupy another person's point of view. In the philosophy of history, the English philosopher R. G. Collingwood (1946) suggested that the inner imitation of thoughts, or what he calls the reenactment of thoughts, is a central epistemic tool for understanding other agents. (For an overview of this tradition, see Stueber 2006.)

In addition to Gordon, Jane Heal (1986) and Alvin Goldman (1989) in the 1980s endorsed the simulation idea. Their core idea is that mind readers simulate a target by trying to create similar mental states of their own as proxies or surrogates of those of the target. These initial pretend states are fed into the mind reader's own cognitive mechanisms to generate additional states, some of which are then imputed to the target. In other words, attributors use their own mind to mimic or "model" the target's mind and thereby determine what has or will transpire in the target.

An initial worry about the simulation idea is that it might "collapse" into theory theory. As Dennett put the problem:

> How can [the idea] work without being a kind of theorizing in the end? For the state I put myself in is not belief but make-believe belief. If I make believe I am a suspension bridge and wonder what I will do when the wind blows, what "comes to me" in my make-believe state depends on how sophisticated my knowledge is of the physics and engineering of suspension bridges. Why should my making believe I have your beliefs be any different? In both cases, knowledge of the imitated object is needed to drive the make-believe "simulation," and the knowledge must be organized into something rather like a theory. (1987, 100–101)

Goldman (1989) responded that there is a difference between *theory-driven* simulation, which must be used for systems different than oneself, and *process-driven* simulation, which can be applied to systems resembling oneself. If the process or mechanism driving the simulation is similar enough to the process or mechanism driving the target, and if the initial states are also sufficiently similar, the simulation might produce an isomorphic final state to that of the target without the help of theorizing.

5. Mirroring and Simulational Mindreading

The original form of simulation theory (ST) primarily addressed the attribution of propositional attitudes. In recent years, however, ST has focused heavily on simpler mental states, and on processes of attribution rarely dealt with in the early ToM literature. We include here the mindreading of motor plans, sensations, and emotions. This turn in ST dates to a paper by Vittorio Gallese and Alvin Goldman (1998), which posited a link between simulation-style mindreading and activity of mirror neurons (or mirror systems). Investigators in Parma, Italy, led by Giacomo Rizzolatti, first discovered mirror neurons in macaque monkeys by using single cell recordings (Rizzolatti et al. 1996; Gallese et al. 1996). Neurons in the macaque premotor cortex often code for a particular type of goal-oriented action, for example, grasping, tearing, or manipulating an object. A subclass of premotor neurons were found to fire both when the animal plans to perform an instance of its distinctive type of action and when it observes another animal (or human) perform the same action. These neurons were dubbed "mirror neurons," because an action plan in the actor's brain is mirrored by a similar action plan in the observer's brain. Evidence for a mirror system in humans was established around the same time (Fadiga et al. 1995). Since the mirror system of an observer tracks the mental state (or brain state) of an agent, the observer executes a mental simulation of the latter. If this simulation also generates a mental-state attribution, this would qualify as simulation-based mindreading. It would be a case in which an attributor uses his own mind to "model" that of the target. Gallese and Goldman speculated that the mirror system might be part of, or a precursor to, a general mindreading system that works on simulationist principles.

Since the mid-1990s the new discoveries of mirror processes and mirror systems have expanded remarkably. Motor mirroring has been established via sound as well as vision (Kohler et al. 2002), and for effectors other the hand, specifically, the foot and the mouth (Buccino et al. 2001). Meanwhile, mirroring has been discovered for sensations and emotions. Under the category of sensations, there is mirroring for touch and mirroring for pain. Touching a subject's legs activates primary and secondary somatosensory cortex. Keysers et al. (2004) showed subjects movies of other subjects being touched on their legs. Large extents of the observer's somatosensory cortex also responded to the sight of the targets' legs being touched. Several studies established mirroring for pain in the same year (Singer et al. 2004, Jackson et al. 2004, and Morrison et al. 2004). In the category of emotions, the clearest case is mirroring for disgust. The anterior insula is well-known as the primary brain region associated with disgust. Wicker et al. (2003) undertook an fMRI experiment in which normal subjects were scanned while inhaling odorants through a mask—either foul, pleasant, or neutral—and also while observing video clips of other people's facial expressions while inhaling such odorants. Voxels in the anterior insula that were significantly activated when a person inhaled a foul odorant were also significantly activated when seeing others make facial expressions arising from a foul odorant. Thus, there was mirroring of disgust.

The critical question for ToM, however, is whether mindreading (i.e., mental attribution) occurs as an upshot of mirroring. In 2005 two similar experiments in the domain of motor intention were performed by members of the Parma group, and are claimed to provide evidence for mirror-based—hence, simulation-based—prediction of motor intentions. One experiment was done with monkeys (Fogassi et al. 2005) and the other with humans (Iacoboni et al. 2005). This section shall sketch the latter study only.

Iacoboni et al.'s study was an fMRI study in which subjects observed video clips presenting three kinds of stimulus conditions: (1) grasping hand actions without any context ("Action" condition), (2) scenes specifying a context without actions (i.e., a table set for drinking tea or ready to be cleaned up after tea) ("Context" condition), and (3) grasping hand actions performed in either the before-tea or the after-tea context ("Intention" condition). The Intention condition yielded a significant signal increase in premotor mirroring areas where hand actions are represented. The investigators interpreted this as evidence that premotor mirror areas are involved in understanding the intentions of others, in particular, intentions to perform subsequent actions (e.g., drinking tea or cleaning up).

This mindreading conclusion, however, is somewhat problematic, because there are alternative "deflationary" interpretations of the findings (Goldman 2008). One deflationary interpretation would say that the enhanced activity in mirror neuron areas during observation of the Intention condition involved only predictions of *actions*, not attributions of *intentions*. Since actions are not mental states, predicting actions does not qualify as mindreading. The second deflationary interpretation is that the activity in the observer's relevant mirror area is a mimicking of the agent's intention, not an intention *attribution* (belief). Reexperiencing an intention should not be confused with attributing an intention. Only the attribution of an intention would constitute a belief or judgment about an intention. Thus, the imaging data do not conclusively show that mindreading took place in the identified premotor area.

However, the Iacoboni et al. study presented evidence of intention attribution above and beyond the fMRI evidence. After being scanned, subjects were debriefed about the grasping actions they had witnessed. They all reported representing the intention of drinking when seeing the grasping action in the during-tea condition and representing the intention of cleaning up when seeing the grasping action in the after-tea condition. Their verbal reports were independent of the instructions the subjects had been given at the outset. Thus, it is quite plausible that their reported intention attributions were caused by activity in the mirror area. So the Iacoboni et al. study does provide positive evidence for its stated conclusion, even if the evidence is not quite as probative as its researchers contend.

Where else might we look for evidence of mirroring-based mindreading? Better specimens of evidence are found in the emotion and sensation domains. For reasons of space, attention is restricted here to emotion. Although Wicker et al. (2003) established a mirror process for disgust, they did not test for disgust attribution. However, by combining their fMRI study of normal subjects with neuropsychological studies of brain-damaged patients, a persuasive case can be made for mirror-caused disgust attribution (in normals). Calder et al. (2000) studied patient NK, who suffered insula and basal ganglia damage. In questionnaire responses NK showed himself to be selectively impaired in experiencing disgust, as contrasted with fear or anger. NK also showed significant and selective impairment in disgust recognition (attribution), in both visual and auditory modalities. Similarly, Adolphs et al. (2003) had a patient B who suffered extensive damage to the anterior insula but was able to recognize the six basic emotions *except disgust* when observing dynamic displays of facial expressions. The inability of these two patients to undergo a normal disgust response in their anterior insula apparently prevented them from mindreading disgust in others, although their attribution of other basic emotions was preserved. It is reasonable to conclude that when *normal* individuals recognize disgust through facial expressions of a target, this is causally mediated by a mirrored experience of disgust (Goldman and Sripada 2005; Goldman 2006).

Low-level mindreading, then, can be viewed as an elaboration of a primitive tendency to engage in automatic mental mimicry. Both behavioral and mental mimicry are fundamental dimensions of social cognition. Meltzoff and Moore (1983) found facial mimicry in neonates less than an hour old. Among adults, unconscious mimicry in social situations occurs for facial expressions, hand gestures, body postures, speech patterns, and breathing patterns (Hatfield, Cacioppo, and Rapson 1994; Bavelas et al. 1986; Dimberg, Thunberg, and Elmehed 2000; Paccalin and Jeannerod 2000). Chartrand and Bargh (1999) found that automatic mimicry occurs even between strangers, and that it leads to higher liking and rapport between interacting partners. Mirroring, of course, is mental mimicry usually unaccompanied by behavioral mimicry. The sparseness of behavioral imitation (relative to the amount of mental mimicry) seems to be the product of inhibition. Compulsive behavioral imitation has been found among patients with frontal lesions, who apparently suffer from an impairment of inhibitory control (Lhermitte et al. 1986; de Renzi et al. 1996). Without the usual inhibitory control, mental mimicry would produce an even larger amount of behavioral mimicry. Thus, mental mimicry is a deep-seated property of the social brain, and low-level mindreading builds on its foundation.

6. Simulation and High-Level Mindreading

The great bulk of mindreading, however, cannot be explained by mirroring. Can it be explained (in whole or part) by another form of simulation? The general idea of mental simulation is the reexperiencing or reenactment of a mental event or process, or an *attempt* to reexperience or reenact a mental event (Goldman 2006, ch. 2). Where does the traditional version of simulation theory fit into the picture? It mainly fits into the second category, that is, *attempted* interpersonal reenactment. This captures the idea of mental pretense, or what we call "enactment imagination" (E-imagination), which consists of trying to construct in oneself a mental state that is not generated by the usual means (Goldman 2006; Currie and Ravenscroft 2002). *Simulating Minds* argues that E-imagination is an intensively used cognitive operation, one commonly used in reading others' minds.

Let us first illustrate E-imagination with intrapersonal applications, for example, imagining seeing something or launching a bodily action. The products of such applications constitute, respectively, visual and motor imagery. To visualize something is to (try to) construct a visual image that resembles the visual experience we would undergo if we were actually seeing what is visualized. To visualize the *Mona Lisa* is to (try to) produce a state that resembles a seeing of the *Mona Lisa*. Can visualizing really resemble vision? Cognitive science and neuroscience suggest an affirmative answer. Kosslyn (1994) and others have shown how the processes and products of visual perception and visual imagery have substantial overlap. An imagined object "overflows" the visual field of imagination at about the same imagined distance from the object as it overflows the real visual field. This was shown in experiments where subjects actually walked toward rectangles mounted on a wall and when they merely visualized the rectangles while imagining a similar walk (Kosslyn 1978). Neuroimaging reveals a notable overlap between parts of the brain active during vision and during imagery. A region of the occipitotemporal cortex known as the fusiform gyrus is activated both when we see faces and when we imagine them (Kanwisher et al. 1997). Lesions of the fusiform face area impair both face recognition and the ability to imagine faces (Damasio et al. 1990).

An equally (if not more) impressive story can be told for motor imagery. Motor imagery occurs when you are asked to imagine (from a motoric perspective) moving your effectors in a specified way, for example, playing a piano chord with your left hand or kicking a soccer ball. It has been shown convincingly that motor imagery corresponds closely, in neurological terms, to what transpires when you actually execute the relevant movements (Jeannerod 2001).

At least in some modalities, then, E-imagination produces strikingly similar experiences to ones that are usually produced otherwise. Does the same hold for mental events such as forming a belief or making a decision? This has not been established, but it is entirely consistent with existing evidence. Moreover, a core brain network has recently been proposed that might underpin high-level simulational mindreading as a special case. Buckner and Carroll (2007) propose a brain system that subserves at least three, and possibly four, forms of what they call "self-projection." Self-projection is the projection of the current self into one's personal past or one's personal future, and also the projection of oneself into other people's minds or other places (as in navigation). What all these mental activities share is projection of the self into alternative situations, involving a perspective shift from the immediate environment to an imagined environment (the past, the future, other places, other minds). Buckner and Carroll refer to the mental construction of an imagined alternative perspective as a "simulation."

So E-imaginative simulation might be used successfully for reading other minds. But what specific evidence suggests that we deploy E-imaginative simulation in trying to mindread others, much of the time or even most of the time? This is what simulation theory concerning high-level mindreading needs to establish. (This assumes that simulation theory no longer claims that each and every act of mindreading is executed by simulation. Rather, simulation theorists are prepared to accept a hybrid approach in which simulation plays a central but not exclusive role.)

Two lines of evidence will be presented here (for additional lines of argument, see Goldman 2006, ch. 7). An important feature of the imagination-based simulation story is that successful mindreading requires a carefully pruned set of pretend inputs in the simulational exercise. The exercise must not only *include* pretend or surrogate states that correspond to those of the target but also *exclude* the mindreader's own genuine states that do not correspond to ones of the target. This implies the possibility of two kinds of error or failure: failure to include states possessed by the target, and failure to exclude states lacked by the target. The second type of error will occur if a mindreader allows a genuine state of his own, which he "knows" that the target lacks, to creep into the simulation and contaminate it. This is called *quarantine failure*. There is strong evidence that quarantine failure is a serious problem for mental-state attributors. This supports ST because quarantine failure is a likely affliction if mindreading is executed by simulation, but should pose no comparable threat if mindreading is executed by theorizing.

Why is it a likely problem under the simulation story? If one tries to predict someone's decision via simulation, one sets oneself to make a decision (in pretend mode). In making this decision, one's own relevant desires and beliefs try to enter the field to "throw their weight around" because this is their normal job. It is difficult to monitor the states that do not belong there, however, and enforce their departure. Enforcement requires suppression or inhibition, which takes vigilance and effort. No analogous problem rears its head under a theorizing scenario. If theorizing is used to predict a target's decision, an attributor engages in purely factual reasoning, not in mock decision making, so there is no reason why his genuine first-order desires or beliefs should intrude. What matters to the factual reasoning are the mindreader's beliefs *about* the target's desires and beliefs, and these second-order beliefs pose no comparable threat of intrusion.

Evidence shows that quarantine failure is in fact rampant, a phenomenon generally known as "egocentric bias." Egocentric biases have been found for knowledge, valuation, and feeling. In the case of knowledge, egocentric bias has been labeled "the curse of knowledge," and it has been found in both children (Birch and Bloom 2003) and adults (Camerer et al. 1989). To illustrate the bias for valuations, Van Boven, Dunning, and Loewenstein (2000) gave subjects Cornell coffee mugs and then asked them to indicate the lowest price they would sell their mugs for, while others who did not receive mugs were asked to indicate the highest price they would pay to purchase one. Because prices reflect valuations, the price estimates were, in effect, mental-state predictions. Both owners and sellers substantially underestimated the differences in valuations between themselves and their opposite numbers, apparently projecting their own valuations onto others. This gap proved very difficult to eliminate. To illustrate the case of feelings, Van Boven and Loewenstein (2003) asked subjects to predict the feelings of hikers lost in the woods with neither food nor water. What would bother them more, hunger or thirst? Predictions were elicited either before or after the subjects engaged in vigorous exercise, which would make one thirsty. Subjects who had just exercised were more likely to predict that the hikers would be more bothered by thirst than by hunger, apparently allowing their own thirst to contaminate their predictions.

Additional evidence that effective quarantine is crucial for successful third-person mindreading comes from neuropsychology. Samson et al. (2005) report the case of patient WBA, who suffered a lesion to the right inferior and middle frontal gyri. His brain lesion included a region previously identified as sustaining the ability to inhibit one's own perspective. Indeed, WBA had great difficulty precisely in inhibiting his own perspective (his own knowledge, desires, emotions, etc.). In nonverbal false-belief tests, WBA made errors in eleven out of twelve trials where he had to inhibit his own knowledge of reality. Similarly, when asked questions about other people's emotions and desires, which again required him to inhibit his own perspective, fifteen of twenty-seven responses involved egocentric errors. This again supports the simulationist approach to high-level mindreading. There is, of course, a great deal of other relevant evidence, which requires considerable interpretation and analysis. But ST seems to fare well in light of recent evidence (for contrary assessments, see Saxe 2005 and Carruthers 2006).

7. FIRST-PERSON MINDREADING

Our last topic is self-mentalization. Philosophers have long claimed that a special method—"introspection," or "inner sense"—is available for detecting one's own mental states, although this traditional view is the object of skepticism and even scorn among many scientifically minded philosophers and cognitive scientists. Most theory theorists and rationality theorists would join these groups in rejecting so-called "privileged access" to one's own current mental states. Theory theorists would say that self-ascription, like other-person ascription, proceeds by theoretical inference (Gopnik 1993). Dennett holds that the intentional stance is applied even to oneself. But these positions can be challenged with simple thought experiments, such as this one:

> I am now going to predict my bodily action during the next twenty seconds. It will include, first, curling my right index finger, then wrinkling my nose, and finally removing my glasses. There, those predictions are verified! I did all three things. You could not have duplicated these predictions (with respect to *my* actions). How did I manage it? Well, I let certain intentions form, and then I detected (i.e., introspected) those intentions. The predictions were based on the introspections. No other clues were available to me, in particular, no behavioral or environmental cues. The predictions must have been based, then, on a distinctive form of access I possess vis-à-vis my current states of mind, in this case, states that were primed to cause the actions. I seem to have similar access to my own itches and memories.

In an important modification of a well-known paper that challenged the existence or reliability of introspective access (Nisbett and Wilson 1977), the coauthor Wilson subsequently provides a good example and a theoretical correction to the earlier paper:

> The fact that people make errors about the causes of their own responses does not mean that their inner worlds are a black box. I can bring to mind a great deal of information that is inaccessible to anyone but me. Unless you can read my mind, there is no way you could know that a specific memory just came to mind, namely an incident in high school in which I dropped my bag lunch out a third-floor window, narrowly missing a gym teacher.... Isn't this a case of my having privileged, "introspective access to higher order cognitive processes"? (2002, 105)

Nonetheless, developmentalists have adduced evidence that putatively supports a symmetry or parallelism between self and other. They deny the existence of a special method, or form of access, available only to the first-person. Nichols and Stich (2003, 168–92) provide a comprehensive analysis of this literature, with the clear conclusion that the putative parallelism does not hold up, and fails precisely in ways that favor introspection or self-monitoring.

If there is such a special method, how exactly might it work? Nichols and Stich present their own model of self-monitoring. To have beliefs about one's own beliefs, they say, all that is required is that there be a monitoring mechanism that, when activated, takes the representation p in the Belief Box as input and produces the representation *I believe that p* as output. To produce representations of one's own beliefs, the mechanism merely has to copy representations from the Belief Box, embed the copies in a representation schema of the form *I believe that* ___, and then place the new representations back into the Belief Box. The proposed mechanism would work in much the same way to produce representations of one's own desires, intentions, and imaginings. (2003, 160–61).

One major lacuna in this account is its silence about an entire class of mental states: bodily feelings. They do not fit the model because, at least in the orthodox approach, sensations lack representational content, which is what the Nichols-Stich account relies upon. Their account is a syntactic theory, which says that the monitoring mechanism operates on the syntax of the mental representations monitored. A more general problem is what is meant by saying that the proposed mechanism would work in "much the same way" for attitude types other than belief. How does the proposed mechanism decide *which* attitude to ascribe? Which attitude verb should be inserted into the schema *I ATTITUDE that* ___? Should it be belief, desire, hope, fear, etc.? Each contentful mental state consists, at a minimum, in an attitude type plus a content. The Nichols-Stich theory deals only with contents, not types. In apparent recognition of the problem, Nichols and Stich make a parenthetical suggestion: perhaps a distinct but parallel mechanism exists for each attitude type. But what a profusion of mechanisms this would posit, with each mechanism essentially "duplicating" the others! Where is nature's parsimony that they appeal to elsewhere in their book?

The Nichols-Stich model of monitoring belongs to a family of self-attribution models that can be called "redeployment" theories because they try to explain self-attribution in terms of redeploying the content of a first-level mental state at a meta-representational level. Another such theory is that of Evans (1982), defended more recently by Gordon (1996), who calls it the "ascent-routine" theory. Gordon describes the ascent routine as follows: the way in which one determines whether or not one believes that *p* is simply to ask oneself the question whether or not *p*. The procedure is presumably to be completed as follows: if one answers the whether-*p* in the affirmative, one then "ascends" a level and also gives an affirmative answer to the question, "Do I think/believe that *p*?"

The ascent-routine theory faces a problem previously encountered with the monitoring theory. The basic procedure is described only for belief and lacks a clear parallel for classifying other attitudes or sensations. How is it supposed to work with hope, for example? Another problem concerns the procedure's details. When it says that a mindreader "answers" the whether-*p* question, what exactly does this mean? It cannot mean *vocalizing* an affirmative answer, because this will not cover cases of self-ascription where the answer is only *thought*, not vocalized. What apparently is meant by saying that one gives the "answer" *p* is that one *judges* the answer to be *p*. But how is one supposed to *tell* whether or not one judges that *p*? Is this not the same question of how one determines whether one (occurrently) believes that *p*? This is the same problem we started with, so no progress appears to have been made.

As we return to an introspectivist approach, notice that it is uncommitted to any strong view about introspection's reliability. Traditionally, introspection was associated with infallibility, but this is an easily detachable feature that few current proponents espouse. Introspectionism is often associated with a perceptual or quasi-perceptual model of self-knowledge, as the phrase "inner sense" suggests. Is that a viable direction? Shoemaker (1996) argues to the contrary. There are many disanalogies between outer sense and introspection, though not all of these should deter a theorist, says Shoemaker. Unlike standard perceptual modalities, inner sense has no proprietary phenomenology, but this should not disqualify a quasi-perceptual analogy. A more serious disanalogy, according to Shoemaker, is the absence of any organ that orients introspection toward its cognitive objects (current mental states) in the manner in which the eyes or nose can be oriented toward their objects. Shoemaker considers but rejects attention as a candidate organ of introspection.

This rejection is premature, however. A new psychological technique called "descriptive experience sampling" has been devised by Hurlburt (Hurlburt and Heavey, 2001) for studying introspection. Subjects are cued at random times by a beeper, and they are supposed to pay immediate attention to their ongoing experience upon hearing the beep. This technique revealed thoughts of which they had not initially been aware, though they were not unconscious. Schooler and colleagues (2004) have made similar findings, indicating that attention is typically required to trigger reflective awareness via introspection. Actually, the term *introspection* is systematically ambiguous. It can refer to a process of inquiry, that is, inwardly directed attention, that chooses a selected state for analysis. Or it can refer to the process of performing an analysis of the state and outputting some description or classification of it. In the first sense, introspection itself is a form of attention, not something that requires attention in order to do its job. In the latter sense, it is an operation that performs an analysis or description of a state once attention has picked out the state to be analyzed or described.

If introspection is a perception-like operation, should it not include a transduction process? If so, this raises two questions: what are the inputs to the transduction process, and what are the outputs? Goldman (2006, 246–55) addresses these questions and proposes some answers. There has not yet been time for these proposals to receive critical attention, so it remains to be seen how this new quasi-perceptual account of introspection will be received. In any case, the problem of first-person mentalizing is as difficult and challenging as the problem of third-person mentalizing, though it has thus far received a much smaller dollop of attention, especially among cognitive scientists.

8. CONCLUSION

Like most topics at the cutting edge of either philosophy or cognitive science, mindreading is awash with competing theories and rival bodies of evidence. The landscape is especially difficult to negotiate because it involves investigations using a myriad of disparate methodologies, ranging from a priori reflection to the latest techniques of contemporary neuroscience. The resulting variety of evidential sources ensures that new and fascinating findings are always around the corner; but it also makes it likely that we will not see a settled resolution of the debate in the very near future. It would be misguided to conclude that the amount of research effort devoted to the subject is disproportionate to its importance. To the contrary, the target phenomenon is a key to human life and sociality. People's preoccupation with mindreading, arguably at multiple levels, is a fundamental facet of human nature, and a philosophico-scientific understanding of how we go about this task must rank as a pre-eminent intellectual desideratum for philosophy of mind and for cognitive science.

REFERENCES

Adolphs, R., Tranel, D., and Damasio, A. R. (2003). Dissociable neural systems for recognizing emotions. *Brain and Cognition* 52: 61–69.
Baron-Cohen, S., Leslie, A., and Frith, U. (1985). Does the autistic child have a "theory of mind"? *Cognition* 21: 37–46.
———. (1986). Mechanical, behavioral, and intentional understanding of picture stories in autistic children. *British Journal of Developmental Psychology* 4: 113–25.

Bavelas, J. B., Black, A., Lemery, C. R., and Mullett, J. (1986). "I show how you feel": Motor mimicry as a communicative act. *Journal of Personality and Social Psychology* 50: 322–29.

Birch, S. A. J., and Bloom, P. (2003). Children are cursed: An asymmetric bias in mental-state attribution. *Psychological Science* 14: 283-86.

Buccino, G; Binkofski, F, Fink, G. R, Fadiga, L, Fogassi, L, Gallese, V, Seitz, R. J, Zilles, K, Rizzolatti, G, and Freund, H.-J. (2001). Action observation activates premotor and parietal areas in a somatotopic manner: An fMRI study. *European Journal of Neuroscience* 13(2): 400–404.

Buckner, R. L., and Carroll, D. C. (2007). Self-projection and the brain. *Trends in Cognitive Sciences* 11: 49–57.

Calder, A. J., Keane, J., Manes, F., Antoun, N., and Young, A. W. (2000). Impaired recognition and experience of disgust following brain injury. *Nature Reviews Neuroscience* 3: 1077–78.

Camerer, C., Loewenstein, G., and Weber, M. (1989). The curse of knowledge in economic settings: An experimental analysis. *Journal of Political Economy* 97: 1232–54.

Carlson, S. M., and Moses, L. J. (2001). Individual differences in inhibitory control and children's theory of mind. *Child Development* 72: 1032–53.

Carruthers, P. (2006). *The Architecture of the Mind*. Oxford: Oxford University Press.

Chartrand, T. L., and Bargh, J. A. (1999). The chameleon effect: The perception-behavior link and social interaction. *Journal of Personality and Social Psychology* 76: 893–910.

Churchland, P. (1981). Eliminative materialism and the propositional attitudes. *Journal of Philosophy* 78: 67–90.

Collingwood, R. G. (1946). *The Idea of History*. Oxford: Clarendon Press.

Currie, G., and Ravenscroft, I. (2002). *Recreative Minds*. Oxford: Oxford University Press.

Damasio, A. R., Tranel, D., and Damasio, H. (1990) Face agnosia and the neural substrates of memory. *Annual Review of Neuroscience* 13: 89–109.

Dennett, D. C. (1987). *The Intentional Stance*. Cambridge, MA: MIT Press.

Dimberg, U., Thunberg, M., and Elmehed, K. (2000). Unconscious facial reactions to emotional facial expressions. *Psychological Science* 11: 86–88.

Evans, G. (1982). *The Varieties of Reference*, edited by J. McDowell. Oxford: Oxford University Press.

Fadiga, L., Fogassi, L, Pavesi, G., and Rizzolatti, G. (1995). Motor facilitation during action observation: A magnetic stimulation study. *Journal of Neurophysiology* 73: 2608–11.

Fodor, J. A. (1983). *The Modularity of Mind*. Cambridge, MA: MIT Press.

———. (1987). *Psychosemantics*. Cambridge, MA: MIT Press.

Fogassi, L., Ferrari, P. F., Gesierich, B., Rozzi, S., Chersi, F., and Rizzolatti, G. (2005). Parietal lobe: From action organization to intention understanding. *Science* 308: 662–67.

Gallese, V., and Goldman, A. (1998). Mirror neurons and the simulation theory of mindreading. *Trends in Cognitive Sciences* 2: 493–501.

Gallese, V., Fadiga, L., Fogassi, L., and Rizzolatti, G. (1996). Action recognition in the premotor cortex. *Brain* 119: 593–609.

Gergely, G., and Csibra, G. (2003). Teleological reasoning in infancy: The naïve theory of rational action. *Trends in Cognitive Sciences* 7: 287–92.

Gergely, G., Nadasdy, Z., Csibra, G., and Biro, S. (1995). Taking the intentional stance at 12 months of age. *Cognition* 56: 165–93.

Goldman, A. I. (1989). Interpretation psychologized. *Mind and Language* 4: 161–85.

———. (2006). *Simulating Minds: The Philosophy, Psychology, and Neuroscience of Mindreading*. New York: Oxford University Press.

————. (2008). Mirroring, Mindreading, and Simulation. In J. Pineda (ed.), *Mirror Neuron Systems: The Role of Mirroring Processes in Social Cognition*. New York: Humana Press.

Goldman, A. I., and Sripada, C. S. (2005). Simulationist models of face-based emotion recognition. *Cognition* 94: 193–213.

Gopnik, A. (1993). How we know our minds: The illusion of first-person knowledge of intentionality. *Behavioral and Brain Sciences* 16: 1–14.

Gopnik, A., Glymour, C., Sobel, D. M., Schulz, L. E., Kushnir, T., and Danks, D. (2004). A theory of causal learning in children: Causal maps and Bayes nets. *Psychological Review* 111: 3–32.

Gopnik, A., and Meltzoff, A. N. (1997). *Words, Thoughts and Theories*. Cambridge, MA: MIT Press.

Gopnik, A., and Wellman, H. (1992). Why the child's theory of mind really *is* a theory. *Mind and Language* 7: 145–71.

Gordon, R. M. (1986). Folk psychology as simulation. *Mind and Language* 1: 158–71.

————. (1996). "Radical" Simulationism. In P. Carruthers and P. Smith (eds.), *Theories of Theories of Mind*. Cambridge: Cambridge University Press.

Hatfield, E., Cacioppo, J. T., and Rapson, R. L. (1994). *Emotional Contagion*. Cambridge: Cambridge University Press.

Heal, J. (1986). Replication and Functionalism. In J. Butterfield (ed.), *Language, Mind, and Logic*. Cambridge: Cambridge University Press.

Hurlburt, R. T., and Heavey, C. L. (2001). Telling what we know: Describing inner experience. *Trends in Cognitive Sciences* 5: 400–403.

Iacoboni, M., Molnar-Szakacs,I., Gallese, V., Buccino, G., Mazziotta, J. C., and Rizzolatti, G. (2005). Grasping the intentions of others with one's own mirror neuron system. *PLoS Biology* 3: 529–35.

Jackson, P. L., Meltzoff, A. N., and Decety, J. (2004). How do we perceive the pain of others? A window into the neural processes involved in empathy. *NeuroImage* 24: 771–79.

Jeannerod, M. (2001). Neural simulation of action: A unifying mechanism for motor cognition. *NeuroImage* 14: S103–9.

Johnson, S. C. (2005). Reasoning about intentionality in preverbal infants. In P. Carruthers, S. Laurence, and S. Stich (eds.), *The Innate Mind: Structure and Contents*. Oxford: Oxford University Press.

Johnson, S. C., Slaughter, V., and Carey, S. (1998). Whose gaze will infants follow? Features that elicit gaze-following in 12-month-olds. *Developmental Science* 1: 233–38.

Kanwisher, N., McDermott, J., and Chun, M. M. (1997). The fusiform face area: A module in human extrastriate cortex specialized for face perception. *Journal of Neuroscience* 17: 4302–11.

Keysers, C., Wicker, B., Gazzola, V., Anton, J-L., Fogassi, L., and Gallese, V. (2004). A touching sight: SII/PV activation during the observation of touch. *Neuron* 42: 335–46.

Kohler, E., Keysers, C., Umilta, M. A., Fogassi, L., Gallese, V., and Rizzolatti, G. (2002). Hearing sounds, understanding actions: Action representation in mirror neurons. *Science* 297: 846-48.

Kosslyn, S. M. (1978). Measuring the visual angle of the mind's eye. *Cognitive Psychology* 7: 341–70.

————. (1994). *Image and Brain: The Resolution of the Imagery Debate*. Cambridge, MA: MIT Press.

Leslie, A. M. (1994). *Pretending* and *believing*: Issues in the theory of ToMM. *Cognition* 50: 211–38.

Leslie, A. M., and German, T. (1995). Knowledge and Ability in "Theory of Mind": One-Eyed Overview of a Debate. In M. Davies and T. Stone (eds.), *Mental Simulation*. Oxford: Blackwell.

Lhermitte, F., Pillon, B., and Serdaru, M. (1986). Human autonomy and the frontal lobes. Part I: Imitation and utilization behavior: a neuropsychological study of 75 patients. *Annals of Neurology* 19: 326–34.

Meltzoff, A. N., and Moore, M. K. (1983). Newborn infants imitate adult facial gestures. *Child Development* 54: 702–9.

Mitchell, P., and Lacohee, H. (1991). Children's early understanding of false belief. *Cognition* 39: 107–27.

Morrison,I., Lloyd, D., de Pelligrino, G., and Roberts, N. (2004). Vicarious responses to pain in anterior cingulate cortex. Is empathy a multisensory issue? *Cognitive Affective Behavioral Neuroscience* 4: 270–78.

Nichols, S., and Stich, S. P. (2003). *Mindreading*. Oxford: Oxford University Press.

Nisbett, R., and Wilson, T. (1977). Telling more than we can know. *Psychological Review* 84: 231–59.

Onishi, K. H., and Baillargeon, R. (2005). Do 15-month-old infants understand false beliefs? *Science* 308: 255–58.

Paccalin, C., and Jeannerod, M. (2000). Changes in breathing during observation of effortful actions. *Brain Research* 862: 194–200,

Perner, J. (1991). *Understanding the Representational Mind*. Cambridge, MA: MIT Press.

De Renzi, E., Cavalleri, F., and Facchini, S. (1996). Imitation and utilization behavior. *Journal of Neurology and Neurosurgical Psychiatry* 61: 396–400.

Rizzolatti, G., Fadiga, L., Gallese, V., and Foggasi, L. (1996). Premotor cortex and the recognition of motor actions. *Cognitive Brain Research* 3: 31–41.

Samson, D., Apperly,I. A., Kathirgamanathan, U., and Humphreys, G. W. (2005). Seeing it my way: A case of a selective deficit in inhibiting self-perspective. *Brain* 128: 1102–11.

Saxe, R. (2005). Against simulation: The argument from error. *Trends in Cognitive Sciences* 9: 174–79.

Scholl, B., and Leslie, A. M. (1999). Modularity, development and "theory of mind". *Mind and Language* 14: 131–53.

———. (2001). Minds, modules and meta-analysis. *Child Development* 72: 696–701.

Schooler, J., Reichle, E. D., and Halpern, D. V. (2004). Zoning-out during reading: Evidence for dissociations between experience and meta-consciousness. In D. Levin (ed.), *Thinking and Seeing: Visual Meta-Cognition in Adults and Children*. Cambridge, MA: MIT Press.

Schulz, L. E., and Gopnik, A. (2004). Causal learning across domains. *Developmental Psychology* 40: 162–76.

Sellars, W. (1956). Empiricism and the Philosophy of Mind. In H. Feigl and M. Scriven (eds.), *Minnesota Studies in Philosophy of Science*, vol. 1. Minneapolis: University of Minnesota Press.

Shoemaker, S. (1996). *The First-Person Perspective and Other Essays*. New York: Cambridge University Press.

Singer, T., Seymour, B., O'Doherty, J., Kaube, H., Dolan, R., and Frith, C. (2004). Empathy for pain involves the affective but not sensory components of pain. *Science* 303: 1157–62.

Spelke, E. (1994). Initial knowledge: Six suggestions. *Cognition* 50: 431–45.

Stich, Stephen (1981). Dennett on intentional systems. *Philosophical Topics* 12: 38–62.

Stueber, K. (2006). *Rediscovering Empathy*. Cambridge, MA: MIT Press.

Tager-Flusberg, H. (2000). Language and Understanding Minds: Connections in Autism. In S. Baron-Cohen, H. Tager-Flusberg, and D. Cohen (eds.), *Understanding Other Minds: Perspectives from Developmental Cognitive Neuroscience*, 2nd ed. Oxford: Oxford University Press.

Van Boven, L., and Loewenstein, G. (2003). Social projection of transient drive states. *Personality and Social Psychology Bulletin* 29: 1159–68.

Van Boven, L., Dunning, D., and Loewenstein, G. (2000). Egocentric empathy gaps between owners and buyers: Misperceptions of the endowment effect. *Journal of Personality and Social Psychology* 79: 66–76.

Wellman, H. M., Cross, D., and Watson, J. (2001). Meta-analysis of theory-of-mind development: The truth about false belief. *Child Development* 72: 655–84.

Wicker, B., Keysers, C., Plailly, J., Royet, J-P., Gallese, V., and Rizzolatti, G. (2003). Both of us disgusted in *my* insula: The common neural basis of seeing and feeling disgust. *Neuron* 40: 655–64.

Wilson, T. D. (2002). *Strangers to Ourselves: Discovering the Adaptive Unconscious.* Cambridge, MA: Harvard University Press.

Wimmer, H., and Perner, J. (1983). Beliefs about beliefs: Representation and constraining function of wrong beliefs in young children's understanding of deception. *Cognition* 13: 103–28.

Woodward, A. L. (1998). Infants selectively encode the goal object of an actor's reach. *Cognition* 69: 1–34.

Zaitchik, D. (1991). Is only seeing really believing? Sources of the true belief in the false belief task. *Cognitive Development* 6: 91–103.

CHAPTER 18

....

BROAD-MINDED: SOCIALITY AND THE COGNITIVE SCIENCE OF MORALITY

....

JOHN M. DORIS
AND SHAUN NICHOLS

LONELINESS is a terrible thing.

This is hardly a novel observation: as Aristotle said, human beings are political animals, and their flourishing is a community affair. Yet however obvious it may be, the importance of human sociality has not been sufficiently appreciated in the study of human reasoning and morality. Much contemporary cognitive science and moral philosophy is Cartesian in spirit: it appears to presuppose that human beings reason best on their own, windows closed and curtains drawn, after the gripping fiction of Descartes' *Meditations*. *Individualism*, the view that *optimal human reasoning is substantially asocial*, pervasively inflects research on human rationality and morality. We will argue that individualism is troubled by much empirical research—research that compels a richly social conception of rationality. This conception, which we call *collaborativism*, maintains that *optimal human reasoning is substantially social*.[1]

....

[1] Thanks to Mark Timmons for suggesting the term *collaborativist*.

1. PRELIMINARIES

We should start by trying for a bit of clarity about the notions we propose to cojoin: reasoning and sociality. Words such as *reason* and *reasoning* cover a multitude of sins, and not all of this sinning is done publically. For instance, the magical moment when one is clobbered by the validity of a deductive inference might be go on entirely "in one's head." But even if there is something private about the workings of deductive inference, it may yet be the case that much theoretical and practical reasoning—"problem solving" broadly construed—is inescapably social. This is what we will argue.

One might reasonably wonder what we mean by *sociality*. Fair enough, though we do not have anything very fancy to say: sociality, in our view, appears in most any circumstance where two or more people interact. While this often involves physical proximity, it need not: distal interactions with technologies such as telephone or e-mail certainly count. Many instances of human sociality involve language use, and the centrality of language in thought is one motivation for collaborativism about reasoning, but much sociality does not conspicuously involve speech: consider the companionable silence of friends, or the unspoken passion of lovers.

There are interesting questions about "virtual" sociality, where a lone person can be said to function socially. The disapproving gaze of the "imagined other" in the experience of shame is familiar in moral philosophy, but examples abound, such as the often solitary, but socially engaged, activity of journal writing. Most broadly, it might be said that anywhere is culture—and that is anywhere there are people—there is quite a lot of virtual sociality about. And virtual sociality might facilitate reasoning; imagined interlocuters, or talking to oneself, may sometimes help in problem solving. But virtual sociality has its limits: too much time talking to oneself and not enough to other selves, and one might go off the rails. In any event, we do not require an expansive understanding of sociality here, and we will not need to much rely on delicate notions of virtual sociality, because there are more than enough obvious examples of sociality in reasoning to motivate our collaborativism.

We are going to ask how individualism fares in light of empirical evidence. In doing so, we might be expected to encounter the familiar philosophical balkiness about mingling descriptive and normative inquiry. Psychologists and cognitive scientists are naturally thought to be concerned with how people *actually* reason, while philosophers are naturally thought to be concerned with how people *should* reason. Of course, the trouble posed by the "fact/value distinction" is trouble that has been around for some time, and while we sympathize with the thought that seeing one's way around this trouble requires seeing that the distinction is highly permeable (Doris and Stich 2005, Railton 2003), we do not need to rehearse the arguments here.

While it is possible that some philosophers aspire to "purely" normative inquiry, articulating standards for ideal reasoning without attending to how people actually reason, philosophers theorizing about reasoning are very often theorizing about *human* reasoning. If so, these philosophers might find it instructive to consider how actual human beings—as opposed to Rufous Spiny Bandicoots, or the Gods of the Pantheon—think things through. So instead of addressing thorny questions about relations between the actual and ideal, we are going ask questions about what we will call *optimal* reasoning: under what circumstances are *actual* human organisms *in fact* able to reason most effectively? We will not presume to say exactly how this should impact normative debate; we will simply assume that answers to such questions will be good to have in hand as discussion proceeds.

Talk such as "most effectively" does invite normative disputation, and without articulating the relevant ideals, one might feel a bit at sea. Fortunately, in many instances, what counts as successful reasoning is not particularly mysterious: a problem might have an uncontroversial solution, so that thinking that arrives there is readily seen to be better than thinking that does not, or the reasoner might reasonably be attributed certain goals, so that determining what is instrumentally rational for her to think or do is relatively unproblematic. This is not the whole story, of course; as many moral philosophers have noted, purely instrumental conceptions of rationality seem unacceptably thin (e. g., Korsgaard 1996; Velleman 2006, 17–19, 349). Often pressing questions of rationality seem to involve "substantive" reasoning about ends rather than instrumental reasoning; people are frequently concerned with what they ought to want, rather than how they ought to go about getting it. Evaluating substantive reasoning is likely to be both complex and contentious, so while we expect, pending powerful arguments to the contrary, that collaborativism extends to substantive matters, we will focus on examples where what counts as good reasoning is more easily pegged. Collaborativism about rationality can be thought of as the view that, very often, human beings solve problems most effectively when the problem solving is a process involving social interaction, and we can motivate this thought without becoming entangled in debates about rationality.

2. A Tourist's History of Individualism

Individualism maintains that optimal human reasoning is substantially asocial, and therefore implies that sociality does not facilitate, and may impede, reasoning. It may immediately be observed that individualism looks to be something of a nonstarter: human beings are ultrasocial animals—perhaps eusocial animals (Foster and Ratnieks 2005)—who do everything in groups, from making love to making war. Who, then, would advocate individualism? The answer, on defensible readings, includes many canonical figures in the Western philosophical tradition. For the limit case, start with Nietzsche (e.g., 1886/1966, 212), who repeatedly warns of the danger "the herd" presents to his "higher type" of man:

> [T]he concept of greatness entails being noble, wanting to be by oneself, being
> able to be different, standing alone and having to live independently.[2]

Call this individualistic perfectionism: human beings can only be at their best when
they are socially unencumbered.

Nietzsche's views are notoriously extreme, but individualist themes are repeatedly sounded by philosophers with more cautious casts of mind. Here is Descartes,
in a famous passage from the *Meditations*:

> I WILL now close my eyes, I will stop my ears, I will turn away my senses from
> their objects, I will even efface from my consciousness all the images of corporeal
> things;....holding converse only with myself, and closely examining my nature,
> I will endeavor to obtain by degrees a more intimate and familiar knowledge of
> myself. (*Meditations* III, § 1)[3]

The problem of the *Meditations* is how knowledge of the external world may be
obtained in the face of skeptical challenge. The solution, it turns out, is not by discussing it with friends, or starting up a lab, but quite the opposite: the isolated enquirer is best able to address the skeptical difficulty. Of course, Descartes in the
Meditations appears to be rather self-consciously spinning a yarn; his method, it
might be said, is very much an ideal method. Once more, there are large questions
about how the ideal relates to the actual, but it is presently enough to note that the
ideal is very much asocial; if the actuality is supposed to diverge from the ideal in
virtue of its sociality, it is obscure where the sociality is supposed to come in.

The *Meditations* looks to be a work of *theoretical* reason (concerning what to
believe), and its relation to the questions of *practical* reason (concerning what to *do*)
that preoccupy moral philosophers may therefore seem obscure. We doubt the distinction between theoretical and practical reasoning is entirely stable, but no matter:
individualism is similarly evident in venerable conceptions of practical reason, such
as Kant's (1784/1963, 3):

> Enlightenment is the release of human beings from their self-incurred tutelage.
> Tutelage is the inability to use one's own reason without direction from someone
> else....Have courage to use your own reason!—that is the motto of enlightenment.

Here again, the idea seems to be that the optimal expression of rationality is effected
when the agent is emancipated from social influence. As with Descartes, people are
supposed to best think things through by themselves.

---

[2] For Nietzsche's individualism, see Leiter (2002, 116–17).

[3] It is perhaps a measure of the seductiveness of this picture that it was drawn by a man
who was an energetic philosophical correspondent, holding converse not only with himself, but
with his interlocutors. Indeed, despite their "official" methodological picture, the *Meditations* are
accompanied by the "Objections and Replies," where Descartes directly engages his critics (for
discussion, see Ariew and Greene 1995; for Descartes's correspondence, see Cottingham et al. 1991;
thanks to Dennis DesChene for help on this point). It is arguable that in the *Regulae*, Descartes
appears as more of a collaborativist. This presents the intriguing possibility, as Eric Brown
has suggested to us, that the *Meditations* are intended as a *reductio* of individualism: embrace
individualism, and you fall prey to skeptical doubt.

Individualism also appears in the thought of a figure that is, in contemporary philosophical taxonomies, rather uncomfortably related to Kant—John Stuart Mill (2002/1863, 65):

> [I]t is only the cultivation of individuality which produces, or can produce, well-developed human beings....what worse can be said of any obstruction to good, than that it prevents this?

While the differences between Mill and Nietzsche are at least as striking as those between Mill and Kant, Mill appears to share with Nietzsche an attachment to individualist perfectionism, where the possibility of human excellence increases as the role of social influence decreases.[4]

Individualist themes in political philosophy are not limited to Mill. Macpherson (1962, 3) argued (with special attention to Hobbes and Locke) that modern liberal-democratic theory is sourced in the "possessive individualism" of seventeenth-century political thought, which conceives of the individual

> as essentially the proprietor of his own person or capacities, owing nothing to society for them....The individual, it was thought, is free inasmuch as he is proprietor of his person and capacities. The human essence is freedom from the dependence on the wills of others, and freedom is a function of possession.

While Macpherson tends to the dramatic,[5] the less flamboyant assessment of Goodin (1998, 531) identifies related themes running from the Enlightenment to contemporary moral and political theory:

> The Enlightenment model of social life...depicts rational (or anyway reasoning) individuals choosing goals and plans and projects for themselves, with those autonomous individuals then coming together, of their own volition, in pursuit of shared interests and common goals....From Pico della Mirandola through Kant and the early Rawls, this vision of modern man as a "sovereign artificer" has reigned supreme throughout mainstream Western moral and political thought.

It is not, of course, that these canonical philosophers somehow failed to notice that human beings live in groups. Rather, it is that human beings' reasoning is best pursued in a way that is somehow prior to, or independent of, social entanglements. As every philosophy graduate student knows, history is a dangerous thing, and hell hath few furies to rival philosophical historians who think the figure on which they specialize has been scorned. Fortunately, we need not risk such fury, because our project retains interest on a wide range of possible histories. Suppose that we are *right* about the history, and canonical philosophers underappreciate the sociality of rationality. Then we have the enticing prospect of showing how contemporary cognitive science finds some of the greatest philosophers in the Western tradition to be

---

[4] Once again, reading a historical figure allows degrees of freedom; as Appiah (2005, 4, 20–21) makes clear, care should be taken not to exaggerate Mill's individualism.

[5] Macpherson's analysis, particularly his history of ideas, is controversial (Dunn 1974). For assessments, see Carens (1993) and Townshend (2000).

guilty of serious error. This would be a largish result. Now suppose —perhaps in the company of many historians—that we are *wrong* about the history. Suppose that our intended targets, when read with proper charity and care, give due weight to sociality; then we have the enticing prospect of showing some of the greatest philosophers in the Western tradition to be even greater than is usually thought, for they anticipated cognitive science that was not produced until ages after their deaths. This, too, would be a largish result. Thus, our endeavor can be seen as either an empirical *revision* of canonical ideas in Western philosophy or an empirical *vindication* of canonical ideas in Western philosophy. Both projects, we submit, are worth undertaking.

3. Individualism in the Cognitive Science of Morality

Whatever one's preferred intellectual history, individualism seems to run far and wide through American folkways, instantiated wherever the "rugged individualist" is celebrated, the "conformist" is castigated, and the child is admonished about the dangers of "peer pressure." Interpretations of folkways, particularly in the absence of systematic empirical study, fraught with peril. But whatever one thinks of past philosophers and present folkways, individualism is evident in the cognitive science of morality.

Research at the intersection of empirical psychology and ethical theory is thriving as never before, and this interdisciplinary work has lately produced a wide range of provocative findings.[6] But this research exhibits a striking feature that we ourselves were oblivious to until recently:*the cognitive science of morality very frequently proceeds with individualist assumptions.* Given that the various sciences of mind are often collected under the heading of *social* science, it is surprising to find that empirical investigation is often no less infected by individualist assumptions than is philosophical rumination. With notable exceptions such as Darley and colleagues' (Latané and Darley 1970; Darley and Batson 1973) studies of "group effects" in helping behavior, and Milgram's (1974) work on destructive obedience, researchers frequently investigate psychological phenomena, particularly morally salient phenomena, by focusing on individual processes at the expense of group dynamics.

Think of the social psychology "vignette" studies that have, in no small measure due to the relative logistical ease and economy of their production, been the paradigm most often pursued by experimental philosophers studying moral cognition: sneak into the department office under cover of darkness, Xerox the needed copies

---

[6] For overviews, see Andreou (2007); Doris and Stich (2005, 2006). For collections giving an extended sense of work in the field, see Doris et al. (2010); Knobe and Nichols (2008); Sinnott-Armstrong (2007a, 2007b, 2007c).

of your surveys, and you are good to go (pending IRB approval), near enough for free. Now look, for a moment, as your dutiful participants quietly (if unnervingly quickly) fill out their questionnaires: they are hunched over their desks, looking neither left or right, saying nothing, as they work through the intricacies of some moral problematic.

If the silence is not deafening, it should be. Instead of this monastic setting, go down to your neighborhood bar, where you might find the regulars discussing some or another moral conundrum: the legitimacy of torture say, or the finer points of marital infidelity. Voices are raised; fingers are pointed. And that is our point: in the wild, folks do not do this stuff quietly—they cajole, they plead, they argue, they abuse. And they do all of this in groups. It is hard to imagine something more unlike experimental participants mutely circling numbers on a Likert scale.

The same is true for another place where cognitive scientists, such as Greene, Moll, Cushman, and Young (e.g., Greene et al. 2001, 2004; Greene 2007; Moll et al. 2002a, 2002b, 2008), have made their way to studying morality: scanner studies of the neural activity associated with moral cognition. Once again, things are pretty quiet (save the considerable din of the fMRI machine); the participants think alone. In both vignette and scanner studies, moral cognition is treated as the activity of solitary individuals. (Indeed, in scanner studies, the variables of interest often involve only *parts* of individuals.) The Legend of the Lonely Thinker, it appears, infuses even the most empirically aggressive philosophical work done in many years.

We do not mean to be dismissive. We are talking about work we admire here, and we would be the first to say it is necessary work. Indeed, it is work we have done ourselves (e.g. Doris et al. 2007; Nichols and Knobe 2007). But we now want to develop an approach to moral psychology that is as garrulously social as lived human life. This will be a noisy undertaking indeed, because this strategy gains psychological realism at the expense of introducing factors—a great mess of factors—inimical to the controlled observation that distinguishes experimental inquiry. But the racket is worth it. We cannot do as much as we would like, because the phenomena we consider are taken over in bits and pieces from a diversity of literatures, and these fragments make for an unruly pastiche. But we think we can do enough to make our thesis plausible, and we hope we can do enough to inspire future research.

4. THE SCIENCE OF SOCIALITY: INDIVIDUALISM AND PATHOLOGY

In the 1950s and 1960s, Harlow (1986; Harlow, Dodsworth, and Harlow 1965) conducted a famous—or infamous—series of experiments documenting the effects of social isolation on rhesus monkeys. Within hours of their birth, infant monkeys were imprisoned in a stainless steel chamber and afforded no contact with any animal, human or nonhuman, for three to twelve months. The consequences were catastrophic. When removed from isolation, two of six monkeys in the three-month isolation group stopped eating entirely, one of these died, and the other had to be force-fed to keep it alive. Monkeys in all isolation groups developed florid psycho-pathologies, including aggression and withdrawal; in some instances females were unable to breed, or properly care for their young. It was extremely difficult to re-store the isolates to approximations of normalcy, but significantly, the only effective treatment was social: Harlow and Suomi (1971, 1537–38) paired damaged monkeys that had been reared in isolation from birth to six months with three-month-old normal monkeys, labeled "therapists," and within six months, the isolates were almost indistinguishable from monkeys that had been raised in a normal social environment.

Studies of this kind using human populations are, thankfully, prohibited on ethical grounds. (Indeed, many of Harlow's experiments with *monkeys* are objec-tionable on ethical grounds.) However, history is scattered with examples of "for-bidden experiments" involving feral children, who have been found living in the wild or among animals, and confined children, who are the victims of egregious abuse and neglect. Given the historical uncertainties surrounding feral children, it is perhaps better to focus on cases of confined children. Among the best known of these dates from the 1970s: "Genie," was discovered by authorities in Los Angeles at age thirteen, having spent most of her life imprisoned alone in a single room (Newton 2002, 208–29). Unsurprisingly, Genie was tragically impaired: such forma-tive social deprivation causes severe cognitive and social disability that, unlike the disability of Harlow's isolated monkeys, appears to be irreversible. Of course, such disability is not solely attributable to social deprivation: confined children are typi-cally subject to abuse and neglect, often horrific abuse and neglect, in addition to isolation. As we have already said, the data we must consider are noisy, and the causal factors at issue are not readily disentangled. But there is little reason to doubt that social isolation is a major pathogen in these cases, just as the Harlow studies would lead one to expect.

Although its etiology is incompletely understood, and there is no reason to think it a sequela of abuse and neglect, autism is another developmental disorder with implications for thinking about sociality. Autistic children present with a broad range of deficits in social reasoning, including disability in attributing mental states to others (see, e.g., Baron-Cohen 1995; Nichols and Stich 2003). At the same time, autistic children present similarly to normally developing children on familiar moral judgment tasks, such as those designed to test facility with the moral/conven-tional distinction (Turiel 1983; Blair 1996). But some researchers have observed that autistic children *do not* present normally when asked to *report* their moral reason-ing. In one study, normally developing and autistic children (ages five to nine) were asked the following question:[7]

---

[7] Shaun Nichols and Trisha Folds-Bennett, unpublished data. James Blair has reported similar findings in personal communication with Nichols.

"Why was it wrong for Johnny to hit Billy?"

Normally developing children gave the expected sorts of answers, exhibiting culturally appropriate instances of moral reasoning:

"Because it hurt Billy."

Autistic children, however, often produced quite inappropriate responses such as:

"I was on an escalator once."

This kind of difficulty is well known from clinical studies in the area: autistic children quite often produce such conversational non sequiturs and present other evidence of social disconnection, such as presentation of strangely inappropriate gifts (e. g., Dawson and Fernald 1987). Apparently, autistic children have trouble engaging in the kind of social dialogue that facilitates normal moral functioning.

Adult clinical populations are similarly suggestive. Consider narcissistic personality disorder, where afflicted individuals present with self-importance and feelings of entitlement (DSM-IV-TR 2000, 715). Unfortunately, the narcissist's affliction is often an affliction on those around him or her: individuals with narcissistic personality disorder may believe that they are exempt from social norms, assuming, for instance, that they do not have to wait in line (DSM-IV-TR, 715). Schwartzberg (2000, 106–88), describes a patient, "Brian," who was compelled into therapy after a series of legal problems, typically resulting "from his belief that rules and laws for other people didn't apply to him." Matters came to a head when Brian arrived at the airport minutes before his scheduled departure to discover that his seat had been reassigned. In an attempt to stop the plane, he claimed that his luggage was aboard with a bomb in it. Apparently, Brian was unable to grasp the fact that his flight was not literally *his* flight. The narcissist behaves, as Mom used to say, like the world revolves around him.

Narcissism is not limited to this sort of adolescent self-absorption; it may take darker turns as well. While the exact relation is unclear, narcissism may be associated with antisocial personality disorders such as psychopathy, with some researchers placing narcissistic and antisocial personality disorders on a continuum where "malignant narcissism" occupies an intermediate (Kernberg 1998; Geberth and Turco 1997).[8] These waters are muddy: neither psychopathy nor narcissism is completely understood. Moreover, some authorities have raised questions about whether such conditions are legitimate mental disorders.[9] And this is a skepticism that lay

[8] Our remarks here concern clinical narcissism. It is worth noting, however, that subclinical narcissistic tendencies do not present as particularly appealing attributes (for some empirical work, see Vazire and Funder 2006, Vazire et al. 2008).

[9] We doubt such skepticism is entirely warranted, particularly for psychopathy. Recent research (e. g., Blair, Mitchell, and Blair 2006) seems to indicate that psychopaths manifest cognitive and neurological abnormalities of the sort that legitimize attribution of mental illness (although some observers have denied that psychopathy is an illness; see Mealey 1995).

people may quite reasonably share, since the line between a narcissist and a jerk may often be a vanishingly fine one. After a fashion, that is the lesson. Both clinical populations present abnormally with regard to social affect and interaction, and whether we think psychopaths and narcissists are legitimately ill or just nasty, both populations fail to reason and behave appropriately on matters of moral concern. As sociality goes—or fails to go—so goes morality.

The evidence indicates that environmental and congenital impoverishments of sociality are implicated in the development of deficits in practical and moral reasoning. But this does not entail that for normally developed adults, optimal reasoning is, in an ongoing way, a social process. Hence, the considerations thus far adduced do not appear to favor collaborativist conceptions of rationality over individualist conceptions; insisting on the sociality of normal development is quite compatible with individualism. The individualism that explicitly acknowledges this observation can be tagged individualism$_a$: optimal human reasoning is substantially asocial, and *(a) sociality is necessary for the development of optimal reasoning.*

Now, the individualist may allow that normal development requires sociality, but deny that optimal reasoning in mature individuals requires it. This suggestion is not implausible, but it will not work out. Rationality is not only socially *developed*, it is socially *sustained*; in many domains, optimal reasoning occurs when the reasoning process occurs as part of a social process.

While the isolation to which confined children are subjected is considered criminal, such treatment of adults is legally sanctioned, in the form of solitary confinement and "security housing units." The terrors of solitary are well-known:

> The air in your cell vanishes. You are smothering. Your eyes bulge out; you clutch
> at your throat; you scream like a banshee. Your arms flail the air in your
> cell....The walls press you from all directions....You become hollow and
> empty....You are dying. Dying a hard death. (Abbott 1991, 25)

If anyone thinks that such effects are limited to the faint of heart, note that this description was penned by Jack Henry Abbott, a hardened recidivist and murderer.

Over the past thirty years or so, American penitentiary practice has, with the aid of technological innovations, refined the practice of solitary confinement into "supermax" institutions, where prisoners are held for extended periods with radically curtailed human contact. At California's Pelican Bay supermax, the "pinky shake," which consists of touching fingers through nickel-sized holes in cell doors, is in some instances the only physical contact individuals have with other human beings for years (Sullivan 2006). The effects of such deprivation now have a name: Security Housing Unit Syndrome (Grassian 1983). To put it mildly, sufferers are not feeling very well.

For example, Haney's (2003, 136) interviews with prisoners at Pelican Bay indicated radically inflated incidences of psychopathology.[10] Below are the percentages of inmates reporting various symptoms; the figure in parentheses indicates the

---

[10] Brodsky and Scogin's (1988) earlier study presents a similar picture for other prisoner populations. For a review of the literature, see Haney (2003).

factor by which this percentage exceeds estimates of base rates in non-incarcerated populations.

Chronic Depression: 77 percent *(3x)*
Trouble Sleeping: 84 percent *(5x)*
Perspiring Hands: 68 percent *(4x)*
Heart Palpitations: 68 percent *(18x)*
Dizziness: 56 percent *(8x)*
Nightmares: 55 percent *(7x)*
Irrational Anger: 88 percent *(30x)*
Confused Thought Processes: 84 percent *(8x)*
Hallucinations: 41 percent *(24x)*

Hallucinations are not common in the population at large. Yet for individuals in "secure" housing the probability of experiencing hallucinations approaches a coin flip—they are pretty nearly as likely to hallucinate as not. Similarly, victims of such treatment are overwhelmingly likely to experience confused thought processes, again a symptom that is (appearances notwithstanding) uncommon in the general population. This is especially salient in the present context because moral agency is plausibly thought to require "cognitive feats" (Doris and Murphy 2007)—moral self-direction, as one might call it, apparently requires that one get quite a lot right about one's circumstances. (*Would that hurt her feelings? Does getting that mean I have to do this?*) A person falling prey to hallucinations, confused thought processes, irrational anger, and the like makes an unlikely candidate for the achievement of such feats.

The prison population is a demographic group exhibiting considerable distress prior to incarceration—for example, substance abuse is considerably elevated among criminal populations (Sinha and Easton 1999)—so it might be thought that little can be ascertained about the pathogenic role of prison isolation. In this regard, Sestoft and associates' (1998, 103) study of Danish prisoners is illuminating: there, the probability of being admitted to the prison hospital for psychiatric reasons after four weeks in solitary confinement was *twenty times higher* than for prisoners held in the general prison population for a comparable period of time.[11] It seems reasonable to suppose that whatever pathologies afflict populations in corrections systems, isolation is an *added* pathogen.

If isolation is implicated in psychopathology, it is reasonable to conclude that sociality is required to *sustain* normal adult reasoning, and is not merely a developmental prerequisite.[12] To acknowledge this, individualism needs to be amended again, to individualism$_{ab}$: optimal human reasoning is substantially asocial, and (a) sociality is necessary for the development of optimal reasoning, and *(b) sociality is necessary for the sustenance of optimal reasoning.*

[11] This data is especially compelling, as it relies on clinical assessment as well as prisoner self-report.

[12] We do not need to deny that some adult human isolates, such as hermits and explorers, may enjoy passable mental hygiene—though some of the anecdotal evidence, such as found in histories of polar exploration (e. g., Bickel 1977, 225–26; cf. 166), gives cause for doubt.

5. SOCIALLY EMBEDDED REASONING

Evidently, optimal cognitive functioning is both developed and sustained through sociality. This appears to trouble individualist accounts of well functioning, at least for variants of those positions that suppose well functioning is something fairly widely accessible, as opposed to the singular achievement of a few übermenschen. But what we have established so far might be taken in the following relatively minimal sense: isolation is causally implicated in psychopathologies of a sort that may impair cognitive functioning. We are suggesting something stronger: the optimal exercise of rationality is a socially *embedded* process. This means that sociality is not just a precondition of rationality, but that even among those with normal cognitive functioning, the optimal exercise of rationality typically occurs as *part* of a social process. But the minimalist reading of the evidence just offered—sociality is required for basic cognitive health—does not entail anything about embeddedness: so long as a person enjoys whatever level of sociality is required to keep madness at bay, perhaps she can reason optimally on her own. However, further evidence indicates that embeddedness is a condition of optimality.

Consider the role of sociality in substantial cognitive achievement, such as scientific and technological discovery. There are various reasons why these attainments tend to be socially engendered and sustained. In the first instance, knowledge is cumulative: the present generation learns from the previous generation, and the steam engine gives way to internal combustion, which in turn (one prays) gives way to some more sustainable technology. In the second instance, knowledge is specialized, especially in cases of large-scale industrial productions: it takes a lot of different engineers, in a lot of different fields and subfields, to make something like a modern commercial aircraft fly. Moreover, as Kitcher (1993, 70; cf. Kitcher 2001, ch. 9; Longino 1990) argues, a diversity of researchers pursuing an assortment of research programs facilitates scientific progress, because in this way a larger percentage of the empirical and theoretical space is investigated, which increases the likelihood that nonobvious but high payoff research programs will be identified.

While the advance of knowledge requires diversity, it equally requires *conformity*. This is due to the importance of expertise and authority in the accumulation and transmission of information. As Boyd and Richerson (2006, 24) observe in their account of cultural evolution, "[t]o get the benefits of social learning, humans have to be credulous, for the most part accepting the ways they observe in their society as sensible and proper." If each generation has to reinvent the wheel, or in Boyd and Richerson's example, the kayak, not a lot of wheels or kayaks will get built. Inevitably, in kayak builders we trust; innovation may modify the shape of the kayak (if not the wheel) from one generation to the next, but a lot has to be taken on the authority of those who have previously worked the field.

This picture contrasts with a certain Enlightenment picture of genius, like that associated with Mill (1863/2002; cf. Brennan 2007), where intellectual achievement is the province of the rogue and the rebel. Surely this is partly true: innovation is driven by those who refuse to see the wisdom in the old wisdom. But it is equally true that science, at least Kuhn's (1962) normal science, is driven by true believers—those able to absorb and exploit the achievements of others. Helping us to appreciate this fact is one service done by the best intellectual history, such as Ellenberger's (1970) monumental *The Discovery of the Unconscious*. Popular culture often appears to credit Freud alone for uncovering (or inventing) the depths of mind, but Ellenberger adduces important elements of this advance in Freud's predecessors who are not today's household names: Mesmer, Charcot, and even the countless unnamed practitioners of shamanistic cures.

Of course, the moral case is different. Skepticism seems to have more bite in ethics than it does elsewhere (Railton 2003, 3), so notions of progress and expertise appear to be much more uncertain in moral domains than they do in technical and (at least some) scientific enterprises. But even on broadly skeptical accounts of moral norms (e. g., Sripada and Stich 2007), it remains the case that basic moral competence, however arbitrary and culturally local one takes that competence to be, requires a substantial measure of social conformity. This applies even to reformers: perhaps it was a moral innovation when abolitionists began to insist that African slaves be regarded as persons, but these same abolitionists were also drawing on a body of cultural norms governing how persons should be treated. The point sharpens when one thinks of religious leaders who have fostered moral dissent: King and Gandhi were not only innovators, but conservatives, drawing on the resources of ancient spiritual traditions. Without a large measure of fealty to tradition, it becomes unclear how people could function morally at all.

The individualist may want to accommodate these observations as well. If so, here is individualism$_{abc}$: optimal human reasoning is substantially asocial, and (a) sociality is necessary for the development of optimal reasoning and (b) sociality is necessary for the sustenance of optimal reasoning, and *(c) sociality is necessary for the transmission of information requisite for optimal reasoning.*

To allow (c) is not yet to admit our point about social embeddedness. The fact that social interactions and institutions transmit *information* requisite for optimal cognitive functioning does not entail that optimal reasoning itself involves social *process*, since once the needed information is in hand, the reasoner may do his own thing. Nevertheless, to say optimal reasoning can typically be asocial in this way appears to bear noteworthy entailments: for instance, it seems to follow that as much can be learned from a book as from a teacher or a research collaboration, and this would seem to make the existence of institutions like universities rather mysterious. We are in substantial sympathy with the suspicion that a primary function of American universities is the provision of socially appropriate mating opportunities for the upper middle class, but it would seem rather odd, given the familiar and plausible views in science studies mentioned above, if collaborative learning and research were only a spandrel of cultural evolution.

There is also direct argument to the effect that the role of sociality in reasoning cannot be merely informational.[13] It is true there is evidence that people tend to believe propositions they entertain (Gilbert et al. 1993), but it is also true that people do not end up believing everything they hear. Reasoning looks to involve not only the passive absorption of information, but also the *selection* of what information to keep and act on. We contend that this selection will often have a social component.

Consider an example. It is likely that most smokers want to quit: 79 percent in a 2002 Gallup poll said they did.[14] In the argot, most smokers are dissonant, rather than consonant (McKennell and Thomas 1967; cf. Kunze 2000). Quitting is notoriously difficult, of course, but there is evidence that public health campaigns decrease rates of smoking (Fiore et al. 2000).[15] It is possible that the mechanism by which such programs work is simply informational; perhaps smokers learn new facts about smoking, and this helps them to quit. But it is entirely likely that smokers *already* possess information to the effect that smoking carries substantial health risks; that is why they want to quit. Instead, anti-smoking campaigns may help people to make better use of the information they already have: for example, to make the information more vivid, which may increase the motivation to act on it. (It is one thing to know that tobacco is implicated in oral cancer, another to see a billboard featuring a face disfigured by oral malignancy.)

The particulars in each case are doubtless complex, but the general point is hard to deny: optimal reasoning involves more than recording information—it involves evaluating information, and this evaluating is socially facilitated. We are not claiming that social processes *always* facilitate good reasoning. The present point is only that the role of the social is not merely informational—even supposing the notion of "merely informational" is one of which we can make adequate sense. We therefore contend that individualism$_{abc}$, while true, does not properly accommodate the importance of social process for optimal reasoning.

As with other aspects of human functioning, the impact of sociality on cognition may be for better or worse, and is everywhere sensitive to variations in task and process. In early studies of group reasoning, research indicated that groups consistently outperformed individuals (e.g., Laughlin 1965; Davis 1969; Maier 1970). But that work compared the output of a group with the output of a single individual, meaning that the effect may have simply been the function of greater numbers (more people may output more ideas, increasing the odds of good ones) rather than any positive effect of the group process itself.[16] In subsequent work comparing the output of a group of *n* individuals with a matched "nominal" group

---

[13] We are indebted here to Eric Brown.

[14] http://www.gallup.com/poll/9910/Tobacco-Smoking.aspx.

[15] Thanks to Ernest Sosa for suggesting this example.

[16] Numerous studies indicate that factual judgments reached by aggregating and averaging the judgments of individuals in a group will be more accurate than the judgment of most individuals in the group (Suroweicki 2005, 3–7), but this does not indicate anything about the efficacy of collaborative reasoning.

of *n* individuals working alone, the results were much more mixed, with numerous findings favoring individuals (Hill 1982; Paulus, Larey, and Ortega 1995). For example, when Taylor and colleagues (1958) compared the output of four individuals "brainstorming" in group discussion to the pooled output of four individuals working separately, they found that individuals who worked on their own produced more overall ideas, more unique ideas, and higher quality ideas than individuals who worked together. Various social processes may impede the effectiveness of group efforts such as brainstorming: free riding, social anxiety (Camacho and Paulus 1995), and "production blocking," where each individual's effective work time is reduced by listening to others (Diehl and Stroebe 1987).

On the other hand, there is experimental evidence for the superiority of groups in working on some problems. Part of the explanation this superiority of groups is the role of error correction; as is evident to anyone who has attended an academic talk, individuals are frequently slower than groups to see the problems in their own work (Hill 1982, 524). Moreover, groups may be quicker to uncover an unobvious solution, as Schwartz (1995) demonstrated by comparing groups and individuals on "turning gear" puzzles. Eight such problems were presented, varying only in how many gears were involved; for example, Schwartz had people imagine five interlocked gears lined up in a row, and asked them which direction the fifth gear would turn if the first gear turns to the right. There is a rule for these problems: if the number of gears is odd, then the last gear will turn in the same direction as the first. Schwartz found that only 14 percent of the individuals discovered the rule, but 58 percent of the pairs did. It is better to appreciate the principle: those who got it were able to solve the problem much faster, as was especially clear on the final trial, which involved over one hundred gears. The large difference in success that Schwartz found between individuals and groups cannot be accounted for by appealing to nominal groups, since the success rate for nominal pairs was 28 percent, less than half that of the actual pairs. Somehow, working together on these problems helped people to recognize the deeper principle.

The potency of group insight looks to have important implications in real-world contexts. Corporations have honed collaborative research into an extremely successful technique; companies such as Gore-Tex, the "guaranteed to keep you dry" fabric manufacturer, use small group research to develop innovations, effecting remarkable market gains (Sawyer 2007). The intellectual benefits of sociality may help explain why some analyses of research productivity and impact indicate that *quantity predicts quality*; researchers who produce *more* research are more likely to produce *impactful* research (Simonton 1997). For many academic philosophers, this result may seem untoward, since much anecdotal evidence from survivors of graduate work in philosophy suggests that the disciplinary culture celebrates being "careful" and not publishing "too much." We have always thought this strategy akin to supposing that the best way to improve one's golf game is by playing fewer golf games, and we wonder whether there are more promising ways to proceed. Subjecting one's work to public (as opposed to in-clique) scrutiny in journals, one may get more, and more diverse, feedback that can help to improve the quality of future research. Furthermore, since it is difficult to anticipate which of one's ideas will "have legs" and generate an influential or provocative research program—just as it may be difficult to predict which mutations will confer reproductive advantage—one is well advised to avail oneself of more, rather than fewer, opportunities for such impact.

Relatedly, people may reason better when a diverse set of views is represented in the population (Zamzow and Nichols 2009, 377–79). A large body of research indicates that motivation plays a crucial role in reasoning; within limits, people will tend to believe what they want to believe (Dawson et al. 2002; Dunning 1999; Kunda 1990). Interestingly, these motivational factors may have asymmetric effects. As you strive for fame and fortune, you are motivated to maintain your own theory and repudiate opposing theories, so you are good at producing support for your view and objections to your opponent's. At the same time, you may be significantly worse at recognizing deficits in your theory and advantages in your opponent's theory. Thus, the quality of your reasoning may be "lopsided" in your favor. In the aggregate, this partiality may not be a bad thing, and may even be a good one, so long as there are people effectively expounding competing views so that the strengths and weaknesses of the contending positions receive balanced exposure. In addition to the aggregate benefits of such "checks and balances," the agonistic character of intellectual discourse in a diverse population of motivated reasoners may provoke individual contestants to reason more compellingly (if not more impartially) as they advocate for their views.

Can the individualist assimilate all this? We are up to individualism$_{abcd}$: optimal human reasoning is substantially asocial, and (a) sociality is necessary for the development of optimal reasoning, and (b) sociality is necessary for the sustenance of optimal reasoning, and (c) sociality is necessary for the transmission of information requisite for optimal reasoning, and *(d) sociality, in the form of socially embedded cognition, is necessary for optimal reasoning*. It is starting to look as though we are losing the plot, for qualification (d) does not look so much like an *emendation* of the original statement of individualism as it does a *rejection* of it. It may be possible to offer readings of "substantially asocial" and "socially embedded" that dispel the appearance of contradiction, or it may be possible to retain individualism as a distinctive brand while jettisoning the "substantially asocial." We certainly do not want to discourage such efforts, but we will not undertake them ourselves. Instead, we will try to ratchet up the tension a little more, by considering specifically moral cognition.

6. SOCIALITY AND MORAL REASONING

Very often, moral reasoning looks to be socially induced. As Haidt (2001, 828–29) develops the point in his "social intuitionist" account, moral judgment,

> is not just a single act that occurs in a single person's mind but is an ongoing process, often spread out over time and over multiple people. Reasons and

arguments can circulate and affect people, even if individuals rarely engage in private moral reasoning for themselves.

Haidt adopts the now familiar "dual process" approach, which distinguishes between *analytic* and *automatic* processing (e.g., Stanovich 2004): conscious, effortful, moral reasoning is supposed to be analytic processing, while "snap" or unreflective judgments, Haidt's (2001) "intuitions," are supposed to be examples of automatic processing. For Haidt, the idea is not that solitary individuals are somehow bereft of moral intuitions, it is that they will relatively seldom go in for moral reasoning without social provocation. Haidt and colleagues' (Haidt, Bjorklund, and Murphy 2000) work on "moral dumbfounding" exemplifies this phenomenon: his participants had no difficulty producing confident moral judgments (such as condemning a case of brother-sister incest as wrong), but only attempted to produce justifications when pressed to do so by experimenters. The kind of elaborated moral cognition that seems distinctive of adult humans, then, will tend to appear *dialogically*, in the context of social interaction. If so, moral reasoning is typically socially embedded reasoning.

While Haidt and colleagues (2000) have produced a variety of provocative evidence for the "intuitionist" part of the Social Intuitionist model, evidence for the "social" is harder to come by. Part of this has to do with methodological/logistical considerations: it is difficult to design well-controlled studies of group interaction, and even in the absence of significant methodological obstacles, doing so requires lots of resources, including larger numbers of participants. So while armchair evidence favors the sociality of morality—just read the paper, or watch the tube—systematic evidence is harder to come by.

Matters are more difficult for us, because we are not only saying people reason morally in groups, we are saying that moral reasoning is *better* when done in groups. And the prima facia case for this is not a promising one, as any number of mass movements like National Socialism would indicate. Experimental study may seem to give the same impression. Indeed, Haidt's (2001) moral dumfounding paradigm induced moral reasoning, but not moral reasoning that was much good. Some of the classic findings in social psychology are similarly dispiriting: in Latané and Darley's (1970, e.g., 101, 111) group effect paradigms the presence of bystanders depressed individual probability of helping in emergency situations, while Milgram's (1974) work on obedience associates sociality and destructive behavior.

We need to slow down. In Milgram's (1974: experiment 5) iconic finding, two-thirds of the subjects were willing to repeatedly shock a vehemently protesting victim in response to the "not impolite" request of the experimenter. Less often remarked on is that obedience dropped to 10 percent in the defiant peers condition (Milgram 1974, experiment 17), where participants were asked to shock the victim in the presence of two experimental confederates who were not complying with the experimenter's requests. Natural contexts also present both sides of this coin: while the Nazis exploited social dynamics to secure compliance, such as the peer "mentoring" of doctors involved in medicalized killing at Auschwitz (Lifton 1986, 193–217), sociality also facilitated *resistance* to the Nazis, as when the extraordinary village of Le Chambon, united under a Huguenot religious tradition by charismatic leadership, cooperated to shelter Jews (Hallie 1979). Experimental evidence also indicates that the *kind* of group matters: Latané and Darley (1970, 63) found that pairs of friends were faster to intervene in a group effect scenario than were pairs of strangers (although pairs did not do so as quickly as individuals).

Good examples matter too, as has been demonstrated in numerous studies of "precedent effects." For example, Reingen (1982) found that students were nearly twice as likely (46 percent versus 25 percent) to donate to the American Heart Association if they were first shown a list of eight students who had previously done so, and participants asked to donate blood were ten times as likely (30 percent versus 3 percent) to sign up after having been shown a list of previous signers. Relatedly, Cialdini and associates (1999) discovered that participants' estimates of how likely they were to take part in an unpaid survey varied dramatically with the number of classmates they had been told agreed to do so: the greater the number of potential co-participants, the greater the estimated likelihood of helping.

The role of social embedding in moral reasoning is especially clear in a paradigmatic instance of moral improvement (evaluative controversy duly noted), the development of more *inclusive* moral attitudes (i.e., less racist, sexist, xenophobic). Such attitudes are not engendered by solitary reflection, but neither are they borne simply of being in the same building with a diversity of people. In the early days of American desegregation, it was hoped that putting African American and white children in the same place would effect increased racial harmony. After all, one of the staple results in social psychology is that familiarity facilitates attraction (e.g., Zajonc 1968, Matlin 1970, Grush 1976). What happened instead was the formation of ethnic in-groups, which eventuated in in-group/out-group hostility and violence (see Rogers et al. 1984; Oskamp and Schultz 1998).

One factor that may have contributed to the divisiveness of desegregated classrooms is the competitive nature of the school environment, an interpretation supported by the classic "Robber's Cave" experiment conducted by Sherif and colleagues (1961) at an Oklahoma State Park. The first part of the experiment generated group competition between two "teams" of children at camp, which led to significant intergroup hostility. The researchers initially tried to reduce friction by bringing the groups together for pleasant activities such as movies and picnics, but this did not work; the activities merely served as occasions for the boys to express hostility with such charming expedients as shouting and throwing food at each other. However, when the experimenters introduced situations in which boys from each team had to work together in pursuit of a common goal, such as dislodging a vehicle trapped in the mud, conflict was significantly reduced. It was only when members of the different teams had to engage in cooperative activity that hostility was alleviated: the critical factor appeared to be not merely living together, but *working* together.

In a similar spirit, liberal theorists advocating "deliberative democracy" promote increased social interaction as a way of developing a more engaged citizenry and a better functioning state (Gutmann and Thompson 2004). But as any viewer of American political debates knows, it is not obvious that making discussion public makes it more intelligent. The practice of "deliberative polling" puts deliberative democracy to the empirical test: participants are surveyed before and after a weekend of structured conversation consisting of both moderated small group discussion and plenary discussion with expert panelists representing different perspectives on the issues. According to Ackerman and Fishkin (2004, 52):

> It is not unusual for deliberation to significantly change the balance of opinion on two-thirds of the policy questions. And more than half of the respondents typically change their positions on particular policy items after sustained conversations.

Changing minds after discussion looks like evidence of reasoning, but is it evidence of *good* reasoning?

At a major deliberative polling event, held in Austin during 1996 (Ackerman and Fishkin 2004, 52), the proportion of participants

favoring "a tax reduction for savings" increased 17 percent;
favoring a "flat tax" decreased 14 percent;
agreeing that "divorce should be made harder to get" (as a way of strengthening the family) increased 21 percent;
agreeing that the "biggest problem facing the American family" is "economic pressure" increased 15 percent;
agreeing that the "biggest problem facing the American family" is "breakdown of traditional values" declined 10 percent.

This is a rather motley assortment of views, and we may again seem plagued by stubborn normative questions. As ever, there is something to that thought, but a bit more can be said.

In the first instance, participants in deliberative polling tend to become much better informed (Luskin et al. 2008, 23), and better-informed reasoning is not implausibly held to be better reasoning. Moreover, there is reason to believe that participants are, with the appropriate structure, able to reason together cooperatively and respectfully (once again, not just any group will do). For example, in a Vermont deliberative polling weekend on energy policy (Luskin et al. 2008, 23), 83 percent of participants agreed that their "small group moderator provided the opportunity for everyone to participate in the discussion"; 71 percent agreed that that the members of their small group did in fact participate "relatively equally in the discussions"; and 83 percent agreed that "the important aspects of the issues were covered in the small group discussions." Participants also report being better able to respect one another's beliefs and circumstances (Luskin et al. 2008, 14): at the end of one weekend, a man was heard remarking to someone he had disparaged at the beginning of discussion, "What are the three most important words in the English language? They are 'I was wrong'" (Ackerman and Fishkin 2004, 54). This sounds pretty good to us: the deliberative polling events have the feel of reasoned discussion, and it is not unlikely that judgments so secured could better withstand critical scrutiny than unconsidered political opinions.

One important feature of group interactions is that they are likely to induce emotional responses. Many familiar emotions such as anger, guilt, and sympathy are characteristically triggered by cues in social interaction: it is the stunned look on your friend's face that makes you feel guilty for slighting her, it is the tone of your colleague debasing a peer that makes you angry, and it is the tears running down the child's face that trigger your sympathy. Guilt, anger, and sympathy are paradigmatically moral emotions (Prinz and Nichols 2010), yet from the perspective of one's short-term material interests, these emotions may seem counterproductive. As serial killer Ted Bundy philosophized:

> Guilt?....It's a kind of social control mechanism—and it's very unhealthy. It does terrible things to our bodies. And there are much better ways to control our behavior than that rather extraordinary use of guilt....I guess I am in the enviable position of not having to deal with guilt. There's just no reason for it. (reported in Michaud and Aynesworth 1989, 288)

In contrast to Bundy's psychopathic insouciance, many people appear liable to guilt when they mistreat someone. And it turns out that this feeling has salutary consequences in the long run. For example, when people feel guilt over hurting someone, this may motivate them to repair the damage that they have caused (Baumeister, Stillwell, and Heatherton 1994). Similarly, when people feel sympathy or empathy at the plight of a person in distress, this may motivate them to help the other person (Batson 1991). The point to emphasize is that such emotions are *socially cued*. Thus, there is another reason to think that sociality will facilitate moral well functioning: moral emotions such as sympathy and guilt may improve moral behavior, and the moral emotions are social emotions that are especially likely to be triggered in a social context.

Moreover, deficiency in the social emotions is plausibly thought to be implicated in moral deficiency, as it was in the case of Ted Bundy (perhaps we may here be indulged in omitting to acknowledge the possibility of normative controversy). If we are right about this, the individualist has a further fact to take on board. Here we go again, with individualism$_{abcde}$: optimal human reasoning is substantially asocial, and (a) sociality is necessary for the development of optimal reasoning, and (b) sociality is necessary for the sustenance of optimal reasoning, and (c) sociality is necessary for the transmission of information requisite for optimal reasoning, and (d) sociality, in the form of socially embedded cognition, is necessary for optimal reasoning, and *(e) sociality, in the form of socially embedded cognition, is necessary for optimal moral reasoning.*

We heartily approve of the qualifications, but we wonder about retaining the individualist brand, since the qualifications follow rather more naturally on a collaborativist tag line, *optimal human reasoning is substantially social.* Indeed, if one begins here, the *qualifications* start to look rather more like *implications.* As we have said, it may be that a suitably qualified individualism might be fashioned into a tidier package, perhaps by offering suitably nuanced readings of "substantially asocial" and "socially embedded" that dispel the awkward appearances. However, we suspect that such efforts are likely to devolve into gerrymandering, and we also suspect that the disagreement is threatening to devolve into a terminological spat.

This is tedium to avoid. The important thing is that the cognitive science of morality accommodate the importance of sociality. We think collaborativism is an efficient way to do so, but we have no objection to a qualified individualism—*so long as that branding does not serve to distract us from the required emphasis on sociality*. There is little in the name, and much in the facts. We think the facts put pressure on the name "individualism," and various understandings of human psychology that have been forwarded in the vicinity of that appellation, but there is perhaps not a great deal that needs be said about nomenclature, once we have reached the appropriate understanding.

7. CONCLUSIONS

We have argued that optimal human reasoning is typically socially embedded reasoning, and any individualism espousing the notion that sociality is inimical to the optimal expression of human rationality is contradicted by the evidence. Thus, classic statements of individualism in the philosophical tradition should be understood as highly—if not explicitly—qualified. On the other hand, if they are not highly qualified, they are badly overstated, in which case they are distinctive, but wrong. This is not to say that people never reason, or reason well, on their own. As we have been at pains to point out, not all cognitive processes are group processes, and not all social processes are felicitous processes. It is to say that an empirically defensible philosophical moral psychology will be more collaborativist in flavor than individualist.

Furthermore, an appropriately developed cognitive science of morality will have to empirically investigate group processes in much more depth than has been done. For example, studies of moral judgment will need to investigate how moral judgment eventuates from group process, and the ways in which these judgments and processes may differ from those identified by the individualistic methodologies that have so far dominated the field. If the cognitive science of morality is to continue to thrive as it has recently, it must move beyond individualism—in our view, considerably beyond it. We hope that both philosophers and scientists will take up our challenge.[17]

[17] Versions of this material were presented at the Ethics Programme and the Seminar in Science Studies, University of Oslo; the 32nd Midwest Philosophy Colloquium, University of Minnesota, Morris; the Experimental Philosophy Meets Conceptual Analysis Conference, Research School of the Social Sciences, The Australian National University; and the Society for Philosophy and Psychology Annual Meetings, York University; we are grateful to these audiences for helpful feedback. Special thanks to James Bohman, Michael Gill, Daniel Haybron, and especially Eric Brown for valuable comments. Doris gratefully acknowledges the National Humanities Center and the American Council of Learned Societies for their support.

REFERENCES

Abbott, J. H. (1991). *In the Belly of the Beast: Letters from Prison.* London: Vintage.

Ackerman, B., and Fiskin, J. S. (2004). *Deliberation Day.* New Haven, CT: Yale University Press.

Alston, W. (1989). *Epistemic Justification: Essays in the Theory of Knowledge.* Ithaca, NY: Cornell University Press.

American Psychiatric Association (2000). *Diagnostic and Statistical Manual of Mental Disorders 4th Ed. Text Revision (DSM-IV-TR).* Washington, DC: Author.

Andreou, C. (2007). Morality and psychology. *Philosophy Compass* 2: 46–55.

Appiah, K. A. (2005). *The Ethics of Identity.* Princeton, NJ: Princeton University Press.

Ariew, R., and Greene, M. (eds.). (1995). *Descartes and His Contemporaries: Meditations, Objections, and Replies.* Chicago: University of Chicago Press.

Aristotle (1984). *The Complete Works of Aristotle,* Vols. 1 and 2: *The Revised Oxford Translation.* J. Barnes (ed.), Princeton, NJ: Princeton University Press.

Aronson, E., and Bridgeman, D. (1979). Jigsaw groups and the desegregated classroom: In pursuit of common goals. *Personality and Social Psychology Bulletin* 5: 438–46.

Aronson, E., Stephan, C., Sikes, J., Blaney, N., and Snapps, M. (1978). *The jigsaw classroom.* Beverly Hills, CA: Sage.

Baron, J. (2001). *Thinking and Deciding.* 3rd ed. Cambridge: Cambridge University Press.

Baron-Cohen, S. (1995). *Mindblindness: An Essay on Autism and Theory of Mind.* Cambridge, MA: MIT Press.

Batson, C. D. (1991). *The Altruism Question: Toward a Social-Psychological Answer.* Hillsdale, NJ: Lawrence Erlbaum.

Baumeister, R. F., Stillwell, A. M., and Heatherton, T. F. (1994). Guilt: An interpersonal approach. *Psychological Bulletin* 115: 243–67.

Bickel, L. (1977). *Mawson's Will: The Greatest Survival Story Ever Written.* New York: Stein and Day.

Bird, C. (1999). *The Myth of Liberal Individualism.* Cambridge: Cambridge University Press.

Blair, R. J. R. (1996). Brief report: Morality in the autistic child. *Journal of Autism and Developmental Disorders* 26: 571–57.

——.(1997). Moral reasoning and the child with psychopathic tendencies. *Personality and Individual Differences* 26: 731–39.

Blair R. J. R., Mitchell, D., and Blair, K. (2006). *The Psychopath: Emotion and the Brain.* Oxford: Blackwell.

Bornstein, G., Kugler, T., and Ziegelmeyer, A. (2004). Individual and group decisions in the centipede game: Are groups more rational players?" *Journal of Experimental Social Psychology* 40: 599–605.

Bornstein, G., and Yaniv, I. (1998). Individual and group behavior in the ultimatum game: Are groups more "rational" players?" *Experimental Economics* 1: 101–8.

Boyd, R., and Richerson, P. J. (2006). Culture, adaptation, and innateness." In P. Carruthers, S. Stich, and S. Laurence (eds.), *The Innate Mind: Culture and Cognition.* Oxford: Oxford University Press, 23–38.

Brennan, J. (2005). Choice and excellence: A defense of Millian individualism. *Social Theory and Practice* 31: 483–98.

Brodsky, S. L., and Scogin, F. R. (1988). Offenders in protective custody: First data on emotional effects. *Forensic Reports* 1: 267–80.

Brown, E. A. (forthcoming). Aristotle on the Choice of Lives: Two Concepts of Self-Sufficiency. In P. Destrée (ed.), *Quel Choix de Vie? Études sur les Rapports Entre Theôri et Praxis chez Aristotle.* Louvain: Peeters.

Brown, J., Dreis, S., and Nace, D. K. (1999). What Really Makes a Difference in Psychotherapy Outcome? Why Does Managed Care Want to Know? In M. A. Hubble, B. L.Duncan, and S. D. Miller (eds.), *The Heart and Soul of Change: What Works in Therapy*. Washington, DC: American Psychological Association, 389–406.

Camacho, L., and Paulus, P. (1995). The role of social anxiousness in group brainstorming. *Journal of Personality and Social Psychology* 68: 1071–80.

Carens, J. H. (ed.). 1993. *Democracy and Possessive Individualism: The Intellectual Legacy of C. B. Macpherson*. Albany: State University of New York Press.

Cialdini, R. B., Wosinska, W., Barrett, D. W., Butner, J, and Gornik-Durose, M. (1999). Compliance with a request in two cultures: The differential influence of social proof and commitment/consistency on collectivists and individualists. *Personality and Social Psychology Bulletin* 25: 1242–53.

Clark, A., and Chalmers, D. J. (1998). The extended mind. *Analysis* 58: 10–23.

Cottingham, J., Murdoch, D., Stoothoff, R., and Kenny, A. (1991). *The Philosophical Writings of Descartes, Volume 3: The Correspondence*. Cambridge: Cambridge University Press.

Cooper, J. M. (1990). *Reason and Emotion*. Princeton, NJ: Princeton University Press.

Cummins, D.D., and Cummins, R. (eds.). (2000). *Minds, Brains, Programs: An Historical Introduction to the Foundations of Cognitive Science*. Oxford: Wiley-Blackwell.

D'Arms, J. (2007). Sentimental Rules and Moral Disagreement. In W. Sinnott-Armstrong (ed.), *Moral Psychology*, Vol. 2: *The Cognitive Science of Morality*. Cambridge, MA: MIT Press.

Darley, J. M., and Batson, C. D. (1973). From Jerusalem to Jericho: A study of situational and dispositional variables in helping behavior. *Journal of Personality and Social Psychology* 27: 100–108.

Darwall, S., Gibbard, A., and Railton, P. (1997). Toward Fin de Siècle Ethics: Some Trends. In S. Darwall, A. Gibbard, and P. Railton (eds.), *Moral Discourse and Practice*. Oxford: Oxford University Press, 3–47.

Dasgupta, N., and Greenwald, A. (2001). On the malleability of automatic attitudes: Combating automatic prejudice with images of admired and disliked individuals. *Journal of Personality and Social Psychology* 81: 800–814.

Davis, J. H. (1969). *Group Performance*. Reading, MA: Addison-Wesley.

Dawes, R. M. (1994). Psychotherapy: The Myth of Expertise. In *House of Cards: Psychology and Psychotherapy Built on Myth*. New York: Free Press, 38–74.

Dawson, E., Gilovich, T., and Regan D. T. (2002). Motivated reasoning and performance on the Wason Card selection task. *Personality and Social Psychology Bulletin* 28: 1379–87.

Dawson, G., and Fernald, M. (1987). Perspective-taking ability and its relationship to the social behavior of autistic children. *Journal of Autism and Developmental Disorders* 17: 487–98.

Descartes, R. (1641/1996). *Meditations on First Philosophy*, translated by John Cottingham. Cambridge: Cambridge University Press.

Diehl, M., and Stroebe, W. (1987). Productivity loss in brainstorming groups: Toward the solution of a riddle. *Journal of Personality and Social Psychology* 53: 497–509.

Doris, J. M., Knobe, J. and Woolfolk, R. L. 2007. Variantism about responsibility. *Philosophical Perspectives: Philosophy of Mind* 7: 337–48.

Doris, J. M., and Murphy, D. (2007). From My Lai to Abu Ghraib: The moral psychology of atrocity. *Midwest Studies in Philosophy* 31: 25–55.

Doris, J. M., and Stich, S. P. (2005). As a Matter of Fact: Empirical Perspectives on Ethics. In F. Jackson and M. Smith (eds.), *The Oxford Handbook of Contemporary Philosophy*. Oxford: Oxford University Press, 114–52.

————. (2006). Moral Psychology: Empirical Approaches. In Edward N. Zalta (ed.)., *The Stanford Encyclopedia of Philosophy* (Winter 2003 Edition), http://plato.stanford.edu/entries/moral-psych-emp/.

Doris, J. M., and The Moral Psychology Research Group (eds.). (2010). *The Moral Psychology Handbook*. Oxford: Oxford University Press.

Dunn, J. (1974). Democracy unretrieved, or the political philosophy of Professor MacPherson. *British Journal of Political Science* 4: 489–99.

Dunning, D. (1999). A newer look: Motivated social cognition and the schematic representation of social concepts. *Psychological Inquiry* 10: 1–11.

Ellenberger, H. (1970). *The Discovery of the Unconscious*. New York: Basic Books.

Fiore, M. C., Bailey, W. C., Cohen, S. J., Dorfman, S. F., Fox, B. J., Goldstein, M. G., et al. (2000). *Treating Tobacco Use and Dependence: A Clinical Practice Guideline*. Rockville, MD: US Department of Health and Human Services. Public Health Service. AHRQ Publication no. 00-0032.

Foster, K. R., and Ratnieks, F. L. W. (2005). A new eusocial vertebrate? *Trends in Ecology and Evolution* 20: 363–64.

Frank, R. (1988). *Passions within Reason*. New York: Norton.

Gantz, T., and Henkle, G. (2002). "Seatbelts: Current Issues." *Prevention Institute*: http://www.preventioninstitute.org/traffic_seatbelt.html.

Geberth, V. J., and Turco, R. N. (1997). Anti-social personality disorder, sexual sadism, malignant narcissism and serial murder. *Journal of Forensic Sciences* 42: 49–60.

Gibbard, A. (1990). *Wise Choices, Apt Feelings: A Theory of Normative Judgment*. Cambridge, MA: Harvard University Press.

————. (2003). *Thinking How to Live*. Cambridge, MA: Harvard University Press.

————. (2006). Moral Feelings and Moral Concepts. In Russ Schafer-Landau (ed.), *Oxford Studies in Metaethics*, Vol. 1. Oxford: Clarendon Press, 195–215.

Gigerenzer, G. (2000). *Adaptive Thinking: Rationality in the Real World*. Oxford: Oxford University Press.

Gigerenzer, G., Todd, P. M., and the ABC Group. (1999). *Simple Heuristics That Make Us Smart*. Oxford: Oxford University Press.

Gilbert, D., Tafarodi, R., and Malone, P. (1993). You can't not believe everything you read. *Journal of Personality and Social Psychology* 65: 221–33.

Goodin, R. E. (1998). Communities of enlightenment. *British Journal of Political Science* 28: 531–58.

Grassian, S. (1983). Psychopathological effects of solitary confinement. *American Journal of Psychiatry* 140: 1450–54.

Greene, J. D. (2007). The Secret Joke of Kant's Soul. In W. Sinnott-Armstrong (ed.), *Moral Psychology, Volume 3: Morality in the Brain*. Cambridge, MA: MIT Press.

Greene, J. D., Nystronm, L. E., Engell, A. D., Darley, J. M., and Cohen, J. D. (2004). The neural bases of cognitive conflict and control in moral judgment. *Neuron* 44: 389–400.

Greene, J. D., Sommerville, R., Nystrom, L., Darley, J., and Cohen, J. (2001). An fMRI investigation of emotional engagement in moral judgment. *Science* 293: 2105–108.

Grush, J. (1976). Attitude formation and mere exposure phenomena: A nonartifactual explanation of empirical findings. *Journal of Personality and Social Psychology* 33: 281–90.

Gutmann, A., and Thompson, D. (2004). *Why Deliberative Democracy?* Princeton, NJ: Princeton University Press.

Haidt, J. (2001). The emotional dog and its rational tail: A social intuitionist approach to moral judgment. *Psychological Review* 108: 814–34.

Haidt, J., Bjorklund, F., and Murphy, S. (2000). *Moral Dumbfounding: When Intuition Finds No Reason*. Unpublished manuscript. University of Virginia.

Hallie, P. P. (1979). *Lest Innocent Blood Be Shed: The Story of the Village of Le Chambon and How Goodness Happened There*. New York: Harper & Row.

Haney, C. (2003). Mental health issues in long-term solitary and "supermax" confinement. *Crime & Delinquency* 49: 124–56.

Hare, R. M. (1952). *The Language of Morals*. Oxford: Oxford University Press.

Harlow, H. F. (1986). *From Learning to Love: The Selected Papers of H. F. Harlow*, edited by C. M. Harlow. New York: Praeger.

Harlow, H. F., Dodsworth, R., and Harlow, M. (1965). Total social isolation in monkeys. *Proceedings of the National Academy of Science USA* 54: 90–97.

Harlow, H. F., and Suomi, S. (1971). Social recovery by isolation-reared monkeys. *Proceedings of the National Academy of Science USA* 68: 1534–38.

Hill, G. (1982). Group versus individual performance: Are N + 1 heads better than one?. *Psychological Bulletin* 91: 517–39.

Horvath, A. O., and Symonds, B. D. (1991). Relation between working alliance and outcome in psychotherapy: A meta-analysis. *Journal of Counseling Psychology* 38: 139–49.

Hume, D. (1777/1975). *Enquiries Concerning Human Understanding and Concerning the Principles of Morals*, 3rd ed. Oxford: Oxford University Press.

Hursthouse, R. (1999). *On Virtue Ethics*. Oxford: Oxford University Press.

Kahneman, D., Slovic, P., and Tversky, A. (1982). *Judgment under Uncertainty: Heuristics and Biases*. Cambridge: Cambridge University Press.

Kahneman, D., and Tversky, A. (1996). On the reality of cognitive illusions: A reply to Gigerenzer's critique. *Psychological Review* 103: 582–91.

Kant, I. (1963). Idea for a Universal History from a Cosmopolitan Point of View. Translated by Lewis White Beck. In L.W. Beck (ed.) *Kant: On History*. New York: Bobbs-Merrill. Originally published in 1784.

Keller, M. B., McCullough, J. P., Klein, D. N., Arnow, B., Dunner, D. L., Gelenberg, A. J., Markowitz, J. C., et al. (2000). A comparison of Nefazodone, the cognitive behavior-analysis system of psychotherapy, and their combination for the treatment of chronic depression. *New England Journal of Medicine* 342: 1462–70.

Kelly, D., Stich, S. P., Haley, K., Eng, S., and Fessler, D. (2007). Harm, affect and the moral/conventional distinction. *Mind and Language* 22: 117–31.

Kennett, J. (2006). Do psychopaths really threaten moral rationalism?. *Philosophical Explorations* 9: 69–82.

Kernberg, O. F. (1998). The Psychotherapeutic Management of Psychopathic, Narcissistic, and Paranoid Transferences. In T. Millon, E. Simonsen, M. Birket-Smith, and R. D. Davis (eds.), *Psychopathy: Antisocial, Criminal, and Violent Behavior*. New York: Guilford, 372–82.

Kitcher, P. (1993). *The Advancement of Science*. Oxford: Oxford University Press.

———. (2001). *Science, Truth and Democracy*. Oxford: Oxford University Press.

Knobe, J., and Nichols, S. (2008). *Experimental Philosophy*. Oxford: Oxford University Press.

Korsgaard, C. M. (1996). *The Sources of Normativity*. Cambridge: Cambridge University Press.

Krupnick, J. L., Sotsky, S. M., Simmens, S., Moyher, J., Elkin, I., Watkins, J., and Pilkonis, P. A. (1996). The role of the therapeutic alliance in psychotherapy and pharmacotherapy outcome: Findings in the National Institute of Mental Health Treatment of Depression Collaborative Research Project. *Journal of Consulting and Clinical Psychology* 64: 532–39.

Kuhn, T. (1962). *The Structure of Scientific Revolutions*. Chicago: University of Chicago Press.

Kunda, Z. (1990). The case for motivated reasoning. *Psychological Bulletin* 108: 480–98.

———. 1999. *Social Cognition: Making Sense of People*. Cambridge MA: MIT Press.

Kunze, M. (2000). Maximizing help for dissonant smokers. *Addiction* 95: 13–17.

Lambert, M. J., and Ogles, B. M. (2004). The Efficacy and Effectiveness of Psychotherapy. In M. J. Lambert (ed.), *Bergin and Garfield's Handbook of Psychotherapy and Behavior Change*. New York: Wiley, 139–93.

Latane, B., and Darley, J. (1970). *The Unresponsive Bystander: Why Doesn't He Help?* New York: Appleton-Century-Crofts.

Laughlin, P. R. (1965). Selection strategies in concept attainment as a function of number of persons and stimulus display. *Journal of Experimental Psychology* 70: 323–27.

Leiter, B. (2002). *Nietzsche on Morality*. London: Routledge.

Lifton, R. J. (1986). *The Nazi Doctors: Medical Killing and the Psychology of Genocide*. New York: Basic Books.

Loeb, D. (1995). Full-information theories of individual good. *Social Theory and Practice* 21: 1–30.

Longino, H. (1990). *Science as Social Knowledge: Values and Objectivity in Scientific Inquiry*. Princeton, NJ: Princeton University Press.

Lowery, B., Hardin, C., and Sinclair, S. (2001). Social influence effects on automatic racial prejudice. *Journal of Personality and Social Psychology* 81: 842–55.

Luborsky, L., Singer, B., and Luborsky, L. (1975). Comparative studies of psychotherapies: Is it true that 'everyone has won and all must have prizes'?. *Archives of General Psychiatry* 32: 995–1008.

Luborsky, L., McLellan, A. T., Woody, G. E., O'Brien, C. P., and Auerbach, A. (1985). Therapist success and its determinants. *Archives of General Psychiatry* 42: 602–11.

Luskin, R. C., Crow, D. B., Fiskin, J. S., Guild, W., and Thomas, D. 2008. Report on the Deliberative Poll on "Vermont's Energy Future". Austin, TX: Center for Deliberative Opinion Research, University of Texas at Austin.

Martin, D. J., Garske, J. P., and Davis, M. K. (2000). Relation of the therapeutic alliance with outcome and other variables: A meta-analytic review. *Journal of Consulting and Clinical Psychology* 68: 438–50.

McKennell, A. C., and Thomas, R. K. (1967). *Adults' and Adolescents' Smoking Habits and Attitudes*. London: British Ministry of Health.

MacIntyre, A. (1984). *After Virtue*. Notre Dame, IN: Notre Dame University Press.

MacPherson, C. B. (1962). *The Political Theory of Possessive Individualism: Hobbes to Locke*. Oxford: Oxford University Press.

Maibom, H. (2005). Moral unreason: The case of psychopathy. *Mind & Language* 20: 237–57.

Maier, N. R. F. (1970). *Problem-Solving and Creativity in Individuals and Groups*. Belmont, CA: Brooks/Cole.

Matlin, M. W. (1970). Response competition as a mediating factor in the frequency-affect relationship. *Journal of Personality and Social Psychology* 16: 536–52.

May, L. (1992). *Sharing Responsibility*. Chicago: University of Chicago Press.

May, L., and Hoffman, S. (1991). *Collective Responsibility*. Lanham, MD: Rowman and Littlefield.

Mealey, L. (1995). The sociobiology of sociopathy: An integrated evolutionary model. *Behavioral and Brain Sciences* 18: 523–99.

Michaud, S., and Aynesworth, H. (1989). *Ted Bundy: Conversations with a Killer*. New York: New American Library.

Milgram, S. (1974). *Obedience to Authority*. New York: Harper and Row.

Mill, J. S. (2002). On Liberty. In D. E. Miller (ed.), *The Basic Writings of John Stuart Mill*, with an introduction by J. B. Schneewind. New York: The Modern Library. Originally published in 1863.

Moll, J., de Oliveira-Souza, R., Bramati, I. E., and Grafman, J. (2002a). Functional networks in emotional, moral, and nonmoral social judgments. *Neuroimage* 16: 696–703.

Moll, J., de Oliveira-Souza, R., Eslinger, P. J., Bramati, I. E., Mourão-Miranda, J., Andreiuolo, P. A., and Pessoa, L. (2002b). The neural correlates of moral sensitivity: A functional magnetic resonance imaging investigation of basic and moral emotions. *Journal of Neuroscience* 22: 2730–36.

Moll, J., de Oliveira-Souze, R., and Zahn, R. (2008). The neural basis of moral cognition: Sentiments, concepts, and values. *The Year in Cognitive Neuroscience* 1124: 161–80.

Murphy, D. (2006). *Psychiatry in the Scientific Image.* Cambridge, MA: MIT Press.

Newton, M. (2002). *Savage Boys and Wild Girls: A History of Feral Children.* London: Faber and Faber.

Nichols, S. (2007). Sentiment, Intention, and Disagreement: Replies to Blair and D'Arms. In W. Sinnott-Armstrong (ed.), *Moral Psychology*, Vol. 2: *The Cognitive Science of Morality.* Cambridge, MA: MIT Press.

Nichols, S., Folds-Bennett, T. Unpublished data.

Nichols, S., and Knobe, J. (2007). Moral responsibility and determinism: The cognitive science of folk intuitions. *Nous* 41: 663–85.

Nichols, S., and Stich, S. P. (2003). *Mindreading.* Oxford: Oxford University Press.

Nietzsche, F. (1966). *Beyond Good and Evil,* translated by Walter Kaufmann. New York: Vintage. Originally published 1886.

Orlinsky, D. E., Rønnestad, M. H., and Willutzki, U. (2004). Fifty Years of Process-Outcome Research: Continuity and Change. In M. J. Lambert (ed.), *Bergin and Garfield's Handbook of Psychotherapy and Behavior Change,* 5th ed. New York: Wiley, 307–90.

Oskamp, S., and Schultz, P. (1998). *Applied Social Psychology.* Englewood Cliffs, NJ: Prentice Hall.

Paulus, P. B., Larey, T. S., and Ortega, A. H. (1995). Performance and perceptions of brainstormers in an organizational setting. *Basic and Applied Social Psychology* 17: 249–65.

Pizarro, D. A., and Bloom, P. (2003). The intelligence of the moral intuitions: A reply to Haidt (2001). *Psychological Review* 110: 193–96.

Prinz, J. J., and Nichols, S. (2010). Moral Emotions. In J. M. Doris and The Moral Psychology Research Group (eds.), *The Moral Psychology Handbook.* Oxford: Oxford University Press.

Railton, P. (2003). *Facts, Values, and Norms: Essays Toward a Morality of Consequence.* Cambridge: Cambridge University Press.

Reingen, P. H. (1982). Test of a list procedure for inducing compliance with a request to donate money. *Journal of Applied Psychology* 67: 110–18.

Rogers, M., Hennigan, K., Bowman, C., and Miller, N. (1984). Inter-Group Acceptance in Classrooms and Playground settings. In N. Miller and M. Brewer (eds.), *Groups in Contact: The Psychology of Desegregation.* New York: Academic Press.

Rosati, C. S. (1995). Persons, perspectives, and full information accounts of the good. *Ethics* 105: 296–325.

Samuels, R., and Stich, S. P. (2004). Rationality and Psychology. In A. Mele and P. Rawling (eds.), *The Oxford Handbook of Rationality.* Oxford: Oxford University Press, 279–300,

Sandel, M. J. (1982). *Liberalism and the Limits of Justice.* Cambridge: Cambridge University Press.

Sawyer, R. K. (2007). *Group Genius: The Creative Power of Collaboration*. New York: Basic Books.

Seligman, M. E. P. (1993). *What You Can Change and What You Can't: The Complete Guide to Successful Self-Improvement*. New York: Knopf.

Shults, R. A., Elder, R. W., Sleet, D. A., Thompson, R. S., and Nichols, J. L. (2004). Primary enforcement seat belt laws are effective even in the face of rising belt use rates. *Accident Analysis & Prevention* 36: 491–93.

Schwartz, D. (1995). The emergence of abstract representations in dyad problem solving. *Journal of the Learning Sciences* 4: 321–54.

Schwartzberg, S. S. (2000). Case 8: Narcissistic Personality Disorder. In *Casebook of Psychological Disorders: The Human Face of Emotional Distress*. Needham Heights, MA: Allyn & Bacon, 106–88.

Sestoft, D., Andersen, H., Lillebaek, T., and Gabrielson, G. (1998). Impact of solitary confinement on hospitalization among danish prisoners in custody. *International Journal of Law and Psychiatry* 21: 99–108.

Sherif, M., Harvey, O., White, B., Hood, W., and Sherif, C. (1961). *Intergroup Conflict and Cooperation: The Robbers Cave Experiment*. Norman: University of Oklahoma Institute of Intergroup Relations.

Smith, M. (1994). *The Moral Problem*. Cambridge: Basil Blackwell.

Smith, G. R., Rost, K., and Kashner, M. A. (1995). A trial of the effect of a standardized psychiatric consultation on health outcomes and costs in somatizing patients. *Archives of General Psychiatry* 52: 238–43.

Sinha, R. & Easton, C. (1999). Substance abuse and criminality. *Journal of the American Academy of Psychiatry and Law* 27: 513–26.

Simonton, D. (1997). Creative productivity: A predictive and explanatory model of career trajectories and landmarks. *Psychological Review* 104: 66–89.

Sinnott-Armstrong, W. (2007 a). *Moral Psychology, Vol. 1: The Evolution of Morality*. Cambridge, MA: MIT Press.

———. (2007 b). *Moral Psychology, Vol. 2: The Cognitive Science of Morality*. Cambridge, MA: MIT Press.

———.(2007 c). *Moral Psychology, Vol. 3: The Neuroscience of Morality*. Cambridge, MA: MIT Press.

Sobel, D. 1994. Full Information accounts of well-being. *Ethics* 104: 784–810.

Sober, E., and Wilson, D. S. (1998). *Unto Others: The Evolution and Psychology of Unselfish Behavior*. Cambridge, MA: Harvard University Press.

Sripada, C., and Stich, S. P. (2007). A Framework for the Psychology of Norms. In P. Carruthers, S. Laurence, and S. Stich (eds.), *The Innate Mind: Culture and Cognition*. Oxford: Oxford University Press, 280–301.

Stanovich, K. E. (2004). *The Robot's Rebellion: Finding Meaning in the Age of Darwin*. Chicago: University of Chicago Press.

Stich, S. P. (1996). *Deconstructing the Mind*. Oxford: Oxford University Press.

Sullivan, L. (2006). "At Pelican Bay Prison, a Life in Solitary." National Public Radio, July 28, 2006, http://www.npr.org/templates/story/story.php?storyId=5584254

Suroweicki, J. (2005). *The Wisdom of Crowds*. New York: Random House.

Taylor, D. W., Berry, P. C., and Block, C. H. (1958). Does group participation when using brainstorming facilitate or inhibit creative thinking? *Administrative Science Quarterly* 3: 23–47.

Thaler, R. H. (1988). Anomalies: The ultimatum game. *Journal of Economic Perspectives* 2: 195–206.

Tollefsen, D. (2002). Collective intentionality and the social sciences. *Philosophy of the Social Sciences* 32: 25–50.

———. (2003). Participant reactive attitudes and collective responsibility. *Philosophical Explorations* 6: 218–34.

———. (2006). The rationality of collective guilt. *Midwest Studies in Philosophy* 30: 222–39.

Townshend, J. (2000). *C. B. Macpherson and the Problem of Liberal Democracy*. Edinburgh: Edinburgh University Press.

Turiel, E. (1983). *The Development of Social Knowledge: Morality and Convention*. Cambridge: Cambridge University Press.

Van der Vegt, G., and Bunderson, J. S. (2005). Learning and performance in multidisciplinary teams: The importance of collective team identification. *Academy of Management Journal* 48: 532–47.

Vazire, S., and Funder, D. C. (2006). Impulsivity and the self-defeating behavior of narcissists. *Personality and Social Psychology Review* 10: 154–65.

Vazire, S., Naumann, L. P., Rentfrow, P. J., and Gosling, S. D. (2008). Portrait of a narcissist: Manifestations of narcissism in physical appearance. *Journal of Research in Personality* 42: 1439–47.

Velleman, J. D. (1988). Brandt's Definition of "Good". *The Philosophical Review* 97: 353–71.

Velleman, J. (2006). *Self to Self: Selected Essays*. New York: Cambridge University Press.

Wampold, B. E., Mondin, G. W., Moody, M., Stich, F., Benson, K., and Ahn, H. (1997). A Meta-Analysis of Outcome Studies Comparing Bona Fide Psychotherapies: Empirically, "All Must Have Prizes". *Psychological Bulletin* 122: 203–15.

Williams, B. (1981). *Moral Luck*. Cambridge: Cambridge University Press.

Wilson, R. A. (2007). A puzzle about material constitution and how to solve it: Enriching constitution views in metaphysics. *Philosophers Imprint* 7: 1–20.

Zajonc, R. (1968). Attitudinal effects of mere exposure. *Journal of Personality and Social Psychology Monograph Supplement* 9: 1–28.

Zamzow, J., and Nichols, S. (2009). Variations in ethical intuitions. *Philosophical Issues* 19(1): 368–88.

CHAPTER 19

CONCEPTUAL DEVELOPMENT: THE CASE OF ESSENTIALISM

SUSAN A. GELMAN AND ELIZABETH A. WARE

THIS chapter concerns conceptual development in children. Although psychologists have commonly studied concepts by focusing on the adult endpoint alone, the facts of development specially inform the study of concepts. Much as grammars are acquired within a tightly constrained timetable, thereby yielding insights about the nature of the mind that is capable of learning them (Chomsky 1968), so too are concepts and conceptual systems acquired within developmental constraints. Relatedly, the study of children clearly raises the issue of whether there are innate concepts, and if so, what those might be. Developmental issues also evoke questions about conceptual change and whether genuine conceptual change is possible in human ontogenesis. These issues are well-known philosophical conundrums and have been discussed by cognitive scientists interested in development.

Advances in the field of cognitive science have made the study of concepts more tractable. One brief example will suffice. Forty years ago, psychologists assumed a certain view of concepts, as arbitrary clusters of features that are jointly necessary and individually sufficient for concept membership (paradigmatic example: "bachelor" = unmarried, marriageable male; see Smith and Medin 1981). This view is rooted in assumptions that favor certain empirical findings and theoretical conclusions (e.g., that concepts are learned, not innate; that concepts are arbitrary, not predictably structured;

that concepts undergo qualitative shifts across development). But then interdisciplinary scholarship undermined this view by showing that natural concepts—those that people spontaneously use—have unexpected structure and principles. From philosophy, it became clear that stipulated features fail to capture natural kinds (Putnam 1975; Quine 1969). From linguistic analyses, one could see that natural language concepts have predictable structure and hierarchical levels (Rosch 1976). From anthropology, universal patterns and principles of categorization emerged (Atran 1998). Thus, natural concepts (that are embodied in language and cross-culturally universal) differ markedly in structure from those created in the lab.

Thanks to an active field of research over the past thirty years, much is known about the nature of early concepts and conceptual development, in content areas that encompass space, time, force, gravity, number, causality, objects, emotions, agency, animacy, theory of mind, etc. Whole volumes have been devoted to reviewing new research findings with infants and children (e.g., Kuhn and Siegler 2006). Obviously an empirical review of this vast literature is beyond the scope of the present chapter. At the same time, a purely descriptive empirical review would fail to engage the issues of greatest philosophical interest.

Therefore, in order to highlight issues of broad theoretical concern, we have opted to focus on a single (yet wide-ranging and fundamental) aspect of conceptual development. Specifically, we focus on the development of *psychological essentialism*: the idea that members of a category have an underlying cause (essence) that makes the category what it is and gives rise to surface commonalities. We focus on essentialism for three reasons. First, it reveals some surprisingly early competencies in children's cognition. Second, it provides central evidence regarding debates about continuity versus conceptual change, and about nativist versus empiricist approaches. And third, essentialism is a way of thinking that emerges early in childhood and is maintained throughout life, yet it makes assumptions about categories that conflict with most scientific characterizations. It is thus a human bias that clashes with reality, thereby raising questions regarding why human thought has the structure that it does, and what functions human reasoning errors have for our species.

We first provide an overview of essentialism in philosophy, evidence for its importance in everyday human thought, and theoretical accounts of its emergence. We use this as a springboard to consider broader issues regarding the origins of concepts and the possibility of conceptual change.

1. Essentialism and Psychological Essentialism: Some Framing Issues

Psychological essentialism is an intuitive folk belief about the nature of (at least some) categories. According to psychological essentialism, an *essence* is an underlying reality or true nature, shared by members of a category, that one cannot observe directly but that gives an object its identity, and is responsible for other similarities that category members share (James 1890/1981; Locke 1959; Medin 1989).[1] A clear example of this belief can be found in a set of experiments conducted by the psychologist Frank Keil (1989). He introduced children and adults to hypothetical scenarios in which animals were superficially transformed so that their appearance and behavior changed dramatically. For example, a raccoon was modified (via surgery to change its body shape, paint to add a white stripe down its back, addition of a smelly sac of liquid, etc.) so that it looked and acted just like a typical skunk. Participants were then asked whether the animal was still a raccoon or had been changed into a skunk. Adults and older children consistently reported that the raccoon had *not* been changed into a skunk. They maintained that its identity could not be modified by means of superficial modifications, and that its outward appearance was irrelevant. This essentialist argument maintains that the animal was born a raccoon and as such would always be a raccoon. The raccoon "essence," as it were, could not be erased or modified.

Thus, there are two primary components to psychological essentialism as an intuitive folk construal: (1) the belief that certain categories are natural kinds (i.e., real, discovered, and rooted in nature), and (2) the belief that there is some unobservable property (part, substance, or quality)—the essence—that causes things to be the way they are and gives rise to the observable similarities shared by members of a category. For example, the essence of a person is whatever quality remains constant as that individual grows and transforms (e.g., baby to child to adult); the essence of water is whatever quality remains constant as that substance changes shape, size, or state (e.g., water to vapor to ice).

Psychological essentialism thus has commonalities with philosophical approaches to essentialism, ranging from Plato to Aristotle, and on to Locke and James (though these approaches are certainly not equivalent to one another; see Gelman 2003 for discussion). It is related to notions of natural kinds (Mill 1843; Quine 1969; Schwartz 1979), as well as to considerations of how to characterize the meaning of words for natural kinds, which cannot be understood in terms of simple perceptual properties (Kripke 1980; Putnam 1975). At the same time, there is an asymmetry between psychological and philosophical approaches to essentialism. One important difference between psychological essentialism and philosophical accounts of essentialism is that the former does not claim to provide a metaphysical account about the structure of the world; indeed, most biologists strenuously deny the existence of essences in the characterization of species (Mayr 1991; Sober 1994; Wilson 1999). Rather, it is a set of claims regarding human conceptual biases (Medin 1989; Medin and Ortony 1989). These could also be considered "heuristics" in the sense that Gigerenzer (2000) uses. Although often psychological essentialism is characterized as a set of "beliefs," this is meant loosely, to include nonconscious, intuitive assumptions, not metacognitive or explicit beliefs.

---

[1] Here we focus on essentialism of *categories*. Although it is also possible to apply essentialism to individuals (e.g., John has a special quality or essence that makes him who he is), we will not consider that type of essentialism in the present chapter.

Whereas psychologists strictly aim to examine the psychological representations of concepts, philosophers have examined essentialism with the goal of addressing a range of issues: psychological, semantic, and metaphysical. For example, Kripke's thought experiments regarding word meaning are analogous to psychological experiments regarding word meaning (e.g., Keil 1989), though Kripke's conclusions extend beyond psychological intuition and into semantics and metaphysics. One point that will be important to keep in mind in our review of the literature on essentialism concerns which claims and evidence speak to psychological issues (of semantic representation), and which speak to broader philosophical issues (concerning the nature of material experience).

It is also important to clarify that essentialism is *not* equivalent to positing the existence of "defining properties." Although this is a common interpretation of essentialism, it is not the sense that we use, for two reasons. First, essentialism is a placeholder concept, with the actual content often unknown. One might think there is *something* deep, nonobvious, and causal that makes humans different from other species well before one learns about DNA, chromosomes, or anything biological. A placeholder is not a definition. For example, to have the concept of tiger, it is not as if the agent has to represent a definition that makes explicit the purported essential properties. Instead, the agent just has to represent that there is a distinct animal kind, with a presumed unique essence, to which a particular instance or set of instances belong. Second, we are interested in the essence as having a causal force, which a purely definitional redescription cannot do (e.g., "featherless biped" may be a serviceable definition of "human," but these properties do not cause one to be a human; see Gelman and Hirschfeld 1999 for more discussion). The causal sense of essence implies that a category has inherent, hidden properties that *determine* observable qualities.

Although discussions of essentialism permeate Western philosophical tradition, recent psychological studies suggest that similar notions pervade commonsense lay reasoning, as well. We turn next to a review of this work.

2. Evidence for Psychological Essentialism in Adults and Children

There is broad evidence for essentialism in everyday thinking, in adults as well as children.

2.1. Adults

Intuitively, essentialism appears to underlie a range of informal observations concerning gender (the former president of Harvard speculated that sex differences in math/science abilities reflect innate differences in ability; Summers 2005), race (most Americans surveyed believe that two people from the same race are more genetically similar to each other than two people from different races; Jayaratne et al. 2006), and biological species (one-half of Americans reject evolutionary theory, reporting instead that species are fixed and unchanging; Evans 2000; Mayr 1991).

More formal scientific investigations use varied tasks (see Gelman 2003, for review), including questionnaire studies in which college students readily agree to statements such as the following when reasoning about a range of human characteristics (Haslam, Rothschild, and Ernst 2000):

- "It is not easy to change this characteristic; it is a fixed attribute." (immutability)
- "It is possible to know about many aspects of X once you become familiar with a few of its basic traits." (inductive potential)
- "The kind of X something is can be largely attributed to its genetic inheritance." (innate potential)
- "Some categories have an underlying reality; although their members have similarities and differences on the surface, underneath they are basically the same." (nonobvious basis)

Furthermore, adults treat underlying causes as more central to category membership than effects (Ahn et al. 2000; Rehder 2006), treat certain properties as inborn or biologically determined (Mahalingam 2003), treat category membership as fixed and unchanging (Prentice and Miller 2006), and treat biological species as having internal, causal natures and as being classifiable into a stable hierarchical taxonomy (Atran 1998).

2.2. Children

As noted earlier, the study of essentialism in children provides not only insights into children's cognition, but information regarding the roots of human concepts. Children—particularly young children who have not yet had formal schooling—provide an ideal population for studying a conceptual system that has been relatively untouched by prior scientific knowledge or Western philosophical traditions. If essentialism grows out of historical, contingent experiences, then children should not (yet) be essentialist. In contrast, to the extent that children also essentialize, this suggests that it may be a fundamental cognitive bias.

As just noted above, essentialism includes several component beliefs, including: that categories have sharp, immutable boundaries; that category members share deep, nonobvious commonalities; and that category membership has an innate, genetic, or biological basis. Elsewhere Gelman has detailed empirical evidence that preschool children expect certain categories (e.g., natural kinds) to have each of these properties (Gelman 2003, 2004). Prentice and Miller (2006, 129) summarize much of this literature concisely:

[A] a wolf remains a wolf even if it is wearing sheep's clothing (Gelman & Markman, 1986, 1987), even if a doctor performs an operation that makes it look like a sheep (Keil, 1989), and even if it eats something that turns it into an object resembling a sheep (Rips, 1989). Moreover, a wolf will develop wolflike characteristics even if it grows up in a community of sheep (Gelman & Wellman, 1991).

To provide a sense of this research, we briefly review two sets of experimental evidence concerning innate potential and underlying structure.

Innate potential. To test the essentialist belief that properties are fixed at birth, researchers have asked children to consider hypothetical scenarios about an organism that had one set of biological parents (or, in the case of plants, a seed that grew from a plant of a particular type) and then was transferred at birth to a new environment and a new set of parents. Children are then asked how the organism will develop, and in particular whether the birth parents or environmental context are more important. For example, in one study, children learned about a newborn tiger that went to live with horses, and were asked whether the tiger would grow to have striped or plain fur, and whether it would neigh or roar (Gelman and Wellman 1991). Preschool children typically reported that it would have striped fur and would roar (i.e., reflecting nativist beliefs). A follow-up study demonstrated that this pattern did not simply reflect associations with the relevant category (e.g., associating tigers with striped fur), because children were much less likely to report that the properties were yet present in the newborn animal. These patterns hold for children's reasoning not only about nonhuman species, but also about human kinds, including race (Hirschfeld 1996) and traits (Heyman and Gelman 2000). Precisely when this understanding emerges is subject to debate (e.g., Solomon et al. 1996; Springer 1996), but even under a conservative estimate this pattern appears by about six years of age.

An interesting additional finding is that children are even more likely than adults to show an innate bias, predicting that birth parents are more determinative than adoptive parents of the language that a child will speak (Hirschfeld and Gelman 1997). Similarly, the youngest children tested (four- and five-year-olds) are most likely to treat gender-linked properties (e.g., a boy wanting to play with trucks; a girl wanting to play with dolls) as innately determined, whereas older children and adults recognize the importance of environmental factors (Taylor 1996).

Nativist beliefs are also not limited to those in the United States. Even in distinct cultures, where beliefs about birth and reproduction vary widely, a nativist bias can be seen, for example, among Torguud adults in Western Mongolia (Gil-White 2001), upper-caste adults in India (Mahalingam 2003), Vezo children in Madagascar (Astuti, Solomon, and Carey 2004), and Itzaj Maya adults and children in Mexico (Atran et al. 2002).

Underlying structure. One primary goal in studies of underlying structure is to examine when in development children consider features that are internal, hidden, nonobvious, and/or non-perceptual. Although classic developmental accounts would argue that such features are beyond the capacity of young children (e.g., Piaget 1983), many researchers have found that preschool children can indeed consider such "underlying" properties. Of greatest relevance to essentialism, preschoolers make inductive inferences about internal features and nonvisible functions on the basis of category membership, even when category membership competes with perceptual similarity (e.g., extending properties from a leaf insect to a typical insect, not to a perceptually similar leaf; Gelman and Markman 1986; Jaswal and Markman 2007). With familiar and/or natural transformations, young children report that an animal cannot be transformed into another kind of thing (for example, a lion in a tiger costume does not become a lion; deVries 1969; Keil 1989). Instead, category membership remains stable over striking transformations as long as the insides remain unchanged (Gelman and Wellman 1991).

Moreover, preschool children can readily consider a striking range of nonobvious features as a basis for their inferences and categorizations, often while disregarding irrelevant perceptual cues. They can reason about internal parts and hidden causes (Diesendruck 2001; Gopnik and Sobel 2000; Gottfried and Gelman 2005). They can reason about functions (Kemler et al. 2000). They can reason about psychological states, including beliefs, desires, and intentions (Wellman 1990).

Indeed, even young infants are sensitive to the significance of nonobvious properties to categories. Before age two, infants generalize domain-specific properties (e.g., sleeping versus driving) exclusively to objects belonging to the same domain (e.g., animals or vehicles; Mandler and McDonough 1996), regardless of whether the objects are perceptually similar. Infants also focus on function information—they expect objects given the same label to exhibit the same nonobvious function (e.g., to rattle when shaken; Graham, Kilbreath, and Welder 2004; Welder and Graham 2001), and are more likely to group objects into a category after seeing them used for the same function (Booth 2006). Infants also show a basic awareness of intention and agency, for example, distinguishing between intentional and unintentional actions (Woodward 1999). All of these capacities likely serve as important precursors to the development of richer beliefs about underlying structure.

3. RECONCILING WITH ARGUMENTS AGAINST ESSENTIALISM

The evidence for essentialism is multipronged and converging. Nonetheless, some scholars have argued that essentialism cannot account for a variety of experimental findings (Braisby, Franks, and Hampton 1996; Rips 2001; Sloman and Malt 2003; Strevens 2000). For example, the extent to which different liquids are judged to be water is independent of the extent to which they share the purported essence of water, H_2O (Malt 1994). A consideration of these arguments highlights broader points about conceptual structure.

One issue is that much of this debate seems to reflect differing ideas of what is meant by *essentialism*. If one considers essentialism to imply that people make use of consistent defining features, such that categories have sharp and absolute boundaries, then it will fall short (Kalish 1995, 1998). The world is a messy place, with individuals crossing even the seemingly firmest of boundaries (e.g., interbreeding species, hybrid plants, intersex babies). Kamp and Partee (1995, 172) suggest that it may be only in abstract domains (e.g., pure mathematics) that we see categories with absolutely sharp boundaries. However, essentialism does not require that categories be treated as absolute; rather, essentialism is the claim that category boundaries are *intensified*—that boundaries between categories are treated as relatively more dichotomous (either/or, discrete, or non-fuzzy) than they truly are, and relatively more dichotomous than one would predict based on typicality or similarity (Diesendruck and Gelman 1999). Boundary intensification is analogous to the categorical perception of speech sounds, where two sounds that fall along a perceptual continuum (e.g., "ba" and "pa") are instead perceived as falling into either of two distinct categories (Wood 1976). Importantly, however, it is not that people perceive no fuzzy boundary between the two ends of the perceptual continuum, but rather that people greatly reduce the fuzzy boundary compared to the variability in the physical stimulus. Thus, if we understand essentialism as implying boundary intensification rather than absolute boundaries, then mixed or hybrid categories need not be opposed to essentialism (Wade 2004).

Disagreement also stems from the kinds of concepts under investigation. Different concepts are construed differently and may be essentialized to greater or lesser degrees (Gelman and Kalish 2006; Medin, Lynch, and Solomon 2000; Prentice and Miller 2007). Indeed, one of the important implications of psychological essentialism is the domain-specificity of how concepts are structured. Animal species are prime candidates for essentialism, simple artifacts (e.g., cups) and certain social categories (e.g., Republicans) less so, and arbitrary constructions (e.g., "objects weighing less than six pounds") unlikely under any account. Thus, the suggestion that artifact concepts are not essentialized need not imply the same about animal kinds. Some of the major arguments against essentialism focus on artifact concepts (e.g., Jones and Smith 1993; Sloman and Malt 2003), yet other scholars have proposed that artifact concepts are poor tests of essentialism because artifact categories do not capture nonobvious inherent commonalities (e.g., Keil 1989).

On the other hand, other scholars (e.g., Bloom 1996; Kornblith 1993) have proposed that essentialism can extend quite broadly to artifacts as well. For example, classifying an object as a cup depends in part on the (nonobvious, causally powerful) intent of the creator. Furthermore, we have the intuition that a work of art, such as a Picasso painting, has a nonobvious essence that does not appear in a reproduction. The question of which domains can be essentialized rests in part on how broad a notion of essentialism one presumes, and remains a question for future research.

Furthermore, even controlling for conceptual content, context and process can have powerful effects. When quick identification is the goal (e.g., in order to avoid a predator), then rapid and automatic processes are likely to take over, with resultant focus on perceptually salient properties; when accuracy in prediction is the goal (e.g., reasoning about genealogy), then people are more likely to make use of hidden, nonobvious properties. Not surprisingly, then, those arguing for the importance of perceptual features in children's concepts have tended to rely on sorting tasks or object-identification tasks (e.g., Smith, Jones, and Landau 1996). In contrast, those arguing for the importance of nonobvious features in children's concepts have tended to rely on tasks that require a deeper level of processing, such as making inductive inferences or reasoning about causal properties (e.g., Ahn et al. 2000; Gelman and Markman 1986). This is not to say, however, that essentialism itself is a deliberative process; to the contrary, we expect it to be unconscious, automatically invoked, etc. The general point here is that, in contrast to the historical assumption that categorization is a single, unitary process (Murphy 2002), categorization serves numerous distinct functions, and we recruit different kinds of information depending on the task at hand (Rips 2001; Smith, Patalano, and Jonides 1998). Thus, essentialist beliefs are recruited when they are most useful, such as when making deeper attributions or inferences.

It is also the case that essentialism itself is likely to have multiple components (Haslam et al. 2000; Rothschild and Haslam 2003), and for certain categories or tasks people may endorse some but not all of these components. Children in particular seem to have difficulty understanding how information about one essentialist assumption (innateness, for example) will have implications for another (such as stability; Gelman, Heyman, and Legare 2007). It is also possible to separate essentialism into a "natural kind" component, which involves treating categories as deeply structured kinds, versus an "essentialist" component, which additionally assumes that there is an underlying essence that *causes* kind membership. Children may not at first appreciate the essentialist component (see Strevens 2000 and Ahn et al. 2001 for debate).

4. Origins and Nativism: Where Does Essentialism Come From?

One of the oldest issues in philosophy and cognitive science is that of innateness, and the extent to which the human mind is a product of nature versus nurture. Yet psychologists are trained to be wary of a strict "nature-versus-nurture" debate (e.g., Pinker 2002; Sameroff 1983) for at least three reasons: (1) all behaviors have both innate and learned components (Elman et al. 1996), (2) learning need not be in conflict with innateness; rather, certain forms of learning are innately prepared (Marler 1991), and (3) genetic structures require appropriate context and experience to develop appropriately; they do not carry within them all the information required for ultimate expression. Nonetheless, despite the dangers associated with considering innateness in overly simplified terms, the question still is timely, leading to questions regarding the nature of the innate mind, the extent to which human cognition is malleable, and the puzzle of where knowledge comes from.

Concepts are internal mental states and as such not directly observable (as we are so keenly aware historically from behaviorist accounts; Skinner 1953). Thus, the nature of conceptual representations is always inferred. When we see data supportive of essence-like construals in children, this raises the question of how to interpret these results, and what they mean about cognition. Debates around essentialism provide a microcosm of more general debates concerning the nature of cognition. As Gelman (2003) notes:

> In any discussion of essentialism, controversies rage. Is essentialism in the world or in the mind? Is essentialism innate in the infant (Atran, 1998) or "a late and sophisticated achievement" (Fodor, 1998, p. 159)? Is essentialism a universal "habit of the mind" (Atran, 1998, p. 551) or limited to certain points in history (Fodor, 1998)? ... Debates about essentialism sit astride debates about the very nature of human cognition.

We can (crudely) divide the arguments about origins into at least two broad sorts (corresponding to nature and nurture, respectively): (1) domain-specific nativist accounts that propose that essentialism is part of a domain-specific module (or modules), and (2) domain-general learning accounts that are bottom-up and empiricist, and attempt to reduce essentialist effects to other factors. We will review each of these possibilities, concluding that essentialism fits neatly into neither framework. Instead, we will propose that essentialism is (3) the convergence of several domain-general capacities that have evolved to render the task of cognitive development tractable.

4.1. Domain-Specific, Modular Accounts of Essentialism

We have seen that essentialism emerges early and consistently, does not require formal schooling, and if anything may be even stronger in early childhood than later. Detailed studies of parental input to children about categories also suggest that parents do not provide explicit instruction about essentialist beliefs (Gelman et al. 1998), and in fact at times try to steer their children away from explicitly essentialized notions (Gelman, Taylor, and Nguyen 2004). Moreover, it cannot be the case that children "read off" essentialism from observing the world around them, because (according to most biologists) essentialism is not in the world, but rather is a human construct (Mayr 1991; but see Bloom 2000; Kornblith 1993). Thus, a poverty-of-the-stimulus sort of argument could be mounted in favor of essentialism as an innate, domain-specific module (Atran 1995; Pinker 2002). In this view, essentialism is part of our evolutionary inheritance—a way of processing the world that has proved to have sufficient survival advantages.

Although this view is appealing for the reasons mentioned above, it encounters problems as well. One difficulty with the biological-module account of essentialism is that there is no evidence that essentialism starts (in childhood) as a biologically specific assumption. Race and other social categories are readily essentialized from an early age (Giles 2003; Hirschfeld 1996), as are concepts that are altogether nonliving (e.g., substances, such as gold, water, etc.; Gelman and Markman 1986). A second difficulty is that essence-like attitudes are found with a wide range of nonbiological phenomena. These attitudes include belief in an inherent, non-visible substance or quality that can be passed along from one individual to the next, and that has vast causal implications for a range of properties, including authenticity. Such attitudes are found not only with essentialism, but with how people respond to metaphorical "contagion," as when one desires to touch objects that belonged to a celebrity (positive contagion), and when one wishes to avoid objects that belonged to a disliked, feared, or stigmatized figure (negative contagion). Rozin and his colleagues have documented the power that such non-rational attractions and repulsions have for adults, even for educated folk in modern society (Nemeroff and Rozin 1994), and initial evidence suggests similar beliefs in young children (Frazier and Gelman 2009). These clearly are nonbiological phenomena, yet seem related to essentialism, essentialist-like, or partaking of similar reasoning strategies. (See Gelman 2003 for a more extended discussion of this argument and alternative accounts.)

4.2. Empiricist Accounts of Essentialism

A wholly different account of essentialism derives from an "attentional learning account" (ALA) of conceptual development, with "dumb attentional mechanisms" or "similarity-based" reasoning as underlying mechanisms (Rakison 2005; Sloutsky 2003; Smith et al. 1996). This account proposes that concepts are built up piecemeal by means of low-level, domain-general associative and similarity-matching mechanisms that take sensory and perceptual data as input and that gradually accrete over developmental time.[2] In this view, in order to learn a concept, children do not come prepared with a set of assumptions about richly-structured kinds, each distinct from the other and underpinned with a placeholder representing a causal force (i.e., the essentialist claim). Instead, to the extent that children even possess concepts (a debated point; Smith 2000), children in reality consider only the most surface aspects available.

Recent years have seen increased attention to empiricist approaches to cognitive development (Sloutsky and Fisher 2004; Yoshida and Smith 2003). This renewal stems in part from subtle and compelling demonstrations of the capacity of infants and young children to track statistical information in the input (Baldwin et al. 2001; Gopnik and Sobel 2000; Kushnir and Gopnik 2005; Saffran, Aslin, and Newport 2004). Children's capacity to extract orderly information from messy input gives reason to expect that they may be able to construct subtle and sophisticated concepts out of low-level cues.

---

[2] Associative and similarity-based mechanisms refer to distinct, but related views. At times we treat them as comparable, but in some cases they make different predictions, and so we separate them.

Empiricist approaches are appealing in their parsimony (one might even say reductionism), making few assumptions about what capacities children start with, and enabling the building up of complex knowledge systems out of the most modest of tools. Certainly domain-general learning processes (including induction, associative learning, and noting of similarity) are important tools throughout the lifetime. However, we would argue that some of the empiricist critiques of essentialism are flawed.

With regard to essentialism specifically, empiricist accounts argue that children may at times *appear* to be essentialist on certain tasks, but this is an illusion, brought about by misleading task features. Consider, for example, one finding argued to favor essentialism: preschool children draw inferences from one category member to another, even when doing so requires going against perceptual similarity (e.g., inferring that a blackbird will have the same diet as a flamingo (when each is labeled as "a bird") rather than the diet of a bat; Gelman and Markman 1986). In contrast, the ALA proposal has argued that the appearance of stimulus items is distorted in a way that is not found in the real world and thus does not reflect how children make inferences in reality (for example, in the real world, children could readily use perceptual similarity, because flamingos really do look more like blackbirds than they do bats; Jones and Smith 1993), or that children's attention is drawn to salient aspects of the procedure (e.g., hearing labels for category members draws children's attention away from visual appearances, because the auditory channel is overall more salient to children; Sloutsky 2003).

In this view, if children are provided with appropriate stimuli, they attend only to the most superficial cues; concept learning in children is best characterized by a focus on the immediate perceptual context; and children's attention is captured by the most salient aspects of that context. One simple illustration of this point is provided by Jones and Smith (1993), who pose the question: why do children refer to a toy bear as a "bear"? Deeper, more essential features unmistakably differentiate toy bears from real bears (e.g., real bears—but not toy bears—are alive, have blood and bones inside, give birth to babies). The only features that toy bears and real bears share and, thus, the only features that children can appeal to label them both as a "bear," are the superficial perceptual features of shape and body parts.

We reject this characterization for several reasons. First, the toy bear example misses the critical point that toy bears are *representations* of real bears, and it is for this reason (we argue) that toy bears and real bears are given the same name. One can test this idea by examining how children label items that are either plausible representations (because they were intentionally created) or cannot be representations (because they came about wholly by accident)—for example, a blob of paint that is shaped like a person, and so could have been either intended as a work of art or the result of an accidental paint spill. In such cases, we find that children as young as two-and-a-half years are more likely to label the item based on its shape when it was described as having been intentionally created (and is therefore a plausible representation; Gelman and Ebeling 1998; see also Bloom and Markson 1998, for related findings). Thus, the subtle and theory-embedded property of creator's intent is more important than outward perceptual features.

Second, the criticisms regarding the stimulus items in essentialist studies are misguided. Thanks to natural selection processes, there are indeed many real-world examples of plants or animals for which outward appearances conflict with adult classification systems and (correspondingly) the labels we provide. Flying mammals, legless lizards, whales, sea anemones, pyrite, etc., not only have served as items in prior experiments, but also have fooled or puzzled adults trying to determine how to classify them.

Third, and most important, the low-level processes proposed to account for concept development encounter serious limitations, both theoretically and empirically, when one considers the nature of young children's concepts. Gelman and Kalish (2006) noted four major limitations of low-level empiricist models: context-sensitivity, role of labels, constraints, and conceptual variation. We briefly review each of these below.

Context-sensitivity is a problem for similarity-based models (though not associative models). This is the phenomenon that even young children weight features differently, depending on the task and items at hand. For example, McCarrell and Callanan (1995) showed that children as young as two years use perceptual similarity cues more when similarity corresponds to function than when it does not. When making an inference about whether an unfamiliar animal can see in the dark, children attended to eye size, but when making an inference about whether the same animal can jump high, they attended to leg length. In other words, children were not using an undifferentiated similarity metric, but rather intuitive notions concerning the causal and functional implications of different parts. Relatedly, two-year-olds use different types of features to extend labels depending on the ontological category of the given objects—they use shape similarities when labeling solid objects, but are more likely to use similarities in material when labeling non-solid substances (Soja, Carey, and Spelke 1991).

As noted above, associationist accounts—unlike similarity accounts—would easily be capable of handling context-sensitivity, and even predict that responses will vary by context (Smith 2000). It is this prediction that leads some scholars to propose that children do not even possess concepts, because concepts vary too freely to be identified as stable entities (Jones and Smith 1993). However, one important difference between an associationist explanation of context-sensitivity and an essentialist explanation of context-sensitivity is that the former assumes that children possess a prior database of associations (built up over direct experiences) that determine children's judgments, whereas the latter assumes that children can make novel inferences that are supported by theory-driven assumptions. Support for the importance of theory-based inferences can be found in several studies with children. For example, preschoolers predict that a wooden pillow will be hard rather than soft, even though all the pillows children have previously encountered have been soft (Kalish and Gelman 1992). Furthermore, Barrett et al. (1993) asked elementary-school children to categorize novel birds into one of two categories, and found that participants noticed correlations that were supported by causal links, and used such correlations to categorize new members (e.g., correlation between brain size and memory). The children did not make use of features that correlated equally well but were unsupported by a theory (e.g., the correlation between structure of heart and shape of beak). Krascum and Andrews (1998) likewise conclude, in their study of category learning in four- and five-year-olds: "the meaningfulness of individual features is not a significant factor in children's category learning, and instead, what is important is that attributes within a category can be linked in a theory-coherent manner" (343).

Role of labels. One of the most striking differences between empiricist and non-empiricist accounts of concepts is in their treatment of language, and labels in particular. According to the similarity view, words function as one more perceptual feature in an undifferentiated similarity judgment, to be combined with other perceptual features (including visual appearances; Sloutsky 2003). In contrast, we maintain that words refer to concepts, and it is those concepts (not the uttered words per se) that mediate children's judgments. A corollary to this point is that children systematically treat different form-classes (e.g., count nouns, proper nouns) as referring to different sorts of concepts (e.g., taxonomic kinds and individuals, respectively), and it is therefore misleading to consider all words to function equivalently, as an undifferentiated aural "feature." Indeed, there is a vast literature demonstrating that young children are sensitive to different word types in their conceptual judgments (see Bloom 2000 for review). For example, although children make kind-based inductive inferences when items are labeled with count nouns (e.g., making inferences from one bird to another, dissimilar bird when each is labeled as "a bird"), they refrain from making such inferences when items are labeled with adjectives (e.g., when each is labeled as "sleepy"; Gelman and Coley 1990). Moreover, children consider the intentions of the speaker (not simply the presence or absence of labels per se); for example, they make use of labels for induction only when they are clearly used intentionally (Jaswal 2004). We would argue, therefore, that children's choices are determined not by labels per se, but rather by the concepts conveyed by those labels.

In further support of this view, prior research has shown that, as early as twenty months of age, children make use of concepts (not words per se) when extending novel count nouns (Booth and Waxman 2002; Booth, Waxman, and Huang 2005). In these studies, young children were taught a novel word, learned some conceptual information about the referent, and then were asked if the label extended to new instances. Importantly, although the perceptual appearances of the objects were identical across conditions, the conceptual information varied—the object was described as having either animal-like properties ("This dax has a mommy and daddy who love it very much…") or artifact-like properties ("This dax was made by an astronaut to do a very special job…"). Participants classified new instances quite differently depending on the conceptual information they had heard. For example, shape similarities were weighted heavily when extending labels in both the animate and artifact conditions. However, texture similarities were weighted *more* heavily in the animate condition relative to the artifact condition. Thus, the properties that mattered most depended on the kind of object being considered, as specified by the given conceptual information.

Constraints. One central problem with associationist and similarity-based accounts is that they are too powerful, able to characterize not only the relevant developmental phenomena, but nearly any possible pattern, no matter how implausible (Keil 1981; Murphy and Medin 1985). Yet children's patterns of acquisition are not unconstrained but rather are typically quite regular and predictable across individuals. Thus, the central challenge for these accounts is to describe constraints such that the hypothesized processes produce only those concepts and developmental changes that actually occur. Empiricist approaches claim that such constraints are provided by the structure of the input. One difficulty with this assumption is that children are not passively shaped by the environment, but at times resist instruction or fail to be influenced by the input. In order to solve this problem, strict associationist models may need to be supplemented with some innate dispositions, such as attentional biases (Rakison 2003).

Conceptual variation. A final challenge for associationist accounts is that much of what we learn seems to require processes of verbally and socially mediated learning, rather than associations provided by direct sensory and perceptual experiences. This is easy to see with certain "advanced" concepts that adults acquire: concepts such as "hegemony," "corporate takeover," or "quark" rest on complex systems of knowledge that are unlikely to be translatable into sensory or perceptual associations. But the same concern holds with even the early-acquired concepts of childhood. With children as well as adults, testimony from expert others is a powerful means of conveying concepts (Koenig, Clément, and Harris 2004).

4.3. Essentialism as the Convergence of Domain-General Capacities Required for Development

We turn now to a position that differs from each of the extremes mentioned above, and in fact is the view we favor. This is the notion that essentialism arises out of a set of domain-general capacities that are invoked differently in different domains depending on the causal structure of each domain (Gelman 2003). Although speculative, this account is most compelling in part because it predicts a broad application of essentialism (applying to concepts beyond the domain of folk biology), yet not promiscuous application (honoring theory-relevant distinctions between concepts of animal kinds versus concepts of simple artifacts, for example). (See also Keil 1994; Leslie 1994; and Sperber 1994 for other accounts of how domain-general mechanisms may interact with properties of each domain.)

We suggest that essentialism is the result of several converging psychological capacities or tendencies. Candidate capacities are provided in Table 1, excluding capacities that simply restate essentialism (e.g., categorical realism: the belief that the world consists of natural kinds) or capacities that are fundamental to any sort of conceptual system (e.g., the capacity to form categories, or the capacity to detect

Table 1. Root capacities underlying essentialism, with corresponding cognitive functions and essentialist instantiations.

Root capacities	Broad cognitive functions	Essentialist instantiations
Appearance versus reality	Categorization	Nonobvious properties as core
Induction from property clusters	Prediction	Inferences about the unknown
Causal determinism	Prediction and explanation	Causal properties as core
Tracking identity over time	Identifying kin, possessions	Importance of origins
Deference to experts	Social learning	Acceptance of category anomalies
ABOVE, COLLECTIVELY		Realist assumptions about categories and names; boundary intensification; category immutability and stability over transformations; importance of nature over nurture

similarity). Importantly, in this framework each capacity is independently motivated by what it provides the developing child's broader cognitive development. Thus, we propose that these capacities did not emerge for the purpose of furthering essentialism, but rather for fundamental developmental tasks. For example, causal determinism motivates children to search for underlying mechanisms and so reach a deeper understanding of events (ultimately enabling greater predictive power). Nonetheless, each capacity also has a direct implication or instantiation for essentialism (e.g., deference to experts leads children to accept category anomalies). Furthermore, these varied, independent capacities combine to yield effects that range beyond those of the individual effects. For example, categorical realism arises not from any individual cause, but rather from the various capacities converging.

Three examples serve to illustrate the framework: induction from property clusters, causal determinism, and deference to experts.

Induction from property clusters. It is often remarked that one of the primary functions of categories is to permit inductive inferences (Smith 1995). For example, armed with the category of "mushrooms," if the mushrooms we have encountered in the past are edible, we can infer that new mushrooms that we come across will be edible. All organisms (from mealworms to humans) appear to have some sort of ability to categorize and use categories as the basis for future action. We propose that a further related capacity is the tendency to assume that property clusters attract other properties. In other words, the more commonalities that a child has learned about a category in the past, the stronger her inferences about that category will be in the future (see Baldwin, Markman, and Melartin 1993 for some evidence from infants). Over time, this capacity could be a key contributor to beliefs about categorical realism. To be clear, the suggestion here is not simply that categories permit induction (although they do). Rather, the claim is that categories also permit second-order inferences, such that they will continue to permit even more inductions into the future (e.g., a newly encountered mushroom will share even unknown properties with other mushrooms).

Causal determinism. Causal determinism—the assumption that properties and events are caused—is a construct that has received much attention in the philosophical literature. There is a variety of indirect evidence suggesting that causal determinism may be operating early in childhood and even infancy (Brown 1990; Bullock, Gelman, and Baillargeon 1982; Gelman and Kalish 1993; Schulz and Sommerville 2006; Shultz 1982). In the case of natural events, causal determinism comes to mean that events without an external cause demand some sort of mediating, inherent cause. When extended to categories, causal determinism could imply that category members with many similarities demand some sort of explanation, and thus lead children to search for hidden, nonobvious, as-yet-unknown causal properties.

Deference to experts. We also suggest that children honor a tacit division of linguistic labor (Putnam 1973), in which they consult parents and other adults to find out what something truly should be called. Just as adults defer to experts in matters of naming natural kinds (Malt 1990; but see Kalish 1995), so too do children (Kalish 1998). Children readily accept experimenter-provided labels, even when such labels are surprising and counterintuitive (Gelman and Coley 1990; Gelman and Markman 1986; Graham et al. 2004; Jaswal and Markman 2007; Welder and Graham 2001). Yet they also recognize subtle social-pragmatic cues as to which labels are reliable and conventional versus arbitrary (Sabbagh and Baldwin 2001). For example, four-year-olds are more willing to request and accept labels for novel objects from a speaker who has accurately labeled objects previously than an inaccurate or ignorant speaker (Koenig et al. 2004). Deference to experts in the matter of naming also implies that there is an objectively "correct" word for items (what we may think of as nominal realism). Such a belief could bolster, and be bolstered by, the essentialist presumption that categories are real.

We suggest that each of these cognitive capacities serves useful functions in its own right, but also (as a by-product) contributes to essentialism. In this view, essentialism is not a single cognitive predisposition but the convergence of several independent capacities or tendencies (see also Kornblith 1993). Although the proposed principles are domain-general, essentializing is not. Domain-specificity emerges because of properties of the world (as we perceive it). For example, the rich property clusters found in animal kinds support rich inductive inferences and causal underpinnings in a manner not possible for simple artifacts such as knickknacks or socks.

5. Conceptual Change

One of the most potent debates in developmental psychology concerns the nature of conceptual change. Do concepts undergo significant change over developmental time, and if so, is this change continuous or discontinuous? Can concepts be translated from one format to another, or are they so wholly incommensurable that translation is not even possible? Consistent with the latter possibility, a classic view of development proposes that children's concepts undergo major qualitative reorganizations with age (Inhelder and Piaget 1964). In this view, young children before age six or seven are preconceptual (unable to use a consistent basis for classifying items; unable to coordinate intensions and extensions of a category); elementary school children (about seven to eleven years) can categorize, but only on the basis of concrete, perceptual cues. Only later (e.g., around age eleven and onward) can children form and reason with abstract concepts. Although few developmental psychologists today would endorse Piaget's proposed stage progression regarding conceptual development, the notion that children's early concepts are perceptually based is an enduring one. Others reject the notion that early concepts are exclusively perceptual, yet nonetheless suggest that children's concepts undergo qualitative changes during childhood (Carey 2000; Gopnik 2003).

In contrast to these views, essentialism suggests an important degree of continuity in children's concepts. Recall that studies of essentialism reveal that children—like adults—are surprisingly able to reason about nonobvious features, to attend to causal structure, and to form abstract generalizations. In all of these respects, the commonalities between preschool children and adults are striking.

Ironically, where conceptual change takes place may not be early in childhood, but later in development. Overcoming essentialism may require quite sophisticated conceptual tools. Even when college students learn about species in a biology course, for example, we suspect it may be difficult to overcome the assumption that species are understood in terms of inherent features that each member possesses (e.g., a "tiger gene"), rather than appreciating that a species is an interbreeding population, characterized by diversity among its members and having no single determinative property. Indeed, in some cases new beliefs are layered on top of old, with not-necessarily-coherent belief systems coexisting (diSessa, Gillespie, and Esterly 2004; Legare and Gelman 2008). However, it would also appear that eventually conceptual change is possible, at least for some concepts and at least for some individuals.

6. CONCLUSION

We focus on the development of psychological essentialism to provide a case study of the issues involved in conceptual development. One important conclusion from the work we have surveyed is that traditional views of children's concepts underestimate the capacity of young children. Such accounts claim that young children begin with only very basic perceptual and attentional mechanisms and then undergo a qualitative shift with age: from concrete to abstract, surface to deep, thematic to taxonomic, or perceptual to conceptual. These findings often derive from studies that ask children to sort an array of items into categories (the "free classification" task), where young children indeed tend to focus on outward appearances to a much greater extent than older children or adults. In contrast, when categories are specified for children (most often by means of labeling routines, e.g., "This is a cat"), we see that categories function much the same way for young children as for adults. By two-and-a half years, children expect categories to reveal deep, nonobvious similarities. Children understand categories as placeholders for rich clusters of information, most of which they have not yet acquired (see also Waxman and Markow 1995).

At the same time, we do not mean to suggest that perceptual features or similarity are unimportant in young children's concepts. Appearances are often the only (or most salient) features available when reasoning about an item, and are important cues to less obvious properties. Children might often appeal to perceptual features in the service of making inferences about underlying properties. Their experiences with categories could lead them to observe that certain perceptual attributes, such as shape or texture, are typically linked to deeper properties, such as function or internal parts. This could lead them to view such perceptual properties as reliable cues to category membership (Ware and Booth 2010). Thus, much of development involves learning to link perceptual cues to nonobvious properties, rather than primarily considering one factor over the other, and exploiting these relations to make new categorizations and inferences.

In this sense, for children (as for adults), a category has two distinct though interrelated levels: the level of observable features and the (hidden) level of causal features. This two-tier structure may motivate children to search for new information, reconcile discrepancies, and make new comparisons—in short, it may be an engine for conceptual growth. There is empirical evidence as well that specifically introducing a nonobvious similarity between dissimilar things can yield dramatic growth in children's concepts (Opfer and Siegler 2004). This view is consistent with many accounts of cognitive development, which assume that children hold multiple contrasting representations that foster growth (e.g., equilibration, competition, or cognitive variability; see Gelman 2003 for review).

More broadly, we have argued that the emphasis on perceptually based, bottom-up processes in empiricist accounts of learning does not adequately capture the nature of concept acquisition in childhood (e.g., identifying the boundaries between words; Saffran et al. 2004; learning to identify the range of shape similarities that constitute a novel shape-based category; Smith 2003). Human concepts are broad and varied (Gelman and Kalish 2006; Medin et al. 2000), and empiricist accounts are not sufficient to account for the rich set of assumptions that guide children's reasoning about the world. In the first few years of life, children appeal to inborn, internal, and nonobvious properties. They link these properties to category membership and use these links to generate predictions about the world around them. Moreover, they do this in the face of competing, readily accessible perceptual similarities and in the absence of any formal instruction. Although the present discussion has specifically targeted questions surrounding essentialism, these conclusions apply to conceptual development in general.

Acknowledgements

Preparation of this chapter was supported by NICHD grant R01 HD36043 to the first author.

REFERENCES

Ahn, W. K., Kalish, C., Gelman, S. A., Medin, D. L., Luhmann, C., Atran, S., Coley, J. D., and Shafto, P. (2001). Why essences are essential in the psychology of concepts. *Cognition* 82: 59–69.

Ahn, W. K., Kim, N. S., Lassaline, M. E., and Dennis, M. J. (2000). Causal status as a determinant of feature centrality. *Cognitive Psychology* 41: 361–416.

Astuti, R., Solomon, G., and Carey, S. (2004). Constraints on conceptual development: A case study of the acquisition of folkbiological and folksociological knowledge in Madagascar. *Monographs of the Society for Research in Child Development* 69(3): 1–135.

Atran, S. (1995). Causal constraints on categories and categorical constraints on biological reasoning across cultures. In D. Sperber, D. Premack, and A. J. Premack (eds.), *Causal Cognition: A Multidisciplinary Debate*. New York: Oxford University Press, 205–33.

Atran, S. (1998). Folk biology and the anthropology of science: Cognitive universals and cultural particulars. *Behavioral and Brain Sciences* 21: 547–609.

Atran, S., Medin, D., Ross, N., Lynch, E., Vapnarsky, V., Ek', E. U., Coley, J., Timura, C., and Baranet, M. (2002). Folkecology, cultural epidemiology, and the spirit of the commons: A garden experiment in the Maya lowlands, 1991–2001. *Current Anthropology* 43: 421–50.

Baldwin, D. A., Baird, J. A., Saylor, M. M., & Clark, M. A. (2001). Infants parse dynamic action. *Child Development* 72: 708–17.

Baldwin, D. A., Markman, E. M., and Melartin, R. L. (1993). Infants' ability to draw inferences about nonobvious object properties: Evidence from exploratory play. *Child Development* 64: 711–28.

Barrett, S. E., Abdi, H., Murphy, G. L., and Gallagher, J. M. (1993). Theory-based correlations and their role in children's concepts. *Child Development* 64: 1595–16.

Bloom, P. (1996). Intention, history, and artifact concepts. *Cognition* 60: 1–29.

Bloom, P. (2000). *How Children Learn the Meanings of Words*. Cambridge, MA: MIT Press.

Bloom, P., and Markson, L. (1998). Intention and analogy in children's naming of pictorial representations. *Psychological Science* 9: 200–204.

Booth, A. E. (2006). Object function and categorization in infancy: Two mechanisms of facilitation. *Infancy* 10: 145–69.

Booth, A. E., and Waxman, S. (2002). Word learning is "smart": Evidence that conceptual information affects preschoolers' extension of novel words. *Cognition* 84: B11–B22.

Booth, A. E., Waxman, S. R., and Huang, Y. T. (2005). Conceptual information permeates word learning in infancy. *Developmental Psychology* 41: 491–505.

Braisby, N., Franks, B., and Hampton, J. (1996). Essentialism, word use, and concepts. *Cognition* 59: 247–74.

Brown, A. L. (1990). Domain-specific principles affect learning and transfer in children. *Cognitive Science* 14: 107–33.

Bullock, M., Gelman, R., and Baillargeon, R. (1982). The Development of Causal Reasoning. In W. J. Friedman (ed.), *Development of Time Concepts*. New York: Academic Press, 209–53.

Carey, S. (2000). The origins of concepts. *Journal of Cognition and Development* 1: 37–41.

Chomsky, N. (1968). *Language and Mind*. New York: Harcourt, Brace, and World.

deVries, R. (1969). Constancy of generic identity in the years three to six. *Monographs of the Society for Research in Child Development* 34(3): 1–67.

Diesendruck, G. (2001). Essentialism in Brazilian children's extensions of animal names. *Developmental Psychology* 37: 49–60.

Diesendruck, G., and Gelman, S. A. (1999). Domain differences in absolute judgments of category membership: Evidence for an essentialist account of categorization. *Psychonomic Bulletin & Review* 6.: 338–46.

Di Sessa, A., Gillespie, N., and Esterly, J. (2004). Coherence versus fragmentation in the development of the concept of force. *Cognitive Science* 28: 843–900.

Elman, J. L., Bates, E. A., Johnson, M. H., and Karmiloff-Smith, A. (1996). *Rethinking Innateness: A Connectionist Perspective on Development*. Cambridge, MA: MIT Press.

Evans, E. M. (2000). Beyond Scopes: Why Creationism Is Here to Stay. In K. S. Rosengren, C. N. Johnson, and P. L. Harris (eds.), *Imagining the Impossible: Magical, Scientific, and Religious Thinking in Children*. New York: Cambridge University Press, 305–33.

Fodor, J. A. (1998). *Concepts: Where Cognitive Science Went Wrong*. New York: Oxford University Press.

Frazier, B. N., and Gelman, S. A. (2009). Developmental changes in judgments of authentic objects. *Cognitive Development* 24: 284–92.

Gelman, S. A. (2003). *The Essential Child: Origins of Essentialism in Everyday Thought*. New York: Oxford University Press.

Gelman, S. A. (2004). Psychological essentialism in children. *Trends in Cognitive Sciences* 8: 404–09.

Gelman, S. A., and Coley, J. D. (1990). The importance of knowing a dodo is a bird: Categories and inferences in 2-year-old children. *Developmental Psychology* 26: 796–804.

Gelman, S. A., Coley, J. D., Rosengren, K. S., Hartman, E., and Pappas, A. (1998). Beyond labeling: The role of maternal input in the acquisition of richly structured categories. *Monographs of the Society for Research in Child Development* 63(1): 1–139.

Gelman, S. A., and Ebeling, K. S. (1998). Shape and representational status in children's early naming. *Cognition* 66: B35–B47.

Gelman, S. A., Heyman, G. D., and Legare, C. H. (2007). Developmental changes in the coherence of essentialist beliefs about psychological characteristics. *Child Development* 78: 757–74.

Gelman, S. A., and Hirschfeld, L. A. (1999). How Biological Is Essentialism? In D. L. Medin and S. Atran (eds.), *Folkbiology*. Cambridge, MA: MIT Press, 403–46.

Gelman, S. A., and Kalish, C. W. (1993). Categories and Causality. In R. Pasnak and M. Howe (eds.), *Emerging Themes in Cognitive Development*. New York: Springer-Verlag, 3–32.

———. (2006). Conceptual Development. In D. Kuhn and R. S. Siegler (eds.), *Handbook of Child Psychology: Vol. 2, Cognition, Perception, and Language*, 6th ed. Hoboken, NJ: Wiley, 687–733.

Gelman, S. A., and Markman, E. M. (1986). Categories and induction in young children. *Cognition* 23: 183–209.

———. (1987). Young children's inductions from natural kinds: The role of categories and appearances. *Child Development* 58: 1532–42.

Gelman, S. A., Taylor, M. G., and Nguyen, S. P. (2004). Mother-child conversations about gender. *Monographs of the Society for Research in Child Development* 69(1): 1–127.

Gelman, S. A., and Wellman, H. M. (1991). Insides and essences: Early understandings of the non-obvious. *Cognition* 38: 213–44.

Gigerenzer, G. (2000). *Adaptive Thinking: Rationality in the Real World*. New York: Oxford University Press.

Gil-White, F. J. (2001). Are ethnic groups biological "species" to the human brain?: Essentialism in our cognition of some social categories. *Current Anthropology* 42: 515–54.

Giles, J. W. (2003). Children's essentialist beliefs about aggression. *Developmental Review* 23: 413–43.

Gopnik, A. (2003). The Theory Theory as an Alternative to the Innateness Hypothesis. In L. Antony and N. Hornstein (eds.), *Chomsky and His Critics*. Oxford: Blackwell Publishing, 238–54.

Gopnik, A., and Sobel, D. M. (2000). Detecting blickets: How young children use information about novel causal powers in categorization and induction. *Child Development* 71: 1205–22.

Gottfried, G. M., and Gelman, S. A. (2005). Developing domain-specific causal-explanatory frameworks: The role of insides and immanence. *Cognitive Development* 20: 137–58.

Graham, S. A., Kilbreath, C. S., and Welder, A. N. (2004). Thirteen-month-olds rely on shared labels and shape similarity for inductive inferences. *Child Development* 75: 409–27.

Haslam, N., Rothschild, L., and Ernst, D. (2000). Essentialist beliefs about social categories. *British Journal of Social Psychology* 39: 113–27.

Heyman, G. D., and Gelman, S. A. (2000). Preschool children's use of trait labels to make inductive inferences. *Journal of Experimental Child Psychology* 77: 1–19.

Hirschfeld, L. A. (1996). *Race in the Making: Cognition, Culture, and the Child's Construction of Human Kinds*. Cambridge, MA: MIT Press.

Hirschfeld, L. A., and Gelman, S. A. (1997). What young children think about the relationship between language variation and social difference. *Cognitive Development* 12: 213–38.

Inhelder, B., and Piaget, J. (1964). *The Early Growth of Logic in the Child: Classification and Seriation*. New York: Harper and Row.

James, W. (1890/1981). *The Principles of Psychology*. Cambridge, MA: Harvard University Press.

Jaswal, V. K. (2004). Don't believe everything you hear: Preschoolers' sensitivity to speaker intent in category induction. *Child Development* 75: 1871–85.

Jaswal, V. K., and Markman, E. M. (2007). Looks aren't everything: 24-month-olds' willingness to accept unexpected labels. *Journal of Cognition and Development* 8: 93–111.

Jayaratne, T. E., Ybarra, O., Sheldon, J. P., Brown, T. N., Feldbaum, M., Pfeffer, C. A., and Petty, E. (2006). White Americans' genetic lay theories of race differences and sexual orientation: Their relationship with prejudice toward blacks, and gay men and lesbians. *Group Processes & Intergroup Relations* 9: 77–94.

Jones, S. S., and Smith, L. B. (1993). The place of perception in children's concepts. *Cognitive Development* 8: 113–39.

Kalish, C. W. (1995). Essentialism and graded membership in animal and artifact categories. *Memory & Cognition* 23: 335–53.

Kalish, C. W. (1998). Natural and artifactual kinds: Are children realists or relativists about categories? *Developmental Psychology* 34: 376–91.

Kalish, C. W., and Gelman, S. A. (1992). On wooden pillows: Multiple classification and children's category-based inductions. *Child Development* 63: 1536–57.

Kamp, H., and Partee, B. (1995). Prototype theory and compositionality. *Cognition* 57: 129–91.

Keil, F. C. (1981). Constraints on knowledge and cognitive development. *Psychological Review* 88: 197–227.

———. (1989). *Concepts, Kinds, and Cognitive Development.* Cambridge, MA: MIT Press.

———. (1994). The Birth and Nurturance of Concepts by Domains: The Origins of Concepts of Living Things. In L. A. Hirschfeld and S. A. Gelman (eds.), *Mapping the Mind: Domain Specificity in Cognition and Culture.* New York: Cambridge University Press, 234–54.

Kemler Nelson, D. G., Russell, R., Duke, N., and Jones, K. (2000). Two-year-olds will name artifacts by their function. *Child Development* 71: 1271–88.

Koenig, M. A., Clément, F., and Harris, P. L. (2004). Trust in testimony: Children's use of true and false statements. *Psychological Science* 15; 694–98.

Kornblith, H. (1993). *Inductive Inference and Its Natural Ground: An Essay in Naturalistic Epistemology.* Cambridge, MA: MIT Press.

Krascum, R. M., and Andrews, S. (1998). The effects of theories on children's acquisition of family-resemblance categories. *Child Development* 69: 333–46.

Kripke, S. A. (1980). *Naming and Necessity.* Cambridge, MA: Harvard University Press.

Kuhn, D., and Siegler, R. S. (eds.). (2006). *Handbook of Child Psychology: Vol. 2, Cognition, Perception, and Language,* 6th ed. Hoboken, NJ: Wiley.

Kushnir, T., and Gopnik, A. (2005). Young children infer causal strength from probabilities and interventions. *Psychological Science* 16: 678–83.

Legare, C. H., and Gelman, S. A. (2008). Bewitchment, biology, or both: The co-existence of natural and supernatural explanatory frameworks across development. *Cognitive Science* 32: 607–42.

Leslie, A. M. (1994). ToMM, ToBY, and Agency: Core Architecture and Domain Specificity. In L. A. Hirschfeld and S. A. Gelman (eds.), *Mapping the Mind: Domain Specificity in Cognition and Culture.* New York: Cambridge University Press, 119–148.

Locke, J. (1959). *An Essay Concerning Human Understanding.* New York: Dover.

Mahalingam, R. (2003). Essentialism, culture, and power: Representations of social class. *Journal of Social Issues* 59: 733–49.

Mahalingam, R., and Rodriguez, J. (2006). Culture, brain transplants and implicit theories of identity. *Journal of Cognition and Culture* 6: 453–62.

Malt, B. C. (1990). Features and beliefs in the mental representation of categories. *Journal of Memory and Language* 29: 289–315.

Malt, B. C. (1994). Water is not H2O. *Cognitive Psychology* 27: 41–70.

Mandler, J. M., and McDonough, L. (1996). Drinking and driving don't mix: Inductive generalization in infancy. *Cognition* 59: 307–35.

Marler, P. (1991). The Instinct to Learn. In S. Carey and R. Gelman (eds.), *The Epigenesis of Mind: Essays on Biology and Cognition.* Hillsdale, NJ: Erlbaum, 37–66.

Mayr, E. (1991). *One Long Argument: Charles Darwin and the Genesis of Modern Evolutionary Thought.* Cambridge, MA: Harvard University Press.

McCarrell, N. S., and Callanan, M. A. (1995). Form-function correspondences in children's inference. *Child Development* 66: 532–46.

Medin, D. L. (1989). Concepts and conceptual structure. *American Psychologist* 44: 1469–81.

Medin, D. L., Lynch, E. B., and Solomon, K. O. (2000). Are there kinds of concepts? *Annual Review of Psychology* 51; 121–47.

Medin, D. L., & Ortony, A. (1989). Psychological Essentialism. In S. Vosniadou and A. Ortony (eds.), *Similarity and Analogical Reasoning*. New York: Cambridge University Press, 179–95.

Mill, J. S. (1843). *A System of Logic, Ratiocinative and Inductive*. London: Longmans Green.

Murphy, G. L. (2002). *The Big Book of Concepts*, Cambridge, MA: MIT Press.

Murphy, G. L., and Medin, D. L. (1985). The role of theories in conceptual coherence. *Psychological Review* 92: 289–316.

Nemeroff, C., and Rozin, P. (1994). The contagion concept in adult thinking in the United States: Transmission of germs and of interpersonal influence. *Ethos* 22: 158–86.

Opfer, J. E., and Siegler, R. S. (2004). Revisiting preschoolers' living things concept: A microgenetic analysis of conceptual change in basic biology. *Cognitive Psychology* 49: 301–32.

Piaget, J. (1983). Piaget's Theory. In P. H. Mussen (ed.), *Carmichael's Manual of Child Psychology*, vol. 1, 4th ed. New York: Wiley.

Pinker, S. (2002). *The Blank Slate: The Modern Denial of Human Nature*. New York: Viking.

Prentice, D. A., and Miller, D. T. (2006). Essentializing differences between women and men. *Psychological Science* 17: 129–35.

———. (2007). Psychological essentialism of human categories. *Current Directions in Psychological Science* 16: 202–6.

Putnam, H. (1973). Meaning and reference. *Journal of Philosophy* 70: 699–711.

Putnam, H. (1975). The Meaning of "Meaning." In *Mind, Language and Reality: Philosophical Papers*, vol. 2. Cambridge: Cambridge University Press.

Quine, W. V. O. (1969). Natural Kinds. In *Ontological Relativity and Other Essays*. London: Columbia University Press, 114, 138.

Rakison, D. H. (2003). Free association? Why category development requires something more. *Developmental Science* 6: 20–22.

———. (2005). The Perceptual to Conceptual Shift in Infancy and Early Childhood: A Surface or Deep Distinction? In L. Gershkoff-Stowe and D. H. Rakison (eds.), 131–158, *Building Object Categories in Developmental Time*. Mahwah, NJ: Erlbaum.

Rehder, B. (2006). When similarity and causality compete in category-based property generalization. *Memory & Cognition* 34: 3–16.

Rips, L. J. (1989). Similarity, Typicality, and Categorization. In S. Vosniadou and A. Ortony (eds.), *Similarity and Analogical Reasoning*. New York: Cambridge University Press.

Rips, L. J. (2001). Necessity and natural categories. *Psychological Bulletin* 127: 827–852.

Rosch, E. (1976). Basic objects in natural categories. *Cognitive Psychology* 8: 382–439.

Rothschild, L., and Haslam, N. (2003). Thirsty for H2o? Multiple essences and psychological essentialism. *New Ideas in Psychology* 21: 31–41.

Sabbagh, M. A., and Baldwin, D. A. (2001). Learning words from knowledgeable versus ignorant speakers: Links between preschoolers' theory of mind and semantic development. *Child Development* 72: 1054–70.

Saffran, J. R., Aslin, R. N., and Newport, E. L. (2004). Statistical learning by 8-month-old infants. *Science* 274: 1926–28.

Sameroff, A. J. (1983). Developmental Systems: Contexts and Evolution. In W. Kessen (ed.), *Handbook of Child Psychology: Vol. 1, History, Theory, and Methods*. New York: Wiley, 237–94.

Schulz, L. E., and Sommerville, J. (2006). God does not play dice: Causal determinism and preschoolers' causal inferences. *Child Development* 77: 427–42.

Schwartz, S. P. (1979). Natural kind terms. *Cognition* 7: 301–15.

Shultz, T. R. (1982). Rules of causal attribution. *Monographs of the Society for Research in Child Development* 47(1): 1–51.

Simpson, T., Carruthers, P., Laurence, S., and Stich, S. (2005). Introduction: Nativism Past and Present. In P. Carruthers, S. Laurence, and S. Stich (eds.), *The Innate Mind: Structure and Contents.* New York: Oxford University Press, 3–19.

Skinner, B. F. (1953). *Science and Human Behavior.* Oxford: Macmillan.

Sloman, S. A., and Malt, B. C. (2003). Artifacts are not ascribed essences, nor are they treated as belonging to kinds. *Language and Cognitive Processes* 18: 563–82.

Sloutsky, V. M. (2003). The role of similarity in the development of categorization. *Trends in Cognitive Sciences* 7: 246–51.

Sloutsky, V. M., and Fisher, A. V. (2004). Induction and categorization in young children: A similarity-based model. *Journal of Experimental Psychology: General* 133: 166–88.

Smith, E. E. (1995). Concepts and Categorization. In E. E. Smith and D. N. Osherson (eds.), *Thinking: An Invitation to Cognitive Science,* vol. 3, 2nd ed. Cambridge, MA: The MIT Press, 3–3.

Smith, E. E., and Medin, D. L. (1981). *Categories and Concepts.* Cambridge, MA: Harvard University Press.

Smith, E. E., Patalano, A. L., and Jonides, J. (1998). Alternative strategies of categorization. *Cognition* 65: 167–96.

Smith, L. B. (2000). Learning How to Learn Words: An Associative Crrane. In R. M. Golinkoff, K. Hirsh-Pasek, L. Bloom, L. B. Smith, A. L. Woodward, N. Akhtar, M. Tomasello, and G. Hollich (eds.), *Becoming a Word Learner: A Debate on Lexical Acquisition.* New York: Oxford University Press, 51–80.

———. (2003). Learning to recognize objects. *Psychological Science* 14: 244–50.

Smith, L. B., Jones, S. S., and Landau, B. (1996). Naming in young children: A dumb attentional mechanism? *Cognition* 60: 143–71.

Sober, E. (1994). *From a Biological Point of View.* New York: Cambridge University Press.

Soja, N. N., Carey, S., and Spelke, E. S. (1991). Ontological categories guide young children's inductions of word meaning: Object terms and substance terms. *Cognition* 38: 179–211.

Solomon, G. E. A., Johnson, S. C., Zaitchik, D., and Carey, S. (1996). Like father, like son: Young children's understanding of how and why offspring resemble their parents. *Child Development* 67: 151–71.

Sperber, D. (1994). The Modularity of Thought and the Epidemiology of Representations. In L. A. Hirschfeld and S. A. Gelman (eds.), *Mapping the Mind: Domain Specificity in Cognition and Culture.* New York: Cambridge University Press, 39–67.

Springer, K. (1996). Young children's understanding of a biological basis for parent -offspring relations. *Child Development* 67: 2841–56.

Strevens, M. (2000). The essentialist aspect of naive theories. *Cognition* 74: 149–75.

Summers, L. H. (2005). Remarks at NBER Conference on Diversifying the Science & Engineering Workforce. Cambridge, MA, January 14, 2005. Retrieved May 7, 2005, from Harvard University, the Office of the President Web site: http://www.president.harvard.edu/speeches/2005/nber.html.

Taylor, M. G. (1996). The development of children's beliefs about social and biological aspects of gender differences. *Child Development* 67: 1555–71.

Wade, P. (2004). Race and human nature. *Anthropological Theory* 4: 157–72.

Ware, E.A, and Booth, A. E. (2010). Form follows function: Learning about function helps children learn about shape. *Cognitive Development* 25: 124–37.

Waxman, S. R., and Markow, D. B. (1995). Words as invitations to form categories: Evidence from 12- to 13-month-old infants. *Cognitive Psychology* 29: 257–302.

Welder, A. N., and Graham, S. A. (2001). The influence of shape similarity and shared labels on infants' inductive inferences about nonobvious object properties. *Child Development* 72: 1653–73.

Wellman, H. M. (1990). *The Child's Theory of Mind*. Cambridge, MA: MIT Press.

Wilson, R. A. (1999). Realism, Essence, and Kind: Resuscitating Species Essentialism? In R. A. Wilson (ed.), *Species: New Interdisciplinary Essays*. Cambridge, MA: MIT Press, 187–207.

Wood, C. C. (1976). Discriminability, response bias, and phoneme categories in discrimination of voice onset time. *Journal of the Acoustical Society of America* 60: 1381–89.

Woodward, A. L. (1999). Infants' ability to distinguish between purposeful and nonpurposeful behaviors. *Infant Behavior & Development* 22: 145–60.

Yoshida, H., and Smith, L. B. (2003). Shifting ontological boundaries: How Japanese- and English-speaking children generalize names for animals and artifacts. *Developmental Science* 6: 1–17.

CHAPTER 20

EVOLUTIONARY PSYCHOLOGY

BEN JEFFARES AND KIM STERELNY

AT its broadest, evolutionary psychology is the study of the evolved mind. It is a branch of cognitive science that takes the evolutionary history—the etiology—of minds as an important component of a complete psychology. Explaining the origins of our cognitive capacities is an important project in its own right, but many evolutionary psychologists think that an evolutionary framework helpfully guides investigations into the cognitive processes of living humans. This chapter will first sketch the historical background against which evolutionary psychology emerged, and then look at two contrasting views of evolved minds. In Section 2, we examine the view that sees minds as collections of evolved task-specific modules: nativist evolutionary psychology. In Section 3, we present the alternative derived directly from evolutionary theory and developmental psychology, a view that takes the role of culture as fundamental. While the truth will lie somewhere between these two views, developing and testing hypotheses will require evidence from the archaeological record. We examine that record and its role in Section 4. Section 5 outlines the implications for cognitive science and the philosophy of mind.

1. EVOLUTIONARY MODELS OF MIND AND BEHAVIOR

The history of evolutionary psychology is as deep as that of serious evolutionary theory itself: both began with Darwin, and others continued his pioneering work on the evolution of emotion and its expression (see especially Richards 1987).

However, evolutionary thinking in psychology essentially disappeared with the development of behaviorism. Moreover, the study of behavior was not yet fully established in zoology. This extinction was not local: the evolution of human cognition disappeared as a topic from serious science. Within the biological sciences, evolutionary psychology only returned as a side effect of the establishment of an evolutionary biology of behavior, and that development took much of the century. Early ethology, the study of animal behavior, did not focus on phenomena that extended naturally to human decision making. Many of the early paradigms were of invariant, quite rigid, often ritualized action patterns. Many examples were drawn from mating and brood care of birds, and experiments were typically designed to show that even quite complex and highly structured act sequences could be controlled by simple releasing stimuli. Action was adaptive despite its simple, cue-driven relation to the world, not because of its cognitive complexity (see, e.g., Tinbergen 1960). There was some speculation within ethology on how human action should be understood within this framework (see Lorenz 1966; Lorenz 1977), but these speculations were not integrated with the Tinbergen-Lorenz experimental program (Burkhardt 2005). Starved of models of cognitive evolution, the study of human evolution concentrated on changes in morphology and behavior. Thus much of the work of archaeologists and physical anthropologists consisted of chronicling the developments leading to full human behavior, with little informed theorizing about the underlying cognitive mechanisms. When archaeologists did speculate about human cognitive evolution, it was more mythical than methodical (Landau 1984).

The connections to human behavior began to be sketched out, as behavioral ecology began to replace ethology (see, e.g., Hamilton 1975). Behavioral ecology had a much greater emphasis on formal evolutionary modeling (including sexual selection and kin selection), and its focus shifted to social and sexual decision making. These connections became explicit in the final chapter of Wilson's synthetic overview, *Sociobiology: The New Synthesis* (1975), and his own subsequent and much less cautious work (1978; Lumsden and Wilson 1981). Wilsonian sociobiology was not *explicitly* a version of evolutionary psychology. Its focus was on supposedly typical forms of human *social behavior*: those to do with mate choice, child care, intergroup relations and the like. The Wilsonians argued that (for example) both hostility toward strangers and male sexual promiscuity are adaptations. But, while the explicit focus was on action, the Wilsonian program was implicitly committed to strong and probably implausible claims about the psychological mechanisms that generated these distinctive patterns of action. For if xenophobia, say, is an adaptation, then it must be a trait whose occurrence and intensity varies independently of the rest of an individual's behavioral phenotype (for otherwise individuals will not differ *just* in their xenophobic tendencies (Lewontin 1978)). For xenophobia to be independent of other behavioral dispositions, the action pattern in question must be generated by an autonomous mechanism (Sterelny 1999). Moreover, this propensity must typically be inherited by an agent's offspring.

This picture of the relations between action and cognition is implausibly crude. An agent's disposition to act is dependent not just on the external environment but on the rest of one's cognitive phenotype: on motivation, emotion, and belief. So on these grounds (and others), the Wilsonian program was subject to brutal criticism (Kitcher 1985), and has essentially vanished, despite the title of Alcock's 2001 work, *The Triumph of Sociobiology*.

Even so, humans are animals, and consequently many of the critics accepted that there must be an account of the evolution of human cognition and behavior that would illuminate our current ways of life and social organization (e.g., Kitcher 2001; although for a sceptic, see Lewontin 1998). Consequently, Wilsonian sociobiology did not vanish without issue. Perhaps its most direct descendant is human behavioral ecology, which maintains the focus on behavior, and like the Wilson program, has imported most of its models from the behavioral ecology of nonhuman animals. Human behavioral ecology focuses on core, fitness-determining decision making (foraging, birth spacing, food sharing, and the like), largely in traditional environments in which the supposed problem of "adaptive lag" (of agents behaving maladaptively because they find themselves in evolutionarily novel environments) is minimized. Human behavioral ecology is focused on individual behavior, and emphasizes the adaptive flexibility of human action: the background assumption is that humans make adaptive choices in just about any social environment. So human behavioral ecologists do not echo Wilsonian sociobiology's claims about genetic constraints on individual action or human social organization, and hence it has not raised the same political suspicions as its predecessor (for reviews of human behavioral ecology that also compare it to other approaches, see Winterhalder and Smith 2000; Smith, Borgerhoff-Mulder, and Hill 2001; Laland and Brown 2002; Smith and Winterhalder 2003).

Human behavioral ecology, like Wilsonian sociobiology, says nothing explicitly about our cognitive mechanisms. Rather, it assumes that those mechanisms allow us to assess the causal structure of our environment and to recognize the likely consequences of our actions. And it presupposes that our subjective utilities track fitness. The outcomes we prefer increase fitness, while those we avoid reduce fitness. In contrast, evolutionary psychologists of sundry varieties focus on cognition and its evolution. There is plenty of disagreement among their ranks, but perhaps the most fundamental divide is on how to incorporate culture within an evolutionary framework. One group, the nativist evolutionary psychologists, minimize the role of culture within human life. While we learn much from others, and while that culturally acquired information is essential for a human life, what we learn is constrained by the innate adaptations of the mind. Just as the space of possible languages is constrained by an innate language faculty, so too the space of possible human mating systems, the space of possible folk psychologies, the space of possible folk biologies, and so forth are constrained by innate specializations. Cultural variation is constrained, and individual acquisition within that constrained space is enhanced, by our adapted mind. This model of evolutionary psychology has seized most of the headlines, and it will be our focus in Section 2. But we will go on to show that it is not the only model of evolutionary psychology: there is a family of alternative models according to which our minds are profoundly shaped by culture and by cultural transmission—alternatives we discuss in Section 3. Methodological issues are the focus of our final section, for it is not obvious that the historical record is rich enough to discriminate between the various alternative models of cognitive evolution.

2. Evolutionary Nativism

Nativist evolutionary psychology is built around a most important insight: ordinary human decision making has a *high cognitive load*. To make good decisions, agents must be sensitive to complex, subtle features of their environment. The information-hungry nature of human action first became apparent in thinking about language. Language makes intensive demands on memory. The different parties to a conversation must remember who said what to whom. It also makes intensive demands on attention. In a conversation, you must do more than recall what has been said: you must *monitor* and *act on* the effects on your utterances and those of others. You need to be alert to the signs that the conversation is going wrong. Moreover, you will often have to do this while also attending to your physical and social world, for often the point of talking is to coordinate joint action: linguistic acts interface, and are smoothly integrated, with the rest of our lives. Famously, Chomskian linguists argued that learning a first language poses an even more formidable challenge. Every child masters a complex, subtle, intricate set of rules (and a huge vocabulary). Children do so (the argument goes) on the basis of impoverished, perhaps even misleading data (this argument is expounded most brilliantly in Pinker 1994; for a sceptical response, see Cowie 1998). No wonder then that language is at the core of the cognitive revolution in psychology and is the model evolutionary nativists use in their attempts to synthesize psychology and evolutionary theory.

For, crucially, the point about the cognitive load on language use holds for human decision making generally. Hence much cognitive psychology—psychology with no professional interest in evolution—has followed Chomskian linguistics into some form of nativism: working with the idea that the human mind is specifically prewired for particular learning tasks. We are cognitively competent in the face of difficult challenges because our minds are specifically structured to meet these challenges. Nativist evolutionary psychology gives this idea of prewiring an evolutionary interpretation. Think, for example, of a social negotiation: deciding who does what in organizing a conference. The participants in such a negotiation must estimate what needs to be done and how the total package can be divided into subtasks. They must understand the relative weight of each chore, and the skill set, reliability, and motivation of their partners. Each must be sensitive to the dynamics of the negotiation itself; they must read one another's moods and intentions. We routinely manage such mixes of mindreading, social negotiation and planning, but that should not blind us to the high cognitive load of such achievements. The evolutionary nativist solution to the cognitive load problem is to appeal to special purpose adaptations. Just as we are specifically adapted for language, so too are we specifically adapted to understand our biological and social environment (Atran 1990, 1998). Our minds are ensembles of special purpose devices, each of which is innately equipped to solve the information-hungry but repeated and predictable problems of human life (the classic source for this view is Barkow, Cosmides, and Tooby 1992; perhaps its most persuasive articulation is Pinker 1997).

In short, evolutionary nativists defend a modular solution to the problem of information load on human decision making. On this view, human minds are ensembles of special purpose cognitive devices: modules. We have a language module; a module for interpreting the thoughts and intentions of others; a "naive physics" module for causal reasoning about sticks, stones, and similar inanimate objects; a natural history module for ecological decisions; a social exchange module for monitoring economic interactions with our peers; and so on. The human mind is not an immensely powerful general-purpose problem-solving engine: it is an integrated array of devices, each of which solves a particular type of problem with remarkable efficiency.

These modules evolved in response to the distinctive, independent, and recurring problems our ancestors faced in their lives as Pleistocene foragers. At some stage in human evolutionary history, language became crucial to human life. In such an environment, the barely lingual would have been under an ever-increasing handicap; hence, there would have been selection for an innate language competence. The lives of human ancestors also depended on cautious cooperation—cautious because free riding would have been an ever-present temptation, and so our ancestors needed to monitor social exchange and to interpret other agents as agents. Our ancestors lived technologically enhanced foraging lives: they needed to understand the causal properties of the raw materials from which they fashioned their tools, and the nature of their biological targets. And so we have folk physics and folk biology modules. And so on.

In this view of cognition, and of cognitive evolution, we cope with the information-processing demands on human life because natural selection has pre-equipped us with both the crucial information and the task-specific processing capabilities we need. Domain specific modules handle information about human language, human minds, inanimate causal interactions, the biological world, and other constant adaptive demands faced by human ancestors (for a recent overview, see Barrett and Kurzban 2006).

So the empirical program of nativist evolutionary psychologists is, first, to use their model of the evolutionary demands on human cognition to develop hypotheses about the set of specific adaptations we should expect to find. Second, they use the methods of experimental cognitive psychology to test the idea that we have distinctive cognitive skills in the domains so identified. If such skills are found, this is treated as a confirmation of the evolutionary model. The results of this empirical program are very controversial. Evolutionary psychologists are very upbeat. Most famously, John Tooby and Leda Cosmides think they have evidence of distinctive reasoning about social interaction and norm violation (Cosmides and Tooby 1992). If the so-called "Wason selection task" is formulated as a problem of policing norm violation, subjects do well. Given logically equivalent tasks with no similar social content, subjects perform poorly. There has also been a good deal of experimental work on mate preference, the results of which have been taken to confirm nativist predictions that males and females have been under divergent selection regimes and look for different qualities in mates (Buss 1994, 2006). As we see it, recent work has undermined Buss's picture in two ways. First, this work highlights the similarities between male and female choice (for example, both sexes weigh intelligence and kindness heavily), and second, mate choice is frequently that expected by behavioral ecologists, demonstrating the adaptive flexibility of mate choice given diverse local conditions (Gangestad and Simpson 2000; Fletcher 2002; Fletcher and Stenswick 2003; Simpson and Orina 2003).

As well as experimental work on adults, nativist evolutionary psychologists also take work in developmental psychology seriously, for innate capacities are expected to have a typical developmental signature, namely, early and uniform appearance. Nativist evolutionary psychologists are especially impressed by evidence of very early theory of mind skills in young toddlers, the development of so-called "theory of mind" capacities (Leslie 2000; Leslie, Friedman, and German 2004; Leslie 2005). Recent studies involve a "differential looking time" test. Subjects look longer at situations that surprise them, so the trick is to devise scenarios that would be surprising if and only if the subject understands mental states and their connections to action. Using these tasks, some developmental psychologists argue that very young children have theory of mind capacities (Gergely and Csibra 2003; Tomasello et al. 2005; Tomasello and Carpenter 2007). Indeed, some very recent work seems to show that children between two and three seem to pass nonverbal tests of sophisticated theory of mind capacities, namely understanding the role of false belief (Baillargeon et al. 2010). That said, these results are difficult to interpret for we need a robust account of why their capacity to understand false belief is masked in some tasks and not others.

Most recently, nativist evolutionary psychologists have turned to moral decision making, arguing that cross-cultural moral judgments are invariant in a subtle and unexpected way. Agents think foreseen but unintended consequences are morally different from foreseen and intended ones: the first, but not the second, can be tolerated as a price paid for avoiding a greater evil (Hauser 2006). There is a good deal of controversy about the robustness of these results and their interpretation. David Buller, in particular, is very sceptical (2005), and his scepticism has provoked a very hostile response (see, e.g., Machery and Barrett 2006). Even some of those broadly sympathetic to the nativist program have doubts about the flagship case of social exchange (see Sperber and Girotto 2003; in response see Cosmides et al. 2005; Cosmides and Tooby 2005).

We do not expect this debate to be over any time soon. We too are sceptical of the nativist program, but for different reasons. Natural selection can build and equip a special purpose module only if the information an agent needs to know is stable over evolutionary time. That sometimes happens: the causal properties of sticks, stones, and bones do not change, hence a naive physics module could well be built into human heads. So there is something to the nativist view, but it is not a general solution to the explanatory challenge posed by our capacities to cope with high cognitive-load problems for many features of human environments have not been stable. The social, physical, and cognitive environments in which we act have been extraordinarily varied. Yet we cope: that is a part of the explanation of our cosmopolitan distribution. Social worlds vary in size, family structure, economic basis, technological elaboration, extent and kinds of social hierarchy, and division of labor. They have varied physically and biologically: the world has changed very dramatically in its physical and biological state over the last few hundred thousand years as it has cycled through ice ages. Indeed, Richard Potts has argued that human evolution has been mainly shaped by increasing, and increasingly intense, climatic variability (1996). Moreover, we are now spread over virtually the whole of the globe. The variability of human environments is especially evident when we bear in mind the fact that we are incorrigible and pervasive ecological engineers: change comes not just from migration and external disturbance, but from our relentless habit of modifying our physical, biological, and social circumstances. Tools, clothes, shelter, fire, and agriculture have changed the world we live in. But so has language, ritual, and extended childhoods in which children live with and learn from their family. If many salient features of human worlds vary across time and space, information about those features cannot be engineered into human minds by natural selection. Our ability to act aptly in such worlds cannot depend on innate modules.

If the nativists are right, and general learning and problem-solving capacities are not powerful enough to solve high-load problems rapidly and accurately, we should suffer from the shock of the new far more brutally than we do. Minds adapted for mid-latitude foraging economies and small-scale social lives should short out when confronted with the cognitive and emotional challenges of (say) New York subways or Mexico City traffic. There are indeed maladaptations of modernity. Favored examples are the diseases of obesity and the restriction of family size by the urban middle class. But we are not, it seems, nearly as hopeless as we should be. Nativist evolutionary psychology seems to predict that the closer an agent's environment is to that of our ancient foraging world, the more reliably adaptive the agent's action should be. But the myth of primitive harmony is indeed a myth: there is no evidence that contemporary small-scale foraging peoples manage their lives more adaptively than, say, urban Mexicans (for a blackly comic catalog of human disaster in traditional societies, see Edgerton 1992; for nativist responses to this apparent paradox, see Sperber 1996; Carruthers 2006).

3. ALTERNATIVE MODELS OF EVOLUTIONARY PSYCHOLOGY

So the situation as we see it is this: The evolutionary nativists are right to think that many routine decision-making situations have a high cognitive load, but they are wrong to think that high-load problems can be confronted successfully only by agents with innately specified, special purpose cognitive capacities. Automatized skills are an alternative means of coping with high-load problems. Automatized skills are phenomenologically rather like modules, but they have very different developmental and evolutionary histories. Skills are slowly built, but once built they are enduring and automatic. As with the nativist psychologist's modules, once they are up and running, they are fast, reliable, automatic, and domain specific. Chess players cannot help but see a chess board and its pieces as a chess position. A birder cannot help but see a particular underwing pattern as a whistling kite. Physical skills such as riding a bicycle might be attenuated by muscular disuse, but they are not forgotten. What is more, automatized skills are often adaptive: they equip agents for the specific features of their environment. The hard-won skills of natural history and bushcraft that enable a forager to move silently, see much that is invisible to others, and find his or her way are plainly critical to survival. So too are the skills of observation, anticipation, and coordination that allow a taxi driver to negotiate the chaos that is Mexico City. But such skills are built into no one's genes. A forager's skills will be very different in an Australian aboriginal in the Pilbara, an Ache hunter-gatherer in a central American rainforest, or an Inuit seal hunter.

Such skills are not only phenomenologically similar to modules, they are also inheritable, for they can be accumulated and transmitted culturally. Children resemble their parents not just because they have inherited their parents' genes, but also because they have inherited their parent's informational resources (and sometimes that of their parents' social partners). Cultural transmission is reliable and of high fidelity in part because we are genetically adapted to pump information from the prior generation (Tomasello 1999; Alvard 2003; Gergely and Csibra 2005; Csibra and Gergely 2006), but also because we structure the learning environment of the next generation. We construct not just our own niche, but that of the next generation. In part, we build not just their physical and biological world, but their learning environment (Avital and Jablonka 2000; Laland et al. 2000; Sterelny 2003, 2006). So there is a family of alternative models of evolutionary psychology (see, e.g., Tomasello 1999; Heyes 2003; Sterelny 2003; Jablonka and Lamb 2005; Richerson and Boyd 2005; Laland 2007). These models emphasize four factors that structure human cognition. These are: (1) Cultural inheritance: we have complex cognitive adaptations—for example, the natural history competences of foragers—that are built by cumulative selection on culturally transmitted variation. (2) We are adapted for cultural learning: Michael Tomasello, for example, thinks that joint attention is a key adaptation underpinning human cultural learning, allowing individuals to monitor (and learn from) the social and technical activities of others. (3) Human cognition is plastic: very different phenotypes emerge from the interaction between environments and inherited resources. (4) We develop in structured learning environments. We now sketch one model from this family (based on Sterelny 2003), and then close with the difficult methodological problem: Can alternative models be tested? Is the historical record rich enough to impose serious evidential constraints on theories of human cognitive evolution?

We begin by contrasting human and chimp culture. Chimpanzee material culture is quite varied, but it is also quite rudimentary: there is no evidence that any chimp tools exist in their current form as the result of a cycle of discovery and improvement (Laland and Galef 2009). Cycles of discovery and improvement depend on reliable and high fidelity transmission between generations. If an innovative australopithecine discovers a more efficient way of flaking stone tools, without reliable social transmission, the technique will disappear at the death of its discoverer. Imitation plays little role in chimp social learning, and they lack language. So chimp social learning is not adapted to a communal data base and a communal skill base that can be ratcheted up over the generations. But for the last couple hundred thousand years the human environment has been the result of a ratchet effect in operation: a cycle in which an innovation is made, becomes standard for the group, and it then becomes a basis for further innovation. So material culture and informational culture is built by cumulative improvements. This process of cumulative improvement depends in part on cognitive adaptations for cultural learning: adaptations that make human children soak up the skills and information of the adults with whom they grow up, but it also depends upon information pooling. Information pooling makes the flow of information across generations much more reliable, as children have access to information controlled by the group as a whole. Moreover, information pooling allows an innovation made by any individual in the group to spread through the group as a whole, and that innovation is then available as a foundation for further improvement. To the extent that information pooling is crucial to the reliable and high fidelity transmission of information across generations, human cooperation and human capacities for high-load decision making are intimately linked.

High fidelity cultural transmission is effective because children are adapted to receive this information, and because information pooling ensures that it is sent reliably, with plenty of redundancy (see Csibra and Gergely 2006, Gergely and Csibra 2005, Gergely et al. 2007); for archaeological support, see Sterelny (2011) and the citations therein). But it also depends on niche construction. We have become cosmopolitan in part because we have learned to take our world with us. We have progressively modified our own physical, social, and biological environment. Tools, clothes, shelters have changed the worlds we live in. But we have modified our learning environment too: reshaping both the information available to children and the access they have to that information. Learning is scaffolded in many ways. Ecological tools have informational side effects. A fish trap can be used as a template for making more fish traps; a toolmaker can be immersed in a social world where environmentally salient tools, tool manufacture, and tool use are ubiquitous. Skills associated with manufacture are demonstrated in a form suited for learning. Completed and partially completed artifacts are used as teaching props. Practice is supervised and corrected. The decomposition of a skill into its components is made obvious; subtle elements will often be exaggerated, slowed down, or repeated. Moreover, skills are often taught in an optimal sequence so that one forms a platform for the next. This makes it possible to learn the otherwise unlearnable. Artifacts also act as props in games, rituals, and storytelling, providing opportunities for motor skill acquisition by novice users, and opportunities for an individual to understand artifact deployment within highly coordinated group activities well before their contribution is crucial to group success.

On this view, the organization of human learning environments, cultural variation in those learning environments, and human developmental plasticity interact to provide a range of human cognitive phenotypes. Humans do not have a single cognitive design; cognitive skills are a fundamental part of our cognitive systems, and these in turn are contingent on our environment. Only highly structured developmental environments make the acquisition of complex automatized skills possible. These skills are phenomenologically like modules: they are fast, automatic, and typically adaptive. Skilled drivers make life-and-death decisions without skipping a conversational beat based on quite subtle information about the physical conditions and the behavior of other drivers. But these skills are developmentally very different from modules: they develop slowly, with much practice and instruction, and with much variation both with and across cultures.

Dan Dennett has suggested a more radical version of the idea that differences in technology result in profoundly different developmental trajectories. He has proposed that our capacity to represent and reason about our own thoughts and those of others depends on prior exposure to public representations. Agents in a culture with enduring public symbols inherit an ability to make those symbols themselves objects of perception and to manipulate them voluntarily. Imagine a group of friends drawing a sketch map in the sand to coordinate a hike. Those representations are voluntary and planned. Dennett suggests that we first learn to think about thoughts by thinking about these public representations. In drafting and altering a sketch map, we are using cognitive skills that are already available—they are just being switched to a new target. Moreover, manipulation of such a public representation makes fewer demands on memory; no one has to remember where on the map the camp site is represented. Rich metarepresentational capacities are developmentally scaffolded by an initial stage in which public representations are objects of thought and action (Dennett 2000); Andy Clark develops a similar picture in his 2002 work.

Nativist evolutionary psychologists think we have a "folk psychology" module for we are very good at estimating what others will think and do. If an arrangement goes wrong, and a friend fails to turn up to a meeting at a café, we are quite good at working how to recoordinate on the fly. We predict others' responses, even taking into account the fact that their response will depend on what they expect us to do.

The nativist explanation of this remarkable capacity is that we have an innate module equipped with a good model of human thought and decision. There is an alternative: we interpret others as the result of having a biologically prepared, culturally amplified, automated skill. That skill is acquired very reliably, both because humans of one generation engineer the learning environment of the next generation, and because the acquisition of this skill is supported by perceptual systems tuned to relevant cues. We are sensitive to facial expression; signs of affect in voice, posture, and movement; the behavioral signatures that distinguish between intentional and accidental action, and the like. These systems make the right aspects of behavior, voice, posture, and facial expression salient to us. They make learning easier because we are apt to notice the right things in other agents. The acquisition of interpretive skills depends on perceptual preadaptation and individual exploration in a socially structured learning environment.

In addition to these perceptual preadaptations, children live in an environment soaked with agents interpreting one another. They are exposed both to third party interpretation, and to others interpreting them. Much of this interpretation is linguistic but there are also contingent interactions in which the child is treated as an agent: imitation games, joint attention, and joint play. It helps that children interact with their developing peers for they have not yet gained the abilities to mask their emotions, inhibit their desires, and suppress their beliefs. Interacting with more transparent agents simplifies the problem of inferring from an action to its psychological root. In adults, the connection between psychological state and action can be very complex and indirect. How could anyone learn that action depends on an agent's beliefs and goals, when those are so hidden? But when children interact with their peers, the connections between desire, emotion, and action will often be very direct. Children are less good at concealing overt signs of their emotion than adults, and less good at resisting the urge to act on those emotions. As three- and four-year-olds are making crucial developmental transitions, this lack of inhibition of their peers simplifies their epistemic environment.

Parents make the interpretive task easier by offering models of their own and their children's actions: they often rehearse interpretations of both their own and their children's actions. Likewise, children's narrative stories are full of simplified and explicit interpretative examples. Language itself scaffolds the acquisition of interpretative capacities by supplying a premade set of interpretative tools. Thus linguistic labels help make differences salient (Peterson and Siegal 1999). Finally, focusing on the concept of belief can make the task of acquiring theory of mind seem even more challenging than it really is. Belief and preference are often hidden, having no overt and distinctive behavioral signature. But many folk psychological concepts—those for sensations and emotions—do have a regular behavioral signature, and these scaffold the acquisition of less behaviorally overt concepts by making available easier examples of inner causes of outer actions (Sterelny forthcoming).

The position outlined agrees with the nativists' scepticism about the power of general purpose learning mechanisms. Nativists think evolution has solved this problem by reducing the amount we need to learn. But the alternative model suggests that evolution has found a different and more flexible strategy: the power of general purpose learning has been augmented by optimizing the learning environment. No doubt the cognitive capacities involved in understanding others would be very hard to acquire by our own unaided efforts, but we do not have to acquire them that way. Our environments have been informationally engineered to circumvent the cognitive limits of individuals on their own. Alternatives to nativist evolutionary psychology present a model of how a fast, automatic, and sophisticated cognitive specialization can develop without it depending on specific innate structures.

Of course many hybrids are possible. Perhaps folk psychology is acquired both through a richly structured environment together with some minimal but specific prewiring: a view that might be very plausible, if the evidence for very early acquisition is further supported. In any case, it is likely that the truth will lie somewhere between the two extremes of nativism and niche construction. The alternatives will be refined and tested within the laboratories of cognitive scientists. But that is not the only evidence that is relevant, for if modules and automatized skills are phenomenologically similar, then one way of choosing between the two hypotheses is the rate of change in behaviors over time. Niche construction and other culture and learning based models predict that human behavior and social organization should be highly variable in space and time (since cultural variation is not constrained by innate modules). Moreover, our distinctive cognitive capacities are coevolutionary products of individual innovation, cultural transmission, and niche construction. This implies that they should appear gradually, as their appearance does not depend on genetic change in the human lineage. The nativist model makes contrary predictions. Change should be more stepwise, as a cognitive module comes on-line, opening up a new set of behaviors. But is there enough information in the historical record for us to tell who is right? We conclude with this pressing methodological problem, beginning with a review of the record and its data.

4. Going Beyond the Evidence? Historical Traces of Cognitive Evolution

As noted in Section 1, there has been speculation about the evolution of human psychology for well over a century. However, within archaeology and paleoanthropology this speculation has typically been naïve about the cognitive sciences. Psychologists' speculations on human evolution have been equally naïve (see for instance Foley 1996). This began to change in the latter decades of the last century, with a small but significant number of publications that were either joint productions of cognitive scientists and archaeologists (Noble and Davidson 1996), or that took cognitive theories more seriously (Wynn 1991; Wynn 1993; Renfrew 1994;

Mithen 1996). This new movement within archaeology and evolutionary studies both tested the claims of evolutionary psychology and morphed into an active branch of "evolutionary psychology" in its own right. So what can this area of evolutionary psychology tell us? Can it discriminate between alternative hypotheses about the history of cognition?

Archaeology can tell us a great deal about our human ancestors. Analysis of a stone tool can show that a hominin manufactured it out of a certain rock type, and thus can also show that hominins traveled long distances, or traded, for the raw material. Biochemical analysis can show that there are residues of animal protein on the tool that indicates its use for butchery. Microscopic analysis can show wear patterns characteristic of certain functions. Fossil evidence provides information about physiological adaptations, such as a shift to bipedalism, and the evolution of precision grip. However, this information is not about cognition as such. To go beyond artifacts and fossils, we have to construct models of behavior, and from these behaviors, models of the associated cognitive skills. So we have evidence for specific cognitive capacities only if these capacities have impacts upon the physical world: impacts that are a consequence of behaviors that are preserved in the historical record. Thus, the best models of cognitive evolution tested in the historical record will take the external environment and its manipulation by actors seriously.

The key, then, to making the physical evidence of the archaeological and fossil record act as a means for testing hypotheses of evolutionary psychology is the construction of models of behavior that make predictions about the interactions of individuals and groups with their physical world. A speech act might leave no direct physical evidence, but a speech act might be necessary for a particular kind of learning, behavior, or activity that in turn leaves some kind of physical evidence. We may not be in a position to determine directly when sophisticated forward planning emerges within the human lineage, but indirectly, the capacity can be inferred from behaviors that require this capacity; behaviors that in turn leave evidential traces.

The notion that we can utilize physical by-products of behavior as tools for understanding minds should not alarm us. We as individuals do this every day when we interpret the desires and beliefs of our fellow agents through the consequences of their actions. We are quite comfortable inferring a set of beliefs and desires about an individual when arriving at a shared office to find an office mate's computer on, and a warm cup of coffee and a scatter of articles on her desk, despite her temporary absence. Forensic scientists routinely reconstruct behaviors and motives from physical evidence in ways that juries find persuasive. So for the historical record to act as a test of evolutionary psychology hypotheses, our models of cognition and behavior should derive predictions about further evidence that can then be detected in the historical record. This information is going to come in two forms: fossil evidence and archaeological evidence.

Behaviors and brains do not fossilize. However, there are endocasts of fossil hominin skulls, which are molds taken from inside a fossil cranium. These can reveal the crude anatomy, hints of surface features, and size of hominin brains. How informative endocasts are, however, remains an open question. It seems unlikely that the folding of the brain surface relates directly to function, as surface folding may well be the result of an allometric process, with increased brain size resulting in increased folding. Sulci, gyrus, and other surface features of the brain, being highly labile and dependent upon body size (Sereno and Tootell 2005), may not then be diagnostic of function. Perhaps the surface features of the brain are only informative where gross anatomy of hominin brains reveal changes in sensory input capacities. Endocasts can reveal differences in the relative size of, say, the frontal lobes versus the cerebellum, but again, how much this reveals about cognitive function is highly dependent upon the extent to which specific, functional aspects are localized, and whether these relations between gross structure and function are stable homologies across species. Certainly, the resolution at this level is not enough to inform us of the emergence of specific functions. However, this gross anatomy, along with overall brain size, may provide clues as to the increasing importance of cognitive strategies in human evolution, even if they are uninformative of specific cognitive adaptations per se (Deacon 1997).

Fossilized physiological traits directly associated with specialized behaviors may provide clues to cognitive developments. Language adaptations provide a good example here. Breathing control necessary for speech requires increased muscle and nerve control and an inevitable widening of the thoracic vertebral canal. Paleoanthropologists have detected this feature in fossils of *Homo neandertalensis*, and *Homo sapiens*, but not in earlier hominins such as *Homo erectus* and the Australopithecines, nor in extant primates (MacLarnon and Hewitt 1999). Other physiological by-products of specialized behaviors include adaptations for precision grip associated with toolmaking (See for instance Marzke 1992; Tocheri et al. 2003). So there is a clear potential for an important subset of behaviors to leave signals in the fossil record, and these are equally signals of the cognitive mechanisms that drive those behaviors.

Archaeological finds of tools, marked and cracked bones, and other manipulated objects are at once highly suggestive of cognitive developments and novel behavior, but they also require a good understanding of the relationship between the evidence, the behavior, and its cognitive basis. Modeling these relationships is not straightforward. For a start, tools and their diversification in the latter stages of human evolution may be in part a response to changing economic requirements (Stiner et al. 2000). So there remains the constant concern that absence of archaeological evidence for cognitive sophistication is not evidence for the absence of cognitive sophistication.

A further problem is that of determining the function of specific tool types, for function tells us about the agent's behavior, and hence the mind responsible for that behavior. Nowhere is this more obvious than in the case of the Acheulean bifaces of the Early Stone Age of Africa, and the Lower Paleolithic of Europe. This technology contravenes many of the expectations we have of a fully human culture: it shows little change over an extended period (approximately two million years ago to .6 mya) and little regional differentiation over a large area (Africa and Eurasia). Quite how hominins used Acheulean handaxes is problematic. The associated cognitive

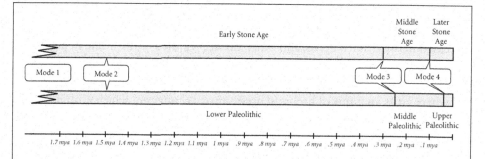

For historical reasons in the development of Archaeology as a discipline, different terminologies exist to describe different "cultures" or stages of tools in Europe and Africa. In Europe the stages are known as Lower, Middle, and Upper Paleolithic. In Africa, similar tool cultures are known as Early, Middle, and Later Stone Age.

The Cambridge Archaeologist Graham Clark proposed an alternative classification based on differing manufacturing methods, which is probably more suitable for evolutionary psychologists.

Mode 1 tools are simple chopping tools and flakes; they emerge approximately 2.6 million years ago in Africa with the *Homo* genus and make a first appearance in Europe some time later. They are typically modified cobbles, and appear to be manufactured by Early *Homo* species in direct response to immediate requirements.

Mode 2 tools are associated with the classic Acheulean Handaxes. These tools are bifacially flaked tools; many seem to be manufactured to a standardized "tear drop" shape and are associated with increased transportation of raw materials. Mode 2 technology makes its first appearance approximately 1.6 to 1.5 million years ago, and it overlaps with *Homo erectus*' long tenure on the planet.

Mode 3 tools are manufactured from a "prepared core." This two-step process has an initial piece of raw material that is shaped, and from this large uniform flakes are removed. These standardized flakes are in turn shaped into different tools. Mode 3 technology is associated with *Homo neandertalensis* and other "Archaic" sapiens.

Mode 4 Tools (Upper Paleolithic and Later Stone Age) represent the emergence of blades and finer worked stone tools, and is generally considered to represent the emergence of a full human suite of toolmaking capacities. (Although not necessarily the emergence of *Homo sapiens* as a species.) Mode 4 tools show regionalization, specialization, and increased use of alternative materials. Symbolic art and other cultural traits are associated with the emergence of Mode 4 technologies. The middle to upper Paleolithic transition represents the sudden arrival of Mode 4 technology in Europe, but appears to have been a gradual transition from middle to later stone age in Africa from approximately one hundred thousand years ago or even earlier (see text).

developments are also difficult to interpret. They have been associated with preferences for symmetry (Wynn 1995, 2002), with sexual selection in hominins (Kohn and Mithen 1999; Kohn 2000), and even as the by-product of a sophisticated hunting strategy utilizing handaxes as "killer Frisbees" (Calvin 1983).

The key to understanding the relevance of such archaeological finds lies in understanding the cognitive requirements and learning strategies of two distinctive behaviors: tool manufacture and tool deployment. Tool manufacture should provide us with information on important planning skills, folk physics (or perhaps more accurately, folk engineering and materials science), and perhaps the learning and teaching skills acquired by hominins. For instance, a crucial development in hominin evolution is decoupling of immediate stimulus from tool manufacture, something of which many primates seem incapable (See for instance Jalles-Filho et al. 2001). The evidence for this comes from manufacturing sites away from kill sites, the emergence of raw material transport and caching as a strategy (Marwick 2003), and higher time investment in tool manufacture. Such analyses suggest increased abilities in forward planning (Suddendorf and Busby 2005; Suddendorf 2006), and possibly in later hominins the evolution of increased working memory (Coolidge and Wynn 2005). While tool manufacture seems implicitly tied to the emergence of technical intelligence, toolmaking is also informative about social intelligence. Modern tool manufacture is frequently a highly social practice, requiring group-level decisions, social pedagogy, and group coordination (for an intriguing example, see Hiscock 2004). Understanding the dynamics of toolmaking, particularly the social skills required for the acquisition of toolmaking competence, is a crucial area of research.

Tool deployment, understood in the light of evidence from bones from kills, tells us about hunting strategies, coordination, exploitation of new habitats, and ability to recognize and exploit seasonal resources. See for instance the shift in game profiles from prehuman to human hunters discussed in (Avery and Cruz-Uribe 1997). In particular, the changes in technology from general purpose cutting devices manufactured and deployed in response to immediately perceived needs, to curated tools manufactured to anticipated needs, through to highly specialized tools with distinct local cultures, suggests changes in planning capacities, learning, and consequent developments in social transmission of environmental information (Jeffares 2010a, 2010b). Archaeological finds thus suggest increases in the sophistication of planning and means-ends reasoning of hominins, plus an increased role for social learning, direct pedagogy, and group-level logistics requiring coordination and cooperation among individuals.

It is important to note here that we can construct from a variety of evidential sources hypotheses about when and why distinctively human cognitive traits appear. Consider Ben Marwick's suggestion that archaeological evidence of long-distance trade networks in raw materials provides evidence for the evolution of language (Marwick 2003). By itself, the claim is highly suggestive but not compelling. However, when combined with physiological evidence of language-related physiological traits such as an increased thoracic vertebrae channel for breathing control (MacLarnon and Hewitt 1999), and increased requirements for directed learning associated with more sophisticated tool types, we may get a temporal cluster of evidence pointing to the emergence of a particular cognitive skill within a particular time frame. In turn, such clusters of evidence can provide us with a chronicle that documents the emergence of a variety of cognitive features.

In short, the archaeological and fossil record can provide information about the tempo and mode of the evolution of particular capacities, and thus it gives us some ability to discriminate between the two models of evolutionary psychology outlined in prior sections. As we noted, the two alternative views of cognitive evolution make quite distinct claims about the timing of the emergence of cognitive capacities. The modular nativist account sees cognition as a set of discrete competences. Consequently, we would expect to see physical evidence for the evolution of these competences as equally discrete units in the physical record. As a new behavior comes "on-line," it would manifest itself in the physical record. However, if human cumulative cultural evolution, niche construction of information, and environments underpin human cognitive evolution, then we should expect to see evidence of an accumulative ratchet of cultural competences in the archaeological record, with technological and behavioral developments building on prior developments.

These alternative views on the tempo of human cognitive evolution have an important echo within paleoanthropology. Much ink has been spilt over the significance of the Middle to Upper Paleolithic transition in tools in Europe approximately fifty thousand years ago. This transition shows the comparatively sudden emergence of new materials for tools. Bone and wood appear for the first time, symbolic art emerges, and we see evidence of other quintessential human cultural traits. In fact, the Middle to Upper Paleolithic transition is so profound and puzzling that some of the hypotheses offered have been wildly speculative. Thus Michael Winkelman suggests that the emergence of symbolic art was a strategy to cope with the integration of a bicarmel consciousness, guided by shamanic cultural practices (Winkelman 2002). Some of the more reasonable suggestions for the cognitive basis for this transition have included the emergence of fully articulate language (wee for instance Lieberman 1998), and (notably for the nativist evolutionary psychologists) Stephen Mithen's suggestion that it marks the final phase of a modular development of human cognition (Mithen 1996). Thus, the Middle to Upper Paleolithic transition could be interpreted as evidence of the emergence of nativist module, with a new set of skills coming "on-line" in the human lineage.

However, recent work suggests that this sudden European transition in tool types represents migration, with culturally developed humans displacing previous populations, and that this transition is consequently an artifact of the European record, and not a sign of a cognitive leap forward. A reevaluation of the archaeological record beyond Europe, particularly Africa, suggests that fully human behaviors are much older, emerging via the cumulative evolution of culture (Mcbrearty and Brooks 2000). The result has been a reevaluation of the emergence of modern humans. There are a range of proposals, some with the emergence of modern humans going back as far as three hundred thousand years ago, and associated with the emergence of Mode 3 tools. (For a brief overview of this debate, see Henshilwood and Marean 2003) Currently, the general picture that emerges is of *Homo sapiens* arising in Africa, slowly developing a sophisticated culture, and then displacing pre-existing hominin populations in Europe and elsewhere. Consequently, archaeological evidence of later hominins favors a cumulative, slow cultural evolution and not a modular, discrete series of punctuated pulses in cultural developments.

5. The Future of the Evolutionary Past

This overview of evolutionary psychology has had two themes. We have underscored the variety of evolutionary approaches to human cognition. There are, of course, various hybrids between evolutionary nativism and the alternatives we have sketched; there is a spectrum of models between undiluted nativism and undiluted constructivism. For example, one might suppose that some or all of the distinctively human cognitive capacities arose and were established in human populations as automated skills (i.e., by the mechanisms discussed in Section 3), and were subsequently modularized by genetic entrenchment, their development becoming less reliant on rich environmental support. The second theme has been to emphasize the constraints history should place on theory. In the existing literature, most of the evidential debates have focused on proximate mechanisms. How do our cognitive capacities develop? How do they operate? These are good questions, for evolutionary psychologists of all stripes make claims about the cognitive capacities of living humans. But they also make claims about the history of those capacities. So it is important to test their models against the record of physical evidence from the past. In the debates on evolutionary psychology, this mode of testing has been underexplored. Despite the obvious limitations of the historical record, we think it is a potentially important source of evidence, as well as an important strand of research in its own right. But that potential will be quite difficult to realize. To test models of cognitive evolution, we need to isolate physical proxies for behaviors: behaviors that are in turn proxies for important cognitive capacities. Those proxies then tell us when the behavior appeared in the hominin repertoire, and the behaviors in turn tell us about the cognitive capacities that underpin those behaviors. That is important, because different models make differing predictions about the tempo and mode of cognitive evolution, so we need a chronology that charts the emergence of our cognitive capacities. As we have just seen, to some extent, and still very tentatively, we can compare different predictions about the construction of the human cognitive suite with its actual construction. The more general point though is that models of cognitive evolution are constrained by physical evidence of our past, not just evidence about current function.

Much of the literature on evolutionary psychology, pro and con, has focused almost wholly on evolutionary nativism. Much of that literature has been more polemical than informative. As we see it, the way forward is to develop and to test specific alternatives to such nativist models (many of which will be hybrids, incorporating elements of those nativist models). There is no novelty in emphasizing testing. We do, however, think that the historical record is richer in information, and hence in the potential to test and refine these alternative methods than has often been supposed. It is true that thoughts do not fossilize. But thoughts drive actions, and these leave traces which, collectively, are surprisingly informative about the lives and minds of ancient humans.

REFERENCES

Alcock, J. (2001). *The Triumph of Sociobiology*. Oxford: Oxford University Press.

Alvard, M. (2003). The adaptive nature of culture. *Evolutionary Anthropology* 12: 136–49.

Atran, S. (1990). *Cognitive Foundations of Natural History*. Cambridge: Cambridge University Press.

———. (1998). Folk biology and the anthropology of science: Cognitive universals and cultural particulars. *Behavioral and Brain Science* 21: 547–609.

Avery, G., and Cruz-Uribe, K. (1997). The 1992–1993 excavations at the Die Kelders Middle and Later Stone Age Cave Site, South Africa." *Journal of Field Archaeology* 24(3): 263.

Avital, E., and Jablonka, E. (2000). *Animal Traditions: Behavioral Inheritance in Evolution*. Cambridge, Cambridge University Press.

Baillargeon, R., Scott, R., and He, Z. (2010). False-belief understanding in infants. *Trends in Cognitive Science* 14(3): 110–18.

Barkow, J. H., Cosmides, L., and Tooby, J. (eds.) (1992). *The Adapted Mind: Evolutionary Psychology and the Generation of Culture*. Oxford: Oxford University Press.

Barrett, C., and Kurzban, R. (2006). Modularity in cognition: Framing the debate. *Psychological Review* 113(3): 628–47.

Buller, D. (2005). *Adapting Minds: Evolutionary Psychology and the Persistent Quest for Human Nature*. Cambridge, MA: MIT Press.

Burkhardt, R. W. J. (2005). *Patterns of Behavior: Konrad Lorenz, Niko Tinbergen and the Founding of Ethology*. Chicago: University of Chicago Press.

Buss, D. M. (2006). Strategies of Human Mating. *Psychological Topics* 15(2): 239–60.

——— . (1994). *The Evolution of Desire: Strategies of Human Mating*. New York: Basic Books.

Calvin, W. H. (1983). A stone's throw and its launch window: Timing precision and its implications for language and hominid brains. *Journal of Theoretical Biology* 104(1): 121–35.

Carruthers, P. (2006). *The Architecture of the Mind: Massive Modularity and the Flexibility of Thought*. Oxford: Oxford University Press.

Clark, A. (2002). *Mindware: An Introduction to the Philosophy of Cognitive Science*. Oxford: Oxford University Press.

Coolidge, F. L., and Wynn, T. (2005). Working memory, its executive functions, and the emergence of modern thinking. *Cambridge Archaeological Journal* 15: 5–26.

Cosmides, J. Tooby, Fiddick, L., and Bryant, G. (2005). Detecting cheaters. *Trends in Cognitive Sciences*, 9(11): 505–6.

Cosmides, L., and Tooby, J. (1992). Cognitive Adaptations for Social Exchange. In
    J. H. Barkow, L. Cosmides, and J. Tooby (eds.), *The Adapted Mind*. Oxford: Oxford
    University Press, 163–227.

Cosmides, L., and Tooby, J. (2005). Neurocognitive Adaptations Designed for Social
    Exchange. In D. Buss, (ed.), *The Handbook of Evolutionary Psychology*. Hoboken, NJ:
    Wiley, 584–627.

Cowie, F. (1998). *What's Within? Nativism Reconsidered*. Oxford: Oxford University Press.

Csibra, G., and Gergely, G. (2006). Social Learning and Social Cognition: The Case for
    Pedagogy. In Y. Munakata and M. H. Johnson (eds.), *Processes of Change in Brain and
    Cognitive Development. Attention and Performance, XXI*. Oxford: Oxford University
    Press, 249–74.

Deacon, T. W. (1997). What makes the human brain different? *Annual Review of
    Anthropology* 26(1): 337–357.

Dennett, D. C. (2000). Making Tools for Thinking. In D. Sperber (ed)., *Metarepresentations:
    A Multidisciplinary Perspective*. New York: Oxford University Press, 17–29.

Edgerton, R. B. (1992). *Sick Societies: Challenging The Myth of Primitive Harmony*.
    New York: Free Press.

Fletcher, G. (2002). *The New Science of Intimate Relationships*. Oxford: Blackwell.

Fletcher, G. J., and Stenswick, M. (2003). The Intimate Relationship Mind. In J. Fitness and
    K. Sterelny (eds.), *From Mating to Mentality: Evaluating Evolutionary Psychology*. Hove,
    UK: Psychology Press, 71–94.

Foley, R. A. (1996). The adaptive legacy of human evolution: a search for the environment
    of evolutionary adaptedness. *Evolutionary Anthropology* 4: 194–203.

Gangestad, S., and Simpson, J. (2000). The evolution of human mating: Trade-offs and
    strategic pluralism. *Behavioral and Brain Sciences* 20(4): 573–644.

Gergely, G., and Csibra, G. (2003). Teleological reasoning in infancy: The naive theory of
    rational action. *Trends in Cognitive Science* 7(7): 287–92.

———. (2005). The social construction of the cultural mind: Imitative learning as a
    mechanism of human pedagogy. *Interaction Studies* 6(3): 463–81.

———. (2006). Sylvia's Recipe: The Role of Imitation and Pedagogy in the Transmission
    of Cultural Knowledge. In N. J. Enfield and S. C. Levenson (eds.), *Roots of Human
    Society: Culture, Cognition and Human Interaction*. Oxford: Berg, 229–55.

Gergely, G., Egyed, K., and Király, I. (2007). On pedagogy. *Developmental Science* 10(1):
    139–46.

Hamilton, W. (1975). Innate Social Aptitudes in Man: An Approach from Evolutionary
    Genetics. In R. Fox (ed.), *Biosocial Anthropology*. London: Malaby Press.

Hauser, M. (2006). *Moral Minds: How Nature Designed Our Universal Sense of Right and
    Wrong*. New York: HarperCollins.

Henshilwood, C. S., and Marean, C. W. (2003). The origin of modern human behavior:
    Critique of the models and their test implications. *Current Anthropology* 44(5): 627–51.

Heyes, C. M. (2003). Four routes of cognitive evolution. *Psychological Review* 110: 713–27.

Hiscock, P. (2004). Slippery and Billy: Intention, Selection and Equifinality in Lithic
    Artefacts. *Cambridge Archaeological Journal* 14(1): 71–77.

Jablonka, E., and Lamb, M. (2005). *Evolution in Four Dimensions*. Cambridge, MA: MIT Press.

Jalles-Filho, E., Da Cunha, R. G. T., and Salm, R. (2001). Transport of tools and mental
    representation: Is capuchin monkey tool behavior a useful model of Plio-Pleistocene
    hominid technology? *Journal of Human Evolution* 40(5): 365–77.

Jeffares, B. (2010a). The Evolution of Technical Competence: Economic and Strategic
    Thinking. In W. Christensen, E. Schier, and J. Sutton (eds.), *ASCS09: Proceedings of the*

9th Conference of the Australasian Society for Cognitive Science. Sydney: Macquarie Centre for Cognitive Science, 162–69.

———. (2010b). The co-evolution of tools and minds: Cognition and material culture in the hominin lineage. *Phenomenology and the Cognitive Sciences* 9(4): 503–20.

Kitcher, P. (1985). *Vaulting Ambition: Sociobiology and the Quest for Human Nature.* Cambridge, MA: MIT Press.

———. (2001). Battling the Undead: How (and How Not) to Resist Genetic Determinism. In R. Singh, K. Krimbas, D. Paul, and J. Beatty (eds.), *Thinking About Evolution: Historical, Philosophical and Political Perspectives: Festschrift for Richard Lewontin.* Cambridge: Cambridge University Press, 396–414.

Kohn, M. (2000). *As We Know It: Coming to Terms with an Evolved Mind.* London: Granta Books.

Kohn, M., and Mithen, S. J. (1999). Handaxes: Products of sexual selection?. *Antiquity* 73: 518–26.

Laland, K. (2007). Niche Construction, Human Behavioral Ecology and Evolutionary Psychology. In R. Dunbar and L. Barrett (eds.), *Oxford Handbook of Evolutionary Psychology.* Oxford: Oxford University Press, 35–48.

Laland, K., and Brown, G. (2002). *Sense and Nonsense: Evolutionary Perspectives on Human Behavior.* Oxford: Oxford University Press.

Laland, K. and Galef, B. (eds.) (2009). *The Question of Animal Culture.* Cambridge, MA: Harvard University Press.

Laland, K. N., Odling-Smee, J., and Feldman, M. (2000). Niche construction, biological evolution and cultural change. *Behavioral and Brain Sciences* 23: 131–75.

Landau, M. (1984). Human evolution as narrative. *American Scientist* 72: 262–68.

Leslie, A. (2000). How to Acquire a Representational Theory of Mind. In D. Sperber (ed.), *Metarepresentations: An Multidisciplinary Perspective.* Oxford: Oxford University Press, 197–224.

———. (2005). Developmental parallels in understanding minds and bodies. *Trends in Cognitive Sciences* 9: 459–62.

Leslie, A. M., Friedman, O., and German, T. P. (2004). Core mechanisms in "theory of mind". *Trends in Cognitive Sciences* 8: 528–33.

Lewontin, R. C. (1978). Adaptation. *Scientific American* 239: 156–69.

———. (1998). The Evolution of Cognition: Questions We Will Never Answer. In D. Scarborough and S. Sternberg (eds.), *An Invitation to Cognitive Science, Volume 4: Methods, Models, and Conceptual Issues.* Cambridge, MA: MIT Press, 107–32.

Lieberman, P. (1998). *Eve Spoke: Human Language and Human Evolution.* New York: W.W. Norton.

Lorenz, K. (1977). *Behind the Mirror.* New York: Harcourt Brace Jovanovich.

Lorenz, K. (1966). *On Aggression.* London: Methuen.

Lumsden, C. J., and Wilson, E. O. (1981). *Genes, Mind, and Culture: The Coevolutionary Process.* Cambridge MA: Harvard University Press.

Machery, E. and Barrett, C. (2006). Debunking adapting minds, *Philosophy of Science* 73(2): 232–46.

MacLarnon, A. M., and Hewitt, G. P. (1999). The evolution of human speech: The role of enhanced breathing control. *American Journal of Physical Anthropology* 109(3): 341–63.

Marwick, B. (2003). Pleistocene exchange networks as evidence for the evolution of language. *Cambridge Archaeological Journal* 13(1): 67–81.

Marzke, M. W. (1992). Evolutionary development of the human thumb. *Hand Clinics* 8(1): 1–8.

Mcbrearty, S., and Brooks, A. S. (2000). The revolution that wasn't: A new interpretation of
    the origin of modern human behavior. *Journal of Human Evolution* 39(5): 453–563.

Mithen, S. J. (1996). *The Prehistory of the Mind: A Search for the Origins of Art, Religion and
    Science*. London: Thames and Hudson.

Noble, W. and Davidson, I. (1996). *Human Evolution, Language and Mind: A Psychological
    and Archaeological Inquiry*. Cambridge: Cambridge University Press.

Peterson, C. C., and Siegal, M. (1999). Insights into Theory of Mind from Deafness and
    Autism." *Mind and Language* 15(1): 77–99.

Pinker, S. (1994). *The Language Instinct: How the Mind Creates Language*. New York:
    William Morrow.

———. (1997). *How The Mind Works*. New York: W.W. Norton.

Potts, R. (1996). *Humanity's Descent: The Consequences of Ecological Instability*. New York:
    Morrow.

Renfrew, C. (1994). Towards a Cognitive Archaeology. In C. Renfrew and E. B. W. Zubrow
    (eds.), *The Ancient Mind: Elements of Cognitive Archaeology*. Cambridge: Cambridge
    University Press, xiv, 195.

Richards, R. (1987). *Darwin and the Emergence of Evolutionary Theories of Mind and
    Behavior*. Chicago: University of Chicago Press.

Richerson, P. J., and Boyd, R. (2005). *Not By Genes Alone: How Culture Transformed Human
    Evolution*. Chicago: University of Chicago Press.

Sereno, M. I., and Tootell, R. B. H. (2005). From monkeys to humans: What do we now
    know about brain homologies?. *Current Opinion in Neurobiology* 15: 135–44.

Simpson, J., and Orina, M. (2003). *Strategic Pluralism and Context-Specific Mate Preferences
    in Humans*. In K. Sterelny and J. Fitness (eds.), *From Mating to Mentality*. Hove, UK:
    Psychology Press, 39–70.

Smith, E. A., Borgerhoff-Mulder, N., and Hill, K. (2001). Controversies in the evolutionary
    social sciences: A guide for the perplexed. *Trends in Ecology and Evolution* 16(3):
    128–34.

Smith, E. A., and Winterhalder, B. (2003). Human Behavioral Ecology. In L. Nadel (ed.),
    Encyclopedia of Cognitive Science. London: Nature Publishing Group. 2: 377–85.

Sperber, D. (1996). *Explaining Culture: A Naturalistic Approach*. Oxford: Blackwell.

Sperber, D., and Girotto, V. (2003). Does The Selection Task Detect Cheater-Detection?. In
    J. Fitness and K. Sterelny (eds.). *From Mating to Mentality: Evaluating Evolutionary
    Psychology*. Hove, UK: Psychology Press, 197–226.

Sterelny, K. (1999). Darwin's dangerous idea: Evolution and the meanings of life. *Philosophy
    and Phenomenological Research* 59(1): 255–62.

———. (2003). *Thought in a Hostile World*. New York: Blackwell.

———. (2006). The evolution and evolvability of culture. *Mind & Language* 21(2): 137–65.

———. (2011). From hominins to humans: How sapiens became behaviorally modern.
    Philosophical Transactions of the Royal Society B: Biological Sciences.

———. (forthcoming). *The Evolved Apprentice: The 2008 Jean Nicod Lectures*.

Stiner, M. C., Munro, N. D., and Surovell, T. (2000). The tortoise and the hare: Small-game
    use, the broad spectrum revolution, and paleolithic demography. *Current
    Anthropology* 41(1): 39–73.

Suddendorf, T. (2006). Foresight and evolution of the human mind. *Science* 312(5776):
    1006–07.

Suddendorf, T., and Busby, J. (2005). Making decisions with the future in mind:
    Developmental and comparative identification of mental time travel. *Learning and
    Motivation* 36(2): 110–25.

Tinbergen, N. (1960). *The Herring Gull's World: A Study of the Social Behavior of Birds.* New York: Lyons and Burford.

Tocheri, M. W., Marzke, M. W., Liu, D., Bae, M., Jones, G. P., Williams, R. C., and Razdan, A. (2003). Functional capabilities of modern and fossil hominid hands: Three-dimensional analysis of trapezia. *American Journal of Physical Anthropology* 122: 101–12.

Tomasello, M. (1999). The human adaptation for culture. *Annual Review of Anthropology* 28: 509–29.

Tomasello, M., and Carpenter, M. (2007). Shared intentionality. *Developmental Science* 10(1): 121–25.

Tomasello, M., Carpenter, M. Call, J., Behne, T., and Moll, H. (2005). Understanding and sharing intentions: The origins of cultural cognition. *Behavioral and Brain Sciences* 28(5): 675–735.

Wilson, E. O. (1975). *Sociobiology: The New Synthesis.* Cambridge, MA: Harvard University Press.

———.(1978). *On Human Nature.* Toronto: Bantam Books.

Winkelman, M. (2002). Shamanism and cognitive evolution. *Cambridge Archaeological Journal* 12(1): 71–101.

Winterhalder, B., and Smith, E. A. (2000). Analyzing adaptive strategies: Human behavioral ecology at twenty five. *Evolutionary Anthropology* 9(2): 51–72.

Wynn, T. (1991). Archaeological Evidence for Modern Intelligence. In R. A. Foley (ed.), *The Origins of Human Behavior.* London: Unwin Hyman, 19.

———. (1993). Two developments in the mind of early homo. *Journal of Anthropological Archaeology* 12(3): 299–322.

———. (1995). Handaxe enigmas. *World Archaeology* 27(1): 10.

———. (2002). Archaeology and cognitive evolution. *Behavioral and Brain Sciences* 25(3): 389–438.

CHAPTER 21

CULTURE AND COGNITION

DANIEL M. T. FESSLER AND EDOUARD MACHERY

HUMANS are unique among animals for both the diverse complexity of our cognition and our reliance on culture, the socially transmitted representations and practices that shape experience and behavior. Adopting an evolutionary psychological approach, in this essay we consider four different facets of the relationship between cognition and culture. We begin with a discussion of two well-established research traditions, the investigation of features of mind that are universal despite cultural diversity, and the examination of features of mind that vary across cultures. We then turn to two topics that have only recently begun to receive attention: the cognitive mechanisms that underlie the acquisition of cultural information, and the effects of features of cognition on culture. Throughout, our goal is not to provide comprehensive reviews so much as to frame these issues in such a way as to spur further research.

1. PSYCHOLOGICAL UNIVERSALS

Psychological universals can be defined as those traits, processes, dispositions, or functions that recur across cultures, with at least a subset of each population (e.g., individuals of a specific gender or at a specific developmental stage) exhibiting the trait. The search for psychological universals has a long tradition, as illustrated by Darwin's (1872) investigation of universal emotions, behaviorists' search for universal laws of learning (Hull 1943), and the Chomskian approach to language and cognition (e.g., Chomsky 1986). This tradition has been in part motivated by the desire to establish the "psychic unity" of humanity.

Because traits may recur across cultures due to cultural influences alone (via common cultural descent, cultural diffusion, or cultural evolutionary convergence), the strongest test of the universality of a given psychological trait is to search for it across maximally disparate cultures. One methodological concern, however, is that whether or not a trait is identified in different cultures will depend in part on how the trait is defined. For instance, if the specific circumstances that trigger shame in the United States are included in the definition of this emotion, shame is unlikely to qualify as a psychological universal; however, if the eliciting conditions are described more abstractly, shame is a good candidate for a universal emotion (e.g., Fessler 2004). It is worth keeping this in mind to avoid empty verbal controversies about the universality of traits (Mallon and Stich 2000). Relatedly, for many traits, similarities and differences across cultures will coexist (Brown 1991; Norenzayan and Heine 2005). Thus, shame might be present in all cultures, yet be triggered by different circumstances, or be expressed differently (see the notion of culture-specific emotion display rules in Ekman and Friesen 1971).

1.1. Generatively Entrenched Homologies

Because cultures vary tremendously with respect to their ecology, social organization, scale, and technology, and because cultural variables affect cognitive development, one might wonder why psychological universals exist at all. The answer is simple for those traits the development or acquisition of which reflects universal properties of physical or social environments (e.g., the belief that water is wet, or the distinction between males and females). The answer is not so straightforward for other psychological universals because cultural variables could plausibly affect their development. One might argue that such universals are the product of evolution by natural selection, and that natural selection tends to select for species-typical traits (Tooby and Cosmides 1992). However, this would be a mistake on two counts. First, much recent research emphasizes that natural selection has favored particular forms of phenotypic plasticity in humans, including the capacity to adapt to, and exploit, parochial cultural information. Second, one cannot presume that natural selection generates homogeneity. In most species, many traits are adaptive polymorphisms, either as a result of frequency-dependent selection or as an adaptive response to environmental variation in the species' range.

So, where do psychological universals come from? Some traits may be psychological universals because they are homologies—features possessed by humans and their relatives by virtue of common descent—that are generatively entrenched. A trait is generatively entrenched if its development is a necessary condition for the development of other traits (Wimsatt 1986). Most modifications of a generatively entrenched trait are selected against because they prevent the development of these other traits. If a psychological trait in humans is homologous to traits in other species, then, given the general absence of culture outside of our own lineage, it follows that the trait originally evolved in a species that had little (if any) capacity for culture. If this trait became generatively entrenched, then natural selection had little scope to act on its development, which remained insensitive to cultural variables.

The approximate numerical sense provides a clear example of a generatively entrenched homology. Research in the United States and in Europe has established that children and adults possess an approximate number sense (Hauser and Spelke 2004; Piazza and Dehaene 2004), being able to approximate the cardinality of sets of visually presented objects or of sequences of sounds without counting, to compare the cardinality of different sets or sequences, and to approximate the results of adding several sets of objects. The accuracy of people's numerical evaluation obeys Weber's law: the mean evaluation is identical to the cardinality of the target set, and the evaluation's standard deviation linearly increases as a function of the cardinality of the target set. Because evaluation is thus increasingly noisy, the accuracy of the numerical comparison between sets or sequences increases as a function of the distance between the cardinality of the sets or sequences to be compared. These are the signature properties of an analogical encoding of the cardinality of sets or sequences.

Recent studies by Pica, Lemer, Izard, and Dehaene (2004) and Gordon (2004) provide strong evidence that this approximate number sense is a psychological universal.[1] Pica et al. studied approximate estimation, comparison, and addition among the Mundurukú, a small-scale society in Brazilian Amazonia having limited contact with nonindigenous people. Most Mundurukú have not received any formal education. Their language has words for only the numbers one to five; above five, the Mundurukú rely on locutions signifying "some" or "many." Strikingly, the numbers three, four, and five are also used to refer to approximate quantities. For instance, in Pica et al.'s experiments, the word for four was used for sets of four and five objects, and the word for five was used for sets of five to nine objects. In spite of the differences between the Mundurukú counting system and that in European (and other) languages, and in spite of the many other differences between the respective cultures, the Mundurukú's approximate number sense is identical to Europeans'. The Mundurukú's performances in estimation, comparison, and addition tasks show the signature properties of the analogical system assumed to encode the cardinality of sets or of sequences. Gordon (2004) found similar results with the hunter-gatherer Pirahã tribe in the Lowland Amazonia region of Brazil. Most strikingly, the language spoken by the Pirahã has words for only one, two, and three. Nonetheless, their performances in tasks tapping into their approximate number sense were very close to Europeans' and Americans', providing further evidence for the universality of the approximate number sense.

The approximate number sense, evident in cultures as diverse as small-scale hunter-horticulturalist societies and modern, technologically complex societies, is

---

[1] These studies also show that cultural counting systems dramatically affect numerical cognition. People with a limited counting system (a few number terms) are unable to count beyond three and cannot accurately conduct arithmetic operations.

also present in numerous animal species (Hauser and Spelke 2004). Thus, in line with our discussion of the origins of psychological universals, the approximate number sense is plausibly a generatively entrenched homology.

1.2. Canalized Traits

Not all psychological universals are generatively entrenched homologies. A number of uniquely human psychological traits are also universal because their development has been canalized during the evolution of human cognition. Natural selection selects against development pathways that rely on specific environmental inputs when these environmental inputs vary, when variation in these environmental inputs cause the development of variable traits, and when there is a single optimally adaptive variant (Waddington 1940).[2] When this happens, natural selection buffers the development of the relevant traits against environmental variation by selecting for developmental pathways that do not depend on these environmental inputs. This phenomenon, known as canalization, likely explains the origins of some psychological universals. Note that, in contrast to the explanation of the evolution of psychological universals examined above, this second account can explain the universality of psychological traits that are not homologies.

Research on so-called folk theories provides some of the best evidence for such universals (Sperber and Hirschfeld 2004; Boyer and Barrett 2005). Folk theories are domain-specific, often implicit bodies of information that people use to reason. Although many folk theories vary across cultures, in some domains, folk theories have a universal core; folk biology and folk psychology are two such cases.

Some aspects of folk biology vary across cultures: in some cultures people have much more extensive biological knowledge than in others, and some reasoning strategies about the biological domain are found only in some cultures (for review, see Medin, Unsworth, and Hirschfeld 2007). Despite such heterogeneity, across cultures people classify animals and plants in a similar way (Berlin, Breedlove, and Raven 1973; Berlin 1992; Atran 1990, 1998; Medin and Atran, 2004; Medin et al. 2007). Plants and animals are organized into hierarchically organized taxonomies of kinds that include (at least) three levels: a "generic species" category (e.g., dogs and cedars), a superordinate category of biological domains (e.g., animals and plants), and a subordinate category of species varieties (e.g., particular breeds or strains).[3] At any

---

[2] By contrast, as will be discussed at length later, selection favors developmental pathways that rely on environmental inputs when the adaptive value of variants is contingent on features of the environment that are stable within an organism's lifespan and home range, yet variable over larger expanses of time and space occupied by the species.

[3] For Atran and colleagues, a generic species is a population of interbreeding individuals that possesses distinctive physiological and behavioral traits. These traits distinguish it from other generic species, and they are easily recognizable by ordinary people. For many organisms with which people are acquainted, generic species correspond to scientific genera (e.g., oak) or species (e.g., cat) or to local variants thereof.

level, membership in a kind is exclusive. For instance, no animal is both a dog and a cat, or a fish and a mammal.

From a cognitive perspective, the generic species level is of particular importance. Atran and colleagues have shown that while Itza' Maya's biological knowledge is much more extensive than American undergraduates', both Itza' Maya and American undergraduates avoid generalizing biological properties to the members of categories whose level is above the generic-species level (Coley, Medin, and Atran 1997). Furthermore, in a range of diverse cultures, membership in generic species is associated with "psychological essentialism" (Medin and Ortony 1989; Gelman 2003; Medin and Atran 2004): people believe that membership in a biological kind is associated with the possession of a causal essence—that is, some property or set of properties that define membership in the kind and cause the members of this kind to possess the kind-typical properties independently of their rearing environment. An essentialist disposition has been found among American children and adults (Keil 1989; Gelman and Wellman 1991), Yucatek adults (Atran et al. 2001), Brazilian adults (Sousa, Atran, and Medin 2002), and among children and adults from Madagascar (Astuti, Solomon, and Carey 2004)

Similarly, in spite of many differences in the way people across cultures explain their and others' behavior (Lillard 1998), psychologists have identified a universal core in folk psychology. In every known culture, people explain behavior in mentalistic terms, that is, by ascribing mental states such as beliefs and desires (Wierzbicka 1992). Furthermore, in their meta-analysis of children's performances in the false belief task, Wellman, Cross, and Watson (2001) have shown that children's understanding of beliefs develops similarly across cultures. They report studies conducted in Western cultures (United States, Austria, Australia), Eastern cultures (Japan, Korea), among the hunter-gatherer Baka (Avis and Harris 1991), and among Quechua-speaking Peruvian Indians (Vinden 1996). Although culture affects how early children come to understand that beliefs can be false, three-year-old children from Western, Eastern, African, schooled, and nonschooled cultures initially fail to understand this, then progressively come to grasp the notion of false belief.

2. CROSS-CULTURAL PSYCHOLOGICAL DIVERSITY

An entirely different approach to the relations between culture and cognition focuses on differences across cultures. Many differences are best described as ethnographic: because across cultures, people live in different social and physical environments, and different cultural framings thereof, and have correspondingly different experiences, their beliefs, concepts, and desires—in brief, the contents of their minds—will often similarly vary. However, looking beyond such differences, scholars have explored the effects of cultural variation on cognitive processes, personality, and perception (for an extensive review, see Kitayama and Cohen 2007). The search for cross-cultural differences is deemed successful to the extent that the psychological differences across cultures are marked and are explained by relevant differences among these cultures.

The search for cross-cultural psychological differences has a long history. Particularly, numerous scholars have addressed the role of linguistic differences in producing psychological differences. The anthropologist Edward Sapir and the linguist Benjamin Whorf famously proposed that the syntax and the vocabulary of different languages promote irreducibly different patterns of thought—what is known as the Sapir-Whorf hypothesis (Whorf 1956). Similarly, Soviet psychologists, particularly Lev Vygotsky and Alexander Luria, argued that languages as well as social activities (e.g., counting routines) constitute tools that allow children to develop symbolic thinking (Vygotsky 1986).

2.1. Proximal Origins of Cross-Cultural Differences: Extended Cognition

Traditionally, psychologists and anthropologists searching for cross-cultural differences have given little thought to the evolutionary origins of this diversity, assuming that evolutionary considerations were only relevant for universal traits, or that evolution was only relevant insofar as it produced an undifferentiated "capacity for culture." As discussed in the next two sections, recent theories and findings belie these assumptions. Here, we successively focus on two proximal causes of cross-cultural psychological diversity.

First, while cognitive science has tended to be methodologically solipsist (Fodor 1980), neglecting the social and physical environment in which cognition takes place, an influential approach, termed *extended cognition*, now insists that social practices, such as counting routines and formal education, as well as physical artifacts, dramatically modify (or, in some formulations of this idea, are constitutive of) people's cognitive processes (e.g., Hutchins 1995; Clark 1997). Because practices and artifacts vary tremendously across cultures, their effect on the mind is a potent source of cross-cultural diversity.

Language is one of the social practices that can potentially cause cognition to differ substantially across cultures. Since the 1990s, a flurry of cross-cultural work in linguistics and psychology has revived interest in the Sapir-Whorf hypothesis (for reviews, see Gentner and Goldin-Meadow 2003; Gleitman and Papafragou 2005; Chiu, Leung, and Kwan 2007). We consider in turn the research on spatial orientation and the more decisive research on color perception.

Levinson and colleagues have shown that languages encode spatial orientation in a variety of ways (Levinson 2003; Pederson et al. 1998). They identify three main ways of describing the location of objects. Speakers who use an intrinsic frame of reference locate objects by describing the relations between these objects (the spoon is beside the plate). Speakers who use a relative frame of reference locate objects by describing the position of these objects in relation to themselves and others (the knife is on my/your right). Speakers who use an absolute frame of reference locate objects by using cardinal directions (the knife is west of the plate).

Levinson and colleagues have argued that these linguistic differences affect people's spatial reasoning. In the rotation experiment (Pederson et al. 1998), subjects are shown an array of objects displayed on a table in front of them. Then, they are asked to turn by 180 degrees. They are then given the objects and asked to recreate the original array on a new table. If speakers of a language with a predominantly relative frame reference also reason relatively, they should preserve the orientation of the objects with respect to their own body: the object that was on the subject's left on the first table should be on her left on the second table, and so on. If speakers of a language with a predominantly absolute frame reference reason absolutely, they should preserve the absolute orientation of objects, thereby changing the orientation of the objects with respect to their own body: the object that was on the subject's left on the first table should now be on her right on the second table. As predicted, Dutch and Japanese subjects, whose languages use a relative frame of reference, preserved the relative orientation of the objects, while Mayans, whose language uses an absolute frame of reference, preserved the absolute orientation of objects. Levinson and colleagues take this and other findings to support the Sapir-Whorf hypothesis.

This finding has, however, been criticized (Gallistel 2002; Li and Gleitman 2002). Li and Gleitman have shown that when a salient object is present in their physical environments, American subjects can be primed to replicate the first array in an absolute manner. Since subjects who replicated the array of objects in a relative manner and those who replicated it in an absolute manner speak the same language (English), it would seem that the linguistic differences between Teztlan and Dutch (or Japanese) do not explain Pederson et al.'s original findings (but see Levinson et al. 2002). Li and Gleitman propose that the task description of the rotation experiment, "Make it the same," is ambiguous, because there are two different ways to reproduce the original array of objects. Subjects use the fact that their language relies predominantly on a relative or on absolute frame of reference to disambiguate the task. Thus, Levinson and colleagues' work does not show an effect of language on thought, but rather an effect of language on the interpretation of a linguistic expression. Furthermore, Li, Abarbanell, and Papafragou (2005) have shown that Teztlan speakers are not only able to use a relative frame of reference to solve spatial problems, but are better at doing it than at using an absolute frame of reference—in clear contrast to the Sapir-Whorf hypothesis.

This is not to say that the Sapir-Whorf hypothesis is unsupported by recent research, as shown by the research on color. The color lexicon varies tremendously across languages (Berlin and Kay 1969). While English has eleven basic color terms, the Dani, a hunter-horticulturalist society in Papua New Guinea, use only two, one for light colors and one for dark ones. Color vocabulary is not entirely arbitrary, however (Kay and Regier 2006). Focal colors constrain languages' color vocabulary: Regier, Kay, and Cook (2005) found that the best examples of color terms for 110 languages from nonindustrialized societies cluster around the focal colors. Furthermore, colors tend to be grouped by similarity. Finally, Kay and Regier (2007) have shown that the boundaries between color terms are not arbitrary, but rather map closely across languages.

Because color vocabulary varies across languages, one might wonder whether people's perception and memory of colors vary across linguistic communities. Heider (1972) and Heider and Olivier (1972) have answered negatively to this question, as the Dani's limited color vocabulary seems to have limited effect on their color perception and memory. Heider and Olivier showed Dani and American subjects a color patch. After a thirty-second interval, subjects were shown an array of similar color patches and were asked to identify the original patch in this array. The pattern of color recognition was very similar between the two groups. Particularly, in both groups, focal colors were recognized more easily (that is, less confused with other colors) than nonfocal colors.

This body of evidence against the Sapir-Whorf hypothesis has been challenged in recent years (Davidoff, Davies, and Roberson 1999; Roberson, Davies, and Davidoff 2000; Roberson et al. 2005; see also Lucy and Shweder 1979; Kay and Kempton 1984). Davidoff and colleagues (1999; Roberson et al. 2000) have focused on color perception among the Berinmo in Papua New Guinea, who have only five basic color terms. They failed to replicate Heider and Olivier's (1972) experiments. In contrast to Americans, the Berinmo were not more accurate at recognizing focal colors than nonfocal colors. Furthermore, when Davidoff and colleagues compared the Berinmo's pattern of color confusion with their pattern of color naming and with the American pattern of color confusion, they found that the Berinmo pattern of color confusion was more similar to the Berinmo pattern of color naming than to the American pattern of color confusion. In line with the Sapir-Whorf hypothesis, this suggests that the Berinmo's color vocabulary affects their color memory.

Additional evidence shows that the effect of color terms on cognition is, at least sometimes, driven by the on-line use of color words (Roberson and Davidoff 2000; Winawer et al. 2007). Similarity judgments and color identification (identifying two patches as being the same color) are affected by the boundaries between color categories drawn by the color terms in the subjects' languages. Winawer et al. simultaneously presented subjects with a target color patch on the one hand and a pair of color patches on the other. Subjects were asked to determine which of the two patches in the pair was identical to the target patch. Reaction times show that this color identification task was easier when two different color terms (in the subject's color vocabulary) applied to the two patches in the pair than when the same term applied to both patches—a form of categorical perception. Importantly, this and other effects of color vocabulary disappear when a verbal dual task prevents subjects from articulating subvocally.

2.2. Proximal Origins of Cross-Cultural Differences: Effects of Environmental Differences

A second approach to the proximal origins of cross-cultural differences notes that people typically have various processes and strategies for fulfilling a given psychological function (for instance, categorizing, reasoning inductively, making decisions under uncertainty, etc.). In any given environment, these strategies do not equally well fulfill their functions. For instance, the different types of spatial orientation do not work equally well in all environments. As a result, people can learn to rely on the processes and strategies that are most efficient in their environments. It is therefore important that social and physical environments vary across cultures. Indeed, culturally transmitted norms directly shape people's social environments, and cultural practices can powerfully modify people's physical environments. Culture is thus a source of diversity in social and physical environments; hence, across cultures, people might come to learn to rely on different processes and strategies, because these are the best ways to fulfill the relevant functions in the environments they inhabit.

It is important to note that the two proximal causes we have considered lead to two different forms of cross-cultural psychological diversity. The idea that artifacts and social practices dramatically modify or are constitutive of people's cognitive processes implies that particular cognitive processes (and other psychological traits) might exist in some cultures, but not in others. In contrast, the idea that people learn to rely on the strategies and processes that are most efficient in their environment suggests that the same processes (and other psychological traits) are present (if only in nascent form) in all cultures, but are differently employed.

Recent work on cultural differences in attention and reasoning provides a good example of the second type of cross-cultural psychological diversity.[4] Nisbett and colleagues distinguish two cognitive styles (Nisbett et al. 2001; Nisbett 2003; Norenzayan, Choi, and Peng 2007). The analytic cognitive style involves detaching focal objects from their context (field independence), focusing on the properties of objects in contrast to relations between objects, relying on rules to classify and reason, and appealing to causal explanation. By contrast, the holistic cognitive style involves paying attention to the context (field dependence), focusing on the relations between objects, and relying on similarity to classify and reason. Nisbett and colleagues have gathered an impressive body of evidence showing that Westerners exhibit an analytic cognitive style, while East Asians display a holistic cognitive style.

Westerners' attention abstracts objects from their context, while East Asians' attention relates them to their context. In the rod-and-frame test, subjects are shown a rod inside a frame and are asked to adjust the rod to a vertical position. People are considered field-dependent to the extent that their judgment is affected by the verticality of the frame. Ji, Peng, and Nisbett (2000) found that Chinese subjects are

---

[4] For a different research program on culture and reasoning, see Medin and Atran (2004); Atran, Medin, and Ross (2005); Medin, Unsworth, and Hirschfeld (2007).

more field-dependent than American subjects. These different patterns of attention affect Westerners' and East Asians' perception (for review, see Nisbett and Miyamoto 2005). Using an eye-tracking method, Chua, Boland, and Nisbett (2005) have shown that Chinese students and American students have different patterns of visual exploration of a scene, with Americans focusing on the main object of the scene, and Chinese paying greater attention to the background. Chua et al. propose that the differences between Westerners' and East Asians' attentional patterns might result from the differences between the visual scenes that are characteristic of the two cultures.

Westerners and East Asians also reason differently in a large number of contexts. According to Norenzayan et al. (2002), when asked to assess the similarity between a target object and members of two different categories, East Asians rely on the family resemblance of the target object to the members of each category, while Euro-Americans look for properties that are necessary and sufficient for belonging to one of the categories. Furthermore, although East Asians are perfectly able to reason according to the rules of propositional logic, they are less disposed than Westerners to do so.

3. The Cognitive Mechanisms Underlying Culture Acquisition

As exemplified by the work reviewed above, a substantial body of research now documents the extent to which cultural information can play an influential, at times even determinative, role in cognitive processes. Missing from much of this literature, however, are ultimate explanations as to why the human mind is so reliant on, or plastic with respect to, cultural information. Some scholars content themselves with the generalization that because much of human social, economic, and even biological life is structured by culture, the general propensity to think in the same manner as those around one evolved because it facilitates coexistence. Congruent with the "extended cognition" perspective described earlier, others argue that human cultures themselves owe their existence to the effectiveness with which cultural concepts, ways of thinking, and artifacts extend basic human information-processing capacities, thereby bootstrapping our innate potential to a higher level of behavioral complexity. Without contesting either of these generalizations, a nascent school of thought adopts a more explicitly mechanistic evolutionary perspective on the relationship between culture and cognition. This perspective begins with the long-recognized observation that, to a much greater extent than is true of other species, humans depend on cultural information to cope with the challenges posed by their physical and social worlds. The adaptive significance of cultural information suggests that natural selection can be expected to have crafted the human mind so as to maximally exploit this resource. We suggest that the evolved mental mechanisms that serve this goal fall into two general categories, reflecting differences in the degree of specificity of the types of information that they acquire (see also Boyer 1998 and Henrich and McElreath 2003 for relevant discussions). We turn first to mechanisms dedicated to the acquisition of specific bodies of knowledge.

3.1. Domain-Specific Cultural Information Acquisition Mechanisms

Two principal obstacles confronting learners who seek to benefit from others' knowledge are the richness of the informational environment and the incompleteness of the discernable information therein. First, human behavior is enormously complex, varying across contexts and persons, while linguistic utterances convey information ranging from the trivial to the life-saving. If, as is often tacitly presumed, learners were indiscriminate sponges, then (1) learners would often fail to understand how to apply what they have learned, and (2) learners would fail to properly prioritize their acquisition efforts, often resulting in both precocity in domains irrelevant to the learner and retardation in relevant domains. Second, much social learning involves the problem of the poverty of the stimulus, as many actions and utterances explicitly present only fractional portions of the information that motivates them (Boyer 1998). We suggest that, for many domains of learning, natural selection has addressed both of these problems by endowing the mind with inborn mechanisms, possessing considerable content, that serve to structure the acquisition of cultural information. Such a system can address the prioritization problem via variable motivational valence, as the attentional resources that various mechanisms command in pandemonium-style competition can be calibrated by natural selection to reflect the relative importance of acquiring cultural information in the respective domains: to foreshadow the discussion that follows, acquiring cultural information about dangerous animals likely has recurrently had a greater impact on children's survival than has learning to perform adult rituals, and, correspondingly, natural selection appears to have crafted the mind such that children find the former much more interesting than the latter. Likewise, by adjusting the developmental timing of such mechanisms, natural selection can ensure that learners acquire the cultural information that is most relevant to the fitness challenges characteristic of their current stage of life (for example, we postulate that young children are more interested in dangerous animals than they are in courtship behaviors). Finally, as has been argued extensively for the case of language acquisition, innate content can help to overcome the poverty of the stimulus problem, as such structure can serve as a foundation that narrows the possible referents or implications of statements and actions.

We believe that investigations aimed at uncovering what we term *domain-specific cultural information acquisition mechanisms* can shed considerable light on how, when, and why knowledge is acquired from others. In the design of such investigations, it is important to recognize that, for any given learning domain, three factors can be expected to constitute necessary conditions for the evolution of a domain-specific cultural information acquisition mechanism. First, the domain must have been of substantial and relatively uniform importance to biological fitness across the diverse socioecological circumstances that characterized ancestral human populations, as this will have provided the steady selection pressure necessary for the evolution of a complex adaptation. Second, the domain needs to involve content that will have varied significantly across said circumstances as, on the one hand, this precludes the evolution of extensive innate knowledge, and, on the other, this maximally exploits culture's ability to effectively compile information of parochial relevance. Lastly, the domain must be one in which individual learning through trial-and-error or direct observation would have been either very costly or impossible much of the time under ancestral circumstances.

As illustration of the above three factors, consider Barrett's (2005) proposal that a dedicated mechanism or set of mechanisms facilitates learning about dangerous animals, which were a persistent threat to ancestral humans. A few classes of dangerous animals will have been ubiquitous, sharing key perceptual features across disparate environments. For example, poisonous snakes are found in most of the ecosystems inhabited by humans. While snakes vary in their morphological details, all snakes share the same basic body plan. This combination of a significant and recurrent source of selection pressure and uniform perceptual features allowed for the evolution of a template-driven learning mechanism that requires only minimal input to produce a fear of snakes, a homologous trait shared with other primates (Öhman and Mineka 2003). Notably, however, in contrast to the case of snakes, many dangerous animals are either confined to discrete geographical areas, do not exhibit categorically distinguishing features, or both. As a consequence, natural selection cannot construct a learning mechanism with the same type of content as that responsible for the fear of snakes—at most, selection can assign innate salience to cues, such as large sharp teeth, that are imperfectly associated with dangerousness. Nevertheless, because cultures can be relied upon to contain information about the identity and attributes of, and strategies for dealing with, locally dangerous animals, natural selection could construct a mechanism dedicated to acquiring this kind of information from others. Importantly, fulfilling the third criterion listed above, learning about dangerous animals from other actors is almost always vastly cheaper than learning through trial and error.

With regard to the informational challenges (described earlier) that confront the human learner, Barrett (2005) suggests that what we would term a *dangerous-animal domain-specific cultural information acquisition mechanism* likely contains conceptual primitives ("animal," "dangerous") that (1) are rapidly mapped onto local lexical terms, (2) have high salience (leading to enhanced attending to, and retention of, any co-occurring information), (3) assist in solving problems of reference and inference (e.g., the learner presumes that statements concerning dangerousness refer to whole species rather than individual instances, that dangerousness will loom large in others' minds as well, allowing for inference as to the topic being discussed, etc.) and

(4) structure the manner in which information is organized and stored (e.g., the learner constructs a danger-based taxonomy of animals, etc.). This mechanism may also be linked to other mechanisms that govern defensive strategies (e.g., freeze, hide, flee, seek arboreal refuge, etc.) such that learning consists of reinforcing one of a number of preexisting potential responses. Next, because the threat posed by dangerous animals begins early in life (and, all else being equal, is often inversely proportional to an individual's size), this mechanism can be expected to begin operating early in development. Finally, children are expected to not only preferentially attend to and retain cultural information about dangerous animals, but to actively pursue such information (e.g., more frequently asking questions regarding dangerous than non-dangerous animals, allocating time to social contexts in which information regarding dangerous animals is likely to be available, etc.).

What other domains of learning might meet the triple criteria of universally high fitness relevance, parochial content, and costly individual learning necessary for the evolution of a domain-specific cultural information acquisition mechanism? Below we list a number of possibilities:

- *Diet.* Humans are generalists, capable of subsisting on a wide variety of foods. However, many ingestible items are inedible or, worse, poisonous. Obtaining nutrition and avoiding toxins are principal determinants of fitness, yet the features of both edible and poisonous substances differ widely across ecosystems. As in the case of predators' sharp teeth, a few perceptual cues (e.g., sweet taste versus bitter taste) offer some guidance in acquiring locally useful information, but, once again, these cues are imperfect and incomplete. Importantly, (1) cultures can be reliably expected to contain locally useful information in this domain, and (2) the costs of individual trial-and-error learning will generally be vastly higher than the costs of cultural learning (consider the two ways in which one might obtain the answer to the question "Is this mushroom poisonous?"). With regard to timing, acquiring competence in this domain first becomes important at weaning, then becomes increasingly important with greater mobility and, eventually, foraging self-sufficiency. Lastly, it is likely that a phylogenetic precursor of a dietary domain-specific cultural information acquisition mechanism set the stage for the evolution of such a mechanism, as social learning plays a role in the dietary behavior of a variety of language-less mammalian generalists (Galef and Giraldeau 2001).
- *Social structure.* All societies operate in part based on a set of interrelated culturally defined roles that demarcate identities and prescribe actions. Roles dictate social behavior in a relational fashion, as the appropriate actions depend not simply on one's own role, but on the interface between that role and the role of the individual with whom one is interacting (e.g., the appropriate actions of a "daughter" depend on with which parent she is interacting). Accordingly, actors must acquire information concerning not only their own roles, but also many that they will never occupy themselves.

Relatedly, coalitions and corporate groups (associations having a collective identity such that the actions of one member affect all) are of critical importance, as (1) they are often the principal sources of aid, and (2) social conflict often occurs along such lines of cleavage. Because both coalitions and corporate groups form along culturally defined lines, actors benefit from acquiring role-related cultural information regarding the nature, composition, and boundaries of such groups. Lastly, on a still larger scale, actors benefit from preferential assortment with those who share a similar cultural background, as this facilitates coordination and cooperation; conversely, it often pays to understand disjunctures or conflicts between those who resemble oneself and others who possess different cultural affiliations.

From roles all the way up to ethnic groups, the scope of social structural knowledge to be obtained is potentially very large; importantly, it is also highly parochial. While social structures share common conceptual elements, such as distinctions based on age, gender, kinship, hierarchy, communality, and group membership (Brown 1991), these core elements map only indirectly onto local forms—most roles, for example, cannot be defined in terms of these primitives. Because social relations would have been a fundamental determinant of fitness in ancestral societies, we can expect selection to have crafted mechanisms specific to the task of acquiring social structural information (Hirschfeld 1996, 2001). Such mechanisms likely contain the aforementioned conceptual primitives, using these as scaffolds to construct simple role concepts early in life that can, in turn, subsequently serve as the foundations for more complex concepts. Consistent with this premise, learners seem predisposed to (1) acquire labels for, and information concerning, a variety of social categories; (2) rapidly recognize coalitions and corporate groups; and (3) apply essentialist reasoning to larger categories of individuals (technically termed ethnic groups) who resemble one another by virtue of their common possession of indices, such as dialects and other markers of ethnicity, of shared cultural knowledge (Hirschfeld 1996, 2001; Gil-White 2001; Machery and Faucher 2005).

- *Tool Use.* Skill in using tools is a determinant of fitness in all known small-scale societies. Some of the informational basis of tool use skill can be obtained though trial-and-error learning, as artifacts' affordances will often bias experimental efforts in the direction of techniques congruent with the tool's design. However, in many instances, trial-and-error learning will be more expensive (in terms of time, and in terms of risk of injury to self, the tool, or the other objects or persons) than social learning; even in technologically simple societies, acquiring mastery of some tool techniques (e.g., flintknapping) is laborious, or even impossible, without cultural information. We can therefore expect natural selection to have crafted domain-specific cultural information acquisition mechanisms dedicated to this domain (see also Csibra and Gergely 2006). These mechanisms may contain conceptual primitives, such as "piercing tool," "cutting tool," "lever," "carrying tool," "container," and so on, that aid in the acquisition of cultural information linking a specific tool, a specific objective, and a specific technique. Testifying to our reliance on socially transmitted information in conceptualizing artifacts, even brief exposure to another's use of a novel artifact suffices to create a categorical representation of its function (Defeyter and German 2003). The predisposition to conceptualize tools in a socially determined fashion is apparently so strong that, even in technologically simpler societies, such categorization reduces the improvisational application of tools to novel purposes, a constraint termed "functional fixedness" (German and Barrett 2005).

The above discussion is intended to be illustrative, not exhaustive; other domains in which domain-specific cultural information acquisition mechanisms likely operate include navigation, fire building (Fessler 2006), disease avoidance, gathering, hunting (Barrett 2005), courtship and mateship, and, perhaps most importantly, morality (Haidt and Joseph 2004). Moreover, research aimed at mapping domain-specific cultural information acquisition mechanisms will likely overlap somewhat with, and should draw on, existing work on psychological universals of the type discussed in Section 1. For example, the postulated dangerous-animal domain-specific cultural information acquisition mechanism is likely either part of, or linked to, broader mechanisms responsible for learning about living things; these mechanisms likely contain considerable innate content, generating the universality of core features of folk biology described earlier. Our goal here, however, is not simply to expand the scope of the search for psychological universals, but rather to direct attention to the means whereby cultural information is acquired.

We hope that, by considering the three criteria of universal ancestral fitness relevance, parochial content variation, and high-cost individual learning, scholars will identify numerous areas in which domain-specific cultural information acquisition mechanisms likely exist, and will then test predictions that ensue. However, even if this enterprise proves successful, it will necessarily capture only a portion of the domains for which cultural learning is important. This is because much of the information that must be acquired to succeed in any given society is parochial in both content and type. For example, although it is true that public performance is universally an avenue for achieving prestige, nevertheless, the nature of such performance is so variable across cultures (and even across roles within a culture) as to likely have made it impossible for natural selection to have provided substantial foundations for the acquisition of the relevant information (e.g., while the category "performance" may be a conceptual primitive, it probably does not subsume more specific concepts, and thus lacks a rich structure that can serve as a scaffold for learning).

3.2. Domain-General Cultural Information Acquisition Mechanisms

Our species' reliance on cultural information that is parochial in both content and type suggests that natural selection may have favored the evolution of *domain-*

general cultural information acquisition mechanisms. At least two classes of such postulated mechanisms are relevant to the present discussion; both address the problems of the complexity and opacity of cultural information and related behavior discussed earlier. First, selection may have favored the evolution of mechanisms dedicated to the complementary tasks of pedagogy and the receipt of pedagogy. Second, whether the agent serving as the source of cultural information is an active pedagogue or a passive target of imitation, in a crowded social world learners must select whom to attend to as a source of cultural information, a task that may be subserved by evolved mechanisms.

Csibra and Gergely (2006) argue that much cultural information transfer is achieved through a goal-directed social process of teaching and learning contingent on ostension, reference, and relevance. *Ostension* denotes the act of indicating that one's current actions are communicative efforts, thereby differentiating such behavior from the stream of potentially observable actions. *Reference* addresses the need to constrain the topic of the communication from the class of all possible topics. Both ostension and reference can be enacted by either the pedagogue (who strives to indicate "now I am *teaching* about *X*") or the learner (who strives to indicate "I *need information* about *X*"). Csibra and Gergely argue that specific cues, and the cognitive mechanisms that process them, have evolved to facilitate, respectively, ostension (e.g., eye contact, eyebrow flashing, and turn-taking contingency by both parties; motherese by pedagogues interacting with infants) and reference (e.g., gaze directing/gaze-direction detection, pointing). *Relevance*, a feature of the inferential process engaged in by the learner once a teaching/learning interaction has been established, involves the presumption that actions lacking an ulterior explanation are designed to convey information in light of the learner's state of knowledge.

To date, the work of Gergely, Csibra, and colleagues has focused primarily on adult/infant interactions. However, ostension, reference, and relevance ought to characterize all teaching/learning interactions, the only principal modification being that bidirectional linguistic communication expands the channels available for ostension and reference. A more substantial difference, however, between infant learners and older learners is that infants have a vastly smaller range of potential pedagogues from which to choose. Older learners therefore face the problem of selecting the targets from whom they hope to learn. This is true both with regard to pedagogical interactions and to imitating a passive (nonteaching) model (while noting that the cognitive mechanisms underlying imitation are deserving of attention from students of the evolved psychology of culture acquisition, we leave this topic for another day—see Tomasello, Kruger, and Ratner 1993). However, the selection criteria differ for the two types of targets. Because pedagogues scale their communication to the competence of the learner, learners can usefully solicit pedagogy from individuals who possess vastly superior skills. In contrast, because cultural information is complex and opaque, in the case of imitation without pedagogy, learners must select targets that are closer to their own current competence, else much of the model's behavior will be subject to misinterpretation.

Differences between the two types of learning are reduced somewhat when the issue of social structure is considered, as the distribution of cultural information over roles is such that individuals will often benefit from seeking to learn from others whose position in the social structure is not too distant from their own, since similar roles entail similar resources, opportunities, and obligations (for seminal empirical work, see Harris 1995). Lastly, the task of target selection is complicated by the bidirectionality of the interaction: even imitation in the absence of pedagogy often has a bidirectional component, as the target must tolerate the presence of the learner. Henrich and Gil-White (2001) argue that, in exchange for access, learners grant higher status to individuals whom they wish to imitate; market forces then influence individual choices, with less-desired targets willing to grant more access. In contrast to Csibra and Gergely (2006), who argue that the costs to the pedagogue are such that pedagogy can be expected to be primarily kin-based (as kin have an interest in the welfare of their relatives), we suggest that the same market model applies to pedagogy: in both cases, the costs that accompany being targeted by a learner can be outweighed by the elevated prestige and power that flows from social support. The learner must thus trade off the value of the knowledge possessed by the target (skill, success, etc.) against the costs of access, keeping in mind that other learners are also competing for access.

The task of selecting targets from whom to learn is characterized by both sufficient importance and sufficient overarching uniformity as to suggest that evolution has created domain-general cultural information acquisition mechanisms for this purpose. We thus expect that actors will be adept at identifying others who possess the optimal combination of superiority and role-relevant knowledge; that observation of such individuals will be more acute, and subsequent information better retained; that ostensive cues will be both displayed toward and sought from such individuals; and that learners will be quite good at optimizing the access/cost ratio in a fluid market.

An alternative strategy to selecting a single individual as the focus of learning is to adopt the prevailing pattern of behavior in the local group. Because a single individual's success may be the product of many factors, raising the dual problems that (1) it may be unclear which aspects should be acquired by the learner, and (2) some of these factors may not be acquirable through learning, the conformist strategy will often prove profitable (Boyd and Richerson 1985; Henrich and Boyd, 1998). Note, however, that this is not so much a method of learning as it is one of deciding which of a number of variants of behavior to adopt. This is because at least some understanding of those variants (and hence some learning) must precede this decision; presumably, this process must be iterated, with the actor coming to recognize finer distinctions among variants as his command of the relevant information increases. Accordingly, many of the same cognitive learning processes must underlie both information acquisition strategies that target particular individuals as sources and conformist strategies that survey larger numbers of individuals. There are also some parallels as regards the task demands of target selection, since the relevance of behavior common across a set of prospective models is in part contingent on their degree of similarity to the learner: behavior that is common among individuals who occupy positions in the social structure similar to the learner's will generally be most relevant to, and hence should be most salient to, the learner. We can thus expect some form of a domain-general cultural information acquisition mechanism to combine information about the respective behaviors of actors with information about their relative social structural similarity to the learner in order to efficiently promote conformist acquisition. Lastly, the conformist strategy is less complex than individually focused learning in regard to calculating the cost/benefit ratio in target selection: because the goal is to acquire the most common variant of behavior, there is no shortage of potential models; hence it is a buyers' market, and the learner should be unwilling to pay much for access to prospective models.

4. THE EFFECTS OF COGNITION ON CULTURE

Thus far, we have explored the extent to which cultural information shapes cognition, and examined how the acquisition of cultural information may be underlain by evolved psychological mechanisms of varying degrees of domain specificity. This may give the impression that cultural information is a static feature of the environment. However, because culture exists in the minds of individuals, the relationship between culture and cognition is bidirectional, and thus dynamic. Specifically, because culture is instantiated through processes of the transmission, retention, and application of information, the composition of culture is subject to the influence of actors' minds, as information that is more likely to be transmitted, retained, and applied will come to predominate, while information for which this is less true will become rarer, and may disappear entirely (Sperber 1996, ch. 5).

4.1. Design Features of Mental Mechanisms Can Influence Cultural Evolution

Anthropologists have long recognized that some ideas are "good to think" (Lévi-Strauss 1962), meaning that they interdigitate with the mind in ways that make them attractive. Originally, this notion was developed with regard to the manner in which the affordances of real-world objects and entities (e.g., animals) facilitate symbolic distinctions that usefully organize human social life. Such nebulous intuitions were later more rigorously reconceptualized by cognitive anthropologists, who observed, for example, that folk taxonomies tend to be structured in ways that complement features of short-term memory (D'Andrade 1995, 42–43). More generally, the relative learnability (ease of acquisition, retention, and use) of cultural information can be expected to influence its persistence and spread. With regard to information acquired via domain-general cultural information acquisition mechanisms, learnability will in part be a function of the way that information is organized (as in the case of folk taxonomies), and will in part be a function of features of social transmission (e.g., ceteris paribus, because of the costs of pedagogy, ideas that can be acquired through imitation alone will spread faster than those that require pedagogy). With regard to information acquired via domain-specific cultural information acquisition mechanisms, learnability will in part be a function of the extent to which ideas contact the evolved content of the respective mechanisms (see also Boyer 1998). For example, Sperber and Hirschfeld (2004) and Barrett (2005) note that the special salience of information regarding dangerous animals is such that ideas concerning dangerous animals are more likely to spread and persist over time, even to the extent that erroneous information about actual creatures (e.g., the belief that wolves often prey on humans; see Sperber and Hirschfeld 2004) and fantastical notions about nonexistent creatures (Sasquatch, the Loch Ness Monster; see Barrett 2005) become widely accepted (for a pressing contemporary example, see Lombrozo, Shtulman, and Weisberg 2006 on the obstacles to learnability of evolutionary theory). Similarly, using both experimental and naturalistic data, Heath, Bell, and Sternberg (2001) have shown that the likelihood that urban legends will be transmitted, and will persist, is in part contingent on the degree to which they elicit disgust. Given that disgust is prototypically elicited by cues associated with pathogen transmission (Curtis and Biran 2001), this pattern is parsimoniously understood as the result of the operation of a domain-specific cultural information acquisition mechanism dedicated to the acquisition of knowledge relevant to disease avoidance.

A general principle that can be expected to apply to many of the mechanisms postulated thus far is that of error management (Haselton and Buss 2000): natural selection will have built biases into information processing systems such that, when these mechanisms err, they do so "on the safe side," making the mistake that, under ancestral conditions, would have been the least costly of the possible errors. One consequence of this is that, as illustrated by the above examples, domain-specific cultural information acquisition mechanisms that address possible hazards will lead to credulity, as actors will accept without evidence, and will not seek to test, socially transmitted information; aggregated over many individuals, the result will often be a proliferation of wholly imaginary dangers. Consistent with this perspective, preliminary evidence from an extensive evaluation of ethnographic materials suggests that around the globe supernatural beliefs tend to depict a world filled with anxiety-provoking dangerous agents and processes (Fessler, Pisor, and Navarrete n.d.). At a more general level still, the content of beliefs that describe hazards in the world will influence the likelihood that those beliefs will persist over time. For example, many proscriptive beliefs include notions of supernatural sanctions. We can expect that the most successful proscriptions will involve sanctions that either (1) are vague, referring to a wide class of possible events (e.g., misfortune, etc.); or (2) though specific, nonetheless refer to a negative event that is relatively common, and for which the objective causes are not readily evident (e.g., infant death in small-scale societies). Given the issue of credulity mentioned earlier, proscriptions enforced by vague supernatural sanctions will be likely to persist because bad things eventually happen to everyone; hence, when a taboo is violated, it is inevitable that sooner or later events will unfold that can be interpreted as evidence of the veracity of the belief. Likewise, proscriptions enforced by more specific sanctions will be likely to persist if the events described therein happen with sufficient frequency that, eventually, they will occur in the lives of rule breakers.

4.2. Misfirings of Other Mental Mechanisms Can Influence Cultural Evolution

In addition to the question of how design features of domain-specific cultural information acquisition mechanisms and domain-general cultural information acquisition mechanisms influence cultural evolution, a growing literature seeks to explain widespread cultural traits in terms of the accidental misfiring of evolved cognitive mechanisms that are not dedicated to the acquisition of cultural information per se. This approach arguably began with Westermarck (1891), who hypothesized that incest taboos result from the accidental triggering by third-party behavior of mechanisms that evolved to reduce inbreeding among close kin. More recently, Boyer (2001) argues that beliefs in supernatural agents are more likely to persist if they are minimally counterintuitive, meaning that the agents possess nearly all, but not all, of the properties expected by mechanisms that serve to detect agents and predict their actions; Fessler and Navarrete (2003) suggest that the centrality of meat in food taboos is an accident of the salience of meat as a stimulus in toxin-detection and pathogen-avoidance mechanisms; and Boyer and Lienard (2006) argue that the combination of constraints on working memory, the operation of mechanisms devoted to avoiding hazards, and the parsing of actions generates a nonfunctional attraction to ritualized behavior. While we feel that efforts such as these are to be encouraged, such hypotheses are necessarily developed on an ad hoc basis, in contrast to the bottom-up predictive potential of a dedicated effort to explore the nature (and consequences for cultural content and evolution) of mechanisms dedicated to the acquisition of cultural information.

5. Conclusion

There are many ways to study the relations between culture and cognition. In this chapter, we have discussed two traditional approaches in psychology and anthropology: searching for psychological universals and for cross-cultural psychological differences. Furthermore, we have attempted to describe what we take to be two of the most exciting contemporary approaches to studying these relations: looking for domain-specific and domain-general cultural information acquisition mechanisms,

and identifying the influences of the structure of our minds on how information is
retained and transmitted.

REFERENCES

Astuti, R., Solomon, G. A., and Carey, S. (2004). *Constraints of Conceptual Development: A Case Study of the Acquisition of Folkbiological and Folksociological Knowledge in Madagascar.* Oxford: Blackwell Publishing.

Atran, S. (1990). *Cognitive Foundations of Natural History: Towards an Anthropology of Science.* Cambridge: Cambridge University Press.

Atran, S. (1998). Folkbiology and the anthropology of science: Cognitive universals and cultural particulars. *Behavioral and Brain Sciences* 21: 547–609.

Atran, S., Medin, D. L., and Ross, N. O. (2005). The cultural mind: Environmental decision making and cultural modeling within and across populations. *Psychological Review* 112: 744–76.

Atran, S., Medin, D. L., Lynch, E., Vapnarsky, V., Ucan Ek', and Sousa, P. (2001). Folkbiology doesn't come from folkpsychology: Evidence from Yukatec Maya in cross-cultural perspective. *Journal of Cognition and Culture* 1: 3–42.

Avis, J. and Harris, P. L. (1991). Belief-desire reasoning among Baka children: Evidence for a universal conception of mind. *Child Development* 62: 460–67.

Barrett, H. C. (2005). Adaptations to Predators and Prey. In D. M. Buss (ed.), *The Handbook of Evolutionary Psychology.* New York: Wiley, 200–23.

Berlin, B. (1992). *Ethnobiological Classification.* Princeton, NJ: Princeton University Press.

Berlin, B., and Kay, P. (1969). *Basic Color Terms: Their Universality and Evolution.* Berkeley: University of California Press.

Berlin, B., Breedlove, D., and Raven, P. (1973). General principles of classification and nomenclature in folk biology. *American Anthropologist* 74: 214–42.

Boyd, R., and Richerson, P. (1985). *Culture and the Evolutionary Process.* Chicago: University of Chicago Press.

Boyer, P. (1998). Cognitive tracks of cultural inheritance: How evolved intuitive ontology governs cultural transmission. *American Anthropologist* 100: 876–89.

Boyer, P. (2001). *Religion Explained: The Evolutionary Origins of Religious Thought.* New York: Basic Books.

Boyer, P., and Barrett, H. C. (2005). Domain Specificity and Intuitive Ontology. In D. M. Buss (ed.), *Handbook of Evolutionary Psychology.* New York: Wiley, 96–188.

Boyer, P., and Lienard, P. (2006). Why ritualized behavior? Precaution systems and action parsing in developmental, pathological and cultural rituals. *Behavioral and Brain Sciences* 29: 595–650.

Brown, D. E. (1991). *Human Universals.* New York: McGraw-Hill.

Chiu, C.-Y., Leung, A. K.-Y., and Kwan, L. (2007). Language, Cognition, and Culture: The Whorfian Hypothesis and Beyond. In S. Kitayama and D. Cohen (eds.), *Handbook of Cultural Psychology.* New York: Guilford, 668–90.

Chomsky, N. (1986). *Knowledge of Language: Its Nature, Origin, and Use.* New York: Praeger.

Chua, H. F., Boland, J. E., and Nisbett, R. E. (2005). Cultural variation in eye movements during scene perception. *Proceedings of the National Academy of Sciences, USA* 102: 12629–33.

Clark, A. (1997). *Being There: Putting Brain, Body, and World Together Again*. Cambridge, MA: MIT Press.

Coley, J. D., Medin, D. L., and Atran, S. (1997). Does rank have its privilege? Inductive inferences within folkbiological taxonomies. *Cognition* 63: 73–112.

Csibra, G., and Gergely, G. (2006). Social Learning and Social Cognition: The Case for Pedagogy. In Y. Munakata and M. H. Johnson (eds.), *Processes of Change in Brain and Cognitive Development. Attention and Performance XXI*. Oxford: Oxford University Press, 249–74.

Curtis, V., and Biran, A. (2001). Dirt, disgust, and disease: Is hygiene in our genes? *Perspectives in Biology and Medicine* 44: 17–31.

D'Andrade, R. (1995). *The Development of Cognitive Anthropology*. Cambridge: Cambridge University Press.

Darwin, C. (1872). *The Expressions of Emotions in Man and Animals*, 1st ed. New York: Philosophical Library.

Davidoff, J., Davies,I., and Roberson, D. (1999). Colour categories in a stone-age tribe. *Nature* 398: 203–4.

Defeyter, M. A., and German, T. P. (2003). Acquiring an understanding of design: Evidence from children's insight problem solving. *Cognition* 89: 133–55.

Ekman, P., and Friesen, W. (1971). Constants across cultures in the face and emotion. *Journal of Personality and Social Psychology* 17: 124–129.

Fessler, D. M. T. (2004). Shame in two cultures: Implications for evolutionary approaches. *Journal of Cognition and Culture* 4: 207–62.

———. (2006). A burning desire: Steps toward an evolutionary psychology of fire learning. *Journal of Cognition and Culture* 6: 429–51.

Fessler, D. M. T., Pisor, A. C., and Navarrete, C. D. (n.d.). The spirits are not your friends: Biased credulity and the cultural fitness of beliefs. Manuscript in preparation.

Fessler, D. M. T., and Navarrete, C.D. (2003). Meat is good to taboo: Dietary proscriptions as a product of the interaction of psychological mechanisms and social processes. *Journal of Cognition and Culture* 3: 1–40.

Fodor, J. A. (1980). Methodological solipsism considered as a research strategy in cognitive psychology. *Behavioral and Brain Sciences* 3: 63–73.

Galef, B. G. J., and Giraldeau, L.-A. (2001). Social influences on foraging in vertebrates: Causal mechanisms and adaptive functions. *Animal Behaviour* 61: 3–15.

Gallistel, C. R. (2002). Language and spatial frames of reference. *Trends in Cognitive Sciences* 6(8): 321–22.

Gelman, S. A. (2003). *The Essential Child: Origins of Essentialism in Everyday Thought*. New York: Oxford University Press.

Gelman, S. A., and Wellman, H. M. (1991). Insides and essences: Early understandings of the nonobvious. *Cognition* 38: 213–44.

Gentner, D., and Goldin-Meadow, S. (2003). *Language in Mind: Advances in the Study of Language and Thought*. Cambridge, MA: MIT Press.

German, T. P., and Barrett, H. C. (2005). Functional fixedness in a technologically sparse culture. *Psychological Science* 16: 1–15.

Gil-White, F. J. (2001). Are ethnic groups biological "species" to the human brain? *Current Anthropology* 42: 515–54.

Gleitman, L., and Papafragou, A. (2005). Language and Thought. In K. H. and R. Morrison (eds.), *Cambridge Handbook of Thinking and Reasoning*. Cambridge: Cambridge University Press, 633–61.

Gordon, P. (2004). Numerical cognition without words: Evidence from Amazonia. *Science* 306: 496–99.

Haidt, J., and Joseph, C. (2004). Intuitive ethics: How innately prepared intuitions generate culturally variable virtues. *Daedalus* 133: 55–66.

Harris, J. R. (1995). Where is the child's environment? A group socialization theory of development. *Psychological Review* 102: 458–89.

Haselton, M., and Buss, D. M. (2000). Error management theory: A new perspective on biases in cross-sex mind reading. *Journal of Personality and Social Psychology* 78: 81–91.

Hauser, M. D., and Spelke, E. S. (2004). Evolutionary and developmental foundations of human knowledge: A case Study of Mathematics. In M. Gazzaniga (ed.), *The Cognitive Neurosciences*, 3rd ed. Cambridge, MA: MIT Press, 853–64.

Heath, C., Bell, C., and Sternberg, E. (2001). Emotional selection in memes: The case of urban legends. *Journal of Personality and Social Psychology* 81: 1028–41.

Heider, E. R. (1972). Universals in color naming and memory. *Journal of Experimental Psychology* 93: 10–20.

Heider, E. R., and Olivier, C. C. (1972). The structure of the color space in naming and memory for two languages. *Cognitive Psychology* 3: 337–54.

Henrich, J., and Boyd, R. (1998). The evolution of conformist transmission and between-group differences. *Evolution and Human Behavior* 19: 215–42.

Henrich, J., and Gil-White, F. J. (2001). The evolution of prestige: Freely conferred status as a mechanism for enhancing the benefits of cultural transmission. *Evolution and Human Behavior* 22: 165–96.

Henrich, J., and McElreath, R. (2003). The evolution of cultural evolution. *Evolutionary Anthropology* 12: 123–35.

Hirschfeld, L. A. (1996). *Race in the Making: Cognition, Culture, and the Child's Construction of Human Kinds*. Cambridge, MA: MIT Press.

———. (2001). On a folk theory of society: Children, evolution, and mental representations of social groups. *Personality and Social Psychology Review* 5: 106–16.

Hull, C. L. (1943). *Principles of Behavior*. New York: Appleton-Century-Crofts.

Hutchins, E. (1995). *Cognition in the Wild*. Cambridge, MA: MIT Press.

Ji, L., Peng, K., and Nisbett, R. E. (2000). Culture, control, and perception of relationships in the environment. *Journal of Personality and Social Psychology* 78: 943–55.

Kay, P., and Kempton, W. (1984). What is the Sapir-Whorf hypothesis? *American Anthropologist* 86: 65–79.

Kay, P. and Regier, T. (2006). Language, thought and color: recent developments. *Trends in Cognitive Sciences* 10: 51–54.

———. (2007). Color naming universals: The case of Berinmo. *Cognition* 102: 289–98.

Keil, F. (1989). *Concepts, Kinds, and Cognitive Development*. Cambridge, MA: MIT Press.

Kitayama, S., and Cohen, D. (eds.). (2007). *Handbook of Cultural Psychology*. New York: Guilford.

Levinson, S. C. (2003). *Space in Language and Cognition: Explorations in Cognitive Diversity*. Cambridge: Cambridge University Press.

Levinson, S. C., Kita, S., Haun, D. B. M., and Rasch, B. H. (2002). Returning the tables: Language affects spatial reasoning. *Cognition* 84: 155–88.

Lévi-Strauss, C. (1962). *Totemism*, translated by R. Needham. Chicago: University of Chicago Press.

Li, P., and Gleitman, L. (2002). Turning the tables: Language and spatial reasoning. *Cognition* 83: 265–94.

Li, P., Abarbanell, L., and Papafragou, A. (2005). Spatial reasoning skills in Tenejapan Mayans, *Proceedings from the 27th Annual Meeting of the Cognitive Science Society.* Hillsdale, NJ: Erlbaum.

Lillard, A. (1998). Ethnopsychologies: Cultural variations in theories of mind. *Psychological Bulletin* 123: 3–32.

Lombrozo, T., Shtulman, A., and Weisberg, M. (2006). The intelligent design controversy: Lessons from psychology and education. *Trends in Cognitive Sciences* 10: 56–57.

Lucy, J. A., and Shweder, R. A. (1979). Whorf and his critics: Linguistic and nonlinguistic influences on color memory. *American Anthropologist* 81: 581–615.

Machery, E., and Faucher, L. (2005). Why Do We Think Racially? In H. Cohen and C. Lefebvre (eds.), *Handbook of Categorization in Cognitive Science.* Amsterdam: Elsevier, 1009–33.

Mallon, R., and Stich, S. P. (2000). The odd couple: The compatibility of social construction and evolutionary psychology. *Philosophy of Science* 67: 133–54.

Medin, D. L., and Atran, S. (2004). The native mind: Biological categorization and reasoning in development and across cultures. *Psychological Review* 111: 960–83.

Medin, D. L., and Ortony, A. (1989). Psychological Essentialism. In S. Vosniadou and A. Ortony (eds.), *Similarity and Anological Reasoning.* New York: Cambridge University Press, 179–95.

Medin, D. L., Unsworth, S. J., and Hirschfeld, L. (2007). Culture, Categorization and Reasoning. In S. Kitayama and D. Cohen (eds.), *Handbook of Cultural Psychology.* New York: Guilford, 615–44.

Nisbett, R. E. (2003). *The Geography of Thought.* New York: Free Press.

Nisbett, R. E., and Miyamoto, Y. (2005). The influence of culture: Holistic versus analytic perception. *Trends in Cognitive Sciences* 9: 467–73.

Nisbett, R. E., Peng, K., Choi, I., and Norenzayan, A. (2001). Culture and systems of thought: Holistic vs. analytic cognition. *Psychological Review* 108: 291–310.

Norenzayan, A., and Heine, S. J. (2005). Psychological universals: What are they and how can we know? *Psychological Bulletin* 131: 763–84.

Norenzayan, A., Choi,I., and Peng, K. (2007). Cognition and Perception. In S. Kitayama and D. Cohen (eds.), *Handbook of Cultural Psychology.* New York: Guilford, 569–94.

Norenzayan, A., Smith, E. E., Kim, B., and Nisbett, R. E. (2002). Cultural preferences for formal versus intuitive reasoning. *Cognitive Science* 26: 653–84.

Öhman, A., and Mineka, S. (2003). The Malicious Serpent: Snakes as a prototypical stimulus for an evolved module of fear. *Current Directions in Psychological Science* 12: 5–9.

Piazza, M., and Dehaene, S. (2004). From Number Neurons to Mental Arithmetic: The Cognitive Neuroscience of Number Sense. In M. Gazzaniga (ed.), *The Cognitive Neurosciences*, 3rd ed. Cambridge, MA: MIT Press, 865–76.

Pederson, E., Danziger, E., Wilkins, D., Levinson, S., Kita, S., and Senft, G. (1998). Semantic topology and spatial conceptualization. *Language* 74: 557–89.

Pica, P., Lemer, C., Izard, V., and Dehaene, S. (2004). Exact and approximate arithmetic in an Amazonian Indigene group. *Science* 306: 499–503.

Regier, T., Kay, P., and Cook, R. S. (2005). Focal colors are universal after all. *Proceedings of the National Academy of Sciences* 102: 8386–91 .

Roberson, D., and Davidoff, J. (2000). The categorical perception of colors and facial expressions: The effect of verbal interference. *Memory and Cognition* 8: 538–51.

Roberson, D., Davies,I., and Davidoff, J. (2000). Color categories are not universal: Replications and new evidence from a stone-age culture. *Journal of Experimental Psychology: General* 129: 369–98.

Roberson, D., Davidoff, J., Davies,I. and Shapiro, L. (2005). Color categories in Himba: Evidence for the cultural relativity hypothesis. *Cognitive Psychology* 50: 378–411.

Sousa, P., Atran, S., and Medin, D. L. (2002). Essentialism and folkbiology: Further evidence from Brazil. *Journal of Cognition and Culture* 2: 195–223.

Sperber, D. (1996). *Explaining Culture: A Naturalistic Approach*. Cambridge, MA: Blackwell.

Sperber, D., and Hirschfeld, L. A. (2004). The cognitive foundations of cultural stability and diversity. *Trends in Cognitive Sciences* 8: 40–46.

Tomasello, M., Kruger, A. C., and Ratner, H. H. (1993). Cultural learning. *Behavioral and Brain Sciences* 16: 495–552.

Tooby, J., and Cosmides, L. (1992). The Psychological Foundations of Culture. In J.H. Barkow, L. Cosmides, and J. Tooby (eds.), *The Adapted Mind: Evolutionary Psychology and the Generation of Culture*. New York: Oxford University Press, 19–136.

Vinden, P. G. (1996). Junin Quechua's children's understanding of mind. *Child Development* 67: 1707–16.

Vygotsky, L. S. (1986). *Thought and Language*. Cambridge, MA: MIT Press.

Waddington, C. H. (1940). *Organisers and Genes*. Cambridge: Cambridge University Press.

Wellman, H. M., Cross, D., and Watson, J. (2001). Meta-analysis of theory-of-mind development: The truth about false belief. *Child Development* 72: 655–84.

Westermarck, E. (1891). *The History of Human Marriage*. New York: Macmillan.

Wimsatt, W. C. (1986). Developmental Constraints, Generative Entrenchment, and the Innate-Acquired Distinction. In W. Bechtel (ed.), *Integrating Scientific Disciplines*. Dordrecht: Martinus-Nijhoff, 185–208.

Whorf, B. L. (1956). *Language, Thought, and Reality*. Cambridge, MA: MIT Press.

Wierzbicka, A. (1992). *Semantics, Culture, and Cognition: Universal Human Concepts in Culture-Specific Configurations*. Oxford: Oxford University Press.

Winawer, J., Witthoft, N., Frank, M. C., Wu, L., Wade, A. R., and Boroditsky, L. (2007). Russian blues reveal effects of language on color discrimination. *Proceedings of the National Academy of Sciences* 104: 7780–85.

CHAPTER 22

....

EXPERIMENTAL PHILOSOPHY

....

JOSHUA KNOBE

THROUGHOUT the cognitive sciences, it is a common practice to engage in the systematic study of people's intuitions about cases. Phonologists study intuitions about nonce words like "tlito" and "foofa"; researchers in decision science study intuitions about bets in which, e.g., one has a 50 percent chance of losing $20 and a 50 percent chance of gaining $25; social psychologists study intuitions about whether an agent in some hypothetical scenario is likely to have committed a particular crime.

Philosophers, too, have long been concerned with the study of people's intuitions, but their methods have often diverged rather dramatically from the ones we typically find in cognitive science. At least traditionally, they did not conduct empirical studies or construct cognitive models. Instead, each philosopher would consult his or her own intuitions from the armchair and then try to capture them in a list of necessary and sufficient conditions.

The new movement of *experimental philosophy* seeks to examine the phenomena that have been traditionally associated with philosophy using the methods that have more recently been developed within cognitive science. That is to say, experimental philosophers examine intuitions about issues that have long been regarded as falling within the province of philosophy, but they approach those intuitions using systematic experimental studies, statistical analyses, and the full array of cognitive hypotheses that have become familiar within the broader world of cognitive science.

The present chapter begins by reviewing some of the major developments within recent work in experimental philosophy, then turns to a discussion of the relationship between this recent work and the sort of inquiry one more typically finds within the analytic tradition.

1. EXPERIMENTAL STUDIES

Perhaps the best way to convey a sense of how research in experimental philosophy proceeds is just by going through a few specific examples. Let us begin, then, by discussing some of the early studies associated with three specific experimental research programs.

1.1. Cultural Differences

Traditional conceptual analysis frequently relies on appeals to intuition, but it is rarely made clear precisely *whose* intuitions are being discussed. In place of a specification of a definite group of people, one typically finds vague uses of the first-person plural. Thus, one may be presented with a case and then told that *we* are inclined to have this or that intuition about it. It seems natural in such a circumstance to ask who exactly is included in the scope of this "we."

Now, it is a striking fact that readers of analytic philosophy papers usually do share the intuitions to which the authors of those papers appeal, and this fact shows that these intuitions are not just idiosyncratic feelings that differ radically from one philosopher to the next. Still, it should be kept in mind that readers of analytic philosophy papers are not exactly a representative sample of the human population. On the contrary, there is a strong tendency for readers of analytic philosophy papers to be highly educated members of Western cultures. A question therefore arises as to whether people of other cultures and socioeconomic statuses share the same intuitions. Some of the most important early work in experimental philosophy was devoted to addressing this question.

Among the most celebrated studies in this vein are Weinberg, Nichols, and Stich's (2001) experiments on intuitions about knowledge. The researchers presented subjects with brief stories about a character who holds a particular opinion, and then asked them whether this character "really knows" or "only believes." So, for example, some subjects were given a vignette modeled on Gettier's famous counterexample to the view that every case of justified true belief counts as knowledge:

> Bob has a friend, Jill, who has driven a Buick for many years. Bob therefore thinks
> that Jill drives an American car. He is not aware, however, that her Buick has
> recently been stolen, and he is also not aware that Jill has replaced it with a
> Pontiac, which is a different kind of American car. Does Bob really know that Jill
> drives an American car, or does he only believe it?

When the researchers gave this question to white undergraduates at Rutgers University, the results were exactly what one would expect. The majority of subjects said that Bob "only believes" and does not really know. But now comes the exciting twist: when the researchers gave this same story to students of East Asian descent, they obtained exactly the opposite results. A *majority* of East Asian subjects actually said that Bob "really knows."

Similar results were obtained by Machery, Nichols, and Stich (2004) in their study of intuitions about reference. They constructed a case modeled closely on Kripke's classic example of Gödel and Schmidt:

> Suppose that John has learned in college that Gödel is the man who proved an important mathematical theorem, called the incompleteness of arithmetic. John is quite good at mathematics and he can give an accurate statement of the incompleteness theorem, which he attributes to Gödel as the discoverer. But this is the only thing that he has heard about Gödel. Now suppose that Gödel was not the author of this theorem. A man called "Schmidt" whose body was found in Vienna under mysterious circumstances many years ago, actually did the work in question. His friend Gödel somehow got hold of the manuscript and claimed credit for the work, which was thereafter attributed to Gödel. Thus he has been known as the man who proved the incompleteness of arithmetic. Most people who have heard the name "Gödel" are like John; the claim that Gödel discovered the incompleteness theorem is the only thing they have ever heard about Gödel. When John uses the name "Gödel," is he talking about:
> (A) the person who really discovered the incompleteness of arithmetic?
> or
> (B) the person who got hold of the manuscript and claimed credit for the work?

Kripke had originally argued that this case could help us to decide between two different theories about the reference of proper names. One sort of theory would say that the correct answer was (A), while another would say that the answer was (B). Kripke then claimed that the correct answer is obviously (B). This claim has played a pivotal role in the debate between the rival theories. Though there has been a great deal of dispute about Kripke's argument as a whole, readers of his work have overwhelmingly concurred with the claim he makes about the correct judgment in this one case.

But, of course, readers of Kripke's work have almost all been members of Western cultures. When the experimental researchers presented this case to a broader sample, they obtained a surprising result. As expected, subjects from Western cultures mostly had the familiar Kripkean intuitions, but subjects from East Asian cultures showed the opposite pattern. In fact, the majority of East Asian subjects who were given the story about Gödel and Schmidt offered response (A), saying that the word "Gödel" referred to the person who really discovered the incompleteness of arithmetic.

In recent years, the emphasis in cross-cultural work in experimental philosophy has been shifting toward the study of *moral* judgments, with papers exploring cross-cultural differences in intuitions about consequentialism (Doris and Plakias 2007; Peng et al., unpublished data) and moral responsibility (Sarkissian et al. 2010; Sommers 2009). This new work ties in nicely with a long tradition in cross-cultural social psychology, and it will be interesting to see how research in this area continues to evolve.

1.2. Folk Psychology and Moral Judgment

It has long been recognized that there are important connections between folk psychology and moral judgment, but one might initially suppose that all of these connections go in the same direction. That is, one might suppose that the connections only arise in cases where we *first* make a folk-psychological judgment and *then* use this initial judgment as input to a process that eventually generates a moral judgment. Recent research in experimental philosophy has called this assumption into question. This research suggests that there might actually be cases in which the chain goes in the opposite direction—so that moral judgments serve as input to folk psychology.

For a particularly striking example, consider the following case:

> The vice-president of a company went to the chairman of the board and said, 'We are thinking of starting a new program. It will help us increase profits, but it will also harm the environment.'
>
> The chairman of the board answered, 'I don't care at all about harming the environment. I just want to make as much profit as I can. Let's start the new program.'
>
> They started the new program. Sure enough, the environment was harmed.

Faced with this vignette, most subjects said that the chairman *intentionally* harmed the environment.

One might think that this judgment was based entirely on certain information about the agent's mental states (e.g., the fact that he specifically knew the policy would harm the environment). But it seems that there is more to the story. For suppose we leave all of the agent's mental states the same but change the moral status of the behavior by simply replacing the word "harm" with "help." The vignette then becomes:

> The vice-president of a company went to the chairman of the board and said, 'We are thinking of starting a new program. It will help us increase profits, and it will also help the environment.'
>
> The chairman of the board answered, 'I don't care at all about helping the environment. I just want to make as much profit as I can. Let's start the new program.'
>
> They started the new program. Sure enough, the environment was helped.

Faced with this second vignette, very few subjects said that the chairman intentionally helped the environment (Knobe 2003). Yet it seems that the only major difference between the two vignettes lies in the moral status of the behaviors described. The results therefore suggest that people's intuitions as to whether or not a behavior was performed intentionally (a folk-psychological judgment) can be influenced by their beliefs as to whether the behavior itself was good or bad (a moral judgment).

A number of subsequent studies have replicated and extended these initial findings (Leslie et al. 2006; McCann 2005; Nadelhoffer 2004a; Nichols and Ulatowski 2007; Young et al. 2006), and at this point, most researchers agree that moral considerations really are affecting people's use of certain terms that one would ordinarily associate with the domain of folk psychology. The remaining discussion is concerned with questions about *why* this effect arises. Thus far, the debate has been dominated by three major views:

- One view holds that moral considerations play no role in people's underlying *concept* of acting intentionally but that something about people's use of language nonetheless allows moral considerations to affect their application of linguistic expressions like the *word* "intentionally" (Adams and Steadman 2004a, 2004b; Nichols and Ulatowski 2007; Sosa 2006).
- A second holds that people's feelings of blame act to distort their folk-psychological judgments and lead them to conclusions that actually violate the criteria set down by their own competence (Alicke and Rose forthcoming; Malle and Nelson 2003; Nadelhoffer 2004b, 2006).
- Finally, a third view holds that moral considerations truly do play a role in the fundamental competence underlying people's concept of intentional action (Knobe forthcoming; Leslie et al. 2006; McCann 2005; Mele 2003).

Although disagreements about these issues continue to persist, new experimental results are coming in at an extremely rapid rate, and participants in the discussion have thus far shown a remarkable willingness to revise their views as new data become available. We will surely see some progress in the years to come.

In the meantime, one of the most exciting developments in research on this topic has been the recent spate of studies showing just how pervasive the relevant phenomena are. These studies have explored the ways in which moral considerations impact not only intuitions about intentional action but also intuitions about *knowledge* (Beebe and Buckwalter forthcoming), *causation* (Hitchcock and Knobe 2009; Menzies forthcoming; Roxborough and Cumby 2009), *action individuation* (Ulatowski 2010), *happiness* (Phillips, Nyholm, and Liao 2010), *valuing* (Gonnerman 2008; Kauppinen 2006; Knobe and Roedder 2009) and the distinction between *doing* and *allowing* (Cushman et al. 2008). Although these papers may not have resolved the deeper question as to why these effects arise in the first place, they do suggest that the effect is a very basic one, which impacts a surprisingly broad swath of ordinary attributions.

1.3. Moral Responsibility

Philosophers have long been debating the relationship between moral responsibility and determinism. One of the key points of contention is whether moral responsibility and determinism are *compatible* or *incompatible*. In pursuing this debate, philosophers frequently appeal to the views of ordinary people. In particular, it is often claimed that ordinary people are incompatibilists:

> Beginning students typically recoil at the compatibilist response to the problem of moral responsibility. (Pereboom 2001, xvi)
> ...we come to the table, nearly all of us, as pretheoretic incompatibilists. (Ekstrom 2002, 310)

> In my experience, most ordinary persons start out as natural incompatibilists...Ordinary persons have to be talked out of this natural incompatibilism by the clever arguments of philosophers (Kane 1999).

> When ordinary people come to consciously recognize and understand that some
> action is contingent upon circumstances in an agent's past that are beyond that
> agent's control, they quickly lose a propensity to impute moral responsibility to
> the agent for that action. (Cover & O'Leary-Hawthorne 1996, 50)

Clearly, these are empirical claims, but it has traditionally been assumed that
there is no need to test them using systematic experimental techniques. After all,
philosophers are continually engaged in a kind of informal polling. They present
material in classes and listen to how their students respond. What they typically
find, it seems, is that students lean strongly toward incompatiblism, and this is
customarily taken to indicate that folk morality is itself incompatibilist. But,
strange as it may seem, when researchers began examining these questions more
systematically, their results did not confirm the claims that philosophy profes-
sors had been making about their students. In fact, the results pointed strongly
in the opposite direction. People's ordinary judgments appeared to be strongly
compatibilist.

The first study to arrive at this surprising conclusion was conducted by Viney
and colleagues (1982, 1988). The researchers used an initial questionnaire to distinguish between subjects who believed that the universe was deterministic and those
who did not. All subjects were then given questions in which they were given an
opportunity to provide justifications for acts of punishment. The key finding was
that determinists were no less likely than indeterminists to offer retributivist justifications. This finding provided some initial evidence that most determinists were
predominately compatibilists.

Woolfolk, Doris, and Darley (2006) arrived at a similar conclusion using a radically different methodology. They ran a series of experiments in which subjects
were given short vignettes about agents who operated under high levels of constraint. In one such vignette, a character named Bill is captured by terrorists and
given a "compliance drug" to induce him to murder his friend:

> Its effects are similar to the impact of expertly administered hypnosis; it results in
> total compliance. To test the effects of the drug, the leader of the kidnappers
> shouted at Bill to slap himself. To his amazement, Bill observed his own right
> hand administering an open-handed blow to his own left cheek, although he had
> no sense of having willed his hand to move. The leader then handed Bill a pistol
> with one bullet in it. Bill was ordered to shoot Frank in the head...

The researchers then manipulated the degree to which the agent was portrayed as
identifying with the behavior he has been ordered to perform. Subjects in one condition were told that Bill did not want to kill Frank; those in the other condition
were told that Bill was happy to have the chance to kill Frank. The results showed
that subjects were more inclined to hold Bill morally responsible when he identified
with the behavior than when he did not. This pattern of responses provides some
evidence for the claim that people's judgments fit with a popular compatibilist
account—namely, the Frankfurtian view that moral responsibility depends in a
crucial way on identification (see Frankfurt 1971).

Of course, subjects in this second study were not directly asked about the problem of determinism, but when the question is posed directly, the compatibilist result comes out even more clearly. Nahmias, Morris, Nadelhoffer, and Turner (2006) ran a series of experiments in which subjects were given stories about agents who performed immoral behaviors in deterministic worlds. Subjects were then asked to say whether these agents were morally responsible for what they had done. In one such experiment, subjects were given the following case:

> Imagine that in the next century we discover all the laws of nature, and we build a supercomputer which can deduce from these laws of nature and from the current state of everything in the world exactly what will be happening in the world at any future time. It can look at everything about the way the world is and predict everything about how it will be with 100% accuracy. Suppose that such a supercomputer existed, and it looks at the state of the universe at a certain time on March 25th, 2150 A.D., twenty years before Jeremy Hall is born. The computer then deduces from this information and the laws of nature that Jeremy will definitely rob Fidelity Bank at 6:00 PM on January 26th, 2195. As always, the supercomputer's prediction is correct; Jeremy robs Fidelity Bank at 6:00 PM on January 26th, 2195.

Subjects were then asked whether Jeremy was morally blameworthy. The vast majority (83 percent) said yes, indicating that they thought an agent could be morally blameworthy even if all of his behaviors were determined by natural laws. The researchers conducted three experiments—using three quite different ways of explaining determinism—and always found a similar pattern of responses.

In light of these results, it may seem mysterious that any philosophers could have thought that ordinary people were incompatibilists. How could philosophers have concluded that people were incompatibilists when they so consistently give compatibilist answers in systematic psychological studies? Are philosophers just completely out of touch with what their undergraduates really think? Perhaps—but it is also possible that something more complex is going on. It might be that people tend to give compatibilist answers to *concrete* questions about particular cases but incompatibilist answers to *abstract* questions about general moral principles. Then the divergence between the findings from psychological studies and the conclusions of philosophers teaching classes might simply be due to a difference between two ways of framing the relevant question.

To test this hypothesis, Nichols and Knobe (2007) ran a simple questionnaire study. All subjects were given a story about a universe ("Universe A") in which events always unfold according to deterministic laws. Subjects in the "abstract condition" were then given the question:

> In Universe A, is it possible for a person to be fully morally responsible for their actions?

Subjects in the "concrete condition" were given the question:

> In Universe A, a man named Bill has become attracted to his secretary, and he decides that the only way to be with her is to kill his wife and 3 children. He

knows that it is impossible to escape from his house in the event of a fire. Before
he leaves on a business trip, he sets up a device in his basement that burns down
the house and kills his family.

    Is Bill fully morally responsible for killing his wife and children?

The results were dramatic. A full 72 percent of subjects in the concrete condition
said that the agent was fully morally responsible, but less than 5 percent of subjects
in the abstract condition said that it was possible to be fully morally responsible in
the deterministic universe.

    If people's intuitions truly do differ depending on whether the case is abstract
or concrete, we will be faced with the difficult task of understanding the internal
processes that lead to different intuitions depending on precisely how the question
is framed. One view would be that people's theory of moral responsibility is fundamentally incompatibilist but that their intuitions can be biased by emotion in certain kinds of concrete cases. Another possibility would be that people's fundamental
capacity for generating judgments of moral responsibility is a compatibilist one but
that they are unable to engage this capacity when faced with purely abstract questions. In any case, one exciting thing about this domain of research is that, although
many of the researchers started out just trying to decide whether ordinary people
were compatibilists or incompatibilists, the resulting experiments appear to be
offering us a glimpse into the most basic mechanisms underlying people's capacity
for moral responsibility judgments.

2. EXPERIMENTAL PHILOSOPHY AND THE ANALYTIC PROJECT

Philosophers have long sought to investigate the ways in which ordinary people
use concepts, but almost all of these investigations have taken place within the
framework of a single basic aim. We can refer to that aim as the *analytic project*.
Its fundamental assumption is that philosophers can look to information about
the way people use words and concepts as a way to gain a better understanding of
the properties and relations that those words and concepts pick out. Thus, it was
assumed that philosophers could look to information about how people use the
word "causation" as a way to gain a better understanding of the true nature of
causation itself, that they could look to information about how people use the
word "knowledge" to gain a better understanding of the true nature of knowledge, and so on.

    In thinking about the significance of experimental philosophy, we can begin by
trying to situate it in relation to this more traditional aim. Is experimental philosophy supposed to advance the analytic project? Or to fight against it? Or just to do
something altogether unrelated?

The answer, it turns out, is that there is no single unified aim that characterizes all work of experimental philosophy. Instead, different experimental philosophy papers appear to be pursuing fundamentally different projects. Some are looking to information about people's use of concepts as a way to learn about the properties and relations those concepts pick out, some are attacking the whole idea of using information about concepts in that way, and some do not really have anything to do with the analytic project and are just pursuing an entirely different aim. Here we discuss each of these approaches in turn.

2.1. Advancing the Analytic Project

The clearest case of an area in which experimental philosophy has served to advance the analytic project is in the debate over free will and moral responsibility. There, experimental philosophers and analytic philosophers have worked closely together, with each group drawing freely on the insights of the other (e.g., Nelkin 2007; Nichols 2006; Sommers 2010; Sosa 2007; Vargas 2006). The guiding assumption is that a collaborative approach, using multiple methodologies, is our best hope of making progress on these difficult questions.

But here it seems that something puzzling is afoot. After all, the evidence being gathered by experimental philosophers is about the percentages of people holding various sorts of intuitions, while the questions under discussion in analytic philosophy are about whether people truly can be free and responsible. How could the former possibly be relevant to the latter? At least at first glance, we seem to be confronted with a yawning gulf that is impossible to cross.

The problems here come out especially clearly when one considers them in concrete detail. Suppose that a philosopher is thinking about a particular case and wondering whether the agent in that case is morally responsible. And now suppose she receives a new piece of information: namely, that precisely 76.3 percent of subjects in a given sample regarded that agent as responsible. How is this information supposed to be relevant to her inquiry? Is she just supposed to adopt whichever position happens to be in the majority?

Absolutely not. If the experimental results are to be relevant here, it must be in some more indirect way. The actual *percentages* do not play any role in philosophical inquiry; rather, we use the percentages to test theories about certain underlying *psychological processes*, and it is these theories that actually bear on the philosophical questions. Thus, we use the experimental results to figure out whether the process that generated people's intuitions relied on reasoning or emotion, whether it was characterized by conscious thought or nonconscious responses, and so forth. It is this understanding of the underlying process that then has implications for the questions discussed within analytic philosophy.

With this framework in place, it should be easy to see how experimental results can be relevant to philosophical problems. Return to our hypothetical philosopher who is working on a question about free will and moral responsibility. She considers a particular case and immediately finds herself thinking that the agent in this case is morally responsible. Now she needs to know whether she can *trust* her initial intuition. That is, she needs to know whether her intuitions are a reliable guide to truth in this domain. It is here that experimental philosophy comes in. The only way to know whether one's intuitions were produced by a reliable process is to know in real, substantive detail what that process was, and the only way to get at the details of these underlying processes is to go out and run systematic experiments.

The real work here lies in using the experimental data to test theories about the underlying processes and using these theories to figure out which intuitions are most reliable. It appears that people have compatibilist intuitions in certain kinds of cases and incompatibilist intuitions in others, but considerable controversy remains about precisely what accounts for the difference. Some say that people's compatibilist intuitions reveal the true nature of their moral views while the incompatibilist intuitions are merely symptoms of a certain kind of confusion (e.g., Nahmias, Coates, and Kvaran 2007; Nahmias and Murray forthcoming), others suggest that people's incompatibilist intuitions reflect their underlying moral theory while their compatibilist responses reflect immediate affective responses (Nichols and Knobe 2007), and still others say that people's compatibilist and incompatibilist intuitions actually reflect different aspects of their theories (Misenheimer 2007). Further research here is desperately needed.

2.2. Against the Analytic Project

Thus far, we have been considering research in experimental philosophy that serves in one way or another to advance the analytic project, but not all work in experimental philosophy fits comfortably within this framework. In fact, some experimental philosophy has been explicitly critical of the analytic project.

For a case of this second type, consider the cross-cultural work in the theory of knowledge. The chief aim of this work is to show that the analytic project in this area has simply been a mistake. That is, the aim is to show that we have little reason to continue with the project of looking at people's intuitions as part of an attempt to figure out which sorts of mental states count as "knowledge."

In responding to this work, a number of philosophers have argued for the claim that we really can use purely armchair methods to determine the extensions of our concepts (Kauppinen 2007; Ludwig 2007; Williamson 2005). These philosophers have suggested that the use of intuitions in the analytic project does not in fact depend on the sorts of assumptions that might be called into question by experimental work. (So, for example, Kauppinen and Ludwig argue for views according to which the intuitions that are relevant here are not the immediate reactions of ordinary folks, but rather the reflective conclusions of trained philosophers.)

These responses raise a number of fascinating questions, but they do not quite address the worry originally posed by the experimental work. That worry was not that *it is not possible* to determine the extension of our concept of knowledge but rather that *it does not even matter* whether we succeed in determining the extension of this concept.

Here it may be helpful to consider the problem in a broader historical context. Traditionally, one of the principal goals of work in the theory of knowledge was to figure out how we ought to go about forming our beliefs. Should we trust the established authorities, or should we try to work everything out for ourselves? Should we believe the testimony of our senses, or should we seek to understand the basic structure of the world using reason alone? The project devoted to answering questions like these might be called the *normative project*. It is clearly a deeply important project, and no one has offered any real arguments against pursuing it.

In the twentieth century, this traditional project was supplemented by another sort of inquiry, which I have been calling the "analytic project." The aim of work in the analytic project was to determine the precise extensions of certain concepts. Research in this project often proceeded by constructing complex hypothetical cases and then asking whether the beliefs described in those cases counted as knowledge. Although work on this project is becoming increasingly entrenched in the world of academic philosophy, it seems that the analytic project does not have quite the same status that the normative project does. That is, it is not simply *obvious* that the analytic project is important and worth pursuing. Instead, the analytic project seems to have a sort of uncertain status, such that reasonable people could disagree about whether it is important or not.

Now, one way to get a better understanding of this disagreement is to imagine various alternative concepts that people might have had. For example, we can consider the concept of *knowledge* and another concept *shmowledge* that has a slightly different extension. As it happens, people turned out to have the concept of knowledge and not the concept of shmowledge, but we can perfectly well imagine a creature of some other sort who makes frequent use of the concept of shmowledge and who never employs the concept of knowledge in any way. What one wants to know now is why it should be any more philosophically significant to determine the extension of the concept of knowledge than to determine the extension of, for example, the concept of shmowledge.

The results of the Weinberg, Nichols, and Stich studies raise this question in a particularly striking way. What they show is that there actually are people out there who have quite different intuitions about the extension of the concept of knowledge. Faced with these results, one might have either of two reactions. The first would be to say:

> Ah, now this is where the going really gets tough! What we need to do now is introduce a sophisticated theory of reference that enables us to decide precisely what the extension of the concept truly is.

But one might also have a rather different sort of reaction, namely:

> Why does it even matter who is truly right about the extension of this concept? And come to think of it, why was I even engaged in this analytic project to begin with? Why don't I just forget about these questions and go back to the more traditional project of asking straightforwardly normative questions about how we should decide what to believe?

It is this latter response that the experimentalists originally hoped to inspire.

2.3. Independent of the Analytic Project

Philosophers working within the framework of the analytic project have long engaged in the study of people's intuitions, but their real interest has not typically been in human beings and the way they think. Instead, the goal has been to understand the true nature of the properties and relations that people's concepts pick out. The study of people's intuitions is regarded only as a means to an end.

The rise of experimental philosophy has led to a change of perspective in this regard. More and more, philosophers are coming to think that they do not need any ulterior motive to engage in the study of people's intuitions. Instead, there is a growing sense that questions about how people think and feel are of deep philosophical importance in their own right.

Nowhere is this approach more apparent than in the debate concerning people's ordinary concept of intentional action. The participants in this debate hardly ever take any interest in the project of using people's intuitions to get a handle on the true nature of intentional action. Instead, the assumption is that it is simply important in and of itself to understand the relation between people's folk-psychological judgments and their moral judgments. Such an investigation can give us important insights into the human capacity for understanding other minds—telling us whether this capacity is something like a purely naturalistic scientific theory or whether it is infused in some fundamental way with normative considerations.

Work within this third type of experimental philosophy is becoming increasingly interdisciplinary. More and more papers are being written as collaborations between philosophers and other cognitive scientists, and many readers do not even bother to keep track of which discipline each researcher officially belongs to. What we see emerging here is not so much a new subfield within philosophy as a new willingness to stop worrying so much about the distinction between philosophy and the rest of cognitive science.

Of course, this interdisciplinary turn is not without its detractors. Some philosophers express concern that the most important and fundamental issues are somehow going overlooked as researchers turn more and more to empirically informed work in cognitive science (Kauppinen 2007; Ludwig 2007). They feel that there is a distinct class of purely philosophical questions that can only be properly addressed using a specific sort of "philosophical method." They then worry that recent work within experimental philosophy is simply failing to address the truly philosophical questions.

In short, contemporary experimental philosophers typically find themselves embedded in two different communities. The members of one community take for granted the importance of empirical research, do not care at all about the distinction between philosophy and the rest of cognitive science, and just want to know whether experimental philosophers have anything important to say about how the mind works. Meanwhile, members of the second community are extremely concerned with the idea that only a specific class of problems can truly be considered "philosophical" and are worried that experimental methods might not be able to contribute anything to the study of those problems. In my view, the best way to address the worries of this second community is just to focus on doing everything possible to address the worries of the first. It probably will not be helpful to search for general theoretical arguments for the claim that experimental methods can help to illuminate philosophical problems. Instead, experimental philosophers should simply go out and try to discover deep and important truths about the workings of the human mind. If these discoveries are profound enough, then experimental philosophers will thereby have proven—in practice, if not in theory—that experimental methods can play a useful role in philosophical inquiry.

3. METAPHILOSOPHICAL QUESTIONS

The chief aim of this chapter has been to review existing work in experimental philosophy, but it may now be felt that, in addition to the questions that arise within each of the separate strands of experimental philosophy we have been discussing, there also exists a range of more general, abstract questions that belong to the field of what might be called *metaphilosophy*. Perhaps the most basic issues here can be captured in the simple question: "If it is supposed to be philosophy, how can it possibly be experimental?" Those who pose this question wish to engage in a general inquiry into the role of empirical methods in philosophy.

It is not clear, however, that this sort of general inquiry is at all worthwhile. What we find in experimental philosophy is not a single unified research program but a loose aggregation of different approaches that are united only by their shared quest to develop adequate empirical theories of people's ordinary intuitions. Each of these separate approaches raises its own distinctive metaphilosophical questions. Thus, some work in experimental philosophy relies on the assumption that the study of people's intuitions about certain questions can help us to find the right answers to those questions, but other experimental work does not make this assumption or even explicitly call it into question. Similarly, some researchers rely on the assumption that purely contingent questions about human nature can have deep philosophical significance, but others use the very same sorts of data without relying on this assumption in any way. It is hard to see how any single metaphilosophical claim could undermine all of these separate projects in one stroke. Instead, it seems that the only way to successfully argue against the whole idea of experimental philosophy would be to take up each strand in turn, arguing that each of the separate approaches we have discussed here is in some way bankrupt or mistaken.

Still, it does seem that the rise of experimental philosophy brings up one important metaphilosophical issue, and that is an issue about a certain kind of *nonexperimental philosophy*—namely, research that pursues the analytic project but rejects any attempt to introduce empirical data. Though this philosophy does not make use of systematic experimental methods, it does purport to tell us something about people's concepts and intuitions. Yet this approach immediately leads to a fundamental worry that one might capture in the question: "If it is about people's concepts and intuitions, how can it possibly be done from the armchair?" For much of the history of analytic philosophy, this question was rarely posed—it was more or less assumed that that was just the way philosophers went about their work—but as more and more researchers begin collecting actual empirical data, the question will become increasingly pressing. It will be interesting to see how those philosophers who foreswear experimental work choose to address it.

REFERENCES

Adams, F., and Steadman, A. (2004a). Intentional action in ordinary language: Core concept or pragmatic understanding?. *Analysis* 64: 173–81.

———. (2004b). Intentional action and moral considerations: Still pragmatic. *Analysis* 64: 268–76.

Alicke, M., and Rose, D. (forthcoming). Culpable control or moral concepts? *Behavioral and Brain Sciences.*

Beebe, J., and Buckwalter, W. (forthcoming). The epistemic side-effect effect. *Mind & Language.*

Cover, J. A., and O'Leary-Hawthorne, J. (1996). Free Agency and Materialism. In J. Jordan and D. Howard-Snyder (eds.), *Faith, Freedom and Rationality.* Lanham, MD: Roman and Littlefield, 47–71.

Cushman, F., Knobe, J., and Sinnott-Armstrong, W. (2008). Moral appraisals affect doing/ allowing judgments. *Cognition* 108(1): 281–89.

Doris, J. M., and Plakias, A. (2007). How to Argue About Disagreement: Evaluative Diversity and Moral Realism. In W. Sinnott-Armstrong (ed.), *Moral Psychology, Vol. 2: The Biology and Psychology of Morality.* Oxford: Oxford University Press, 303–32.

Ekstrom, L. (2002). Libertarianism and Frankfurt-style Cases. In R. Kane (ed.), *The Oxford Handbook of Free Will* (New York: Oxford University Press).

Frankfurt, H. (1971). Freedom of the will and the concept of a person. *Journal of Philosophy* 68: 5–20.

Gonnerman, C. (2008). Reading conflicted minds: An empirical follow-up to Knobe and Roedder. *Philosophical Psychology* 21: 193–205.

Hitchcock, C., and Knobe, J. (2009). Cause and norm. *Journal of Philosophy* 11: 587–612.

Kane, R. (1999). Responsibility, luck, and chance: Reflections on free will and indeterminism. *Journal of Philosophy* 96: 217–40.

Kauppinen, A. (2006). Lovers of the good: Comments on Knobe and Roedder. *Online Philosophy Conference.*

———. (2007). The rise and fall of experimental philosophy. *Philosophical Explorations* 10(2): 95–118.

Knobe, J. (2003). Intentional action and side effects in ordinary language. *Analysis* 63: 190–93.

———. (forthcoming). Person as scientist, person as moralist. *Behavioral and Brain Sciences.*

Knobe, J., and Roedder, E. (2009). The ordinary concept of valuing. *Philosophical Issues* 19(1): 131–47.

Leslie, A., Knobe, J., and Cohen, A. (2006). Acting intentionally and the side-effect effect: "Theory of mind" and moral judgment. *Psychological Science* 17: 421–27.

Ludwig, K. (2007). The epistemology of thought experiments: first vs. third person approaches. *Midwest Studies in Philosophy* 31: 128–59.

Machery, E., Mallon, R., Nichols, S., and Stich, S. Semantics, cross-cultural style. *Cognition* 92 (2004): B1–B12.

Malle, B. F., and Nelson, S. E. (2003). Judging mens rea: The tension between folk concepts and legal concepts of intentionality. Behavioral Sciences and the Law 21: 563–80.

McCann, H. (2005). Intentional action and intending: Recent empirical studies. *Philosophical Psychology* 18: 737–48.

Mele, A. (2003). Intentional action: Controversies, data, and core hypotheses. *Philosophical Psychology* 16: 325–40.

Menzies, P. (forthcoming). Norms, causes, and alternative possibilities. *Behavioral and Brain Sciences*.

Misenheimer, L. (2007). Predictability, Causation, and Free Will (unpublished manuscript) University of California–Berkeley.

Nadelhoffer, T. (2004a). Praise, side effects, and intentional action. *The Journal of Theoretical and Philosophical Psychology* 24: 196–213.

———. (2004b). Blame, badness, and intentional Action: A reply to Knobe and Mendlow. *Journal of Theoretical and Philosophical Psychology* 24: 259–69.

———. (2006). Bad Acts, blameworthy agents, and intentional actions: Some problems for juror impartiality. *Philosophical Explorations.* 9: 203–19.

Nahmias, E., Coates, J., and Kvaran, T. (2007). Free will, moral responsibility, and mechanism: Experiments on folk intuitions. *Midwest Studies in Philosophy* 31: 214–42.

Nahmias, E., Morris, S., Nadelhoffer, T., and Turner, J. (2006). Is incompatibilism intuitive?. *Philosophy and Phenomenological Research* 73(1): 28–53.

Nahmias, E., and Murray, D. (forthcoming). Experimental Philosophy on Free Will: An Error Theory for Incompatibilist Intuitions. In J. Aguilar, A. Buckareff, and K. Frankish (eds.), *New Waves in Philosophy of Action*. Basingstoke, UK: Palgrave-Macmillan.

Nelkin, D. (2007). Do we have a coherent set of intuitions concerning moral responsibility? *Midwest Studies in Philosophy* 31: 243–59.

Nichols, S. (2006). Folk intuitions on free will. *Journal of Cognition and Culture* 6: 56–86.

Nichols, S., and Knobe, J. (2007). Moral responsibility and determinism: The cognitive science of folk intuitions. *Nous* 41: 663–85.

Nichols, S., and Ulatowski, J. (2007). Intuitions and individual differences: The Knobe effect revisited. *Mind & Language* 22: 346–65.

Nyholm, S. (2007). Moral Judgments and Happiness. (unpublished manuscript). University of Michigan.

Peng, K., Doris, J., Nichols, S., and Stich, S. (unpublished data). University of California–Berkeley.

Pereboom, D. (2001). *Living Without Free Will.* Cambridge: Cambridge University Press.

Phillips, J., Nyholm, S., and Liao, S. (2010). The Ordinary Concept of Happiness. (unpublished manuscript). Yale University.

Roxborough, C., and Cumby, J. (2009). Folk psychological concepts: Causation. *Philosophical Psychology* 22: 205–13.

Sarkissian, H., Chatterjee, A., De Brigard, F., Knobe, J., Nichols, S., and Sirker, S. (2010). Is belief in free will a cultural universal? *Mind & Language*. 25: 346–58.

Sommers, T. (2009). The two faces of revenge: Moral responsibility and the culture of honor. *Biology and Philosophy* 24: 35–50.

——. (2010). Experimental philosophy and free will. *Philosophy Compass* 5: 199–212.

Sosa, E. (2007). Experimental philosophy and philosophical intuition. *Philosophical Studies* 132: 99–107.

Stich, S. P. (1990). *The Fragmentation of Reason*. Cambridge, MA: MIT Press, Bradford Books.

Sytsma, J., Livengood, J., and Rose, D. (2010). Two Types of Typicality: Rethinking the Role of Statistical Typicality in Ordinary Causal Attributions. (unpublished manuscript). University of Pittsburgh.

Tannenbaum, D., Ditto, P. H., and Pizarro, D. A. (2007). Different Moral Values Produce Different Judgments of Intentional Action (unpublished manuscript). University of California-Irvine.

Ulatowski, J. (2010). Action under a Description. (unpublished manuscript). University of Nevada, Las Vegas.

Vargas, M. (2006). Philosophy and the folk: On some implications of experimental work for philosophical debates on free will. *Journal of Cognition and Culture* 6: 249–64.

Viney, W., Waldman, D., and Barchilon, J. (1982). Attitudes toward punishment in relation to beliefs in free will and determinism. *Human Relations* 35: 939–49.

Viney, W., Parker-Martin, P., and Dotten, S. D. H. (1988). Beliefs in free will and determinism and lack of relation to punishment rationale and magnitude. *Journal of General Psychology* 115: 15–23.

Weinberg, J., Nichols, S., and Stich, S. (2001). Normativity and epistemic intuitions. *Philosophical Topics* 29: 429–60.

Williamson, T. (2005). Armchair philosophy, metaphysical modality and counterfactual thinking. *Proceedings of the Aristotelian Society* 105: 1–23.

Woolfolk, R., Doris, J., and Darley, J. (2006). Attribution and alternate possibilities: Identification and situational constraint as factors in moral cognition. *Cognition* 100: 283–301.

Young, L., Cushman, F., Adolphs, R., Tranel, D., and Hauser, M. (2006). Does emotion mediate the effect of an action's moral status on its intentional status? Neuropsychological evidence. *Journal of Cognition and Culture* 6: 291–304.

INDEX

Abarbanell, L., 509
Abbott, Jack Henry, 434
abduction, 75, 341
Abramson, Fred, 161
absolute nonconceptual content, 296n7
abstracta, 265
abstract condition, 534
accelerating Turing machine, 164n30
access consciousness, 23
accidie, 193
Ackerman, B., 443
action-based framework, 279
action condition, 413
action individuation, 532
action potential, 226, 281
activation of concepts, 352
active attention, 101
Adams, Fred, 118n, 142n29
  on extended cognition, 141
  on Glenberg, 132n19
adaptation, 97
adaptive lag, 482
adaptive processes, 97
Adolphs, R., 413
Adrian, Edgar, 236
adults, psychological essentialism in, 457–58
affect program, 188
affordances, 121–22, 259n14
Aizawa, K., 118n, 142n29, 253n3
  on extended cognition, 141
ALA. *See* attentional learning account
Alcock, J., 482
allocation devices, 88
allocation problem, 81
  Fodor's version of, 81–82
Allport, D.A., 209, 211, 216
Alzheimer's disease, 141
amae, 185, 193
ambiguity, 365
American Council of Learned Societies, 445n17
amnesia, 37
amok, 194
analog computation, 235
analytical behaviorism, 340
analytic cognitive style, 511
analytic processing, 441
  advancing, 536–37
analytic project, 535–40
  arguments against, 537–38

independent of, 539–40
anaphoric relations, 369
Andreou, C., 430n6
Andrews, S., 467
animal cognition, 49–50
A-not-B error, 137
anterior insula, 412
anthropomorphism, 172
apodeictic, 158–59, 162
appraisal, 184
a priori knowledge, 320n1
archaeology, 492–96
Ariew, A., 324, 325, 326, 428n3
Aristotle, 286, 319, 456
Armon-Jones, C., 184
Armstrong, Katherine, 216
arousal, 21
artificial intelligence, 11, 147–73
  human-level, 147–48, 151, 166–67
  narrow mechanism and, 161–62
ascent-routine theory, 419
associationism, 241
  essentialism and, 466
asymmetric dependency analysis, 345
Atran, S., 506n3, 507
attention, 35, 201–18
  active, 101
  passive, 101
attentional learning account (ALA), 464
attenuation, 134
attributions, 413
audition, 102
auditory cues, 98
auditory-location, 106
Australian Aboriginal languages, 390
Australopithecines, 493
autism, 329n12, 432
automatic mimicry, 414
automatic processing, 441
autonomous mobile robot, 135
awareness, 22
  first-person, 32–33

Baars, Bernard, 28
backward masking, 34

CPSIA information can be obtained
at www.ICGtesting.com
Printed in the USA
BVOW08s0527230517
484901BV00003B/3/P